Western Cordillera and adjacent areas

Edited by

Terry W. Swanson
Department of Earth and Space Sciences
University of Washington
Seattle, Washington 98195
USA

THE
GEOLOGICAL
SOCIETY
OF AMERICA

Field Guide 4

3300 Penrose Place, P.O. Box 9140 ▪ Boulder, Colorado 80301-9140 USA

2003

Published by The Geological Society of America, Inc.
3300 Penrose Place, P.O. Box 9140, Boulder, Colorado 80301
www.geosociety.org

Printed in U.S.A.

Library of Congress Cataloging-in-Publication Data

Western cordillera and adjacent areas / edited by Terry Swanson.
    p. cm. -- (GSA field guide ; 4)
  Includes bibliographic references.
  ISBN 0-8137-0004-3 (pbk)
    1. Geology--West (U.S.)--Guidebooks. 2. Geology, Stratigraphic--Guidebooks. 3. West (U.S.)--
Guidebooks. I. Field guide (Geological Society of America) ; 4.

QE79.W47 2003
557.8--dc22

                                   2003049537

Cover: The Wallula Vineyard is situated on south-facing bluffs overlooking the Columbia River along the border of Washington and Oregon. Flat-lying flows of the 15–17 Ma Grande Ronde Basalt of the Columbia River Basalt Group crop out on both sides of the river and are overlain by a thin veneer of flood gravels and loess. See "Wine and geology—The terroir of Washington State," by Busacca and Meinert, p. 69–85.

Photo by: Lawrence D. Meinert.

Cover design by: Heather L. Sutphin.

10 9 8 7 6 5 4 3 2 1

Contents

Geological Society of America
Field Guide 4
2003

Glacial Lake Missoula, Clark Fork ice dam, and the floods outburst area: Northern Idaho and western Montana

Norman B. Smyers
USDA-Forest Service, Lolo National Forest, Bldg. 24, Fort Missoula, Missoula, Montana 59804, USA

Roy M. Breckenridge
Idaho Geological Survey, University of Idaho, Moscow, Idaho 83843, USA

ABSTRACT

The first day begins in Spokane and follows Interstate 90 to Missoula, Montana. Stops are at the Rathdrum Prairie ice-dam outburst area east of Spokane; flood outwash deposits in and around Coeur d'Alene, Idaho; Glacial Lake Missoula deposits west of Missoula, Montana; and Glacial Lake Missoula shorelines that are etched into the hillsides surrounding the city of Missoula. The second day's travel includes sites in the Missoula Valley that display features associated with the catastrophic emptying of the lake and the subsequent refilling. The remainder of the day's travel will be north and west of Missoula to sites along U.S. Highway 93 and State Routes 28, 382, and 200 to view the relationship of valley glaciers and Glacial Lake Missoula, pingo scar terrain, large terminal moraines, Joseph T. Pardee's classic giant current ripple forms and gulch fills in the Camas Prairie and along the Flathead River, and Eddy Narrows, ending in Thompson Falls. The third day's route continues west on Montana State Route 200 with stops to study glacial gravels associated with ice-dam advances up the Clark Fork River and the features associated with the ice dam that repeatedly occupied the Purcell Trench, dammed the Clark Fork River, and created multiple Glacial Lakes Missoula. Lastly, we will stop at the outburst area in Farragut State Park and the Rathdrum Prairie breakout area to view notable examples of current ripples and outburst features.

Keywords: Clark Fork ice dam, current ripples, glacial, glacial lake, Lake Missoula, floods, scabland, pingo.

INTRODUCTION

During the last ice age, a lobe of the Cordilleran Ice Sheet advanced from Canada into the Idaho Panhandle, blocked the mouth of the Clark Fork River, and created Lake Missoula (see Fig. 1 for the regional details of these late Pleistocene features). At its maximum extent, this glacial lake was up to 2000 ft (610 m) deep and covered 3000 m² (7680 km²) of western Montana. Catastrophic failure of the Clark Fork ice dam released 500 mi³ (2167 km³) of water at a rate 10 times the combined flow of all the present-day rivers on Earth. The torrent of water and ice thundered across the states of Montana, Idaho, Washington, and Oregon at speeds more than 65 mi (105 km) per hour to the Pacific Ocean. The enormous energy of the flood water carved the land into canyons and cataracts, excavated more than 50 mi³ (205 km³) of soil and rock, piled boulders in huge gravel bars, and drowned entire valleys in beds of mud. The continuous southward flow at the ice front repeatedly blocked the Clark Fork River and refilled Lake Missoula. This cycle was repeated until ca. 12,000 yr ago, at the end of the ice age. The latest episode of flood outbursts occurred from ca. 17,000 to 12,000 yr ago.

BRETZ AND PARDEE

In 1923, Professor J Harlen Bretz (Fig. 2) of the University of Chicago began publishing a series of papers explaining the origin of the Channeled Scabland in eastern Washington. He attributed this system of dry channels, coulees, and cataracts to an episode of

Smyers, N.B., and Breckenridge, R.M., 2003, Glacial Lake Missoula, Clark Fork ice dam, and the floods outburst area: Northern Idaho and western Montana, *in* Swanson, T.W., ed., Western Cordillera and adjacent areas: Boulder, Colorado, Geological Society of America Field Guide 4, p. 1–15. For permission to copy, contact editing@geosociety.org. © 2003 Geological Society of America.

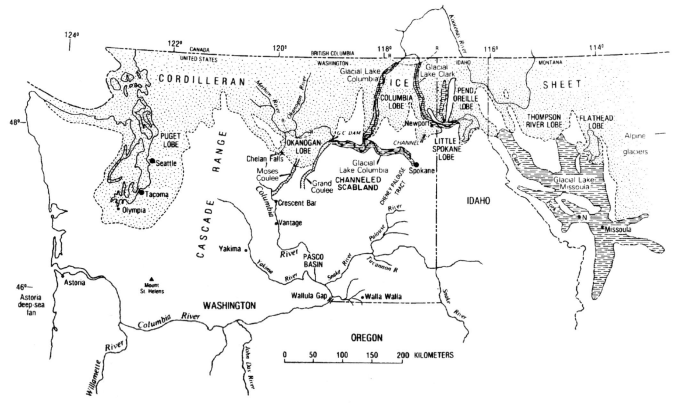

Figure 1. Regional index map. From Waitt (1980).

Figure 2. J Harlen Bretz, ca. 1910.

flooding on a scale larger than geologists had ever recognized. His hypothesis was disputed by prominent geologists, and the resulting controversy is one of the most famous in geologic literature. Bretz's ideas for such large-scale flooding were viewed as a challenge to the uniformitarian principles then ruling the science of geology. Joseph T. Pardee (Fig. 3) of the U.S. Geological Survey first studied Glacial Lake Missoula in 1910, but it wasn't until the early 1940s that he presented evidence for rapid drainage of the large ice-dammed lake. He estimated the lake contained 500 mi³ of water (2100 km³) and drained at a rate of 9.46 mi³ (39.7 km³) per hour. Pardee's explanation of unusual currents and flood features in the lake basin provided Bretz with the long-awaited source for flooding in the Channeled Scabland. More recent calculations have increased the original estimate. Recently, geologists' attention has mostly focused on the recognition of evidence for multiple floods and timing of the ice age events.

Today the effects of the ice age floods can be observed in an area covering 16,000 mi² (40,960 km²) in four states (Fig. 3). The field trip will explore deposits and features of Glacial Lake Missoula, the Clark Fork ice dam, and the "break out" from Lake Pend Oreille across the Rathdrum Prairie and Spokane Valley.

The trip begins and ends in Spokane, Washington (Fig. 4), and consists of 19 stops in Idaho and Montana. Along the route and between stops some geological features are also noted with mileage between stops. Trip mileage is shown in miles and kilometers. Unit measurements are shown in English and metric units.

Figure 3. Joseph T. Pardee, ca. 1900.

FIELD TRIP DESCRIPTION

Day 1. Spokane to Missoula

Spokane, the "Lilac City," has a history tied to mining in the Silver Valley of Idaho. Natural resources first attracted French, English, and U.S. citizens to this area. The fur trade was established in 1810, and gold was discovered in 1860. The town of Spokane Falls was platted by James Glover in 1878, and the Northern Pacific Railroad reached the town site in 1881.

Geologically, Spokane is near the margin of the Columbia Plateau with the Northern Rocky Mountains. The Columbia Plateau is underlain by lava flows of the Columbia River Basalt Group erupted during the Miocene epoch between 17 and 6 million years ago. These basalts are among the largest terrestrial basalt flows known on Earth. The deposits between the separate flows are interbeds of the Latah Formation and contain the fossils of a warm temperate Miocene flora similar to those of central China and the southeastern United States today. Quaternary-age loess of the Palouse Formation blankets much of the Columbia Plateau in this region. This rich deposit is the basis of dry farming on the rolling Palouse country. The loess was stripped and the underlying basalt was scoured during catastrophic ice age floods from Glacial Lake Missoula.

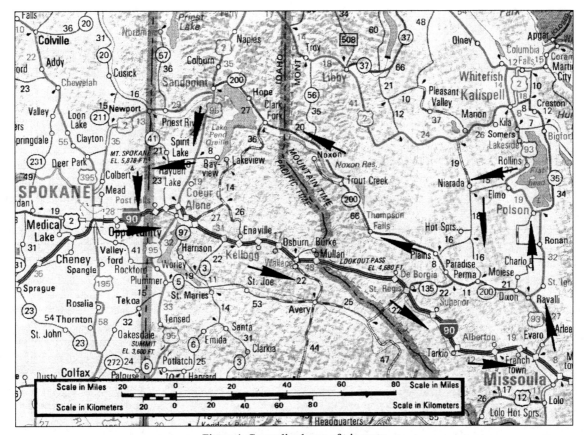

Figure 4. Generalized map of trip route.

The Spokane or Missoula floods, as these events are called, formed landscape that has been referred to as "scabland" by J Harlen Bretz. South and west of Spokane is the beginning of the Cheney-Palouse scabland tract of anastomosing channels that stretch 85 mi (136 km) south to the Snake River. The landscape between Spokane and Coeur d'Alene is dominated by erosional and depositional landforms of the latest floods.

Cumulative		
Miles	*(km)*	*Description*
0.0	(0.0)	Enter I-90 eastbound at interchange of I-90 and Division Street, Spokane, Washington.
1.0	(1.6)	Near the Hamilton exit, Columbia River Basalt Group outcrops can be seen in road cuts of the interstate and in Liberty Park on the south. Lava flows filled the Spokane Valley to an elevation of at least 2400 ft (732 m), but from here to Coeur d'Alene only small remnants of the flows have survived erosion. The elevation at the top of Spokane Falls is ~1860 ft (567 m).
4.0	(6.4)	Near the Sprague exit, the rounded summit of Mount Spokane, a granitic pluton (elevation 5878 ft, 1792 m), is visible to the northeast on the skyline. The bedrock on the north side of the valley is the Precambrian Priest River Crystalline Complex that is over 1.5 billion years old. The Precambrian rocks south of the valley are mostly metasediments of the Proterozoic Belt Supergroup.
5.0	(8.0)	Broadway exit. The aquifer gravels in the valley are as deep as 500 ft (152 m) and are also an important source of construction aggregate. Several large gravel pits intercept the water table at ~75 ft (23 m).
6.4	(6.8)	Argonne exit. On the south, the Dishman Hills natural area, a north-trending ridge of Precambrian bedrock, narrows the Spokane Valley. Missoula flood waters exceeded an elevation of 2700 ft (823 m) here and left huge boulders scattered on the scoured bedrock of Dishman Hills.
8.4	(13.5)	Near Pines exit. North of I-90 a large flood pendant bar trails to the west on the lee side of the scoured bedrock knobs. The knobs are Precambrian Newman Lake gneiss, part of the Priest River Complex. A prominent terrace of flood gravel to the south parallels I-90 for several miles.
14.5	(23.3)	Liberty Lake exit. Gravel deposits that impound Liberty Lake were originally thought to be glacial moraines, but are now recognized as catastrophic flood bars. Most tributary valleys of the Rathdrum Prairie contain similar

flood-dammed lakes, including Newman, Hauser, Spirit, Hayden, and Twin.

18.5	(29.8)	Port of entry, Washington border. The state line is the northward extension to the Canadian border of a meridian located at the confluence of the Snake and Clearwater Rivers ~115 mi (185 km) to the south. I-90 crosses the Spokane River, which originates from Coeur d'Alene Lake.
23.0	(37.0)	Post Falls. Treaty Rock is the site where Frederick Post contracted with Chief Seltice in 1871 to purchase the waterfall from the Coeur d'Alene tribe. For years electricity from Post Falls supplied power to the Coeur d'Alene mining district 30 mi (48 km) to the east. The present-day dam controls the flow of the Spokane River from Coeur d'Alene Lake.
25	(40.2)	Take the eastbound exit from I-90. Turn left at light on Seltice then turn north on Idaho 41 at the intersection. Continue north on Highway 41 to the traffic light at Mullan Avenue. Turn right on Mullan, then right to the Highlands on Sterling Ave. At the top of the grade turn left on Inverness and park in the parking lot behind the building on the left.
25.8	(41.5)	Stop 1.

Stop 1. Ross Point–Highlands

The Purcell Trench, a major topographic depression in British Columbia and northern Idaho, is incised into the margin of a metamorphic and granitic complex that formed during Mesozoic convergent tectonics, but later was subjected to Eocene extension. To the north the trench is bounded on the west by the Selkirk Range and on the east by the Cabinet Mountains. Here it is bounded by the Mount Spokane upland on the west and the Coeur d'Alene Mountains on the east (Lewis et al., 2002). The Purcell Trench has been the site of several drainage reversals. Prior to the Miocene, an ancestral river flowed within the Purcell Trench from the Canadian border south to the Coeur d'Alene area in a meandering pattern following the least resistant rock exposures and fault zones. This pattern is still apparent in the present-day meandering shape of Lake Pend Oreille. During the Miocene, Columbia River Basalt Group flows filled drainage systems and reversed the drainage in the Purcell Trench to the north. Some of the younger valley-filling basalts are exposed in Hoodoo Channel west of Lake Pend Oreille.

From this view the main path of the flood outbursts from the Pend Oreille Lake basin can be followed across the trench, south toward Coeur d'Alene, and west to Spokane. Another floodpath was to the north down the Hoodoo Valley to the Pend Oreille River and Little Spokane River via Newport, Washington. The water scoured as high as 3450 ft (1067 m) on Round Mountain. Flood bars filled the side valleys of the prairie and formed Spirit, Twin, Hayden, Newman, Liberty, and Coeur d'Alene Lakes (Breckenridge and Othberg, 1999b).

Here one can visualize the size of the catastrophic floods moving across Rathdrum Prairie and west through the Spokane Valley. The trimline of flood-water erosion on the bedrock south of Post Falls is above 2600 ft (792 m) in elevation. That is a depth of over 400 ft (122 m) across the prairie between Rathdrum and Post Falls and 250 ft (76 m) above Ross Point. O'Connor and Baker (1992) computed that the peak discharge through the Spokane Valley exceeded 17 ± 3 million m^3/s. Multiple episodes of flood erosion and deposition have left a complicated record in the Rathdrum gravels. Ross Point is a remnant of one of the oldest floods that has been dissected by later and mostly smaller floods. Figure 5 shows the geomorphic signature of the floods in the Rathdrum Prairie. It is likely that the evidence for even older floods and ice advances has been removed by the latest flood events.

Some early workers did not accept the catastrophic flood hypothesis and considered the deposits and landforms in the Spokane Valley and eastern Washington as glacial in origin. Features like those impounding Liberty Lake, Newman Lake, Coeur d'Alene, and Twin Lakes were interpreted as lateral moraines. Even Bretz did not consider the gravels as flood in origin. Most geologists today consider the deposits in the valley to be of flood origin, but it is not known how much of the fill represents reworked till or glacial outwash. In either case, the gravels are an important sole-source aquifer for the region. Several sand and gravel operations produce commercial aggregate from the flood deposits near this intersection. The suite of rock types contains clasts derived from the basin of Glacial Lake Missoula in Montana as well as from the Purcell Trench in Idaho and British Columbia. Boulders of granitic rock types and metasedimentary rocks of the Precambrian Belt Group are the most common. Intact basalt columns are derived locally from the rimrock nearby. Large-scale foreset beds tens of feet thick are common. The high-energy clast-supported gravels are matrix-poor and have high porosity.

CaCO$_3$ content and bulk density of samples taken from the cemented gravel unit were determined in the laboratory. Measured CaCO$_3$ content within the gravel unit is variable and averages ~10% by weight. This openwork fabric results in the remarkable water capacity of the Rathdrum Prairie–Spokane aquifer that serves nearly 400,000 people in the two-state area. The hydrology of the aquifer is not well understood. In some reaches the Spokane River recharges the aquifer, while in others the aquifer discharges to the river. Little subsurface stratigraphy has been documented, mainly because most wells have been so productive there has been no incentive to study the stratigraphy. Gravity studies of the Rathdrum Prairie and Spokane Valley indicate the bedrock may be as much as 1000 ft (305 m) deep at Highway 41 (Adema et al., 1998).

Return to Idaho 41. Turn left on 41 to the I-90 interchange and re-enter I-90 eastbound.

Cumulative Miles	(km)	Description
31	(49.9)	Coeur d'Alene, the "City by the lake," has a history of mining, logging, and recreation. It began as a military fort established by General William Tecumseh Sherman in 1879. Coeur d'Alene was the name French fur traders gave early in the nineteenth century to the shrewd natives in the region. It means "heart as pointed as an awl" in respect of their sharp trading savvy. In the 1880s, the discovery of gold and silver in the Coeur d'Alene Mountains kindled a mining boom. The Silver Valley has more recorded silver production than any district on Earth. At the turn of the century dozens of steamboats plied the lake carrying tourists on popular excursions; by the 1930s the steamers had all but disappeared.
33.6	(54.1)	Take Sherman Avenue exit off I-90. Proceed through the intersection on Lake Drive to Rutledge Point Park.
35.1	(56.5)	Stop 2.

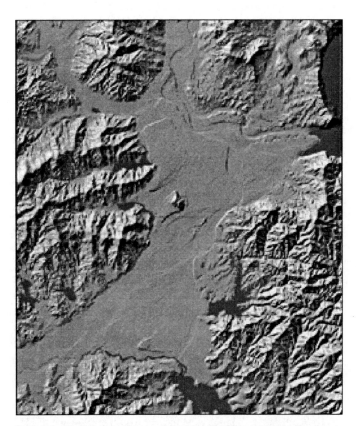

Figure 5. Digital elevation model of the Rathdrum Prairie flood outburst area showing morphology of flood deposits and lakes dammed by flood deposits. Lake Pend Oreille is at top right and Coeur d'Alene Lake at bottom right.

Stop 2. Rutledge Point–Centennial Trail

Coeur d'Alene Lake is the second largest lake in Idaho and is dammed by Lake Missoula flood gravels here on the north. The lake basin is mostly developed in an embayment of the Columbia River basalts (Lewis et al., 2002). Due to its origin as a drowned river system it has numerous bays and an extensive shoreline. The lake is fed by the St. Joe and Coeur d'Alene river systems and is the source of the Spokane River, a tributary of the Columbia River. Glacial Lake Missoula floods inundated the area occupied by the present lake basin and drained west through Rathdrum Prairie and the Spokane Valley except for one overflow outlet at Setters, Idaho, at 2560 ft (780 m) that crossed the Coeur d'Alene drainage divide into Latah Creek. The high water level marked by ice rafted erratics is at least as high as 2600 ft (792 m) around the lake (Dort, 1960). Scabland features can be recognized in the basalt rimrock around the lake. Some early geologists extended the southern ice limit as far as the city of Coeur d'Alene and considered Tubbs Hill as ice scoured. A seismic profile of the lake bottom between Tubbs Hill and Arrow Point by the U.S. Geological Survey reveals spectacular giant current ripples (Breckenridge and Othberg, 1999a).

   Return to Sherman Avenue via Lake Drive. Turn right at the stoplight and take the I-90 east entrance. Take I-90 east.

Cumulative Miles	(km)	Description
41.5	(66.8)	East end of Coeur d'Alene Lake and Fourth of July Pass. East and north of Coeur d'Alene, the preponderance of the bedrock along most of the trip route consists of Belt Supergroup rocks, a Precambrian rock assemblage that was deposited 1.5 billion to 800 million years ago in a slowly subsiding structural basin that occupied much of northeastern Washington, northern Idaho, and western Montana. In northwestern Montana, Harrison et al. (1986) estimated the greatest thickness of this assemblage of sandstone, silt, dolomite, and limestone at more than 20,000 ft (6098 m). The heat and pressure associated with deep burial has slightly metamorphosed portions of this rock assemblage into argillite, quartzite, schist, and marble. The Belt Supergroup rocks are the rock units that form much of the spectacular mountain ranges of Glacier National Park and the Canadian Rockies.
69	(111.0)	Smelterville area. The Coeur d'Alene mining district is located in northern Idaho, with the town of Wallace at its center. In February 1985, only a few months after the Coeur d'Alene district's 100th birthday, miners wrested the billionth ounce of silver from the district's mines and Coeur d'Alene laid claim to being the largest silver mining district in the world. The other major silver districts, including Pachuca and Guanajuato, Mexico, and Potosi, Bolivia, have all been mined since the 1500s and each has produced about a billion

ounces of silver. The district also claims records for silver mines in the United States: the deepest—the Star-Morning mine (7900 feet, 2408 m deep); the richest—the Sunshine mine (over 300 million ounces of silver); and the biggest—the Bunker Hill (over 180 mi, 290 km of underground workings). The Sunshine is also the deepest mine below sea level (3300 ft, 1006 m). Major production for the district is from veins in metasedimentary rocks of the Belt Supergroup. Several active mines have followed these veins well over a mile deep. For a more complete guide to the district, see Bennett et al., 1989.

94	(151.2)	Lookout Pass, Idaho-Montana state line.
100	(161)	Stop 3.

Stop 3. Taft Interchange

   The Taft interchange is the site of a railroad construction town as well as the intersection of three transportation routes—Mullan Road, the Milwaukee Road rail line, and the Northern Pacific Railroad rail line. The town of Taft was built in 1906 to support the building of the 8771 ft (2647 m) long Taft Tunnel, a railroad tunnel for the Milwaukee Road that was cut-through the Bitterroot Mountain range (McCarter, 1992). Mullan Road is named for U.S. Army Captain John Mullan; he constructed the first road across Montana in the late 1850s (Van West, 1986). The Northern Pacific followed his route some 30 years later, as did Milwaukee Road in the early 1900s.

   In 1910 Pardee put the highest stand of Glacial Lake Missoula at 4200 ft (1280 m) above sea level and then lowered it to 4150 ft (1265 m) in 1942. Others (Alt, 2000; Waitt, 1984) have suggested the maximum elevation to have been as much as 4250 ft (1296 m) to 4300 ft (1311 m), and Chambers (1984) suggested an elevation of 4352 ft (1340 m). It seems, however, that the most frequently used height is 4200 ft (1280 m), the one that will be used for the purposes of this trip.

   Using 4200 ft (1280 m), the interchange at an elevation of 3600 ft (1098 m) would have been covered with as much as 600 ft (183 m) of water, while the east portal or entrance to the Taft Tunnel, which lies ~2.5 mi (4 km) south at 4200 ft (1280 m), would have been at the very edge of the lake. If the lake, railroad, and tunnel had all been here 12,000–17,000 yr ago during the highest stand of Glacial Lake Missoula, an eastbound Milwaukee Road train emerging from the tunnel would have had to travel more than 150 mi (241 km) east before it would once again be above the waters of the lake.

Cumulative Miles	(km)	Description
110.7	(171.1)	Haugan, Montana. An 80-ft (924-m) thick Glacial Lake Missoula delta deposit is 1 mi (1.6 km) south, where Forest Road 386 crosses Big Creek.

137.0 (220.4) Tarkio Flat. The rolling hill in the large flat are giant current ripples created by fast-draining lake waters.

138.1 (222.2) Exit the interstate, turn right and then left, and follow the road (the old Milwaukee Road rail-bed) that leads to the Tarkio fishing access site.

138.9 (223.5) Take the left-hand route along old rail-bed.

139.1 (223.8) Stop 4.

Stop 4. Tarkio Flood Gravels and Glacial Lake Missoula Silts

At an elevation of 3000 ft (915 m), this location at the highest lake stand would have been covered by 1200 ft (366 m) of water. Here the Clark Fork River valley widens, flood water velocity decreased, and a thick layer of sand, gravel, and cobbles was deposited on the west and southwest flanks of Martel Mountain. The cobbles found at this site consist chiefly of quartzite, argillite, dolomite, and siltstones derived from various Belt Supergroup units. However, igneous intrusive rocks can be found here as well, granites and diorites that may have come from the Bitterroot Mountains to the south and the Garnet Range immediately east of Missoula. Unconformably overlying the gravel and cobbles are nearly horizontal beds of Glacial Lake Missoula clay and silt.

Cumulative Miles	(km)	Description
140.1	(225.4)	Enter Interstate 90 for eastbound travel.
154.3	(248.3)	Alberton, Montana. Exit I-90 at the west interchange and turn left (north) onto north frontage road (Railroad Avenue) and travel east though the main part of town. The large rocky knobs to the north of the interstate are highly erosion-resistant masses of Belt rock that the rushing flood waters left behind in what would have been the middle of the flood channel.
155.4	(250.0)	Turn left onto U.S. Highway 10.
161.5	(259.8)	Stop 5.

Stop 5. Glacial Lake Missoula Rhythmites

The elevation is ~3080 ft (939 m); at the highest lake stand this location would have been covered by 1120 ft (341 m) of water. The light pink clay, silt, and sand-sized materials here represent a thick accumulation of the glacial flour that clouded the waters of Glacial Lake Missoula. Chambers (1984) described the sand and silt deposits as being composed chiefly of quartz mixed with some mica and other dark opaque minerals. The red-to pink-colored Bonner Quartzite, an older Belt rock unit, is a likely source rock for the material at this location. Pardee (1942) and Alden (1953) referred to these sediments as being finely laminated and resembling varves. Chambers (1984, p. 190), who considers this area as the type section for Lake Missoula clays and silts, prefers the term rhythmites because, on close inspection, they appear as "…individual units …with no connotation of time, whereas varves are couplets that may represent annual deposits." Unfortunately, in terms of establishing a chronology, none of the rhythmite sequences associated with Glacial Lake Missoula have yet to yield material that can be used to produce a reliable radiometric date. Therefore, these deposits provide few clues regarding their time of deposition and their relationship to ice-dam failures, flood events, and lake-filling rates.

Cumulative Miles	(km)	Description
162.9	(262.2)	Turn left onto the eastbound I-90 onramp.
178.2	(286.7)	The city of Missoula lies at the north end of the Bitterroot Valley where the north-flowing Bitterroot River enters the west-flowing Clark Fork River. The population of the city and the surrounding area is ~86,000 and growing. The surface of the valley is covered by modern alluvium that hides the relics—silts, gravels, cobbles, and boulders—of Glacial Lake Missoula and the flood waters that tore through this area. McMurtrey et al. (1965) estimated the valley fill to be as much as 3000 ft (915 m) thick, an unconfined aquifer that is the area's primary water source.
184.5	(296.9)	Junction of I-90 and Reserve Street. Rising above the east side of the valley and guarding the west entry to the Clark Fork River's Hellgate Canyon are, north to south, the rounded masses of Mounts Jumbo (4768 ft, 1453 m) and Sentinel (5158 ft, 1573 m). Superimposed on these mountains are a series of horizontal lines, the multiple shorelines of Glacial Lake Missoula. The valleys and depressions on the upper slopes of these two mountains may represent erosion associated with permanent ice fields that existed on these slopes during the more frigid glacial episodes. Alt (2000, p. 32) has suggested that the somewhat straight vertical lines or gullies on the front of the mountains are due to mudflows that occurred when the "Sudden drainage of the lake would have left the slope saturated with water and without buoyant support."
189.0	(301.1)	Junction of I-90 and Van Buren Street. Exit for south travel on Van Buren Street.
189.1	(304.3)	Stop 6.

Stop 6. Glacial Lake Missoula Shorelines

Park in the Eastgate Shopping Center parking lot and walk south across the Madison Street pedestrian bridge to the south side of the Clark Fork River and the north edge of the University of Montana campus.

The elevation is ~3200 ft (976 m); at the highest lake stand this spot would have been covered by 980 ft (299 m) of water.

The shoreline features etched into Mounts Jumbo and Sentinel were described by Pardee in his 1910 *Journal of Geology* article titled "The Glacial Lake Missoula." It is an interesting artifact of history to note that while the existence of Glacial Lake Missoula was know many years before the Bretz scabland controversy erupted in the mid to late 1920s, it was left to Pardee in 1940 and 1942 to make the connection between the two.

Cumulative Miles	(km)	Description
189.1	(304.3)	Eastgate Shopping Center parking lot. Leave the parking lot and turn left (west) onto Broadway and proceed west through downtown Missoula.
193.0	(310.5)	Junction of Broadway and Reserve Street. Turn left onto the onramp for northbound Reserve Street travel.
194.1	(312.3)	Ruby's Inn Motel. End of Day 1.

Day 2. Missoula to Thompson Falls, Montana

Cumulative Miles	(km)	Description
0.0	(0.0)	Interstate 90 and Reserve Street. Enter I-90 for westbound travel. The slump on the north side of the Best Western–Grant Creek Inn reveals the unstable nature of the Eocene sedimentary units that make up the hills to the immediate north of the interstate. These sediments are, for the most part, an assemblage of well-sorted and well-rounded cobbles, gravel, sand, silt, clay, and volcanic ash (Lewis, 1998).
6.3	(10.1)	Turn right for northbound travel on U.S. Highway 93.
13.3	(21.4)	Evaro Hill summit. At an elevation of only 3956 ft (1206 m), Evaro Hill summit would have been covered with more than 200 ft (61 m) of water at the highest lake stand. During episodes of catastrophic draining, water would have initially flowed in two directions—south and west down the Clark Fork River, and north and west down Finely Creek and the Jacko River to the Flathead River.
19.0	(30.6)	View of the Jacko Valley and northeast to the Mission Range. The dark tree-covered mass that sits in the middle of the Jacko River floodplain, ~2 mi (3.2 km) to the northeast of the highway, is a terminal moraine.
23.3	(37.5)	Junction of U.S. 93 and Montana Route 559, Jacko Road. Smith et al. (2000) noted the presence of wave-cut shorelines on the hills to the northeast of the community of Arlee, but they also stated that these hills were too

low to have been glaciated. On the other hand, they do contain depression sand drainage features very similar to and at nearly the same elevation as those seen on the slopes of Mounts Jumbo and Sentinel. Perhaps these depressions were also the loci of permanent icefields during the more frigid episodes of the ice age?

30.3	(48.8)	Stop 1.

Stop 1. Glacial Lake Missoula Clays and Silts

This site contains numerous small-scale cross laminations in the rhythmites. Chambers (1984) found that the ripple-drift laminations fall into two categories: an in-drift form and an in-phase form. The in-drift forms are characterized by climbing sets of leeside laminae and stoss-side (windward-side) laminae, with continuity of laminae from one ripple to the next; whereas the in-phase (sinusoidal-ripple laminations) are characterized by superposition of the laminae and are generally symmetrical in profile with lee and stoss sides of equal thickness (Chamber, 1984).

Chambers (1984) also noted the presence of deformation structures; e.g., load casts and associated flame structures, and varves with small-scale décollement that he believed might be due to intrastratal sliding from overburden pressure or possibly the grounding of icebergs. Aside from the small structural detail, the color of the sediments is markedly different from those seen at Ninemile Creek—here white to light gray as opposed to light pink.

Cumulative Miles	(km)	Description
39.4	(63.4)	Turnoff to the St. Ignatius Mission.
45.1	(72.6)	Post Creek. Some investigators consider the underpinnings of the mile (1.6 km) long rampart the road climbs to be a moraine. But, as noted by Smith et al. (2000), evidence collected in the 1990s by a number of investigators now indicates that the rampart is made up of Glacial Lake Missoula lacustrine deposits, evidence that will be reviewed further at Stop 2.
48.2	(77.6)	Turn left (west) onto the road that leads to the U.S. Fish and Wildlife Service viewing area for the Ninepipe Reservoir National Wildlife Refuge.
48.4	(77.9)	Stop 2.

Stop 2. Pingo Scars and the Mission Mountains

The elevation here is ~3045 ft (919 m); at the highest lake stand this site would have been covered by ~1185 ft (361 m) of water. The numerous round depressions in this area of fine-grained sediments have been described as kettles by some (Alt and Hyndman, 1986) and as pingo scars by others (Levish et al., 1993); however, kettles are normally associated with coarse glacial till and drift, and not areas underlain by fine-grained sediment types. The process leading to the development of a pingo scar is well illustrated by Figure 6. In the opinion of Levish et al. (1993), the ponds seen around Ninepipe Reservoir are water-filled pingo remnants or pingo scars since they occur in fine-grained sediment—lake clays and silts—as opposed to the cobbles, pebbles, and coarse sands normally associated with till and drift.

The Mission Mountains to the east are a fault-block mountain range that owes its relief to repeated movement along the Mission Valley fault. McDonald Peak is the range's highest point at 9820 ft (2993 m). Levish et al. (1993), discussing the preliminary findings of a Bureau of Reclamation dam safety investigation, reported the discovery of single-event fault scarp displacements of as much as 20–23 ft (6–7 m) along the western edge of the Mission Mountains. Trenching studies showed the most recent event occurred ca. 4000 yr B.P. Clearly, this is a fault zone that has the potential to produce a significant seismic event.

Cumulative Miles	(km)	Description
64.3	(103.5)	Turn left (west) on Caffery Road. To the right is the Polson Moraine, which on its south side bears evidence of Glacial Lake Missoula shorelines.
65.3	(105.1)	Turn right (north) onto Skyline Drive and proceed up the southern slope of the Polson Moraine.
66.7	(107.3)	Turn east onto the dirt road that leads to a fenced radio tower.
66.8	(107.5)	Stop 3.

Stop 3. Overview of the Polson Moraine and Flathead Lake

During the Pleistocene, large glaciers repeatedly occupied the Rocky Mountain Trench. This long and narrow linear structure extends north from Polson more than 400 mi (645 km) to Valemount, British Columbia, Canada. In that area, the glaciers that occupied the trench began as outflow from the Cordilleran Ice Sheet that covered the central mountains of British Colombia and western Alberta. The highest point on the Polson Moraine is 3487 ft (1063 m) above sea level. The moraine has a general elevation of between 3300 and 3400 ft (1000–1040 m). Only a few large boulders dot the surface of the moraine, a characteristic it shares with many of the moraines in western Montana; this is a characteristic that may indicate that Belt rocks apparently do not weather well so as to persist as large erratics. Smith et al. (2000) speculate that the core of the moraine was deposited somewhere between 19 ka and 15 ka, the time of the last glacial maximum in the area. This is also the time Smith et al. (2000) believe that many of the mountain-valley glaciers that descended into the Mission Valley terminated in the waters of Glacial Lake Missoula, an explanation for the apparent lack of distinctive terminal moraines for many of these glaciers.

Cumulative Miles	(km)	Description
66.9	(107.6)	Leave the radio tower area, turn right and drive west on Skyline Drive to downtown Polson.

69.1	(111.2)	Turn left (west) onto 2nd Avenue (Highway 93).
69.4	(111.7)	Polson Bridge. Flathead Lake has the greatest surface area, 192 mi^2 (496 km^2), of any freshwater lake in the United States west of the Mississippi River. It has a maximum length of 27.3 mi (43.9 km), a width of 15.5 mi (24.9 km), and a maximum depth of 371 ft (113 m). The lake's greatest depth occurs along a long narrow trench on the east side. In many places, the eastern wall of this is the Mission Valley fault.
73.0	(117.5)	Prominent shoreline features can be seen on either side of the highway as it climbs the grade to Jette Meadows. Those shorelines higher than the lowest point on the Polson Moraine belong to Glacial Lake Missoula, whereas those lower could be either from a lower stand of Lake Missoula or due to wave action on the lake, ancestral Flathead Lake, tucked behind the Polson Moraine.
86.4	(139.0)	Elmo and junction of U.S. 93 and State Route 28. Turn left (west) onto State Route 28.
89.7	(144.3)	Stop 4.

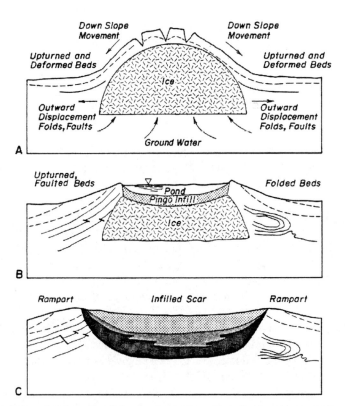

Figure 6. Schematic of pingo scar development. A: During pingo growth material forced outward by expanding ice core. B: Pingo decay and collapse. C: Infilled pingo scar and ramparts. From Levish et al. (1993).

Stop 4. Big Arm Moraine

When the Flathead lobe of the Rocky Mountain Trench glacier reached the Flathead Valley it pushed out in all directions and as far west as the crest of this hill between Elmo and the head of Big Draw, the valley to the immediate west. This splinter of the Flathead lobe has left behind an assemblage of igneous, metamorphic, and well-indurated sedimentary rocks, many of which originated in central British Colombia and bear evidence of their long journey south—faceting and striae.

Cumulative		
Miles	(km)	Description
94.0	(151.0)	Big Draw. Alt and Hyndman (1986) suggest that Big Draw was an ancestral route of the Flathead River, but Larry Smith (Montana Bureau of Mines and Geology [MBMG], 2002, personal commun.) says that recent work by MBMG personnel indicates the that landform and water-well drilling log evidence to support a major drainage way is lacking. The highway skirts the southern edge of the Hog Heaven Hills, the remains of a mineralized Tertiary volcanic complex (Harrison et al., 1986). This 25 mi² (64 km²) assemblage of light-colored quartz latite, latite, andesitic tuff, and basalt is the host for copper, silver, and lead ores that have been extracted in the Hog Heaven mining district, which lies ~6 mi (9.7 km) north (Johns, 1970).
105.1	(169.1)	Turn off to the town of Hot Springs. The hot springs in this area are of the low temperature variety, averaging ~45 °C, hot enough for swimming and bathing, but not hot enough for energy generation (Sonderegger and Bergantino, 1981).
108.0	(173.8)	Junction of State Routes 28 and 382. Turn left (southeast) onto State Route 382.
109.9	(176.8)	Stop 5.

Stop 5. Markle Pass and Camas Prairie Current Ripples

The elevation at this location is ~3300 ft (1006 m); at the highest lake stand this site would have been covered by ~900 ft (275 m) of water. A review of the floods debate history shows that one critical piece of information, more than any other, seems to have done more to tip the balance of belief in Bretz's favor regarding his theories about the formation of the Channeled Scabland—information that was, perhaps, pivotal. That one piece of information was presented by Pardee at a 1940 meeting in Seattle and later in a 1942 Geological Society of America Bulletin article. In that article, Pardee (1942) successfully demonstrated that a series of elongated narrow ridges occupying the Camas Prairie were giant current ripples that he theorized could only have been created by deep and fast-moving water. In places, the giant current ripples seem to flow south down the slopes of

Markle Pass and to spread out across the Camas Prairie. Some of these wave forms obtain a height of 30 ft (9.1 m) and a wavelength of nearly 250 ft (76 m).

A critic of Bretz would be faced with a great dilemma if he or she were to accept Pardee's explanation for these Camas Prairie landforms, because to accept them as current ripples one also has to accept the fact that the responsible waters had to flow westward and through that portion of the Columbia River drainage system now occupied by the Channeled Scabland.

Cumulative		
Miles	(km)	Description
122.0	(196.3)	Junction of State Routes 382 and 200. Turn right (west) onto Highway 200.
124.4	(205.0)	Stop 6.

Stop 6. Gulch Fills

Pardee (1942, p. 1589–1594) described the occurrence of deposits of coarse gravel and small boulders filling many of the higher reaches of gullies, gulches, and drainages adjacent to the Clark Fork and Flathead Rivers, deposits he termed "high eddy deposits" or "gulch fills." These deposits are analogous to pebble, sand, and silt-sized materials being deposited between the crevices of the boulders and rocks by a modern-day stream or rivers during periods of high flow.

Cumulative		
Miles	(km)	Description
159.1	(256.0)	Stop 7

Stop 7. Eddy Narrows

Pardee (1942) calculated the volume of water that he believed moved through Eddy Narrows during the largest of the flood events at 9.5 mi³ (39.9 km³) of water per hour; this at a flow rate of ~44 mi (71 km) per hour. He also noted that the fast-moving water had stripped the surrounding canyon walls of much of their topsoil and talus up to the level he estimated to be the highest stand of the lake (Pardee, 1942).

Cumulative		
Miles	(km)	Description
170.0	(273.5)	Thompson Falls, Montana. End of Day 2.

Day 3. Thompson Falls, Montana, to Spokane, Washington

Cumulative		
Miles	(km)	Description
0.0	(0.0)	Thompson Falls, Montana. Continue west on Highway 200.
10.2	(16.4)	Turn left on Beaver Creek Road and proceed 2.1 mi (3.4 km) to the gravel pit adjacent to the road.
12.3	(19.8)	Stop 1.

Stop 1. Beaver Creek Gravel Pit

This site exposes a section deposited at the mouth of Beaver Creek within the Clark Fork valley. The bedding sets are tens of feet thick and the foresets dip *upriver* to the east (Fig. 7). The clasts are poorly rounded, and some cobbles are striated. Although most of the clasts are of Precambrian Belt metasediments derived locally, some are from granite and diorite (Purcell sills) from rocks exposed in the Purcell Trench. The granules are cemented, and the pit walls must be ripped before the aggregate can be excavated. This gravel probably was deposited in latest-glacial time from the ice margin into Lake Missoula in a pro-glacial Gilbert-type delta. It is not to be confused with the so-called gulch fills or eddy bars of Pardee that formed in the side-valleys during Missoula floods. Most of the drainages on the south side of the Clark Fork valley from Lake Pend Oreille upstream to nearly Thompson Falls contain these features. Thus, late glacial ice advanced upstream much farther than had been previously recognized. Furthermore, late phases of Glacial Lake Missoula associated with this ice dam drained quiescently enough to leave the deposits preserved.

Return to Highway 200 and proceed west.

Cumulative		
Miles	(km)	Description
32.9	(53.0)	Bull River. An ice lobe moved south down the Bull River valley. Although the latest Wisconsin end moraines do not appear to show the ice reached the Clark Fork valley, at glacial maximum Bull River ice would have terminated in Glacial Lake Missoula.
		Just past the Montana-Idaho state line, take entrance to Cabinet Gorge Dam and proceed to the visitors overlook at the dam.
51.2	(82.4)	Stop 2.

Stop 2. Cabinet Gorge Dam Overlook

Cabinet Gorge Dam on the Clark Fork River was completed in 1952 by Washington Water Power (WWP) Company, now AVISTA. The 600 ft (83 m) long and 200 ft (63 m) high true arch dam is constructed in the Libby Formation of the Precambrian Belt Supergroup, a Belt unit formed ca. 1400 Ma. This dam, in coordination with the Albeni Falls Dam downstream, controls the water level of Pend Oreille Lake, normally 2062 ft (628 m). The prominent terrace south of the river is mostly Missoula flood deposits, but logs from monitor wells drilled in 1952 by WWP show cycles of clay till and interbedded lake deposits indicating multiple episodes of ice damming (H.T. Stearns, 1986, personal commun.). Glacial erosion as well as till deposits indicative of an ice margin are found in this area, so many geologists have shown the ice lobe terminus near here. The bedrock bench on the north side of the valley has abundant till cover, interpreted to be ice-marginal deposits; flood drainage was possibly diverted to the south side of the valley by the ice from the north, at least in the waning stages of smaller late floods. The bench may represent the edge of the ice dam failure or the margin of subglacial flow. Just to the north, ice flowed through cols as high as 6000 ft (1830 m) in elevation across the Cabinet Mountains between the Bull River and the Purcell Trench.

Glacial Lake Missoula was determined by Pardee (1942) to contain 500 mi^3 (2167 km^3) of water with a surface area of 2900 mi^2 (7424 km^2). These figures were based on the elevation of the highest evidence for a lake level from evidence of old shorelines. He determined the highest shorelines on the mountain slopes in the Missoula Valley at an elevation of at least 4250 ft (1295 m). An old aerial photograph of the lower Clark Fork valley (Fig. 8) reveals spectacular shorelines along the valley directly south of Cabinet Gorge at least as high as at Missoula.

Return to Highway 200 and turn west toward Clark Fork.

Figure 7. Photograph of Beaver Creek gravel pit showing Gilbert-type foresets dipping upstream. These deposits are interpreted to be proglacial delta building into Glacial Lake Missoula.

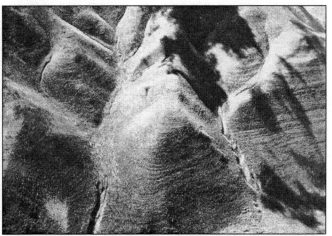

Figure 8. Oblique aerial photograph ca. 1930. Shorelines of Lake Missoula near Dry Creek in this postfire photograph are evidence for a maximum stand of the lake near the mouth of the Clark Fork River. Photograph courtesy of the Washington Air National Guard.

Cumulative		
Miles	*(km)*	*Description*
57.9	(93.2)	Immediately east of Clark Fork large glacial grooves and ice striations can be seen along the road in the Wallace and Striped Peak Formations. This is a very hazardous site to visit due to highway traffic; a drive-by is recommended. Continue west through Clark Fork. Continue through Hope on Highway 200 and across the bridge at the boat basin. Turn out on the left at the highway sign.
69.8	(112.4)	Stop 3.

Stop 3. Purcell Trench and Cordilleran Ice–Highway Overlook

Lake Pend Oreille in northern Idaho is a fascinating geomorphic feature. This lake, the deepest by far in the region, lies in a basin formed by Cordilleran glaciation immediately below the site of the ice dams that repeatedly formed Pleistocene Lake Missoula. The Cordilleran Ice Sheet extended farthest along major south-trending valleys and lowlands, forming several composite lobes segregated by highlands and mountain, leaving behind distinctive landforms, weathering, and soils in these valleys. This view provides a vista of the intersection of the Clark Fork valley and the Purcell Trench as well as the area of the Clark Fork ice dam and Lake Pend Oreille trough (Fig. 9). The Purcell Trench, a major topographic depression in British Columbia and northern Idaho, is incised into the margin of a metamorphic and granitic complex, formed during Mesozoic convergent tectonics, but later subjected to Eocene extension. The trench is bounded on the west by the Selkirk Range and on the east by the Cabinet Mountains. The Purcell Trench has been

the site of several drainage reversals. Prior to Miocene time (22 Ma), ancestral river ranges flowed within the Purcell Trench from the Canadian border south to the Coeur d'Alene area in a meandering pattern following the least resistant rock exposures and fault zones. This pattern is still apparent in the present day meandering shape of Lake Pend Oreille. During Miocene time, Columbia River Basalt Group invaded the southern part of the trench and formed a lava-dammed lake with water levels up to ~2400 ft (800 m) above sea level (Savage, 1967). However, the basalt dam eventually was eroded, restoring southward drainage. During the Pleistocene, Cordilleran ice advancing into Lake Pend Oreille repeatedly blocked the Clark Fork drainage and formed Glacial Lake Missoula (Pardee, 1910). The terminus of the glacial advance extended to the southern end of the lake and upon each failure of the ice dam, much of the catastrophic flood discharge was directed through the lake basin and into the Rathdrum Prairie (O'Connor and Baker, 1992). Paleomagnetic analysis of lake filling and emptying cycles near the dam and in Lightning Creek shows secular variation in each of the lake filling cycles, but that analysis shows none within the instantaneous flood cycles (Breckenridge and Othberg, 1998). The Cordilleran late Wisconsin ice was north of the Canadian border as late as 17,500 yr B.P. and the maximum in the Purcell Trench was reached ca. 15,000 yr B.P. (Clague et al., 1980). Glacial lobes moved south from Canada again blocking the south end of the trench and impounding Lake Pend Oreille basin, this time with glacial debris. In eastern Washington, Wisconsin glacial lobes dammed the Columbia River and its tributary drainages, forming Pleistocene Lake Columbia (Richmond, 1986; Waitt and Thorson, 1983) with a water level up to 2400 ft (800 m) above sea level. Lake Columbia covered the Rathdrum Prairie and cre-

Figure 9. Aerial view, southwest from above Clark Fork, Idaho, of Green Monarch Mountain and the location of the ice dam at the eastern end of Pend Oreille Lake.

ated proglacial waters in front of the ice sheets advancing down the Purcell Trench. The present outlet of Lake Pend Oreille is at the northwest arm of the lake issuing into the Pend Oreille River, which then flows to Washington and north into British Columbia. No modern surface drainage of the northern Rathdrum Prairie exists; all water flow is subsurface.

One of the most intriguing questions about the catastrophic flooding is how the ice dam failed. Various mechanisms for glacial outburst floods have been proposed: Ice erosion by overflow water, subglacial failure by flotation, deformation of ice by water pressure, and erosion of subglacial tunnels by flowing water. One popular model suggests a self-dumping phenomenon. In this mechanism, floodwaters are released when the lake level reaches nine-tenths the height of the ice. At this depth the increasing hydrostatic pressure makes several things happen: The ice becomes buoyant, subglacial tunnels form and enlarge, and drainage occurs until hydrostatic pressure is decreased and the ice again seals the lake. The self-emptying model is used to explain the numerous cycles in the rhythmite deposits and to interpret each cycle as a separate flood. Even so, only the total collapse of the ice dam can explain the largest of the catastrophic floods. Sub-glacial tunneling and enlargement due to thermal erosion progressing to collapse have also been proposed as well as catastrophic failure due to water pressure. All are dependent on the configuration of the ice dam and structure of the ice.

After the latest flood, the Pend Oreille basin was again reoccupied by a glacier but this advance did not result in catastrophic lake drainage. Terminal deposits of till at the southern end of Lake Pend Oreille are undisturbed. Proglacial deltas and kame terraces in the Clark Fork valley left by this advance are intact and therefore postdate the catastrophic Missoula floods. Furthermore, giant current ripples, expansion bars, and other flood deposits in the lower Clark Fork valley are mantled by lacustrine silts, indicating that the last Glacial Lake Missoula did not drain catastrophically. Waitt (1985) interprets ^{14}C evidence to show that Glacial Lake Missoula existed only between ca. 17,200 and 11,000 yr ago. The Purcell Trench was free of Cordilleran ice by the time of two closely spaced Glacier Peak eruptions ca. 11,200 yr ago (Carrara et al., 1996). Latest glacial alpine moraines formed by valley glaciers in the Selkirk and Cabinet Ranges occupy the tributary valleys of the Purcell Trench. These deposits must postdate retreat of the Cordilleran ice lobe and are evidence for a late Wisconsin alpine glacial advance. An uncalibrated radiocarbon age of 9,510 ± 110 yr B.P. (Mack et al., 1978) from a peat bog within the glacial limit on the west slope of the Selkirk Mountains is a minimum-limiting age on the last alpine ice.

Continue west on Highway 200 to Sandpoint.

| Cumulative | | |
Miles	(km)	Description
84	(135.2)	In Sandpoint, turn south on U.S. 95.
107.8	(173.5)	Turn right on Homestead Road, cross the railroad tracks, and park near the large boulders.
107.9	(173.6)	Stop 4.

Stop 4. Gravel Bar–Railroad Cut

This railroad cut provides a view through the middle of a huge gravel expansion bar deposited by floods from the outburst area. The huge boulders stacked here were removed from the cut, which was recently widened to accommodate double track. The position of the boulders in the bar indicates that they were not ice-rafted but carried in the bedload of the high-energy floodwaters.

Return to U.S. 95 and turn south to the junction with Highway 54 in Athol. Turn left on Highway 54 to Farragut State Park. Check in at the visitor center. Fees may apply. Take the South Road to the Sunrise-Willow day-use areas.

| Cumulative | | |
Miles	(km)	Description
118.6	(190.9)	Stop 5.

Stop 5. Farragut State Park, Sunrise-Willow Day-use Area

Farragut State Park is located at the "breakout" of Glacial Lake Missoula floods. Failure of the ice dam in the Clark Fork valley fractured and broke apart the 20-mi-long tongue of ice occupying the lake basin and a torrent of water and ice burst from the lake. Churning flood waters flowed 2000 ft (610 m) deep across Farragut State Park.

Lake Pend Oreille, the largest lake in Idaho, is located ~80 km south of the British Columbia border in the Purcell Trench. The lake level is held at 2062 ft (629 m) above sea level, with the surrounding terrain as high as 6002 ft (1830 m). The maximum depth of the lake is an impressive 1150 ft (351 m). This depth makes it the deepest lake, by far, in the region. The location of the lake is probably related to an old river valley controlled by faults. Lake Pend Oreille was carved repeatedly by a lobe of Pleistocene ice, scoured by ice age floods and filled with glacial outwash and flood deposits. The lake is now dammed at the south end by thick glacial and flood deposits underlying Farragut State Park. Little is known about Lake Pend Oreille basin because the lake bottom has never been cored. The great depth of the lake and the thickness of the lake bottom sediment make it unlikely that such a core will be obtained in the foreseeable future. However, the Idaho Geological Survey obtained data from United States Navy seismic reflection surveys performed on the lake. Although these surveys were conducted primarily for geotechnical studies of the lake-bottom surface, the data show distinct sub-bottom reflections. The seismic reflection surveys show that the bedrock lake basin has been glacially overdeepened to a depth more than 500 ft (152 m) below present-day sea level. The seismic sections are interpreted to show a record of subglacial erosion, Missoula flood deposition, and a postflood glacial readvance (Breckenridge and Sprenke, 1997).

Seismic studies of the bedrock morphology and sedimentary facies of inland linear valleys, such as the Okanagan and Kalamalka valleys in British Columbia, have revealed bedrock erosion to depths well below sea level and subsequent rapid infilling by sediment (Gilbert, 1975; Fulton and Smith, 1978; Mullins et al., 1990; Eyles et al., 1990; Desloges and Gilbert, 1991). These valleys, which are similar to Lake Pend Oreille basin, contain substantial sediment thicknesses and owe their locations to structural lineaments and their morphology to large-scale glacial erosion beneath Cordilleran Ice Sheets. Shaw et al. (1999) proposed that flood releases beneath the British Columbia Ice Sheet were a contributing source of scabland flooding. These are all elongate, deep valleys, controlled by preexisting fluvial channels, preferentially carved along preexisting structural features at the southern margins of the Cordilleran Ice Sheet.

Return to Athol and at the intersection of U.S. 95, proceed west on State Highway 54 to Ramsey Road.

Cumulative Miles	(km)	Description
131.6	(211.8)	Stop 6.

Stop 6. Giant Current Ripples–Ramsey Road

One of Bretz's most important pieces of evidence for catastrophic flooding was the "giant current ripples." These large-scale bedforms appeared as patterns of parallel ridges and swales on many aerial photographs in the flood channels in the scabland of Washington but had escaped recognition from the ground because of their large size. The ripples form transverse to the current direction and form cusps that are convex upstream. As in dunes, the arms point downstream. Furthermore, the size of the cusps appears to decrease in the direction of lower velocity. Internally the ripples consist of gravel and pebble foresets. Giant current ripples exhibit an asymmetrical profile with the downstream (lee) slope steeper than the upstream slope. Crests range from 60 to 600 ft (20–200 m) apart and heights range from 3 to 45 ft (1–15 m) (Baker, 1978) and are among the largest measured throughout the flood area. The height of the crests and distance between crests as well as the particle size show consistent relationships that can be related to flow parameters such as the depth, velocity, and stream power in which they formed. The Spirit Lake current ripples can also be easily recognized from the air by their characteristic pattern accentuated by vegetation. This ripple field is immediately in the path of the breakout from Lake Pend Oreille and experienced some of the highest energy flows. The number of crests between highway mile markers can be readily counted while driving this section of Highway 54.

Continue west on Highway 54 to junction Highway 41.

Cumulative Miles	(km)	Description
134.6	(216.6)	Junction of State Highways 54 and 41. Turn south on Highway 41 toward Rathdrum.
140.0	(225.0)	Several clusters of giant boulders are located along the highway between Spirit Lake and Twin Lakes. The boulders are mostly a granodiorite, similar to an outcrop near Bayview. The rock type, plus the fact that geophysical surveys of the immediate area reveal an ancestral valley, shows the boulders are not

outcrop but have been transported. The largest boulder measures 49 × 40 ft (15 × 12 m) and weighs over 1600 tons (J. Browne, 1999, personal commun.). Their size seems anomalously large even for catastrophic floods. Perhaps they were deposited by a tremendous base surge or were rafted by a great berg of the disintegrated ice lobe.

At Rathdrum follow the signs for State Highway 41 through Rathdrum. Continue south across the prairie.

152.7   (245.7)   Junction State Highway 41 and Interstate 90.

End of Field Trip and Trip Log–Return to Spokane. End of Day 3.

REFERENCES CITED

Adema, G.W., Sprenke, K.F., and Breckenridge, R.M., 1998, Bed Morphology of the Spokane Valley/Rathdrum Prairie Aquifer from a Detailed Gravity Survey: Geological Society of America Abstracts with Programs, v. 30, no. 6, p. 1.

Alden, W.C., 1953, Physiography and glacial geology of western Montana and adjacent areas: U.S. Geological Survey Professional Paper 231, 200 p.

Alt, D., and Hyndman, D.W., 1986, Roadside Geology of Montana: Mountain Press Publishing Company, Missoula, Montana, 427 p.

Alt, D., 2000, The catastrophic drainage of Glacial Lake Missoula, *in* Roberts, S., and Winston, D., eds., Geologic field trips, western Montana and adjacent areas: Rocky Mountain Section of the Geological Society of America, University of Montana, p. 31–39.

Baker, V.R., 1978, Large-scale erosional and depositional features of the Channeled Scabland, *in* Baker, V.R., and Nummedal, D., eds., The Channeled Scabland: National Aeronautics and Space Administration, p. 81–115.

Bennett, E.H., Siems, P.I., and Constantopoulos, J.T., 1989, The geology and history of the Coeur d'Alene Mining District, *in* Chamberlain, V.E., Breckenridge, R.M., and Bonnichsen, B., eds., Guidebook to the Geology of Northern and Western Idaho and Surrounding Area: Idaho Geological Survey Bulletin 28, p. 137–156.

Breckenridge, R.M., and Othberg, K.L., 1998, Chronology of Glacial Lake Missoula Floods–Paleomagnetic Evidence from Pleistocene Lake Sediments near Clark Fork, Idaho: Geological Society of America Abstracts with Programs, v. 30, no. 6, p. 5.

Breckenridge, R.M., and Othberg, K.L., 1999a, Surficial Geologic Map of the Coeur d'Alene Quadrangle, Kootenai County, Idaho: Idaho Geological Survey SGM-7, 1:24,000.

Breckenridge, R.M., and Othberg, K.L., 1999b, Surficial Geologic Map of the Post Falls Quadrangle and Part of the Liberty Lake Quadrangle, Kootenai County, Idaho: Idaho Geological Survey SGM-5, 1:24,000.

Breckenridge, R.M., and Sprenke, K.F., 1997, An overdeepened glaciated basin, Lake Pend Oreille, northern Idaho: Glacial Geology and Geomorphology, rp01/1997.

Carrara, P.E., Kiver, E.P., Stradling, D.F., 1996, The southern limit of Cordilleran ice in the Colville and Pend Oreille valleys of northeastern Washington during the late Wisconsin glaciation: Canadian Journal of Earth Science, v. 33, p. 769–788.

Chambers, R.L., 1984, Sedimentary evidence for multiple Glacial Lakes Missoula: Montana Geological Society 1984 Northwestern Montana Field Conference Guidebook, p. 189–199.

Clague, J.J., Armstrong, J.E., and Mathews, W.H., 1980, Advance of the late Wisconsin Cordilleran ice sheet in southern British Columbia since 22,000 yr. BP: Quaternary Research, v. 13, 322–326.

Desloges, J.R., and Gilbert, R., 1991, Sedimentary record of Harrison Lake: implications for deglaciation in southwestern British Columbia: Canadian Journal of Earth Sciences, v. 28, p. 800–815.

Dort, W., Jr., 1960, Glacial Lake Coeur d'Alene and berg-rafted boulders: Idaho Academy of Science Journal, v. 1, p. 82–92.

Eyles, N., Mullins, H.T., and Hine, H.C., 1990, Thick and fast: Sedimentation in a Pleistocene fiord lake of British Columbia, Canada: Geology, v. 18, p. 1153–1157.

Fulton, R.J., and Smith, G.W., 1978, Late Pleistocene stratigraphy of south-central British Columbia: Canadian Journal of Earth Sciences, v. 15, p. 971–980.

Gilbert, R., 1975, Sedimentation in Lillooet Lake, British Columbia: Canadian Journal of Earth Sciences, v. 12, p. 1697–1711.

Harrison, J.E., Griggs, A.B., and Wells, J.D., 1986, Geologic and structure of the Wallace 1° × 2° Quadrangle, Montana and Idaho: Montana the Bureau of Mines and Geology Montana Atlas 4-A, 1:250,000.

Johns, W.M., 1970, Geology and mineral deposits of Lincoln and Flathead Counties, Montana: Montana Bureau of Mines and Geology Bulletin 79, 182 p.

Levish, D., Ostenaa, D., and Klinger, R., 1993, Quaternary geology of the Mission Valley, Montana: Guidebook—Friends of the Pleistocene 1993 Rocky Mountain field trip, September 10–12, 1993: Seismotectonic and Geophysics Section, Geology Branch, Geotechnical Engineering and Geology Division, Denver Office, U.S., Bureau of Reclamation, Seismotectonic Report 93-7, 23 p.

Lewis, R.S., 1998, Geologic map of the Missoula west 30′ × 60′ quadrangle: Montana Bureau of Mines and Geology Open File Report 373, scale 1:100,000.

Lewis, R.S, Burmester, R.F., Breckenridge, R.M., McFadden, M.A., and Kauffman, J.A., 2002, Geologic Map of the Coeur d'Alene 30′ × 60′ Quadrangle, Kootenai and Shoshone Counties, Idaho: Idaho Geological Survey Geologic Map 33, scale 1:100,000.

Mack, R.N., Rutter, N.W., Bryant, V.M., Jr., and Valastro, S., 1978, Re-examination of postglacial vegetation history in northern Idaho: Hager Pond, Bonner County: Quaternary Research, v. 10, p. 241–255.

McCarter, S., 1992, Guide to The Milwaukee Road in Montana: Helena, Montana, Montana Historical Press, 104 p.

McMurtrey, R.G., Konizeski, R.L., and Brietkrietz, 1965, Geology and ground-eater resources of the Missoula Basin, Montana: Montana Bureau of Mines and Geology Bulletin, v. 47, 36 p.

Mullins, H.T., Eyles, N., and Hinchey, E.J., 1990, Seismic reflection investigation of Kalamalka Lake: a "fiord lake" on the Interior Plateau of southern British Columbia: Canadian Journal of Earth Sciences, v. 27, p. 1225–1235.

O'Connor, J.E., and Baker, V.R., 1992, Magnitudes and implications of peak discharges from Glacial Lake Missoula: Geological Society of America Bulletin, v. 104, p. 267–279.

Pardee, J.T., 1910, The Glacial Lake Missoula: Journal of Geology, v. 18, p. 376–386.

Pardee, J.T., 1942, Unusual currents in Glacial Lake Missoula, Montana: Geological Society of America Bulletin, v. 53, p. 1569–1599.

Richmond, G.R., 1986, Tentative Correlation of deposits of the Cordilleran Ice-Sheet in the Northern Rocky Mountains, *in* Richmond, G.M., and Fullerton, D.S., eds., Quaternary Glaciations in the Northern Hemisphere: Quaternary Science Reviews, v. 5, p. 99–128.

Savage, C.N., 1965, Geologic history of Pend Oreille Lake region in north Idaho: Idaho Bureau of Mines and Geology pamphlet, 18 p.

Shaw, J., Munro-Stasuik, M., Sawyer, B., Beaney, C., Lesemann, J-E., Musacchio, A., Rains, B., and Young, R.R., 1999, The Channeled Scabland: Back to Bretz: Geology, v. 27, p. 605–608.

Smith, L.N., Blood, L., and LaFave, J.I., 2000, Quaternary geology, geomorphology, and hydrogeology of the upper Flathead River valley area, Flathead County, Montana, *in* Roberts, S., and Winston, D., eds., Geologic field trips, western Montana and adjacent areas: Rocky Mountain Section of the Geological Society of America, University of Montana, p. 41–63.

Sonderegger, J.L., and Bergantino, R.N., 1981, Geothermal resources map of Montana: Montana Bureau of Mines and Geology Hydrologic Map 4, 3 p., 1 sheet, scale 1:1,000,000.

Van West, C., 1986, A Traveler's Companion to Montana History: Montana Historical Society Press, Helena, Montana, 239 p.

Waitt, R.B., Jr., 1980, About forty last-Glacial Lake Missoula jokulhlaups through southern Washington: Journal of Geology, v. 88, p. 653–679.

Waitt, R.B., Jr., 1985, Case for periodic, colossal jokulhlaups from Pleistocene Glacial Lake Missoula: Geological Society of America Bulletin v. 96, p. 1271–1286

Waitt, R.B., Jr., and Thorson, R.M., 1983, The Cordilleran Ice Sheet in Washington, Idaho, and Montana, *in* Wright, H.E., Jr., ed., Late-Quaternary Environments of the United States, Volume 1: The Late Pleistocene, Porter, S.C., ed.: University of Minnesota Press, p. 53–70.

Printed in the USA

Geological Society of America
Field Guide 4
2003

Sequence stratigraphy of the Sauk Sequence:
40th anniversary field trip in western Utah

Kevin R. Evans
James F. Miller
Department of Geography, Geology, and Planning, Southwest Missouri State University, Springfield, Missouri 65804, USA

Benjamin F. Dattilo
Geosciences Department, University of Nevada Las Vegas, Las Vegas, Nevada 89154, USA

ABSTRACT

The Sauk Sequence comprises more than 5 km of mixed carbonate and siliciclastic strata on the Paleozoic miogeocline of the eastern Great Basin. Rapid, post-rifting subsidence was the single most important factor for providing accommodation for accumulation of sediments. Despite the enormous thickness of strata and the tendency for unconformities to die out toward the margin of the continent, bounding surfaces of the Sauk Sequence and several sequence boundaries within this interval are preserved in mountain ranges of western Utah. The base and top of the Sauk Sequence are thick sandstones. The development of microkarst or truncation surfaces associated with major facies disclocations and deposition of major influxes of siliciclastics are the hallmarks of sequence boundaries and correlative conformities in this setting. The style of sequence boundary development was mostly a function of magnitude and duration of sea-level fall but was also influenced by tectonic features such as the House Range Embayment.

Keywords: sequence stratigraphy, Sauk Sequence, Cambrian, Ordovician, Utah, tectonics.

INTRODUCTION

The year 2003 marks the fortieth anniversary of publication of Larry Sloss's seminal paper on interregional unconformities and large-scale depositional sequences of North America (Sloss, 1963). The Sauk Sequence was named in an earlier publication (Sloss et al., 1949), but Sloss (1988) acknowledged that the ideas set forth in his 1963 paper were the ones geologists most commonly cite. Combined with the development of seismic stratigraphy, it was Sloss's 1963 paper that provided the fundamental concepts for interpreting the history of ancient sea-level changes.

The year 2003 also is the hundredth anniversary of Charles Walcott's expedition to the House Range, and it marks the fiftieth anniversary of publication of Lehi Hintze's paper on the trilobites and biostratigraphy of the Ordovician Pogonip Group. These authors named many stratigraphic units in the Cambrian part (Walcott, 1908) and Ordovician part (Hintze, 1951) of the

Sauk Sequence. These and other strata were mapped recently at a 1:100,000 scale by Hintze and Davis (2002a, b).

Consequently, we observe these anniversaries and celebrate this unique interval in Earth's history with a field trip in an area that records a remarkably complete saga of Cambrian-Ordovician sea-level fluctuations. Western Utah (Fig. 1) has an arguably nearly continuous record of the Sauk Sequence that was deposited on the subsiding western margin of Laurentia. Lower Cambrian to lower Triassic strata in the House Range and Confusion Range to the west total ~32,000 ft (9.8 km), and of this the Sauk Sequence accounts for more than half, ~16,900 ft (5.1 km).

SEQUENCE STRATIGRAPHY

As sequence stratigraphy developed and evolved over the last quarter of the twentieth century, so too has the scale of observation. Many sequences constitute the Sauk Sequence. In

Evans, K.R., Miller, J.F., and Dattilo, B.F., 2003, Sequence stratigraphy of the Sauk Sequence: 40th anniversary field trip in western Utah, *in* Swanson, T.W., ed., Western Cordillera and adjacent areas: Boulder, Colorado, Geological Society of America Field Guide 4, p. 17–35. For permission to copy, contact editing@geosociety.org. © 2003 Geological Society of America.

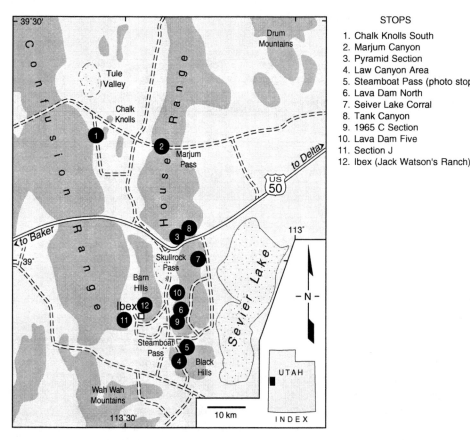

STOPS
1. Chalk Knolls South
2. Marjum Canyon
3. Pyramid Section
4. Law Canyon Area
5. Steamboat Pass (photo stop)
6. Lava Dam North
7. Seiver Lake Corral
8. Tank Canyon
9. 1965 C Section
10. Lava Dam Five
11. Section J
12. Ibex (Jack Watson's Ranch)

Figure 1. Location of field trip stops in western Utah.

cratonic settings it is not uncommon to recognize sequences that range from a few meters to tens of meters in thickness, whereas sequences typically translate to tens or hundreds of meters on the rapidly subsiding miogeocline. During the Cambrian and Ordovician, the western margin of Laurentia was a carbonate platform that had lesser amounts of siliciclastics, and keep-up sedimentation was the norm. Sea-level drops that may have produced unconformity on the craton are recorded as lowstand strata. As a result, in this miogeoclinal setting it is difficult to recognize the characteristic features of an ideal sequence, such as maximum flooding surfaces or zones (Fig. 2), in part because ideal sequence models were developed in very different depositional settings. Consequently, we do not utilize systems tract terminology. Our use of "lowstand" or "highstand" simply refers to conditions or sediments deposited during relatively low or high sea level.

In our detailed studies of the Orr Formation through the House Limestone (Fig. 3), we recognize sequence boundaries as actual surfaces of unconformity or erosional truncation, although some boundaries can be shown to correlate with such surfaces in other areas. We also recognize smaller units, packages, and package boundaries that usually are conformable but are associated with facies shifts that may or may not correlate with cratonic sequence boundaries. Our detailed sequence stratigraphy in the Orr Formation–House Limestone interval is controlled by exten-

sive, detailed biostratigraphy and insoluble residue data from conodont samples (Miller et al., 2003).

SAUK SEQUENCE IN WESTERN UTAH

The central House Range has long been considered a Cambrian Mecca (Hintze and Robison, 1987), and spectacular exposures of gently dipping lower Ordovician strata in the southern House Range are classics since the work of Hintze (1951, 1953). These strata are now the reference sections for North America's upper Cambrian Millardan Series (Palmer, 1998) and lower Ordovician Ibexian Series (Ross et al., 1997).

Sloss (1963) argued that profound unconformities bounding his sequences graded into more continuous deposition on continental margins. In western Utah, the base of the Sauk Sequence is the base of the Prospect Mountain Quartzite. An unconformity at that level is exposed in the Drum and Wah Wah Mountains, where the Prospect Mountain disconformably overlies the Mutual Formation (Fig. 3). In the southern Great Basin, this surface is a disconformity.

Palmer (1981) divided the Sauk Sequence into three parts, identified as Sauk I, II, and III. Golonka and Kiessling (2002) recognized a Sauk IV division. Osleger and Read (1993) identified several third-order sequences in upper Cambrian strata in several parts of the United States, including western Utah. Ross

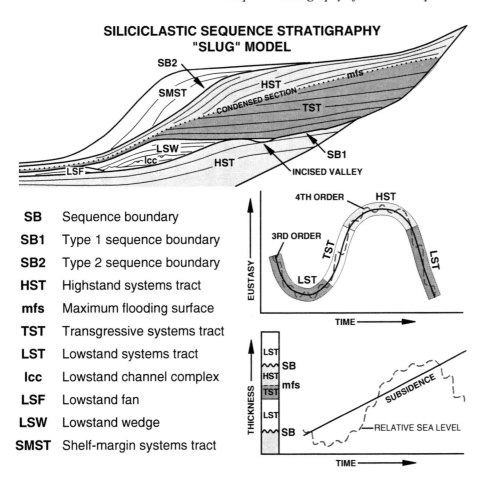

SILICICLASTIC SEQUENCE STRATIGRAPHY "SLUG" MODEL

SB	Sequence boundary
SB1	Type 1 sequence boundary
SB2	Type 2 sequence boundary
HST	Highstand systems tract
mfs	Maximum flooding surface
TST	Transgressive systems tract
LST	Lowstand systems tract
lcc	Lowstand channel complex
LSF	Lowstand fan
LSW	Lowstand wedge
SMST	Shelf-margin systems tract

Figure 2. Traditional model for siliciclastic sequence stratigraphy.

and Ross (1995) have identified third-order sequences in Ordovician strata in the United States, including western Utah. Parts of the interval covered by those authors were subdivided into smaller sequences by Miller et al. (2003).

We will not examine the Prospect Mountain Quartzite, which is exposed in the northern part of the House Range. The top of the Sauk Sequence in the Great Basin is the top of the widespread Eureka Quartzite, which is overlain by the Ely Springs Dolomite. Hintze (1988, fig. 26) shows a large hiatus at this contact.

INFLUENCE OF REGIONAL TECTONICS

Depositional facies of strata from the Wheeler Shale to the Notch Peak Formation were influenced by three regional tectonic features: the Tooele Arch on the north, the House Range Embayment in the central House Range, and the Wah Wah Arch (Miller et al., 2003) in the southern House Range and Wah Wah Mountains. Disjunct facies distribution and regional thickness patterns (strata are thicker in the central House Range) provided evidence for the existence of a persistent depositional basin that Rees (1986) identified as the House Range Embayment. This tectonic feature trends generally NE–SW across western Utah, Nevada, and to the California-Nevada border. This relatively narrow embayment along the margin of Laurentia brought deep marine

environments far into the miogeocline. The southern boundary apparently was a growth fault. Deep water conditions graded northward into a shallow ramp over the Tooele Arch. Detailed study (Miller et al., 2003) of facies changes in the Notch Peak Formation indicates that the south margin of the House Range Embayment was a little north of Skull Rock Pass.

Sequences change thickness and facies and some sequence boundaries die out into lateral conformities as they are traced from shallow-marine settings on the Wah Wah Arch into the deep, rapidly subsiding House Range Embayment. During the trip we will demonstrate such lateral changes in a general way for the Marjum-Weeks/Pierson Cove–Trippe–Wah Wah Summit interval and in more detail for the upper Notch Peak Formation (Fig. 3).

DAY 1

Leave motel at 7:30 a.m. and drive west from Delta, Utah, on U.S. Highway 6 (also U.S. 50) for 32 mi (52 km). Turn right onto Long Ridge Reservoir Road at the U-Dig Fossils sign. Drive northwest ~10 mi (16 km) to the junction with Marjum Canyon Road. Turn left, and follow Marjum Canyon Road 12 mi (19 km) to the top of Marjum Pass. Proceed 5.8 mi (9.4 km) to the mouth of the canyon. Drive west 2.6 mi (4.2 km) to Tule Valley Road, and proceed west 7.4 mi (11.9 km) to a five-road intersection;

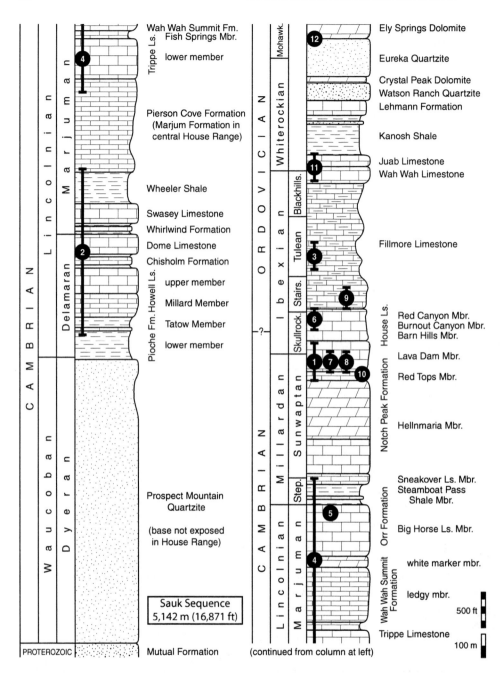

Figure 3. Generalized composite strati-graphic sections of Sauk Sequence strata in the House Range and Confusion Range, western Utah. Strata examined at numbered field trip stops are indicated by black-filled circles and bars. Chronostratigraphy is based on Palmer (1998) and Ross et al. (1997). Stratigraphic thicknesses of units is after Hintze (1988).

turn hard left. Proceed 4.5 mi (7.3 km), turn left onto the Lake Bonneville gravel terrace. Drive 0.2 mi (0.3 km) to the end of the terrace.

Stop 1. Chalk Knolls South

Uppermost Cambrian and Lowermost Ordovician Sequences in the House Range Embayment

We stop here first so as to catch the morning light on these exposures. This painted section is in the middle of the House Range Embayment. Exposed here is the top of the Hellnmaria Member, all of the Red Tops and Lava Dam members of the Notch Peak Formation, and most of the Barn Canyon Member of the House Limestone (Fig. 3). Miller et al. (2003) divided the Sauk III interval into 13 sequences (Fig. 4). Exposures here include Sequence 3 to the middle of Sequence 11.

The lowest exposure is peritidal peloid grainstone at the top of Sequence 3, which forms the top of the Hellnmaria Highstand, an extended period of generally high sea level. The Sequence 3–4 boundary is a double karst truncation surface at 0 and 3 ft that can be traced to near the southern end of the House Range; this doublet also occurs in central Texas. Microbiolites occur here in Sequences 4, 5, 8, 9, and 10. A widespread microbialite is truncated at the top of Sequence 4, at 20 ft in the section. The Red Tops Lowstand in lower Sequence 5 contains multiple microbial units. Package 5C and Sequence 6 are transgressive to highstand lime mudstone in the lower Lava Dam Member.

Here, Sequence 7 (Lange Ranch Lowstand) cannot be divided into the three packages that are separated by truncation surfaces near Skull Rock Pass; we will see these packages tomorrow afternoon at Stops 7 and 8. Rapid subsidence in the Chalk Knolls area produced a record of continuous sedimentation (lateral conformity). The base of the Ibexian Series (247 ft) is slightly above the base of the Lange Ranch Lowstand in other sections. The top of Sequence 7 should be slightly above the lowest *Fryxellodontus inornatus*, at 317 ft, but the Sequence 7–8 boundary is unclear here because of the lateral conformity compared with sections farther south.

Sequences 8 and 9 are also conformable and consist of a thick microbiolite pile from 317 to 581 ft in the upper half of the Lava Dam Member. Stromatolite growth began at the base of Sequence 8, and they kept growing as sea level rose. A thin break in the continuity of these stromatolites at 430 ft is the Sequence 8–9 boundary, but stromatolites fill Sequence 9. Packages 9B and 10A (Fig. 4) are missing at a karst surface at the top of the interval, but these packages are present in sections to the south. This interval, the Basal House Lowstand, records a major change in local depositional patterns. The Notch Peak Formation thickens northward into the House Range Embayment, but the basal House Limestone is condensed here and thickens to the south. This change in depositional pattern is at the karst surface at the Notch Peak–House contact, and Sequence 10 is unusually thin.

The top of Sequence 10 at 619 ft is a thin but widespread microbialite that is truncated at the base of the Drum Mountains Lowstand, which correlates with the *Acerocare* Regressive Event in Europe (Figs. 4 and 5). This level is a significant unconformity in central Texas. Sequence 11 begins at 619 ft and consists of a series of thin packages that comprise the Stonehenge Transgression (Tremadocian Transgression). The widespread olenid trilobite *Jujuyaspis borealis* at 660 ft indicates correlation with the base of the Tremadocian Series of Europe. The lowest occurrence of the conodont *Iapetognathus fluctivagus* is at the same footage.

The lowest local occurrence of this conodont was chosen to characterize the base of the Ordovician System at the Global Stratotype Section at Green Point in Newfoundland. This section is a deep slope facies, and conodonts near the boundary point have been eroded from the shelf or upper slope and redeposited at the base of the continental slope in an inverted succession (Miller et al., 2003). Thus, it is impossible to correlate this stratotype point to sections where conodonts occur in a normal succession. The highest exposures at 742 ft are in the *Cordylodus angulatus* Zone.

Retrace the route to Marjum Pass Road and drive into Marjum Canyon; stop just above the mouth of the canyon.

Stop 2. Marjum Canyon

Sequences in Lower to Middle Cambrian Strata

Strata exposed in Marjum Canyon range from the Pioche Formation to the Pierson Cove Formation (Fig. 3). The interval is too thick for us to see it all, but we will make a series of stops to see representative intervals. Figure 5 shows inferred sequences and a sea-level curve for the entire Sauk Sequence.

Stop 2a. Pioche Formation

Top of Sauk I Transgressive Siliciclastic Complex

The lower Pioche Formation is the top of ~1200 m of siliciclastics at the base of the Sauk Sequence. The Pioche consists of micaceous shales, siltstones, and sandstones that overlie the Prospect Mountain Quartzite, which is exposed for miles north of Marjum Canyon. This stratigraphic unit is radically different from any other shaly unit in the House Range and more closely resembles the Proterozoic Inkom Formation. Gamma-ray spectroscopy of the Pioche reveals its highly radioactive nature, and readings range upward of 300 counts per second. The lower member probably includes sediments derived by deep erosion of the igneous craton and deposition of shales on the shoulders of the continent. This erosion may be related to the Sauk I–II regression.

The contact between the lower member and the Tatow Member of the Pioche Formation can be identified by a change from green to yellow or rusty-brown shale with silty thin carbonate beds. Strata of the Tatow Member are the first carbonates in the Cambrian succession and mark a deepening event. The Tatow may record the Sauk II transgression that followed the Sauk I–II regression of Palmer (1981), or that deepening may be near the base of the overlying Millard Member.

Drive up Marjum Canyon Road for 0.4 mi (0.6 km); turn left into the side canyon and park. Walk up the side canyon.

Stop 2b. Howell Limestone

Highstand Carbonates and Chisholm Lowstand

The Howell is divided into two members, a lower, dark gray Millard Member and a light-colored upper member. The Millard Member has many oncoids and burrows. The upper member is fine grained and light in color, typical of peritidal to supratidal carbonates. The Howell is a highstand deposit that shallows upward. Above it are lowstand deposits of the Chisholm Formation. A characteristic feature of the thick carbonate succession in the House Range is that lowstand deposits are present and often are very distinctive. The Chisholm Formation is one of several lowstand intervals that include a lower shale, a middle carbonate, and an upper shale. This three-part succession can be seen in slopes above the Howell cliffs.

Return to the canyon and drive uphill for 1.1 mi (1.8 km); park near a bench mark set in the Dome Limestone cliffs opposite large landslide blocks north of the road.

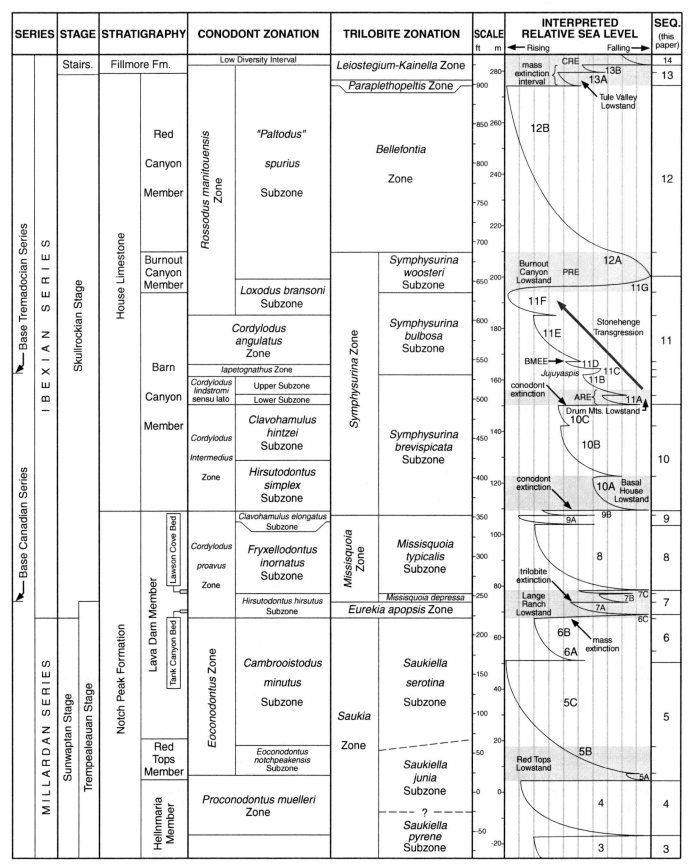

Figure 4. Chronostratigraphic, biostratigraphic, and sequence stratigraphic classification for Notch Peak, House, and basal Fillmore Formations in western Utah. Number-letter combinations (eg., 5A, 5B, 5C) are divisions of sequences, referred to in text as Packages. Lowstand intervals are shaded. SEQ.—sequences; CRE—*Ceratopyge* Regressive Event; PRE—*Peltocare* Regressive Event; BMEE—Black Mountain Eustatic Event; ARE—*Acerocare* Regressive Event. (From Miller et al., 2003, fig. 3.)

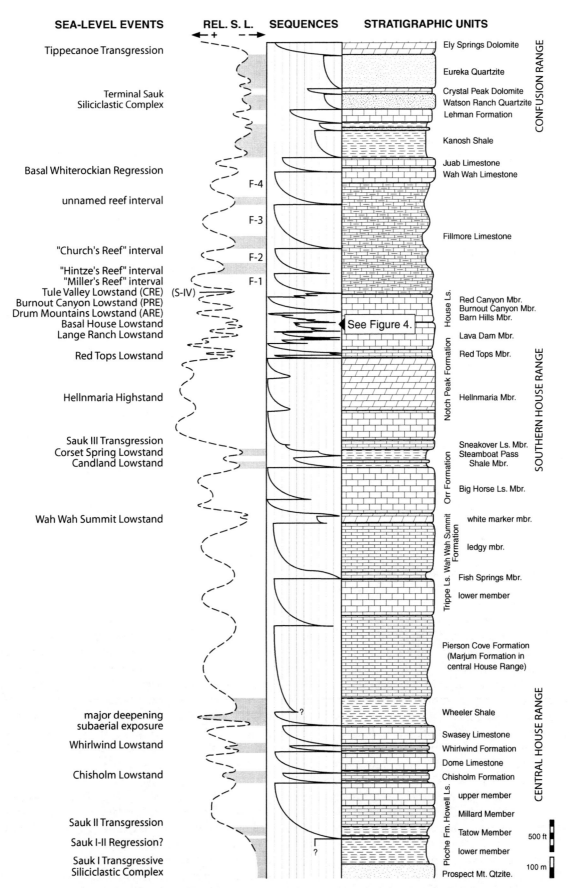

Figure 5. Stratigraphic units of Sauk Sequence in western Utah showing inferred sequences and changes in relative sea level. Shaded intervals are siliciclastics. F-1 to F-4 are sequences in the Fillmore Formation discussed in text. Compare with Figure 3. CRE—*Ceratopyge* Regressive Event; PRE—*Peltocare* Regressive Event; ARE—*Acerocare* Regressive Event; S-IV—base of Sauk IV Supersequence; REL. S.L.—relative sea level.

Stop 2c. Dome Limestone

Carbonate Platform Shoal Facies

The nearly vertical cliff on the south side of the canyon wall exposes a spectacular ooid grainstone shoal with extensive fore-set bedding. The Dome Limestone is interpreted as a highstand deposit above the lowstand deposits of the Chisholm Formation.

Continue driving up the canyon for 0.3 mi (0.5 km); stop near the upper end of the narrow part of the canyon.

Stop 2d. Whirlwind Formation

Lowstand Shales with a Middle Limestone Unit

The Whirlwind Formation comprises the Whirlwind Low-stand and includes lower and upper shaly intervals with lime-stone in the middle, similar to the Chisholm Formation below. The lower and middle units are juxtaposed here by faulting. The Whirlwind has common specimens of the trilobite *Ehmaniella*, a widespread index fossil. Cliffs above the Whirlwind are the Swasey Limestone, another highstand carbonate. We interpret the Chisholm and Whirlwind formations as lowstand intervals. Influxes of shale supressed carbonate sedimentation, resulting in slight relative deepening, so that subsidence provided enough new accommodation to outstrip sea-level falls. Evans and Gold-stein (2001) published this hypothesis as the "Carbonate Supres-sion-Accommodation Generation" (SAGE) model.

Continue driving up the wide part of the canyon for 0.8 mi (1.3 km); stop near where power lines cross the road and the cut stump of a juniper tree is seen to the south.

Stop 2e. Swasey–Wheeler Contact

Karst Sequence Boundary and Flooding Event

Ledges across the gully to the north are near the top of the Swasey Limestone. After examining them briefly, cross the road and walk south along a gully to exposures of the Swasey Limestone–Wheeler Shale contact. The top of the Swasey is an ooid grainstone that is terminated by a red-stained, low-relief karst surface that we interpret as a sequence boundary. The lowest beds of the Wheeler are atypical thin, dark limestones that contain concretions (look near the tree roots) that have abundant specimens of the agnostid *Ptychagnostus gibbus*, an important index species that is typical of deep, offshore facies. Slightly higher, the dark limestones grade into gray clay shales that are more typical of the Wheeler Shale. These lower strata of the Wheeler record an abrupt, rapid relative rise in sea level that is widely recognized. This abrupt rise may be related to fault move-ment on the south edge of the House Range Embayment, where this deep-water facies was deposited. Correlative strata in the Wah Wah Mountains to the south are dominantly tidal flat and carbonate shoal facies that were deposited on the Wah Wah Arch (Rees, 1986).

Return to vehicles and drive 1.1 mi (1.8 km) to the crest of Marjum Pass. Turn right on the dirt track opposite a sign indicating

this pass; drive to the power line access road and park. Walk south over Wheeler Shale to Marjum contact.

Stop 2f. Wheeler Shale and Base of Marjum Formation

Deep-Water Deposits of House Range Embayment

Exposed here are most of the Wheeler Shale and the base of the Marjum Formation. These are deep marine strata that were deposited in the House Range Embayment. Many whole trilobites are preserved in these deep-water strata, especially in the Wheeler Shale at nearby Antelope Springs. This interval is a completely different, shallow-marine facies on the Wah Wah Arch.

The Marjum Formation displays a facies transition with the Pierson Cove Formation. Slopes and ledges south of the pass are deep-water limestones of the Marjum Formation. Massive cliffs at the top are shallower limestones of the Pierson Cove, which may have been deposited near the southeast margin of the House Range Embayment. The Marjum thins to the northeast as the Pierson Cove thickens. The overlying Weeks Formation has platy, shallow-water limestone. The Wheeler to Weeks interval records filling of accommodation and gradual shallowing in the House Range Embayment. Tomorrow we will see some of the equivalent shallow deposits on the Wah Wah Arch.

Retrace the route down Marjum Canyon to the junction with Tule Valley Road; turn left toward Painter Spring. Make a brief photo stop to view west side of House Range, then proceed south ~5 mi (8 km) to Painter Spring Road. We will have a brief stop for water from the large tank at the roadside. Proceed ~3 mi (4.8 km) south to the next brief photo stop for sequence stratig-raphy exposed on the west face of Notch Peak. Proceed ~9.5 mi (15.3 km) to U.S. Highways 6 and 50. Turn left on the pavement and proceed east to the downhill end of the Skull Rock Pass area; turn left on the dirt track between mile posts 39 and 38.

Stop 3. Pyramid Section

Church's Reef in the Fillmore Formation

The name of the section is based on the shape of the hill where it was measured. The Fillmore Formation is a thick (~550 m) unit of interbedded limestones, flat-pebble conglomerates, and shales of early Ordovician age (Tulean Stage, Fig. 3). We tentatively divide the Fillmore into four 100-m-scale sequences (Fig. 5), F-1 (sequence 14 of Fig. 4), which starts just above the base of the formation, through F-4, which ends in the Wah Wah Formation. The Pyramid section exposes most of Sequence F-2, and the 1965 C section at Stop 9 on Day 3 exposes the base of Sequence F-1.

Several meter-scale cycles consist of basal shale grading upward into nodular, interbedded laminae (~1 cm) of silty shale and lime mudstone or calcisiltite (Fig. 6). This material grades upward into laminated lime mudstone or sponge-microbialite reef mounds. The upper surface of the reef mound or lime mud-stone is frequently marked by a truncation surface with a silici-fied rust-stained rind covered by a thin layer of echinoderm-rich grainstone (encrinite) with rounded, silicified, lime mudstone lithoclasts. The truncation surface/encrinite caps the meter-scale cycle. Flat pebble conglomerate and fine-grained grain-stone (calcarenite) beds appear to be event deposits that occur randomly within the meter-scale cycles. Meter-scale cycles occur in the Fillmore Formation and in the middle Cambrian Pierson Cove, Wah Wah Summit, and in the upper Cambrian Sneakover Limestone Member, Hellnmaria Member, and lower Lava Dam Member. Such cycles are commonly thought to be related to Milankovich forcing.

The entire Pyramid with the underlying flat cuesta surface displays the same lithologies in a typical 100-m-scale sequence, F-2. Here the meter-scale cycles are expressed differently throughout the succession with (1) a basal interval of shaly meter-scale cycles with thin sandy units, and flat-pebble conglomerates forming the recessive area around the base of the Pyramid, interpreted as a lowstand deposit; this grades upward to (2) an interval of meter-scale cycles with increasingly thick nodular-bedded limestones and thin reef horizons with well developed truncation surfaces and thin flat-pebble conglomerates and calcarenites forming the sides of the pyramid; this is capped by (3) an interval of thick flat-pebble conglomerates and occasional sandy units forming the steep top of the Pyramid, interpreted as a highstand deposit. The upper, highstand deposit of Sequence F-1 and the lower, lowstand deposit of Sequence F-2 can also be seen in the series of roadcuts from the crest of Skull Rock Pass westward (upward stratigraphically) to the turnoff to the Pyramid, which we will pass on the way back to Delta.

There are four prominent sponge-microbialite patch reef horizons in the Fillmore; three have been named for geologists by Jim Sprinkle (University of Texas—Austin): Church's Reef, exposed here and named for Steve Church (1974), who described these reefs; Hintze's Reef, below this level and named for Lehi Hintze (1951), who first found these reefs and named the Fillmore Formation; and Miller's Reef, named for Jim Miller (1969) because "it tells a good story." Hintze's Reef is well exposed in the nearby roadcut at the crest of Skull Rock Pass, and Miller's Reef will be seen at Stop 9. These names have come into common usage and are shown on Figure 5. Church's reef consists of long, loaf-shaped mounds containing sponges and *Calathium* in a cryptomicrobial micritic matrix. The mounds are a meter in diameter, spaced several meters apart, and aligned in a NE–SW direction, parallel to paleo currents (Church, 1974; Dattilo, 1993). Of particular interest is the truncated and stained upper surface of this reef, which has the rough scalloped appearance of karst and may indicate a minor lowstand.

Return to the highway; turn left (east), passing the exposure of Hintze's Reef at the top of the Skull Rock Pass roadcuts. Proceed to Delta and the Rancher Motel, ~50 mi (80 km). End of Day 1.

DAY 2

Leave for the field at 7:30 a.m. Drive west from Delta on U.S. Highways 6 and 50 through Skull Rock Pass. Turn left (south) at

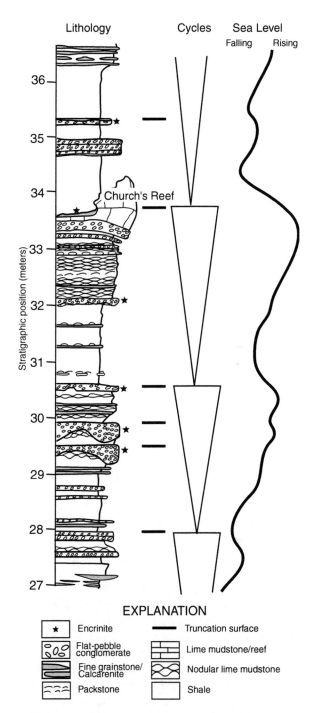

Figure 6. Graphic lithologic column showing the interval near Church's Reef, meter-scale cycles, and interpreted sea-level fluctuations. Strata are in Sequence F-1 of Fillmore Formation. Compare with Figure 5.

mile post 39 onto Tule Valley Road and proceed ~23 mi (37 km) to the south end of House Range. Turn left at the junction with Garrison–Black Rock Road; proceed east 2.2 mi (3.5 km). Turn left on the narrow gravel road that traverses the canyon west of Mount Law; proceed 2.3 mi (3.7 km) north; park on the road.

Stop 4. Mount Law Canyon

Middle to Upper Cambrian on the Wah Wah Arch

We will make a series of three stops to examine Marjuman and Steptoean strata (Fig. 3), many of which are peritidal equivalents of strata we saw in Marjum Canyon. The succession here is different despite essentially comparable rates of subsidence on the Wah Wah Arch and the House Range Embayment. Thick sediments accumulated in the House Range Embayment during sea-level lowstands, whereas thick peritidal carbonates accumulated on the Wah Wah Arch during sea-level rises.

Stop 4a. Upper Pierson Cove Formation

Cyclic Tidal-Flat Deposits of the Wah Wah Arch

Limestones at this stop are tidal-flat deposits in the upper Pierson Cove Formation (Fig. 3), the shallow-water equivalent of the Marjum Formation, which was deposited in the House Range Embayment. The Pierson Cove has many meter-scale carbonate cycles made of subtidal thrombolitic mounds and supratidal rhythmites. Cliffs to the north are carbonate shoal deposits of the Trippe Limestone, a shallow facies equivalent to the lower part of the Weeks Limestone, which overlies the Pierson Cove at the top of the cliffs at Marjum Pass.

Along the low ridge to the east is a dark gray stromatolitic bed with unusual irregular cavities filled with orange silty carbonate grainstone. This stromatolite marks the end of obvious meter-scale cycles in the uppermost Pierson Cove Formation and the beginning of deepening of the depositional environment.

Drive north 0.8 mi (1.3 km); park to the left of the road.

Stop 4b. Wah Wah Summit Formation

Thin Peritidal Sequences of the Wah Wah Arch

Exposures west of the road include the Wah Wah Summit and Orr formations (Fig. 3). The lower, dark, ledgy member and the upper, light, white marker member of the Wah Wah Summit form the lower part of the ridge to the west, where there is a painted section in the upper member. The Wah Wah Summit is laterally equivalent to the upper part of the Weeks Limestone in the Marjum Pass area. The top of the Wah Wah Summit is separated from the Big Horse Member of the Orr Formation by a truncation surface that we interpret as a sequence boundary.

The Wah Wah Summit is a series of many meter-scale cycles. In the lower, ledgy member, cycles are subtidal lime mudstones, but they are peritidal to supratidal in the upper, white marker member. Lower parts of cycles in the upper member consist of light gray to pink, burrowed, peloid grainstones with small thrombolitic mounds deposited in shallow subtidal to intertidal settings. Upper parts of cycles are made of yellow-orange laminae of silty, dolomitic, lime mudstone deposited in intertidal to supratidal settings.

The Big Horse Member is ~225 m thick and represents a return to deeper subtidal conditions with dark burrowed lime mudstone. The Big Horse appears to be one large sequence with three major lithofacies: (1) thick beds of burrowed lime mudstone in the lower one-third; (2) thin-bedded, silty, dolomitic mudstone that weathers recessively, ~10–15 m thick; and (3) a thick, upper complex of carbonate shoal deposits and thrombolitic to stromatolitic bioherms, to be seen at the next stop.

Proceed north ~1.2 mi (1.9 km), turn right onto flat area, and drive as close as possible to exposures.

Stop 4c. Mount Law Section

Corset Spring Lowstand and Sauk II–III Boundary

The upper Big Horse Limestone, Steamboat Pass Shale, and Sneakover Limestone members of the Orr Formation and part of the Hellnmaria Member of the Notch Peak Formation are exposed on Mount Law. The Big Horse Limestone is light-colored, microbial limestone at the base of the mountain; the top of this unit is truncated and overlain by the Steamboat Pass Shale Member. Although the section is faulted, it exposes the mixed shale and carbonate facies of the Steptoean Stage.

The Steptoean is bounded by biomere boundaries, major trilobite extinctions (Fig. 7). The lower extinction is the Marjumiid–Pterocephaliid biomere boundary, and the upper one is the Pterocephaliid–Ptychaspid biomere boundary, just below the *Irvingella major* Zone (Palmer, 1998). In the southern House Range and Wah Wah Mountains, these strata are ~130 m thick (Evans, 1997). Five trilobite zones characterize the Steptoean Stage, including *Aphelaspis, Dicanthopyge, Prehousia, Dunderbergia*, and *Elvinia* zones. Trilobites from each zone have been reported from the southern House Range and Wah Wah Mountains, but lowest Steptoean strata contain few fossils, and a discrete biomere boundary cannot be recognized. It is likely that erosion has removed some strata from the top of the underlying stage, at the top of the Big Horse. The Pterocephaliid–Ptychasid biomere boundary is the base of the Sunwaptan Stage. This boundary can be identified within a few tens of centimeters in the relatively deep-water limestones of the Sneakover Limestone Member.

Interpretations of the sea-level history of the Steptoean Stage have been controversial for more than 30 years. Most investigators have agreed that facies deepened upward during the *Elvinia* chron. Most of the controversy surrounds interpretations of the timing and direction of sea-level changes during the *Aphelaspis* through *Dunderbergia* chrons. Facies deepen upward during the *Aphelaspis* chron in the Great Basin, but in Texas this interval is regressive, high-energy carbonates (Osleger and Read, 1993). The key to interpreting this interval is the antithetical relationship between carbonates and siliciclastics on rapidly subsiding margins. Although facies generally deepen upward in the Great Basin, the deposition of shales, which potentially could be interpreted as the product of sea-level rise, correlates with hiatus on the craton (Evans, 1997). If subsidence outstrips the rate of sea-level fall, deepening-upward successions can accumulate (Fig. 8).

Evans (1997) considered that Steptoean strata record two sequences that are separated by a microkarst surface (Fig. 7). A thin carbonate bed in the upper part of the Steamboat Pass Shale

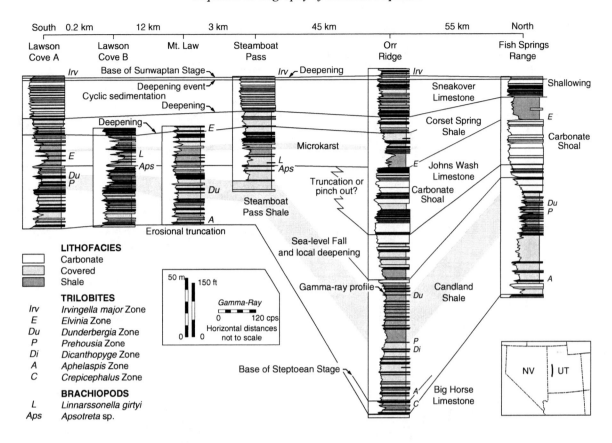

Figure 7. Correlation of western Utah sections of Steptoean Stage. Left edge of each section is an outcrop gamma-ray log. Compare with Figure 3. (Modified from Evans, 1997.)

and its equivalent, the Corset Spring Shale, contains microkarst fractures and fillings. Evans (1997) interpreted this interval as the Sauk II–III boundary. On Mount Law this interval is cryptic, but elsewhere there is offset along fractures, autoclastic breccia, in situ dissolution features, and a generally disrupted fabric recording subaerial exposure. Although an influx of shale may have recorded sea-level fall prior to subaerial exposure, this bed is a record of the lowest stand of sea level within the lower 25 m of *Elvinia* Zone strata. Other investigators have placed the Sauk II–III boundary at the top of the Johns Wash Limestone or below it (Osleger and Read, 1993; Thomas, 1994).

The Johns Wash Limestone is a 60 m carbonate shoal deposit in the central House Range. It is not present on the Wah Wah Arch; rather, the entire *Dunderbergia* interval is abbreviated in this area, and there is no indication that the Johns Wash is represented by hiatus. The base of the Corset Spring Shale and its equivalents appear conformable above the *Dunderbergia* Zone in the southern and central House Range.

The Sauk II–III boundary was described by Palmer (1981) as near the base of the *Elvinia* Zone. Thomas (1994) recognized an unconformity above the *Aphelaspis* Zone and below the *Dunderbergia* Zone in Wyoming. He considered that this level was

the Sauk II–III boundary. The only karstification in Steptoean Stage strata is in the lower *Elvinia* Zone in the Great Basin, so we interpret the Sauk II–III boundary to be at the upper level.

Two non-paleontological tools have potential for correlation of Steptoean strata. Stable isotope studies may provide a means for global chronostratigraphic correlation. The Steptoean Positive Carbon Isotope Excursion (SPICE) event is found in Kazakhstan, Australia, China, and Laurentia (Saltzmann et al., 2000). Resolution may be limited by the large volume of shale through part of this interval in the eastern Great Basin. Gamma-ray profiles and logs provide an inexpensive alternative for high-resolution regional correlation in shaly facies (Fig. 7).

Blocks of limestone and dolomite from the lower part of the Hellnmaria Member have fallen from cliffs above the Orr Formation. Some contain unusual black and white banded dolomite, known as zebra rock, that is widespread at the base of the middle map unit of the Hellnmaria Member. It is quarried extensively in this area and sold to lapidaries who form it into spheres, eggs, obelisks, and bear fetishes. (Look for these at vendor displays at the conference.) This rock is interpreted as recrystallized microbial mats deposited on tidal flats; rarely, small stromatolites are found growing up from white dolomite

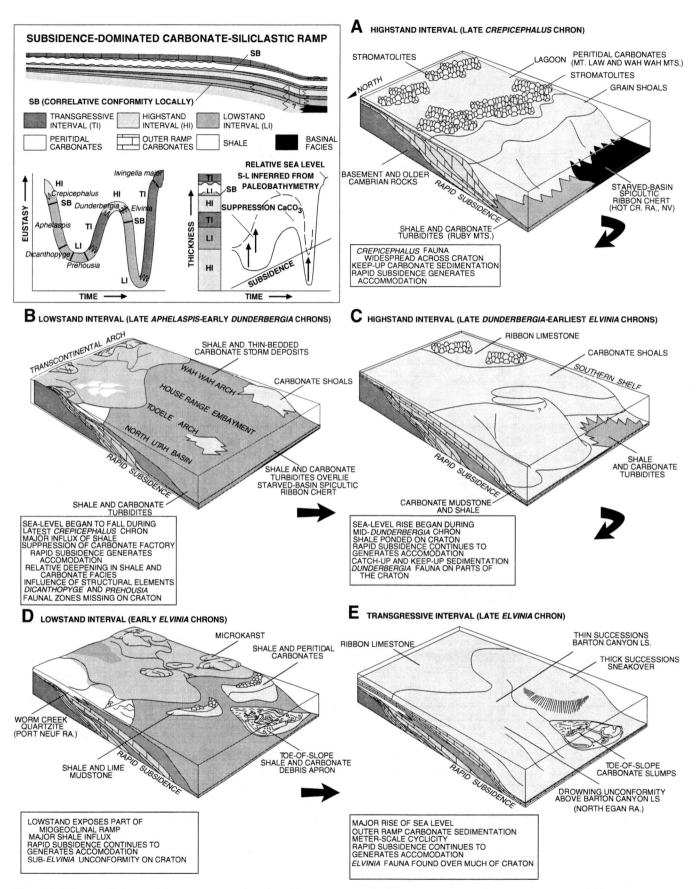

Figure 8. Summary diagram of the sequence stratigraphy of Steptoean strata in the eastern Great Basin. Sequence stratigraphic model in upper left stands in contrast to generalized sequence stratigraphic model for siliciclastic strata illustrated in Figure 2. The carbonate Suppression-Accommodation Generation (SAGE) model considers the capacity for thick accumulation of mixed siliciclastic and carbonate strata during drops in sea level. Carbonate deposition essentially kept pace during most rises of sea level. Interpretive block diagrams A–E illustrate the type of sedimentation during highstands and lowstands of sea level during the Steptoean Stage. SB—sequence boundary; S-L—sea level.

laminae. This lithology appears to be a stillstand deposit, and it marks the base of Sequence 2 (Fig. 4).

Proceed north ~2 mi (3.2 km); turn right onto Steamboat Pass Road for 0.6 mi (1.0 km). Turn right on a dirt track into a north-south canyon, and proceed ~0.3 mi (0.5 km).

Stop 5. Steamboat Pass Photo Stop

Strata exposed in this canyon are in the upper Big Horse Member of the Orr Formation, and microbial mounds occur on both sides of the road. These strata may record filling of accommodation prior to the prolonged lowstand recorded in the overlying Steamboat Pass Shale Member, which forms Steamboat Pass.

A light, peritidal limestone near the top of the Big Horse Member marks a regression at the base of the Millardan Series. North of the Steamboat Pass Road are exposures of the upper Steamboat Pass Shale. The karst surface (Sauk II–III boundary) seen at the previous stop is in this slope, which is part of the Corset Spring Lowstand. The upper part of the slope is the base of the Sauk III Transgression. Above this slope is the ledge-forming Sneakover Limestone Member at the top of the Orr Formation (Figs. 3 and 5). The Sneakover consists of 14 meter-scale cycles; Evans (1997) was able to trace these cycles as far north as the Fish Springs Range and westward into Nevada (Fig. 7). These meter-scale cycles continue into the lower map unit of the Hellnmaria Member of the Notch Peak Formation, which forms the limestone cliffs of Steamboat Mountain.

A few ledges below the top of the Sneakover Limestone is the base of the *Irvingella major* Zone, which marks a deepening event at the base of the Sunwaptan Stage (Fig. 7). Cyclic strata on Steamboat Mountain and the overlying dolomites comprise the Hellnmaria Highstand, which we infer was an extended time of generally high sea level. The Hellnmaria Member is ~320 m thick here and includes the upper part of Sequence 1 and all of Sequences 2–4 (Fig. 4). A painted section of the entire Notch Peak Formation is in and at the head of Steamboat Canyon.

Return to Steamboat Pass Road; turn right and proceed 1.8 mi (2.9 km); turn left on Red Canyon Road. Red-brown exposures farther north are the Oligocene Needles Range Tuff, for which the canyon is named. Proceed 2.8 mi (4.5 km) and turn left onto an indistinct dirt track. Drive 0.1 mi (0.2 km), turn right on a track that follows a small, dry canyon; follow this track 0.3 mi (0.5 km), and park at the end of the track on the skyline.

Stop 6. Lava Dam North Section

Burnout Canyon and Tule Valley Lowstands

This section was painted and described by Hintze (1973) and redescribed by Miller et al. (2003). Exposed below the brown cliffs is the top 50 ft (15.2 m) of the Notch Peak Formation and a complete section of the Barn Canyon Member of the House Limestone (248 ft, 75.6 m). The brown cliff at the parking area is the Burnout Canyon Member (64 ft, 19.5 m). Above the cliff is the type section of the Red Canyon Member (229 ft, 69.8 m).

The lower 5 ft (1.5 m) of the Fillmore Formation is exposed at the top of the section.

This section and the next two we will visit were sampled in detail for conodonts, and the concentration of HCl-insoluble residue (mostly fine quartz sand) was calculated for each conodont sample. Miller et al. (2003) graphed insoluble residue concentration for these and several other sections of the Notch Peak and House formations (Fig. 9). The model they used is that sea-level drops caused quartz sand to prograde into this area as source areas were exposed to erosion. Rising sea level covered the source areas, cutting off the supply of quartz sand. Variations in the insoluble residue curves thus track sea-level changes, and these variations, together with other lithologic data, are useful for delineating sequence and package boundaries.

A Tertiary fault uplifts the east side of the exposure ~30 m, making the Burnout Canyon Member accessible without having to traverse the cliff it forms. This member comprises the Burnout Canyon Lowstand (Fig. 4). The lower part is regressive, the upper part is transgressive, and the pinning point marks the Sequence 11–12 boundary. These beds have up to 100 per mil quartz sand and are characterized by abundant white to brown chert nodules and beds that are interpreted as diagenetically altered sandy limestone. The Burnout Canyon Lowstand appears to correlate with the *Peltocare* Regressive Event in northern Europe (Figs. 4 and 5). The upper part of Sequence 12 records transgressive to highstand conditions that includes most of the Red Canyon Member. Insoluble residues decrease rapidly in the lower part of the Red Canyon.

Sequence 13, the Tule Valley Lowstand, begins with a thin, brown, laminated dolomitic bed at 585 ft and continues above the top of the section. This brown bed records an abrupt drop in sea level, an increase in quartz sand, and a major change in trilobites at the *Bellefontia–Paraplethopeltis* Zone boundary. This bed has the distinction of being the only limestone bed in the House Limestone that was sampled for conodonts (twice!), but none was recovered (Miller et al., 2003, their Table 7). Several conodont taxa typical of the thick *Rossodus manitouensis* Zone disappear at or slightly above this brown bed, and several taxa appear for the first time just above it. Three feet higher is a trilobite-brachiopod grainstone that is just below a widespread 3 m cliff at the top of the House Limestone. This cliff contains trilobites of the *Leiostegium–Kainella* Zone; the base of this zone has been identified as a biomere boundary. Samples from this cliff contain very little quartz sand; however, a 3.2 kg sample of dark lime mudstone near the top of this cliff at 600 ft contained ~1500 acrotretid and lingulid brachiopods (Popov et al., 2002). The top 5 ft of the Lava Dam North section is assigned to the Fillmore Formation and the *Rossodus manitouensis* Zone.

Return to Red Canyon Road and turn left. Proceed along a narrow, winding road for ~7 mi (11.3 km) to a junction; turn right. Proceed ~1.5 mi (2.4 km) to a better north-south road; turn left toward Skull Rock Pass. Drive ~8.5 mi (13.7 km), and turn left just past a sheep corral on the left. Park near the base of the section located in the gully at the north end of the hill.

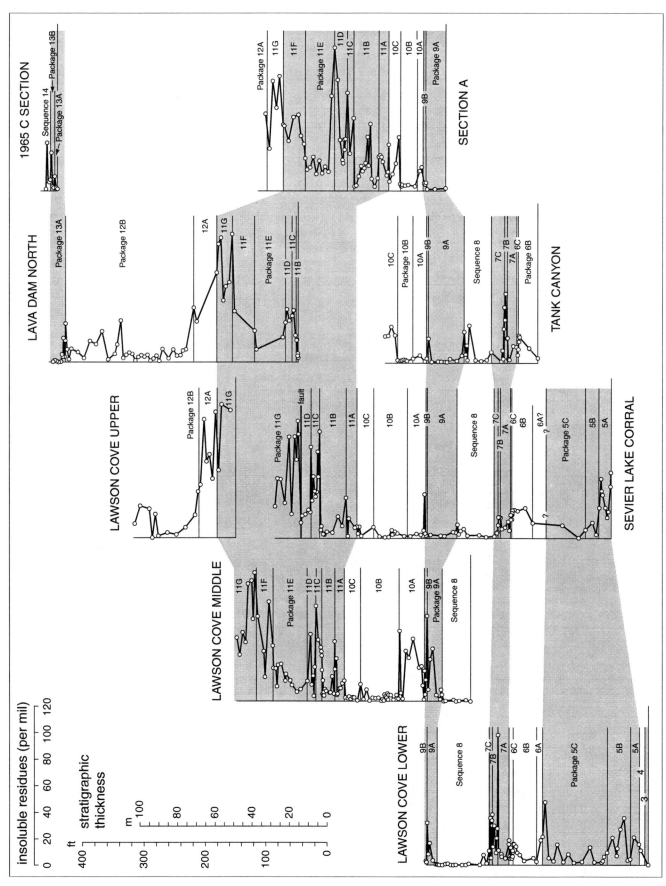

Figure 9. Correlation of measured sections in western Utah using variations in HCl-insoluble residues from conodont samples. Compare with Figure 4. (From Miller et al., 2003, fig. 17.)

Stop 7. Sevier Lake Corral Section

Red Tops and Lange Ranch Lowstands in the Wah Wah Arch–House Range Embayment Transition Zone

Exposed strata at this painted section are the upper approximately one-third of the Red Tops Member (31 ft, 9.4 m) and the entire Lava Dam Member of the Notch Peak Formation. The Barn Canyon Member of the House Limestone was measured 0.3 mi (0.5 km) to the south. The insoluble residue graph for this section is on Figure 9.

A thin stromatolite bed in the middle of the Red Tops Lowstand marks the base of Package 5B (Fig. 4) at 18–20 ft in the section. The Red Tops Lowstand is mostly high-energy grainstone that rapidly grades upward into lime mudstone at the contact with the Lava Dam Member. This contact is the base of Package 5C and marks a rapid relative rise in sea level. A thick, covered interval in the lower Lava Dam Member probably conceals the upper part of Package 5C and all of Package 6A. Lime mudstones above ~100 ft in the Lava Dam Member are assigned to Package 6B.

The base of the Ibexian Series coincides with the base of Package 6C, a thin shallowing-upward interval at the base of the Lange Ranch Lowstand. The base of 6C is an iron-rich hardground at 129 ft that coincides with a major conodont and trilobite faunal change at the base of the *Hirsutodontus hirsutus* Subzone of the *Cordylodus proavus* Zone (Fig. 4). The base of Sequence 7 is the base of the Tank Canyon Bed, a mottled, gray, ooid grainstone that marks a regressive minimum at 130.5 ft. This bed can be traced from the northern Wah Wah Mountains to several miles north of Skull Rock Pass. Package 7A, a thin transgressive-regressive cycle, includes strata from 130.5–147 ft. Package 7B is a second cycle that includes the interval 147–150.5 ft; the top coincides with the top of the *Hirsutodontus hirsutus* Subzone. Package 7C includes the interval from 150.5–157 ft. At the top of 7C is the Lawson Cove Bed, a thin stromatolite that can be traced over the same large area as the Tank Canyon Bed. The top of this marker bed is a karst exposure surface, and along strike to the south there are dissolution pockets (grikes) filled with brown sand at and just below the top of this bed. The sand can be traced along strike as a thin bed. Overlying strata are assigned to Sequence 8.

Packages 6C through 7C can be traced as far south as Lawson Cove in the Wah Wah Mountains and as far north as Sneakover Pass, ~6.6 mi north of here. Near our next stop at Tank Canyon, planar truncation surfaces occur at the bases of 7B and 7C. In the House Range Embayment farther north (Chalk Knolls, Stop 1 yesterday), the named marker beds and truncation surfaces are absent, which we interpret as an example of lateral conformity in the rapidly subsiding House Range Embayment. In the Chalk Knolls area, growth of stromatolites began at about the level of the Lawson Cove Bed and continued through Sequences 8 and 9, and the interval is much thicker than here. This must have been a time of rapid subsidence in the House Range Embayment, perhaps due to movement on the fault along its southeast edge, near this area.

A microbial bed with a karst dissolution surface at its top occurs at the level of the Lawson Cove Bed in the cratonic Lange Ranch section in central Texas and in the Dayangcha section in North China. Such wide distribution of a microbial bed at this level, just above the lowest occurrence of *Fryxellodontus inornatus*, suggests that deposition of these thin packages in Sequence 7 was controlled by global eustatic events and not by simply local influences.

A thin stromatolite interval occurs at the base of Sequence 8 at this section, but it appears to have been drowned during rapid sea-level rise. None of the rest of Sequence 8 is stromatolitic in the cliffs above us; neither is it in sections to the south nor at our next stop to the north. Only the upper strata of Sequence 9 are stromatolitic here, but at our next stop, Sequence 9 is completely filled with stromatolites.

Return to the road; turn left, and proceed ~3 mi (4.8 km) to U.S. Highway 6 (also Hwy. 50). Turn left and proceed 2.2 mi (3.5 km) west; turn right on the gravel road toward Tank Canyon. Proceed 1.4 mi (2.3 km); pull off to the right of the road and park near the mouth of a small side canyon.

Stop 8. Tank Canyon Section

Transition into the House Range Embayment

Strata exposed in this painted section include most of the Lava Dam Member and the lower Barn Canyon Member of the House Limestone. This location is close to the south edge of the House Range Embayment. Facies changes in Sequences 8 and 9 (upper Lava Dam Member) are used to delineate a transition zone with the Wah Wah Arch. The insoluble residue graph for this section is on Figure 9.

The section begins in Package 6B. Packages 6C to 7C are comparable to what we saw at Stop 5, but no exposure surface is seen at the top of the Lawson Cove Bed. Planar truncation surfaces occur at the base of 7B and 7C several hundred meters down the canyon, but they are not observed here. Sequence 8 lacks stromatolites and consists of relatively deep subtidal lime mudstone. The base of Package 9A is at 118 ft, which coincides with the base of the *Clavohamulus elongatus* Subzone (Fig. 4). A brief sea-level drop is recorded by an influx of quartz sand at this level. Waters shallowed enough to permit stromatolite growth, which continued up section as sea level rose, and Package 9A is all stromatolites. In the upper part of 9A, stromatolites prograded southward and are found at Sevier Lake Corral but not farther south. At Sneakover Pass, 2.5 mi northwest, and at Chalk Knolls South (Stop 1 yesterday), all of Sequence 8 and Package 9A are stromatolites, and the interval is thicker than here.

Facies changes in this stratigraphic interval are used to delineate a transition zone between the Wah Wah Arch to the south and the House Range Embayment to the north. Sequences 8 and 9 are filled with microbial boundstone as far west as Sawmill Canyon in the Egan Range, near Lund, Nevada. The top of Sequence 9 in the House Range and the Egan Range is at the base of the *Cordylodus intermedius* Zone.

Stromatolites end and are truncated at the base of Package 9B, which is an invariably thin interval (3 ft thick here) that ends with an informal marker bed, the *Symphysurina* coquina. This bed at 178 ft is the top of the Notch Peak Formation and also the base of the *Cordylodus intermedius* Zone. Strata of Package 10A contain 30%–50% white to brown chert and are assigned to the Barn Canyon Member of the House Limestone. Package 10A contains abundant fine quartz sand, some of which is diagenetically altered to chert. This interval records a relative drop in sea level, the Basal House Lowstand. Package 10A is thicker to the south and is relatively thin here (178–192 ft). These cherty strata can be seen on canyon walls to the south as a brown banded interval above the massive stromatolitic cliffs. Packages 9B and 10A are missing at Chalk Knolls South, where the Basal House Lowstand is recorded by a karst surface that separates stromatolites of Packages 9A from chert-free lime mudstone in Package 10B.

Package 10B is the interval between the cherty beds and the lowest microbialites that characterize Package 10C (192–225 ft). Package 10C typically has one or more stromatolitic or thrombolitic units, which record shallowing and filling of accommodation in the upper part of Sequence 10. Here the thrombolites at 225 ft are a little above the base of the *Clavohamulus hintzei* Subzone, which is at 220 ft. This subzone has abundant silicified burrows, probably *Thalassinoides*.

Retrace the route to U.S. Highway 6 (50); turn left, and drive ~45 mi (73 km) to the Rancher Motel in Delta. End Day 2.

DAY 3

Leave the motel at 7:30 a.m., and drive west on U.S. Highway 6 (50) to mile post 39. Turn left onto Tule Valley Road, and drive south ~3.5 mi (5.6 km) past the Ibex well. Turn left on a dirt track and follow it ~0.3 mi (0.5 km) to its end. Walk southeast into the mouth of a dry canyon.

Stop 9. 1965 C Section

Sauk III–Sauk IV Supersequence Boundary

This section was described by Hintze (1973); the lower 50 ft was described again by Miller et al. (2003). Strata include the top 3-m ledge of the House Limestone and about the lower half of the Fillmore Formation. The upper part of Sequence 13 (Fig. 4) is exposed, as well as Sequence F-1 (= Sequence 14) and F-2 in the Fillmore. At this stop we will only be able to examine Sequence 13 and Package F-1A in detail, but the turnoff to Section C provides an excellent view of alternating shale-rich and limestone-rich intervals that demarcate the sequences in the Fillmore Formation. Sequence F-1 is particularly well displayed, starting in the upper ledges of the House Limestone near the base of the succession. Moving upward, the shaly lowstand interval is interrupted by a thin but prominent ledge capped by Miller's Reef at 170–183 ft. Above this, the shale resumes but then it gives way to more nodular-bedded limestone that is capped by Hintze's Reef at 367–379 ft (Hintze, 1973), which is expressed as a distinct, thin, bluish band. The sequence and the ridge are capped by a

thick series of massive flat-pebble conglomerates at 380–485 ft, which is visible as a thick, dark ledge.

At closer inspection, Package 13A includes the upper ledge of the House Limestone and up to the 4 ft level in the Fillmore. The base of Package 13B at 4 ft is a coquina of the orthid *Nanorthis hamburgensis*; this package extends to 11.5 ft to a thin, sandy layer that is a major sequence boundary and marks the base of Sequence F-1. Conodonts are present in beds up to 10 ft but are absent at 12 ft. A sparse fauna at 14 ft is assigned to the *Rossodus manitouensis* Zone, although conodonts in that sample may be redeposited in clasts. The diverse, long-ranging fauna that characterizes the *R. manitouensis* Zone disappears above 14 ft. Younger strata are referred to the Low Diversity Interval, a conodont zonal unit characterized by both low conodont abundance and taxonomic diversity. The Tule Valley Lowstand and Low Diversity Interval correlate with the *Ceratopyge* Regressive Event in northern Europe (Figs. 4 and 5).

Golonka and Kiessling (2002) proposed a new Sauk IV supersequence, split off from the upper part of the Sauk III interval of Palmer (1981). Their Sauk IV begins at the upper Tremadocian unconformity, which is a well-understood horizon in Great Britain but not in western Utah. One advantage of the Ibex succession is that it is quite complete and lacks the large hiatus present in Great Britain at this level. The most important Ibexian sequence boundary in this general interval is the top of Sequence 13, at the 11.5 ft level in the 1965 C Section. Overlying strata of the Low Diversity Interval record an extended lowstand that correlates with part of the Roubidoux Formation in Missouri (Repetski et al., 2000) and the coeval New Richmond Sandstone in Wisconsin. These siliciclastic intervals may be equivalent to the hiatus below the Arenigian Transgression in Great Britain and also to the *Ceratopyge* Regressive Event recognized in northern Europe. Sequence boundaries may be diachronous where there is a large hiatus, so the top of Sequence 13 (Fig. 4) is probably equivalent to the Tremadocian-Arenigian unconformity in Great Britain.

We place the Sauk III–IV boundary at the top of Sequence 13 (Fig. 4), which is less than one meter from the boundary between the *Rossodus manitouensis* Zone and the Low Diversity Interval. Sauk IV continues to the top of the Sauk Sequence at the top of the Eureka Quartzite.

Miller's Reef is a short hike above the sequence boundary and is within the Low Diversity Interval, and it caps Package F-1A within the F-1 Sequence. Miller's Reef differs from Church's Reef (Stop 3) in its biotic composition and gross geometry; its framework is a continuously distributed, meter-thick layer of columnar stromatolites with attached sponges but no *Calathium*. On the other hand, this horizon resembles Church's Reef in being capped by a deeply scalloped truncation surface that cuts into stromatolites and sponges, marking a minor regressive event.

Return to Tule Valley Road; turn left and proceed ~2.9 mi (4.7 km). Near the south end of an Oligocene volcanic feature called the Lava Dam, turn left on a distinct dirt track, and follow it ~0.5 mi (0.8 km) to its end. Walk to the base of the section in a dry stream bed at the north end of the hill.

Stop 10. Lava Dam Five Section

Sabkha Evaporites in the Red Tops Lowstand

This painted section exposes approximately two-thirds of the Red Tops Member and all of the Lava Dam Member of the Notch Peak Formation, as well as the lower one-third of the Barn Canyon Member of the House Limestone. The base of the Ibexian Series is defined halfway up the side of this hill (Ross et al., 1997). We will examine only the base of the section, in the lower part of the Red Tops Member; these strata are assigned to the *Eoconodontus notchpeakensis* Subzone (Fig. 4).

At the base of the section is a rippled ooid-intraclast grainstone overlain by very thin-bedded grainstone with interlaminated shale and evaporites. This is a fortuitous exposure where the stream occasionally washes these lithologies clean; usually, they form a covered slope. The evaporites are secondarily remobilized into joints. Overlying lithologies are ooid and skeletal grainstone. We interpret these shales and evaporites as a sabkha environment and as the lowpoint of the Red Tops Lowstand, at the Package 5A–5B contact. At Lawson Cove in the northern Wah Wah Mountains, this level is at the base of a stromatolite unit that coincides with the HERB event, a distinctive negative peak on the carbon-isotope curve (Ripperdan, 2002). We have no isotope data here, however. An eolian deposit, the Lily Creek Sandstone, occurs at this level in the Black Mountain section in the cratonic Georgina Basin in Australia.

Return to vehicles and to the Tule Valley Road; turn left. Proceed 0.2 mi (0.3 km), turn right on Snake Pass Road, and follow this circuitous route for ~6.5 mi (10.5 km) across Tule Valley. Turn right toward Ibex and Fossil Mountain. The ridge to our left is capped by Oligocene tuffs, and the north-dipping strata below the tuffs are the Fillmore Formation. Overlapping sections G–J of Hintze (1951, 1973) are measured along this ridge. The road crosses a fence; ~0.6 mi (1.0 km) past this fence, turn left onto a dirt track that follows a dry stream bed. Follow this track ~0.5 mi (0.8 km) to Section J.

Stop 11. Section J

Upper Ibexian and Lowest Whiterockian Cyclic Beds

Strata exposed in this painted section include the Wah Wah Limestone, Juab Limestone, and Kanosh Shale. The Wah Wah Limestone comprises eight sedimentary cycles marked by ledges of resistant carbonates interbedded with recessive mixed carbonates and siltstone to shale. These cycles resemble those found in the Fillmore Formation below. The carbonate ledges mark cycle tops and consist of tan burrowed lime mudstone with pockets of fine bioclast-intraclast-peloid grainstone interposed with rusty-brown-weathering, wispy siltstone laminae. Discrete calathid reef mounds are found in ledges four and six. These reef mounds are mostly light gray; they are lenticular in cross section and round in plan view in a rare three-dimensional exposure in the gully east of the line of section.

The cycle bases are recessive intervals of nodular- to wavy-bedded light gray to tan lime mudstone to fine grainstone within gray to yellow-weathering siltstone interbedded with thin beds of intraclast packstone. Commonly, these intervals are covered with slopewash and are best exposed below the resistant ledges. An olive-green shale unit with interbedded carbonates, ~4 m thick, is found between ledges seven and eight. Carbonates in this interval vary from rusty-brown-weathering silt-rich fine bioclastic grainstone to medium gray lime mudstone to intraclast-peloid grainstone to brachiopod grainstone to packstone composed mostly of the orthid *Hespernomiella minor*. A red to rusty-brown-weathering fine sandstone is found in the upper part of this shale unit, just below ledge seven. This sandstone represents the most regressive part of this succession, and we interpret it as a sequence boundary.

The *Hespernomiella minor* coquina marks the base of the Whiterockian Series, so this important chronostratigraphic change is nearly coincident with (but a little below) the sequence boundary. Whiterockian faunas are among the best known in middle Ordovician strata in the western United States. Strata of this age are found only in a few places around the world. The uppermost sequence of the Wah Wah Limestone is a prominent limestone ledge, and overlying silty carbonates of the Juab Limestone form a recessive bench. This contact can barely be distinguished in gamma-ray profiles (Fig. 10). The Juab Limestone is notably cyclic with clean carbonates interbedded with silty carbonates. Exposed above Section J is a 3-ft-thick interval of shale that can be correlated with the subsurface. A laterally persistent thrombolitic boundstone bed crops out at the top of the Juab.

The Kanosh Shale is among the thickest clay shales in the Sauk Sequence in this area. It contains numerous hardground surfaces and abundant fauna. Residual total organic carbon ranges up to 5% (P. Lillis, 2000, personal commun.), but with a conodont alteration index of 3–3.5 regionally, the level of maturation has passed the window for oil generation. This influx of shale, like the siliciclastics of the Steptoean Stage, suppressed carbonate productivity, marking the beginning of a major phase of regression at the top of Sauk Sequence (Fig. 5).

Cycles in the Fillmore Formation through Kanosh Shale can be traced laterally for many miles using outcrop gamma-ray profiles. A composite log of strata exposed at the Square Top section (north of the 1965 C section), at Section J and at the nearby K-South section is shown on Figure 10. The composite outcrop profile can be correlated confidently with a gamma-ray log from a borehole drilled ~9 mi (~14.5 km) north of here, on the crest of the Confusion Range.

On the steep slope above the Kanosh Shale, a thick sandstone marks the base of the Lehman Formation, but it consists mostly of fossiliferous limestones that represent a short-lived return to normal marine sedimentation. The overlying Watson Ranch and Eureka quartzites culminate deposition of the Sauk Sequence. A short-lived cycle of restricted bank-top carbonate sedimentation is recorded by the Crystal Peak Dolomite, a dark gray-brown unit that lies between the Watson Ranch and Eureka quartzites.

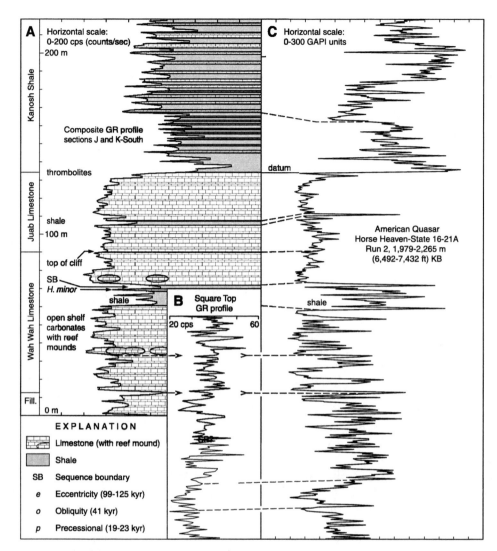

Figure 10. Outcrop gamma-ray (GR) profile of composite Section J and K-South (A) and uppermost Square Top (B) sections can be correlated confidently with the gamma-ray log of the American Quasar Horse Heaven–State 16–21A oil well (C), located ~9 mi north of Section J. GAPI log is the corrected gamma-ray scale in API units. Run 1 in log suite covered an interval higher in the borehole.

Return to the main north-south road and turn left. Proceed ~100 m to the crossroads, and continue north 0.2 mi (0.3 km); turn oblique right. Drive 0.8 mi (1.3 km), and turn right to Ibex; continue 0.3 mi (0.5 km) to the ruins of a building at Ibex.

Stop 12. Ibex (Jack Watson's Ranch)

Eureka Quartzite, Ely Springs Dolomite, and Top of Sauk Sequence

A century ago, Ibex was the homestead of Jack Watson, who lived alone here from the 1890s to the 1930s and tended cattle. The land is still privately owned and is a registered historic landmark. This place is the namesake for the Watson Ranch Quartzite, which can be seen across the valley on Fossil Mountain, below the dark Crystal Peak Dolomite and the Eureka Quartzite. This place is also the namesake for the Ibexian Series (Ross et al., 1997).

The Eureka is exposed near small dams that Jack Watson built to catch rainwater that runs off the impermeable quartzite. Field geolo-gists in the area often find pieces of vitreous Eureka that were fashioned into stone implements by prehistoric inhabitants. The Eureka is strongly burrowed by *Skolithos* and represents the culmination of the Whiterockian regression. The Sauk Sequence has a kind of symmetry in that it begins with one thick sandstone (Prospect Mountain Quartzite) and ends with another (Eureka Quartzite).

We will walk north to the contact between the Eureka Quartzite and the Ely Springs Dolomite. Interbedded sandstone and dolomite at the base of the Ely Springs is known as the Ibex Member, but most of the Ely Springs is dark dolomite. Sandstones in the Ibex Member presumably were reworked from weathered Eureka Quartzite. There is no sharp Eureka–Ely Springs contact, but this rather nebulous level is the Sauk–Tippecanoe sequence boundary. Fossils in this interval are rare, so age relationships are not known in detail, but Hintze (1988, fig. 26) shows a large hiatus at this level.

Retrace the route to the main north-south road; proceed south. At the crossroads near Warm Point, turn left, and drive ~6 mi (9.7 km) to Tule Valley Road. Turn left and drive north to

U.S. Highway 6 (50). Turn right onto the pavement and drive ~50 mi (81 km) to Delta for luggage.

Follow U.S. Highway 6 out of Delta. At Lynndyl, turn right on Utah Highway 132 and follow it to Nephi. At Nephi, turn north on I-15 to Salt Lake City. Follow I-80 to Salt Lake International Airport. After turning in vehicles at Enterprise Car Rental, we will use company facilities for transportation to the terminal. End of trip.

REFERENCES CITED

Church, S.B., 1974, Lower Ordovician patch reefs in western Utah: Brigham Young University Geology Studies, v. 21, no. 3, p. 41–62.

Dattilo, B.F., 1993, The Lower Ordovician Fillmore Formation of western Utah: Storm-dominated sedimentation on a passive margin: Brigham Young University Studies, v. 38, p. 71–100.

Evans, K.R., 1997, Stratigraphic expression of Middle and Late Cambrian sea-level changes: Examples from Antarctica and the Great Basin, USA [Ph.D. dissertation]: Lawrence, Kansas, University of Kansas, 177 p.

Evans, K.R., and Goldstein, R.H., 2001, The carbonate suppression-accommodation generation (carbonate SAGE) model: sea-level lowstands can mimic maximum flooding zones on rapidly subsiding margins: Geological Society of America Abstracts with Programs, v. 33, no. 3, p. 73.

Golonka, J., and Kiessling, W., 2002, Phanerozoic time scale and definition of time slices, *in* Kiessling, W., Flügel, E., and Golonka, J., eds., Phanerozoic reef patterns: Society for Sedimentary Geology Special Publication 72, p. 11–20.

Hintze, L.F., 1951, Lower Ordovician detailed stratigraphic sections for western Utah: Utah Geological and Mineralogical Survey Bulletin 39, 99 p.

Hintze, L.F., 1953, Lower Ordovician trilobites from western Utah: Utah Geological and Mineralogical Survey Bulletin 48, 249 p.

Hintze, L.F., 1973, Lower and Middle Ordovician stratigraphic sections in the Ibex area, Millard County, Utah: Brigham Young University Geology Studies, v. 20, no. 4, p. 3–36.

Hintze, L.F., 1988, Geologic History of Utah: Brigham Young University Geology Studies, Special Publication 7, 202 p.

Hintze, L.F., and Davis, F.D., 2002a, Geologic map of the Wah Wah Mountains North 30′ × 60′ quadrangle and part of the Garrison 30′ × 60′ quadrangle, northwest Millard County and part of Beaver County, Utah: Utah Geological Survey, Map 182.

Hintze, L.F., and Davis, F.D., 2002b, Geologic map of the Tule Valley 30′ × 60′ quadrangle and parts of the Ely, Fish Springs, and Kern Mountains 30′ × 60′ quadrangles, northwest Millard County, Utah: Utah Geological Survey, Map 186.

Hintze, L.F., and Robison, R.A., 1987, The House Range, western Utah: Cambrian mecca, *in* Beus, S.S., ed., Rocky Mountain Section of the Geological Society of America: Boulder, Colorado, Geological Society of America, Centennial Field Guide Volume 2, p. 257–260.

Miller, J.F., 1969, Conodont fauna of the Notch Peak Limestone (Cambro-Ordovician), House Range, Utah: Journal of Paleontology, v. 43, p. 413–439.

Miller, J.F., Evans, K.R., Loch, J.D., Ethington, R.L., Stitt, J.H., Holmer, L., and Popov, L.E., 2003, Stratigraphy of the Sauk III interval in the Ibex area, western Millard County, Utah and central Texas: Brigham Young University Geology Studies (in press).

Osleger, D.A., and Read, J.F., 1993, Comparative analysis of methods used to define eustatic variations in outcrop: Late Cambrian interbasinal sequence development: American Journal of Science, v. 293, p. 157–216.

Palmer, A.R., 1981, Subdivision of the Sauk Sequence, *in* Taylor, M.E., ed., Short Papers for the Second Symposium on the Cambrian System: U.S. Geological Survey Open-File Report 81-743, p. 160–162.

Palmer, A.R., 1998, A proposed nomenclature for stages and series for the Cambrian of Laurentia: Canadian Journal of Earth Science, v. 35, p. 323–328.

Popov, L.E., Holmer, L.E., and Miller, J.F., 2002, Lingulate brachiopods from the Cambrian-Ordovician boundary beds of Utah: Journal of Paleontology, v. 76, p. 211–228.

Rees, M.N., 1986, A fault-controlled trough through a carbonate platform: The Middle Cambrian House Range Embayment: Geological Society of America Bulletin, v. 97, p. 1054–1069.

Repetski, J.E., Loch, J.D., Ethington, R.L., and Dresbach, R.I., 2000, A preliminary reevaluation of the stratigrapy of the Roubidoux Formation of Missouri and correlative Lower Ordovician units in the southern Midcontinent, *in* Johnson, K.S., ed., Platform carbonates in the southern Midcontinent, 1996 symposium: Oklahoma Geological Survey Circular 101, p. 103–106.

Ripperdan, R.L., 2002, The HERB Event: End of the Cambrian carbon cycle paradigm?: Geological Society of America, Abstracts with Programs, v. 34, no. 6, p. 413.

Ross, R.J., Jr., Hintze, L.F., Ethington, R.L., Miller, J.F., Taylor, M.E., Repetski, J.E., Sprinkle, J., and Guensburg, T.E., 1997, The Ibexian, lowermost series in the North American Ordovician, *in* Taylor, M.E., ed., Early Paleozoic Biochronology of the Great Basin, Western United States: U.S. Geological Survey Professional Paper 1579, p 1–50.

Ross, C.A., and Ross, J.R.P., 1995, North American Ordovician depositional sequences and correlations, *in* Cooper, J.D., ed., Ordovician Odyssey: Short Papers for the Seventh International Symposium on the Ordovician System: Pacific Section of Society for Sedimentary Geology, Book 77, p. 309–313.

Saltzmann, M.R., Ripperdan, R.L., Brasier, M.D., Lohmann, K.C., Robison, R.A., Chang, W.T., Peng, S., Ergaliev, E.K., and Runnegar, B., 2000, A global carbon isotope excursion (SPICE) during the late Cambrian; relation to trilobite extinctions, organic-matter burial and sea level: Palaeontology, Palaeoclimatology, Palaeoecology, v. 162, no. 3-4, p. 211–223.

Sloss, L.L., 1963, Sequences in the cratonic interior of North America: Geological Society of America Bulletin, v. 74, p. 93–113.

Sloss, L.L., 1988, Forty years of sequence stratigraphy: Geological Society of America Bulletin, v. 100, p. 1661–1665.

Sloss, L.L., Krumbein, W.C., and Dapples, E.C., 1949, Integrated facies analysis, *in* Longwell, C.R., Chairman, Sedimentary facies in geologic history: Boulder, Colorado, Geological Society of America Memoir 39, p. 91–124.

Thomas, R.C., 1994, The timing of the Sauk II-Sauk III (Upper Cambrian) unconformity in the western United States: Geological Society of America Abstracts with Programs, v. 26, no. 6, p. 122.

Walcott, C.D., 1908, Nomenclature of some Cambrian Cordilleran formations: Smithsonian Miscellaneous Collections, v. 53, no. 1, p. 1–12.

Geological Society of America
Field Guide 4
2003

Tectonic geomorphology and the record of Quaternary plate boundary deformation in the Olympic Mountains

Frank J. Pazzaglia
Department of Earth and Environmental Sciences, Lehigh University, Bethlehem, Pennsylvania 18015, USA

Glenn D. Thackray
Department of Geosciences, Idaho State University, Pocatello, Idaho 83209, USA

Mark T. Brandon
Department of Geology and Geophysics, Yale University, New Haven, Connecticut 06520-8109, USA

Karl W. Wegmann
Washington Department of Natural Resources, Division of Geology & Earth Resources, Olympia, Washington 98504-7007, USA

John Gosse
Department of Earth Sciences, Room 3006, Life Sciences Centre, Dalhousie University, Halifax, Nova Scotia B3H 4J1, Canada

Eric McDonald
Desert Research Institute, Division of Earth and Ecosystem Sciences, 2215 Raggio Parkway, Reno, Nevada 89512, USA

Antonio F. Garcia
Department of Physics, California Polytechnic State University, San Luis Obispo, California 93407, USA

Don Prothero
Department of Geology, Occidental College, 1600 Campus Road, Los Angeles, California 90041-3314, USA

ABSTRACT

We use Quaternary stratigraphy to reconstruct landscape evolution and measure tectonic deformation of the Olympic Mountains section of the Pacific Northwest Coast Range. An important motivation for understanding orogenesis here, and throughout the Coast Range, is the concern about the relationship of active deformation to seismic hazards associated with the Cascadia subduction zone. There is also much interest in apportioning the nature of the deformation, whether cyclic or permanent, whether it involves mainly shortening parallel or normal to the margin, and how the deformation on the pro- versus retrowedge sides of the orogen compare. Pre-Holocene stratigraphy and structure provide the only records of sufficient duration to separate long-term permanent deformation from earthquake-cycle elastic deformation. For this reason, active-tectonic studies have focused on deformation of Quaternary deposits and land-forms, which are best preserved along the Pacific Coast and offshore on the continental shelf. At least four major glacial advances are recorded in the valley and coastal deposits along the western margin of the Olympic Peninsula. Both numeric and relative dating, including soils of these deposits, establish a stratigraphic anchor that is used to document the relationship between margin parallel and margin normal deformation in

Pazzaglia, F.J., Thackray, G.D., Brandon, M.T., Wegmann, K.W., Gosse, J., McDonald, E., Garcia, A.F., and Prothero, D., 2003, Tectonic geomorphology and the record of Quaternary plate boundary deformation in the Olympic Mountains, *in* Swanson, T.W., ed., Western Cordillera and adjacent areas: Boulder, Colorado, Geological Society of America Field Guide 4, p. 37–67. For permission to copy, contact editing@geosociety.org. © 2003 Geological Society of America.

the Olympic Mountains, which, on a geologic time scale (>10³ yr), seems to be the fastest deforming part of the Cascadia forearc high. The glacial stratigraphic framework is extended to fluvial terraces of the Clearwater drainage, which remained unglaciated during the late Pleistocene and Holocene, preserving a record of river incision, with each terrace recording the shape and height of past long profiles. We assess how fluvial terraces are formed in this tectonically active setting and then use features of the terraces to estimate incision rates along the Clearwater long profile. The long fluvial history preserved in the Clearwater ensures that the unsteady deformation associated with the earthquake cycle is averaged out, leaving us with a record of long-term rock uplift as well as horizontal shortening. We show, however, that the earthquake cycle may play an important role in terrace genesis at the millennial time scale.

Keywords: geomorphology, active tectonics, subduction wedges, glacial geology, soils, terraces.

INTRODUCTION

We designed a 4-day field trip to exhibit the geology, geomorphology, and active tectonics of the Olympic Peninsula. The trip is the culmination of over a decade of collaborative research in the Olympic Mountains by the authors and their colleagues. The following text and figures are culled from numerous published papers and theses that report the findings of this research, namely, Brandon and Vance (1992), Thackray (1996, 1998, 2001), Wegmann (1999), Brandon et al. (1998), Pazzaglia and Brandon (2001), Wegmann and Pazzaglia (2002), and Tomkin et al. (2003). It is organized around four major topics that should generate lively discourse on how to use and interpret basic field relationships in tectonic geomorphology research: (1) How are complexly juxtaposed glacial, fluvial, and eolian surficial deposits and their corresponding soils assembled into a dated stratigraphic framework in a tectonically active setting?; (2) What is a river terrace, how is it made, and what do river terraces tell us about active tectonics?; (3) What is driving orogenesis for the Olympic Mountain segment of the Cascadia subduction zone? Is it shortening parallel to the direction of plate convergence, shortening normal to the direction of plate convergence, or some combination of both? Are there any geomorphic or stratigraphic field relationships that can actually be used to track the horizontal movement of rocks and thus interpret the shortening history over geologic time scales? How do the tectonics on the prowedge versus the retrowedge sides of the Olympic Mountains compare?; (4) We know that uplift along Cascadia includes the effects of cyclic earthquake-related deformation and long-term steady deformation. How do these different types of uplift influence incision and aggradation in the rivers of the Olympic Mountains?

The trip begins by laying out the big-picture tectonic setting. The concept of a pro- and retrowedge is introduced and exhibited on the eastern flank and core of the range. Geologic and thermochronologic evidence for margin parallel versus margin normal shortening is introduced. We then begin building a Quaternary stratigraphic foundation anchored along the western coast of the Olympic Peninsula that will be extended landward into the Clearwater drainage. As far as possible, we will present the deposits in

stratigraphic order, from oldest to youngest. Throughout the trip, we will show the data and reasoning for the spatial correlation of deposits, their numeric age, and the resulting tectonic implications. An important consideration in understanding deformation in this setting is how rocks move horizontally through the subduction wedge. We present geomorphic and stratigraphic data to help resolve the horizontal translation of rocks and thus provide some constraints for shortening over geologic time scale.

As noted, the glacial-interglacial stratigraphy is critical to the analysis of tectonism, and additionally has implications for paleoclimatic processes. Glaciers have descended repeatedly into the coastal lowlands, constructing extensive glacial and glacial-fluvial landforms and depositing a variety of sediments. Abundant organic matter facilitates detailed radiocarbon dating of glacial events, and magnetostratigraphy provides additional age control on older deposits. In the Queets and Hoh River valleys, the glaciogenic sediments include lacustrine-outwash-till sequences. In sea-cliff exposures, a last interglacial wave-cut platform separates early and middle Pleistocene sediments from late Pleistocene sediments. Two older glacial units—the Wolf Creek and Whale Creek drifts—predate the last interglaciation. The Whale Creek drift is of middle Pleistocene age and the Wolf Creek drift is associated with magnetically reversed sediments and is therefore of probable early Pleistocene age. Three additional stratigraphic units—the Lyman Rapids, Hoh Oxbow, and Twin Creeks drifts—document six late Pleistocene glacial advances. The maximum late Pleistocene advance occurred between 54,000 yr B.P. and the last interglacial sea-level highstand (ca. 125,000 yr B.P.), most likely between 55,000 and 75,000 yr B.P. The second glacial maximum occurred ca. 33,000–29,000 cal. yr B.P., while the advance correlative with the ca. 21,000 cal. yr B.P. ice-sheet maximum was far less extensive. Correlation of glacial fluctuations with pollen fluctuations determined from Kalaloch sediments by Heusser (1972) indicates that the glacial advances were driven dominantly by sustained moisture delivery and may reflect insolation-modulated variations in westerly atmospheric flow.

Fluvial terraces are the main source of geologic and Quaternary stratigraphic data used in our tectonic interpretations. Terraces are landforms that are underlain by an alluvial deposit, which in turn sits on top of a strath, which is an unconformity of variable lateral extent and local relief. Typically, the strath is carved into bedrock, but it can also be cut into older alluvial deposits. At the coast, we recognize straths and their accompanying overlying alluvial deposits, and then show how those features continue upstream into the Clearwater River drainage. The straths and terraces are exposed because there has been active incision of the river into the rocks of the Olympic Peninsula. The most obvious conclusion is that river incision is a response to active rock uplift. But straths and terraces indicate that the incision history of at least one river has not been perfectly steady. There has been variability in external factors, such as climate or tectonics, which has modulated the terrace formation process. What we hope to demonstrate is that the variability in incision process and rate is primarily attributed to climate, but that continued uplift provides the means for long-term net incision of the river into the Olympic landscape.

The first day will involve traveling from Brinnon, on the eastern side of the peninsula, to Hurricane Ridge, in the core, and will end at Kalaloch, on the Pacific Coast. The next day will be mostly dedicated to understanding the coastal stratigraphy in and around Kalaloch, where many of the age constraints for surficial

deposits are located and there are good exposures of the accretionary wedge rocks. The field relationships for permanent shortening of the Olympic wedge will also be explored. The third day will be devoted to the Clearwater drainage and an investigation of terraces of various size, genesis, and tectonic implication. We will consider the myriad of processes that have conspired to construct and preserve the terraces and the possible contributions of both cyclic and steady uplift. The fourth and final day will be spent briefly visiting three geologic features on the way back to Seattle.

GEOLOGIC SETTING

Tectonics

The Olympic Mountains are the highest-standing part of the Oregon-Washington Coast Ranges occupying a 5800 km^2 area within the Olympic Peninsula (Fig. 1). The central part of the range has an average elevation of ~1200 m, and reaches a maximum of 2417 m at Mount Olympus (Fig. 1C). The Olympics first emerged above sea level ca. 18 Ma (Brandon and Vance, 1992), and they then seem to have quickly evolved into a steady-state

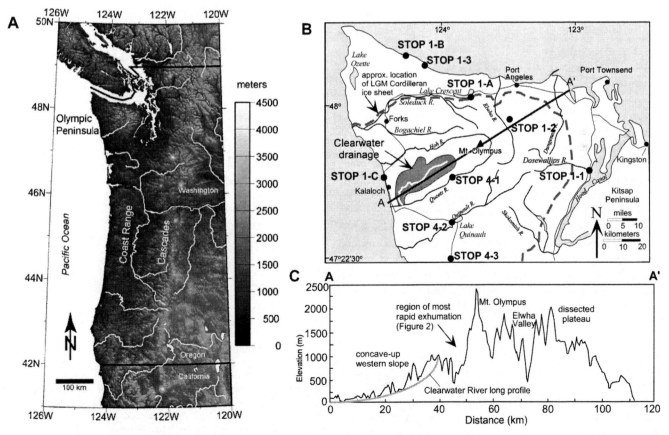

Figure 1. A: Shaded relief digital elevation model of the Pacific Northwest showing the Olympic Mountains in the context of the Coast Ranges. B: Major drainages of the Olympic Peninsula. The gray dashed line marks the southern boundary of the Last Glacial Maximum of the Cordilleran Ice Sheet. Note stops for Days 1 and 4 of the field trip. C: Topographic section across the Olympic Peninsula, parallel to the modern convergence direction (A–A′ in B).

mountain range, defined here by rock uplift rates that are closely balanced by erosion rates (Brandon et al., 1998). Fission-track cooling ages indicate that the fastest erosion rates, ~0.8 m/k.y., are localized over the highest part of the range (Fig. 2A). Rocks exposed there were deposited and accreted in the Cascadia Trench during the late Oligocene and early Miocene, and then exhumed from a depth of ~12–14 km over the past 16 m.y. Present-day rugged relief and high-standing topography are consistent with ongoing tectonic activity, which includes shortening both normal to and parallel to the margin (Fig. 2B). The Cascadia subduction zone underlies a doubly vergent wedge (in the sense of Koons, 1990, and Willett et al., 1993). The change in vergence occurs at the crest of the Oregon-Washington Coast Range, which represents the forearc high. The doubly vergent system includes a prowedge (or proside) that overrides oceanic lithosphere and accretes turbidites of the Cascadia drainage, and a retrowedge (or retroside) that underlies the east-facing flank of the Coast Range (Willett, 1999; Beaumont et al., 1999) (Fig. 3). This usage emphasizes the asymmetry of the underlying subduction zone, defined by subduction of the proplate (Juan de Fuca) beneath the retroplate (North America).

Much of the Cascadia forearc high is underlain by the Coast Range terrane, a slab of lower Eocene oceanic crust (Crescent Formation and Siltez River Volcanics), which occurs as a landward-dipping unit within the Cascadia wedge (Fig. 3A) (Clowes et al., 1987). Accreted sediment that makes up the proside of the wedge reaches a thickness of 15–25 km at the present Pacific Coast (Fig. 3B) and locally extends landward beneath the Coast Range terrane. The Coast Range terrane is clearly involved in subduction-related deformation, even though the rate of deformation is relatively slow when compared with the accretionary deformation occurring at the toe of the seaward wedge. Nonetheless, the Cascadia wedge, by definition, includes all rocks that are actively deforming above the Cascadia subduction zone. Thus, the Coast Range terrane cannot be considered a rigid "backstop," but instead represents a fully involved component of the wedge.

In the Olympic Mountains, the Coast Range terrane has been uplifted and eroded away, exposing the Hurricane Ridge thrust and the underlying Olympic structural complex (Brandon and Vance, 1992; Stewart and Brandon, 2003) (Fig. 3). The Olympic structural complex is dominated by relatively competent and homogeneous assemblages of sandstone and mudstone, with minor conglomerate, siltstone, and basalt (Tabor and Cady, 1978a, b). A large part of the Olympic structural complex was formed by accretion of seafloor turbidites into the proside of the wedge, starting ca. 35 Ma (Brandon et al., 1998). Where exposed in the Olympics, those accreted sediments are now hard, well-lithified rocks. The steep, rugged topography of the Olympics is supported by both basalts of the Coast Range terrane and accreted sediment of the Olympic structural complex, which suggests that there is little difference in their frictional strength. Uplift in the Olympic Mountains has been driven by both accretion and within-wedge deformation (Fig. 3) (Brandon and Vance, 1992; Willett et al. 1993, see stage 2 of their Fig. 2; Brandon et al., 1998; Batt et

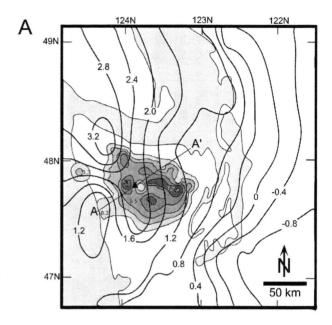

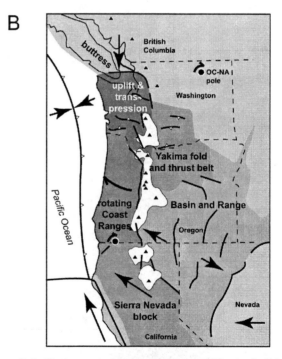

Figure 2. A: Contour map showing short-term uplift rates (solid contour lines) as determined by geodetic measurements (Savage et al., 1991; Dragert et al., 1994) and long-term erosion rates (shaded contour intervals) as determined by fission-track thermochronology (Brandon et al., 1998). Rates are in mm/yr. The preservation of the Quinault and other adjacent near-shore units indicates slow long-term uplift and erosion along the west coast. B: Relative motion of major tectonic blocks in the Pacific Northwest calculated from geodetic data with respect to the Olympic Cascade–North American (OC-NA) pole. Arrows indicate relative block motion. Onshore white polygons and black triangles are volcanic deposits and volcanoes respectively. In this interpretation, uplift of the Olympic Mountains is accomplished by transpression between the Oregon Coast Ranges and a proposed Vancouver Island buttress (modified from Wells et al., 1998).

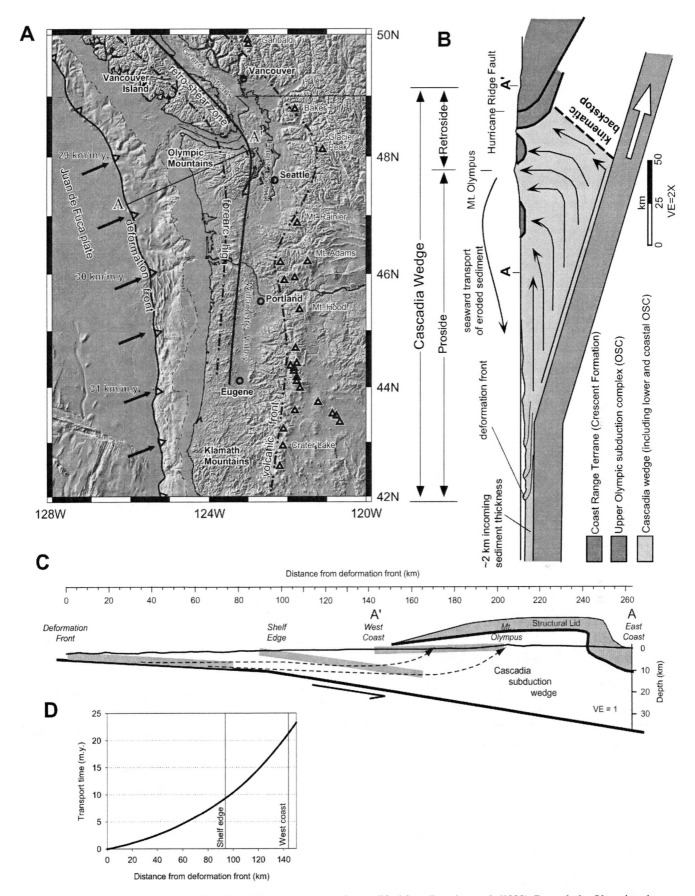

Figure 3. A: Simplified geologic map of the Cascadia convergent margin, modified from Brandon et al. (1998). Beneath the Olympics, the convergence velocity of the Juan de Fuca plate relative to North America is 36 mm/yr at an azimuth of 54°, which is nearly orthogonal to the modern subduction zone (option 2 for Juan de Fuca/Pacific in DeMets et al., 1990, and "NA-PA Combined" in DeMets and Dixon, 1999). B: Schematic section (A–A′ in part A) showing the regional-scale structure of the Cascadia accretionary wedge (after Brandon et al., 1998). VE—vertical exaggeration. C: Inferred displacement path for early Miocene accreted sediments in the Cascadia wedge (Stewart and Brandon, 2003, their Fig. 12). D: Transport time needed to move from the site of initial accretion at the front of the wedge to a location rearward in the wedge (Stewart and Brandon, 2003, their Fig. 13).

al., 2001). Accretion occurs entirely on the proside of the wedge, resulting in decreasing material velocities toward the rear of the wedge. In the Olympics, retroside deformation is marked by folding of the Coast Range terrane into a large eastward-vergent structure (Tabor and Cady, 1978a, b). The upper limb of that fold, which underlies the eastern flank of the Olympics (Fig. 4), is steep and locally overturned, in a fashion similar to the folding illustrated in Willett et al. (1993, stage 2 in their Fig. 2). We infer from the steep topographic slope on the retroside of the wedge that folding is being driven by a flux of material from the proside of the wedge, and that the wedge has not yet begun to advance over the retroside plate (Willett et al., 1993).

Deep erosion and high topography in the Olympics are attributed to an arch in the subducting Juan de Fuca plate (Brandon and Calderwood, 1990; Brandon et al., 1998). The subducting plate is ~10 km shallower beneath the Olympics relative to areas along strike in southwest Washington and southern Vancouver Island (Crosson and Owens, 1987; Brandon and Calderwood, 1990). Stated in another way, the shallow slab beneath the Olympics means that less accommodation space is available to hold the growing Cascadia wedge (Brandon et al., 1998). This situation, plus higher convergence rates and thicker trench fill along the northern Cascadia Trench, has caused the Olympics to become the first part of the Cascadia forearc high to rise above

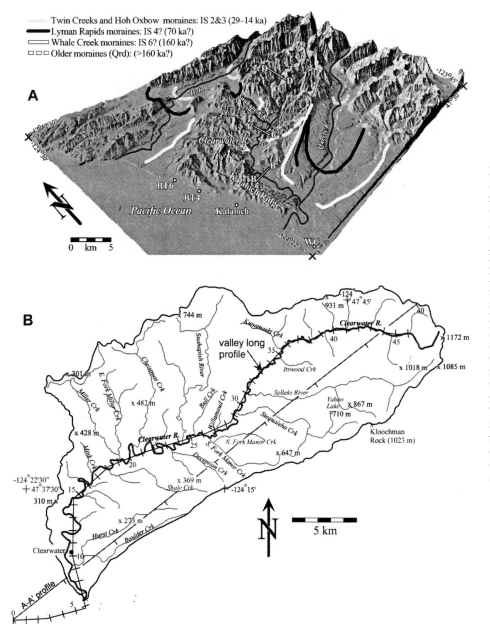

Figure 4. A: Digital shaded-relief image (30-m-resolution digital elevation model) showing the relation of the Clearwater drainage to adjacent drainages and glacial deposits in the western Olympics (glacial data are from Thackray, 1996, 2001; Easterbrook, 1986). BT6, BT4, and WC mark Beach Trail 6, Beach Trail 4, and Whale Creek, where important stratigraphic relationships are exposed along the coast. Profiles B–B′, C–C′, and D–D′ mark cross-valley sections of the Clearwater Valley, as presented in Figure 8. B: Map of the Clearwater drainage. The valley profile is shown as a crooked thin line with tics marking valley kilometers from the mouth of the Queets River at the coast. The straight section (A–A′ in Figs. 1 and 3) lies along the southeast side of the drainage. Final results were projected into A–A′, which parallels the local convergence direction for the Cascadia subduction zone. IS—isotope stage.

sea level. The early development of subaerial topography, plus continued accretion and uplift, account for the deep erosion observed in the Olympics. The corollary to this interpretation is that adjacent parts of the forearc high will evolve in the same way, although more slowly because of lower accretionary fluxes and a larger accommodation space for the growing wedge.

Active Tectonics

There is particular interest in Cascadia regarding the evidence of cyclic deformation related to large earthquakes at or adjacent to the subduction zone (Savage et al., 1981, 1991; Thatcher and Rundle, 1984; Dragert, 1987; Rogers, 1988; Atwater, 1987, 1996; Holdahl et al., 1987, 1989; West and McCrumb, 1988; Darenzio and Peterson, 1990; Atwater et al., 1991; Bucknam et al., 1992; Hyndman and Wang, 1993; Dragert et al., 1994; Mitchell et al., 1994). Fundamental to these studies is the distinction between short-term (10^2–10^3 yr) cyclic elastic deformation adjacent to the seismogenic subduction thrust and long-term (10^4–10^5 yr) permanent deformation associated with growth and deformation of the overlying Cascadia wedge. The earthquake cycle is probably partly decoupled from the permanent deformation, so we cannot easily integrate the effects of numerous earthquake cycles and arrive at the final long-term deformation. Furthermore, aseismic ductile flow, occurring within the deeper parts of the Cascadia wedge, probably also contributes to deformation manifested over long time spans.

Holocene deposits preserved in locally subsiding estuaries along the west coast of the Olympics provide good evidence of cyclic deformation related to large prehistoric earthquakes (Atwater, 1987, 1996). Seismogenic slip associated with these earthquakes, both on the subduction thrust and also on upper-plate faults, contributes to long-term deformation of the margin. However, it is difficult to separate elastic deformation, which is created and then recovered during each earthquake cycle, from the permanent deformation associated with fault slip.

Pre-Holocene stratigraphy and structure provide the only records of sufficient duration to separate long-term permanent deformation from earthquake-cycle elastic deformation. For this reason, local active-tectonic studies have focused on deformation of Quaternary deposits and landforms, which are best preserved along the Pacific Coast and offshore on the continental shelf (Rau, 1973, 1975, 1979; Adams, 1984; West and McCrumb, 1988; Kelsey, 1990; Bockheim et al., 1992; Kelsey and Bockheim, 1994; Thackray and Pazzaglia, 1994; McCrory, 1996, 1997; McNeill et al., 1997, 2000; Thackray, 1998). Mud diapirism, which is widespread beneath the continental shelf and along the west coast of the Olympics (Rau and Grocock, 1974; Rau, 1975; Orange, 1990), may be a local factor contributing to the observed deformation of Quaternary deposits.

In contrast, much less is known about the long-term deformation of the coastal mountains that flank the Cascadia margin (Fig. 2A). The development and maintenance of the Oregon-Washington Coast Range as a topographic high suggests that it is an actively deforming part of the Cascadia plate boundary. Diverse

geologic and geodetic data sets seem to indicate shortening and uplift both parallel (Wang, 1996; Wells et al., 1998) and normal (Brandon and Calderwood, 1990; Brandon and Vance, 1992; Brandon et al., 1998) to the direction of convergence (Fig. 2). This relationship is best documented in the Olympic Mountains (Figs. 1, B and C), which, on a geologic time scale (>10^3 yr), seem to be the fastest deforming part of the Cascadia forearc high.

Geodetic and tide-gauge data (Reilinger and Adams, 1982; Holdahl et al., 1989; Savage et al., 1991; Mitchell et al., 1994) indicate that short-term uplift is very fast on the Olympic Peninsula, ranging from 1.2 to 3.2 m/k.y., with the highest rates along the west side of the peninsula (Fig. 2). These high rates probably include a significant component of earthquake-cycle elastic deformation, given that the Cascadia subduction thrust is presently locked. This conclusion is supported by geologic evidence, which indicates insignificant long-term uplift or growth in coastal regions around the peninsula over the past 10 m.y. For instance, exposures of upper Miocene to lower Pliocene shallow-marine deposits locally crop out near modern sea level (Rau, 1970; Tabor and Cady, 1978a; Armentrout, 1981; Bigelow, 1987; Palmer and Lingley, 1989; Campbell and Nesbitt, 2000). These units currently sit within ~200 m of their original depositional elevation, which implies rock-uplift rates less than ~0.05 m/k.y. Slow long-term rock and surface uplift is also consistent with the preservation of extensive middle and lower Pleistocene deposits and constructional landforms along much of the west coast (Thackray and Pazzaglia, 1994; Thackray, 1998).

We use fluvial terraces to examine the pattern and rates of long-term river incision across the transition from the relatively stable Pacific Coast to the actively uplifting interior of the Olympic Mountains. We focus on the Clearwater drainage (Figs. 1B and 4), which remained unglaciated during the late Pleistocene and Holocene, and thus was able to preserve a flight of fluvial terraces, with each terrace recording the shape and height of past long profiles, with the oldest record extending back into the middle Pleistocene.

Quaternary Stratigraphy

The Quaternary stratigraphy in the lower Queets and Hoh Valleys provides a key framework for understanding terrace age relationships in the Clearwater Valley (Fig. 1). Coastal stratigraphy has been correlated with inland stratigraphy exposed in stream cuts and gravel pits (Thackray, 2001; Pazzaglia and Brandon, 2001), providing detailed age and stratigraphic control. Fill terraces in the lower Clearwater Valley merge with outwash terraces in the lower Queets Valley, and outwash from the lower Hoh Valley breached a low divide in the Snahapish Valley, further influencing the Clearwater terrace stratigraphy. Major outwash terraces in the lower Queets and Hoh Valleys include the Wolf Creek (early Pleistocene?), Whale Creek (middle Pleistocene), Lyman Rapids (55,000–125,000 yr B.P.) and Hoh Oxbow 2 (ca. 33,000–29,000 cal. yr B.P.) terraces. Each outwash terrace surface is underlain by thick (4–15 m) gravel of dominant cobble-pebble sizes, locally containing boulders. Relationships with underlying till, glacial-lacustrine, and bedrock units are exposed locally.

FIELD RESULTS

Day 1. Brinnon to Hurricane Ridge to Kalaloch (Fig. 2)

Start. Stop 1-1. Intersection of Route 101 and Dosewallips Valley Road, Brinnon, Washington

The purpose of this stop is to observe the rocks exposed in the retrowedge of the Olympic accretionary wedge. We will also have the opportunity to observe a Gilbert-delta and discuss glacial and fluvial stratigraphy in the Dosewallips drainage.

A discussion of the Crescent Formation and deformation in the retrowedge is presented above in the Tectonic section. From Brinnon, we gain a good vantage point to envision shortening east of the Hurricane Ridge fault and the topographic crest of the Olympic Mountains.

The fluvial stratigraphy of the Dosewallips drainage has been mapped and incision rates for the eastern part of the Olympic Peninsula are known for a few spot locations (Garcia, 1996). The delta exposed here is related to the Vashon stage glaciation (ca. 14 ka) and is thought to have become emergent by ca. 12.7 ka as a result of both draining of the ancestral Puget Sound, which the delta was building into, as well as isostatic rebound from the retreating Vashon Ice Sheet (Thorson, 1989). Here at Brinnon, there has been 40 m of incision, most of it through Vashon glaciofluvial deposits, which translates to a rate of 3.36 mm/yr. This rate is probably not representative of longer-term tectonic rates as it includes a significant amount of glacio-isostatic rebound. Farther upstream, the glacially polished valley bottom, which is genetically related to drift and other glacial deposits, is deeply incised by the Dosewallips River. At the confluence of the West Fork and main fork of the river (20 km upstream), there has been 19 m of incision and at the confluence with Silt Creek (7 more km upstream), there has been 22 m of incision. Assuming all of this incision is post-Vashon and post-emergence of the delta, the bedrock incision rates are 1.5 and 1.7 mm/yr respectively. Furthermore, a Holocene terrace 15 km upstream near the Elkhorn campground, with a base 5.8 m above the channel was radiocarbon dated at 3570 ± 60 yr B.P., with a corresponding incision rate of 1.6 mm/yr. Accordingly, the postisostatic rebound incision rate for the eastern part of the Olympic Peninsula is between 1.5 and 1.7 mm/yr. There does not appear to be any appreciable increase or decrease in incision rate upstream, nor is there any appreciable change in incision form or process where the Dosewallips River crosses the Hurricane Ridge fault.

Leave parking lot and proceed north on U.S. Rt. 101. The highway cuts through steep topography of the Quilcene Range, all of it underlain by Crescent Formation basalt.

Cumulative Miles	(km)	Description
11.0	(17.7)	Pass through Quilcene, stay north on U.S. Rt. 101.
20.5	(33.1)	Intersection of U.S. Rt. 101 and Rt. 104. Continue on U.S. Rt. 101 toward Port Angeles.
21.2	(34.2)	Crescent Formation exposed in the low outcrop to the right.
22.4	(36.1)	Hills on the left were eroded and smoothed by the continental ice sheet.
41.5	(66.9)	Enter Sequim.
42.2	(68.1)	Intersection with Sequim-Dungeness Way. Proceed straight.
51.1	(82.4)	Low road cut on right and gravel pit on left expose coarse ice sheet outwash.
57.5	(92.7)	Enter Port Angeles.
58.6	(94.5)	Turn left on Race Street, following signs for Olympic National Park south toward Hurricane Ridge.
64.4	(103.9)	Pass entrance station to the park.
68.6	(110.6)	View to the northeast of the Strait of Juan de Fuca.
70.0	(112.9)	Exposures of pillow basalt and pillow basalt breccia, interbedded with bright red, fossiliferous limestone.
71.3	(115.0)	Blue Mountain overlook.
75.5	(121.8)	Cross the Hurricane Ridge fault, which separates the Olympic structural complex from the structurally overlying Crescent basalt and associated forearc basin strata.
76.0	(122.6)	Stop 1-2.

Stop 1-2. Hurricane Ridge Visitor Center, Olympic National Park

The purpose of this stop is to observe the core of the Olympic Mountains and to discuss thermochronologic data (summarized above) and the feedback between topography, tectonics, and erosion. The Olympic Mountains have been proposed as a range that is in or near flux steady state (Brandon et al., 1998; Pazzaglia and Brandon, 2001; Willett and Brandon, 2002). The topography of the range that is displayed from this vantage point (on a clear day) allows us to think about the roles that elevation, relief, and erosion play in maintaining that flux steady state. The extensive thermochronologic data set for the Olympic Mountains, river suspended sediment data, and river incision rates allow for a comparison of erosion as a function of relief across the range. The Olympic Mountains have a nonlinear relationship between slope gradient and erosion rate which supports the emerging view that erosion rates in tectonically active mountain ranges are adjusted to the rate of uplift, rather than the slope steepness (Brozovic et al., 1997; Montgomery and Brandon, 2002; Fig. 5). Furthermore, the mean slopes calculated for a 10-km-diameter circle across the Olympic Mountain core are relatively invariant despite significant differences in erosion rate or rock type (core vs. peripheral rocks) (Montgomery, 2001). Cross-valley profiles extracted from valleys that have been glaciated, partially glaciated, and nonglaciated show significant differences (Montgomery, 2002). Glaciated valleys draining more than 50 km² have two to four times the cross-sectional area and up to 500 m of greater relief than comparable fluvial valleys. These results argue that climate can affect the overall mass flux out of tectonically active ranges like

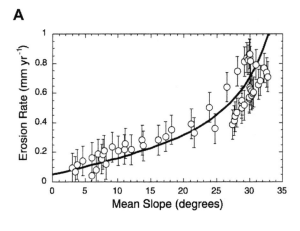

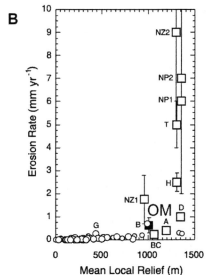

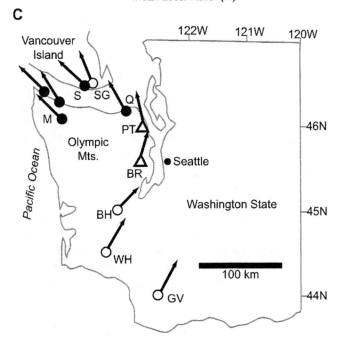

Figure 5. A: Plot of erosion rate data from thermochronometry versus corresponding mean slope determined for a 10-km-diameter circle operating on a 10-m-grid digital elevation model. Error bars represent an uncertainty in the erosion rates of 0.1 mm/yr. The solid line curve shows the predicted nonlinear response of erosion rate to slope angle predicted for landscapes where landslides are important (Roering et al., 2001). B: Plot of erosion rate vs. mean local relief for large, mid-latitude, tectonically inactive basins (the Ahnert, 1970 data, open circles), and large tectonically active basins (filled squares). G—Ganges; B—Brahmaputra; NZ—New Zealand; H—central Himalaya; NP—Nanga Parbat; OM—Olympic Mountains; T—Taiwan; D—Denali; A—Alps. Note location of the Olympic Mountains on this plot. From Montgomery and Brandon (2002). C: Map of western Washington showing paleomagnetic directions and rotations (arrows). Black circles show recently completed results; open symbols show previously published results. BH—Black Hills Volcanics; BR—Bremerton basalts; GV—Goble Volcanics; M—Makah-Hoko River Formations; PT—Port Townsend basalts; Q—Quimper and Marrowstone Formations; S—Sooke Formation; SG—Sooke Gabbro of Metchosin Volcanics; WH—Willapa Hills Volcanics. SG after Irving and Massey (1990); rest of figure modified from Beck and Engebretson (1982, Fig. 1).

---

the Olympic Mountains. Lastly, the elevation of Mount Olympus, or the central massif that contains Mount Olympus, might be partially attributed to an isostatic response to this glacially enhanced erosion (Montgomery and Greenberg, 2000). Calculation of the isostatic rebound at Mount Olympus attributable to valley development ranges from 500 to 750 m (21%–32% of its height) for a 5–10 km effective elastic thickness of the crust. This isostatic effect is one explanation for why the Olympic massif stands above a regional topographic envelope map constructed across the entire range. Alternatively, the transition from fluvial to glacial valleys was more gradual, occurring over 5–10 m.y. as the topography of the Olympic Mountains grew. During the same time, rock uplift would have kept the mean elevation constant, which is limited more by wedge taper than prevailing climate.

Return to Hurricane Ridge Road and retrace route back to Port Angeles.

Cumulative Miles	(km)	Description
87.6	(141.3)	Proceed through entrance gate and down into Port Angeles.
93.4	(150.6)	Turn left onto U.S. Rt. 101 south.
95.8	(154.5)	Leave Port Angeles, begin driving west toward Lake Crescent.
103.2	(166.5)	Cross the Elwha River and pass through the town of Elwha. Rough estimates of the sediment trapped behind the dam, combined with negligible estimates for the rate of chemical denudation, indicate a basin-wide erosion rate of 0.14 mm/yr (Dethier, 1986; B. Stoker, 1991, personal commun.; Brandon et al., 1998).
109.2	(176.1)	Passing Lake Sutherland, notice the predominantly hummocky topography surrounding the lake and ahead. Lake Sutherland has an

outlet to the east that flows into Indian Creek and ultimately to the Elwha River. Indian Creek is small and underfit in its valley.

111.7 (180.2)	The road swings to the south and ascends a low, hummocky ridge defining the eastern boundary of Lake Crescent. Enter Olympic National Park and follow the southern shoreline of Lake Crescent.

113.9 (183.7)	**Optional Stop 1-A. Sledgehammer Point.** The purpose of this optional stop is to observe the Crescent basalts and discuss the origin of Lake Crescent. Lake Crescent is the largest lake on the peninsula. Currently, it has an outlet to the north, through the town of Crescent, directly to the Strait of Juan de Fuca. Its former outlet was back to the east, down the Indian Creek valley. The hummocky topography we passed between Lake Sutherland and Lake Crescent is the debris from a large landslide, derived from the hillside to the north. This natural dam raised the lake level to its present elevation of 176 m (579 ft). The age of the landslide is thought to be early Holocene or late Pleistocene (Tabor, 1975). Lake Crescent is clear, cold, and deep; soundings indicate depths of at least 180 m (600 ft). The lake basin probably reflects both a structural control as well as deep gouging by the continental ice sheet. The pillow basalts exposed in the road cut opposite Sledgehammer Point are vertically bedded with top to the east (Muller et al., 1983).

Return to U.S. Rt. 101 south.

Cumulative Miles	(km)	Description
122.4	(197.4)	Cross a low divide mantled by glacial deposits but cored by bedrock between Lake Crescent and Soleduck Valley.
126.8	(204.5)	Pass the entrance station to the National Park Service (NPS) Soleduck Valley.
140.4	(226.5)	Sappho, turn right on Rt. 112, head north toward Clallam Bay.
155.5	(250.8)	Clallam Bay. Turn left, remaining on Rt. 112, and head west toward Sekiu.
157.8	(254.5)	Sekiu. Follow signs to Olsen's Resort. Park at the resort and access the coastal exposures.

Stop 1-3. Physt Conglomerate and Forearc Basin Rocks of the Olympic Peninsula

The purpose of this stop is to discuss the forearc basin rocks and present paleomagnetic data related to the direction and magnitude of shortening across the Olympic Peninsula (Fig. 5C). The forearc basin rocks are thick packages of Eocene-Oligocene deep marine sediments, which apparently were deposited in deep troughs during and after the accretion and docking of the Crescent terrane (Niem et al., 1992). The upper Eocene-Oligocene Lincoln Creek Formation, on the southern flank of the Olympics, spans over 1000 m in thickness (Prothero and Armentrout, 1985). On the north flank of the Olympics, the Pysht Formation (Durham, 1944; Snavely et al., 1978) may be as much as 2000 m thick, although there may be fault repetition of the section. These thick Eocene-Oligocene marine sequences not only record the history of events in the region in the middle Cenozoic, but they also constrain the tectonic models for the Olympics.

Over the past thirty years, numerous paleomagnetic studies have been conducted on many of these rocks. Nearly all the rocks south and east of the Olympic core show a significant post-Oligocene clockwise rotation (Wells and Coe, 1985; Wells, 1990). Until recently, only limited paleomagnetic studies were conducted to the north of the Olympics (Beck and Engebretson, 1982; Symons, 1973; Irving and Massey, 1990), and the results were ambiguous. In the past few years, however, extensive paleomagnetic studies of these rocks have greatly modified our picture of the mid-Cenozoic tectonics of the Olympics.

The most striking result is that the majority of the rocks north of the Olympic core complex show a post-Oligocene *counterclockwise* tectonic rotation, the opposite of the clockwise rotations reported east and south of the Olympics. On the northwest corner of the Olympic Peninsula, the upper Eocene-Oligiocene Hoko River and Makah Formations show a counterclockwise rotation of ~30 ± 3° (D.R. Prothero, 2003, personal commun.). On the northeastern corner of the Olympic Peninsula, the upper Eocene-Oligocene Quimper sandstone and Marrowstone shale show a counterclockwise rotation of 42 ± 9° (D.R. Prothero, 2003, personal commun.). Across the Strait of Juan de Fuca, on the southern tip of Vancouver Island, the upper Oligocene Sooke Formation yields a counterclockwise rotation of ~25 ± 10° (D.R. Prothero, 2003, personal commun.), consistent with earlier results reported by Symons (1973) and Irving and Massey (1990) for the underlying Eocene Mechosin Volcanics. Although the Sooke Formation is north of the Canadian border, it is tectonically part of the Olympics, since it lies to the south of the Leech River fault.

The only exceptions to this trend are the result reported from the Pysht and Clallam Formations along the central coast of the Olympics. The Oligocene Pysht Formation produces a clockwise tectonic rotation of ~49 ± 6° (Prothero et al., 2001), and the overlying lower Miocene Clallam Formation yields a clockwise rotation of 45 ± 15° (Prothero and Burns, 2001). This suggests that the tectonics of the north coast of the Olympics are more complex that previously expected, and we are currently analyzing these data to find an explanation for this apparent contrary rotation.

However, the overwhelming counterclockwise sense of rotation on rocks from the Quimper Peninsula in the northeast to Makah Bay on the northwest demands some sort of explanation. Tabor and Cady (1978b) commented on the horseshoe-shaped pattern of the Tertiary rocks wrapped around the Olympic core, and postulated that they had undergone some sort of oroclinal bending around the Olympics as they pushed eastward. We favor a model similar to this, where the eastward movement of the Olympic core "bulldozes" the rocks in its path into a curved pattern. Such bulldozing would rotate the rocks north of the core in a counterclockwise sense, and those south of the core in a clockwise sense.

Return to Olsen's Resort and Rt. 112. Proceed west on Rt. 112 to optional stop 1-B, otherwise, retrace route out to Sappho and U.S. Highway 101.

Cumulative Miles	(km)	Description
157.8	(254.5)	Return to Rt. 112 and proceed toward Neah Bay.
165.7	(267.3)	**Shipwreck Point.** This is ***Optional Stop 1-B.*** The purpose of this stop is to observe turbidites and olistostromes in the forearc sedimentary rocks. A walk to the east along the coastal outcrops from Shipwreck Point leads to exposures of the Janssen Creek Member of the Makah Formation, with its huge turbidites and olistostromes plainly visible. These rocks are also rich in early whale fossils, the mysticetes type that are transitional between tooth and baleen groups. Return to Sekiu and then to Sappho and U.S. Highway 101.
191.0	(308.1)	Intersection with U.S. Rt. 101, turn right.
202.6	(326.8)	Cross Soleduck River and the southern boundary of Vashon-age drift from the Juan de Fuca lobe of the continental ice sheet.
205.3	(331.1)	Enter Forks, the largest town on the western Olympic Peninsula.
210.9	(340.2)	Cross the Bogachiel River.
216.5	(349.2)	Ascend a low divide between the Bogachiel and Hoh River drainages underlain by a large lateral moraine complex of the Hoh Valley. The moraine here forms the northern divide of the lower Hoh River valley. A minimum age for the moraine of 30,000 ± 800 yr B.P. was obtained from organic material in a bog ~0.5 km to the southwest (Heusser, 1974). Approximately 6 km upstream from this point, another well-defined moraine ridge has yielded a basal bog minimum age of 18,800 ± 800 yr B.P. (Heusser, 1974). A nearly identical age from peat clasts in overbank silt was obtained in the lower Bogachiel drainage (Heusser, 1974, 1978). The interstadial peat site (Heusser, 1978), 60 m above the Bogachiel River, exhibited 85 cm of peat underlain by more than 2 m of unweathered diamicton and 2 m of brown, compact, thoroughly weathered till. The top of the peat has a reported age of 59,600 yr B.P. (essentially infinite). Pollen in the interstadial peat indicated subalpine forests

being succeeded by tundra and park-tundra type floras. The pollen in the bog suggests a rise in arboreal species including hemlock and pine at 30 ka, followed by tundra and park-tundra types by 20 ka, a rise of Sitka spruce and alder between 20 and 10 ka, maximum percentages of alder in the early Holocene, and finally to a modern palynoflora dominated by western hemlock. The glacial stratigraphy and palynostratigraphy suggests (1) a pre-Wisconsinan (>59,600 yr B.P.) and early Wisconsinan (<59,600 yr B.P.) glaciation that crossed the divide between the lower Hoh and Bogachiel drainages, extending all the way to the present coast, and (2) a late Wisconsinan (Fraser, ~18 ka) glaciation that extended to within 6 km east of this point in the Hoh Valley. The upper Hoh Valley currently supports active cirque glaciers. A goal of the Day 2 itinerary is to revisit the glaciation history since Heusser's important contributions 30 years ago.

219.0	(353.2)	Pass entrance road for the Hoh Rain Forest.
220.8	(356.1)	Cross the Hoh River. The Hoh River supports one of the last remaining salmon runs on the peninsula. The road now bends to the southwest and heads for the coast.
232.2	(374.5)	U.S. Rt. 101 bends directly south, enters Olympic National Park, and parallels the coast.
234.0	(377.4)	***Optional Stop 1-C. Destruction Island viewpoint.*** The purpose of this stop is to observe the Pacific Coast and set the stage for the next day's tour of glacial and tectonic topics.
240.5	(387.9)	Turn right into Kalaloch Lodge parking lot. This is the end of Day 1, and headquarters for the remainder of the field trip.

Day 2. Coastal Exposures Near Kalaloch, Olympic National Park, Moses Prairie Paleo–Sea Cliff and the Record of Horizontal Shortening, Hoh Formation, Glaciation in the Hoh Valley (Fig. 6)

Start, Kalaloch Lodge

Walk to the beach overlook for a brief overview. Kalaloch and the entire field-trip route lie in the west-central part of the Olympic Peninsula, between two large drainages, the Hoh and Queets Rivers, that drain the northwest, west, and southwest flank of Mount Olympus (Figs. 1B and 3A). The Clearwater River is tucked away between these two master drainages, and we use the fluvial and glacial deposits of the Hoh and Queets Rivers to constrain the ages of terraces in the Clearwater drainage (Fig. 7). The Olympic coast here is a constructional feature underlain by glaciofluvial deposits. It lacks the distinct, uplifted marine terraces characteristic of Cascadia in Oregon and northern California, although the effects of glacio-eustasy and coastal tectonics cause a major unconformity

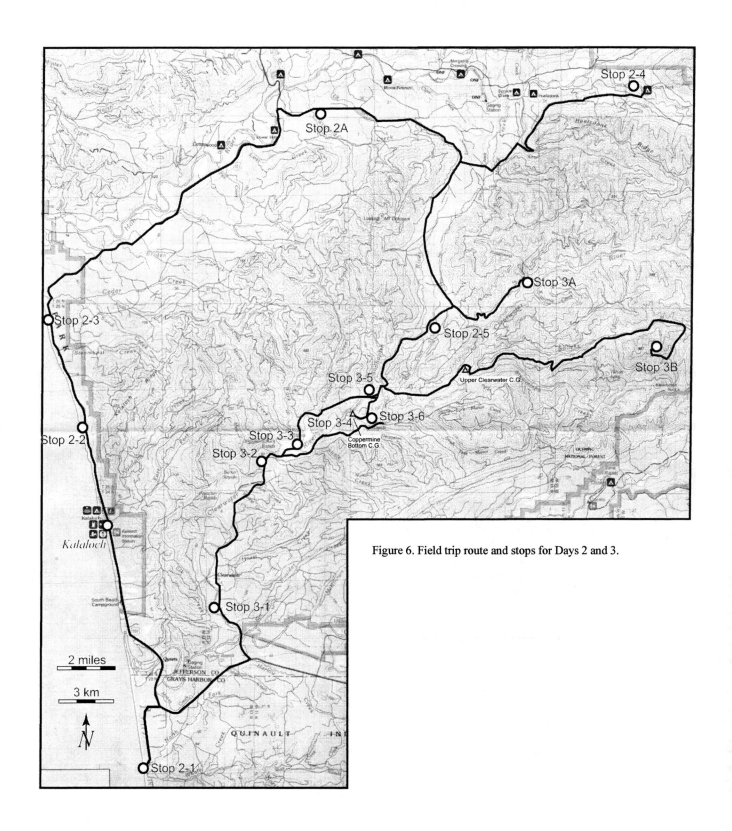

Figure 6. Field trip route and stops for Days 2 and 3.

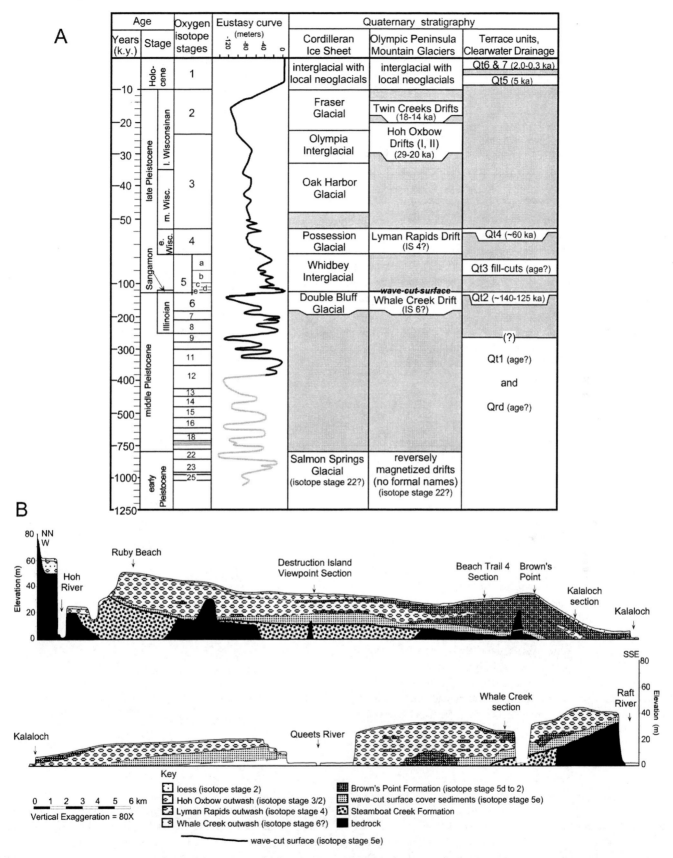

Figure 7. A: Regional stratigraphic correlations for glacial and fluvial deposits in the western Olympic Peninsula and surrounding regions. Note the variable scaling in the time axis. Eustasy curve is from Chappell et al. (1996) for 0–140 ka, and Pillans et al. (1998) for 140–300 ka. The next column shows deposits associated with advances and retreats of the Cordilleran Ice Sheet in Puget Sound, as synthesized by Easterbrook (1986). The next column shows the stratigraphic record of alpine glaciation in the western Olympics, based on the work of Thackray (1996, 2001). We have excluded his drift unit, called Hoh Oxbox 0 and estimated to be 39–37 ka, because it is based solely on an isolated sequence of lake sediment. The final column shows terrace stratigraphy in the Clearwater Valley as determined by work presented here. B: Compiled stratigraphy of the Olympic Peninsula west coast sea cliff from Hoh Head to Whale Creek (from Thackray, 1996).

in the coastal stratigraphy. Both stratigraphy and correlation of the widespread unconformity indicate active tectonic deformation in the form of a broad (tens of km) fold with an axis oriented roughly northeast or parallel to the direction of convergence (Rau, 1973; McCrory, 1996; Thackray, 1998). The mouth of the Queets River marks the approximate southern limb of a syncline centered on Kalaloch, and the mouth of the Hoh River marks the northern limb of the syncline. The significance of this syncline and all associated broad folding of Quaternary deposits at the coast is in the general lack of significant rock uplift and the comparatively small amount of northerly shortening in comparison to the large amount of shortening and uplift in the direction of plate convergence. The relative tectonic stability of the coast is also supported by preservation of early Pleistocene glaciofluvial deposits, as well as Miocene-Pliocene neritic shelf-basin deposits like the Quinault and Montesano Formations that outcrop at or near sea level. However the precise paleoelevation history of such ancient deposits, like the Quinault and Montesano Formations, must be viewed in the context of the large but unknown degree of horizontal translation they have experienced because of wedge shortening.

Exit Kalaloch and turn right.

Cumulative Miles	(km)	Description
2.7	(4.4)	South Beach campground is to the right. The 20 m tread here has been dated as 4570 ± 60 radiocarbon yr B.P. The dated material unconformably overlies late Pleistocene alluvium (Lyman Rapids outwash).
4.9	(7.9)	Cross the Queets River. Highway 101 is following a big meander loop of the Queets River. To the left are several late Pleistocene to Holocene terraces and sloughs. To the right are treads of the 30 m terrace. Engineering borings for the bridge show that the alluvium is thin (6 m); there is no deep, filled thalweg beneath the river channel. In other words, the river is essentially running over a low-relief bedrock strath.
6.6	(10.6)	Turn right and stay straight on the dirt road. This is now Quinault Nation land, and access permission is needed from the tribal government in Tahola. Moses Prairie is on the left, following the valley of the North Fork Whale Creek.
7.1	(11.5)	Stay left on the main dirt road, proceed south, remain on the 30 m terrace.
8.9	(14.4)	Stop 2-1.

Stop 2-1. Whale Creek

There are two objectives at this stop. The first is to observe the southern limb of the Kalaloch syncline and deformed Quaternary stratigraphy. Follow the old track along the north bank of the creek and cross driftwood field to the beach. Walk north ~200 m to the first exposures lying parallel the coast (Fig. 8).

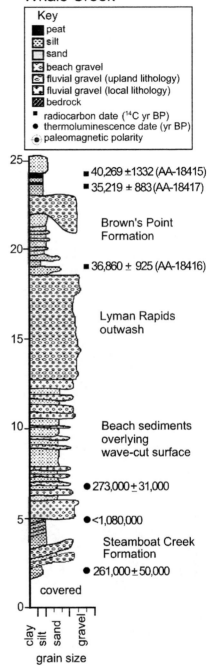

Figure. 8. Whale Creek section.

The stratigraphic sequence in this area is broadly similar to that at Destruction Island viewpoint and similarly fills a 1.5-km-wide area seaward of a paleoshoreline. The wave-cut surface here is not marked by a boulder lag, but is obvious as an angular unconformity. It has an apparent dip of ~3° north in this immediate area. Strata underlying the wave-cut surface have an apparent dip of 11° north. The wave-cut surface descends to beach level a short distance north of this first exposure. Horizontally bedded beach deposits overlie the wave-cut surface and form the base of the exposed sequence north of that location. Approximately 1 km farther north, the beach deposits abut against a 5-m-deep gravel-filled channel. The gravel fill is capped by ~1 m of silt and peat, which thickens northward. Two kilometers north of Whale Creek, the silt/peat sequence is 7.5 m thick, including a 2-m-thick gravel bed in the middle of the sequence. These sediments likely accumulated during latter substages of the last interglaciation.

The sequence above the wave-cut platform at the first exposure north of Whale Creek consists of 6.3 m of outwash and 6.3 m of fine-grained sediments. The latter consist of interbedded fine sand and silt, fining generally upward to silt with peat interbeds. A pebble-gravel bed cuts that fining upward sequence. Pollen spectra from this fine-grained sequence indicate cooling climatic conditions, from stable, mild conditions (at base) to cold-climate conditions represented by an alpine assemblage at the top (Florer, 1972). The sequence yielded three radiocarbon dates of 35–40 ka.

If time and tides permit, return to Whale Creek and cross to the south side. Walk ~0.5 mi south to the prominent exposure. On the way, note prevalence of slumping. The green, beach-side cabin (if still in existence) formerly sat on a level patch of ground about half way up the cliff. In the early 1990s the hillside mobilized, lowering the cabin to beach level and rotating it 90°. This spectacular outcrop provides clues to the style of deformation on the south limb of the Kalaloch syncline. Bedding attitudes in the older sequence (below the wave-cut surface) steepen progressively from north to south across the outcrop, from ~18° north to nearly 90°. Apparent unconformities separate beds of different attitudes. These relationships are suggestive of a fault-propagation fold at the tip of a buried thrust fault, the upper portion of the fold having been removed during wave-cut surface formation. The progressively steeper dips suggest that the fault was active through the period in which the strata were deposited. Such a structure may be responsible for the relatively steep inclination of the wave-cut surface on this limb of the Kalaloch syncline.

The second objective is to observe the paleo–sea cliff, relate the 30 m and 60 m coastal treads to glaciations in the Hoh and Queets River valleys, and discuss how the paleo–sea cliff tracks the horizontal motion of rocks and shortening of the wedge (Fig. 9). For a detailed discussion, see Willett et al. (2001) and Pazzaglia and Brandon (2001).

Retrace route out to U.S. Rt. 101.

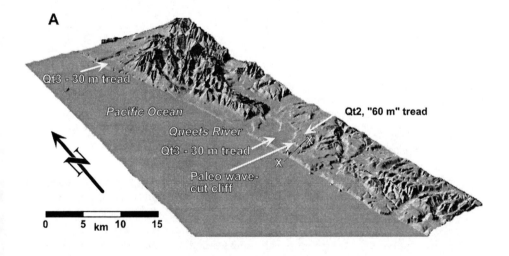

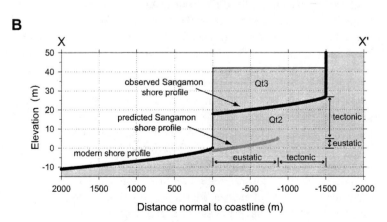

Figure 9. A: Digital shaded relief showing the buried Sangamon sea cliff. The cliff shows up as subdued scarp that parallels the coast at a distance 1500 m inland. The feature is visible in the image from ~15 km south to ~5 km north of the Queets River. B: A schematic cross section of the modern shoreface and sea cliff and the buried Sangamon sea cliff at the mouth of the Queets River. The section X–X' follows A–A'; it has an azimuth of 54°, parallel to the plate convergence direction (Pazzaglia and Brandon, 2001).

Cumulative		
Miles	(km)	Description
11.2	(18.1)	Turn left on U.S. Rt. 101.
18.4	(29.7)	NPS Kalaloch campground is on the left. Highway 101 hugs the coast, traveling on a tread ~20 m in elevation. This tread is underlain by late Pleistocene alluvial and eolian deposits as well as Holocene marsh deposits.
20.3	(32.7)	Pass Brown's Point (Beach Trail 3) and rise onto a 30 m terrace tread. This 30 m terrace is well preserved along the coast and will figure significantly into the stratigraphic story.
21.1	(34.0)	Stop 2-2.

Stop 2-2. Beach Trail 4

The purpose of this stop will be to see the rocks of the Cascadia wedge and begin developing the stratigraphic framework of the Quaternary deposits by observing the 122 ka wave-cut unconformity. A key point is that rocks exposed at the coast may have moved northeastward into the coast, with little to no uplift. Park in the NPS parking lot and descend the trail leaving from the southeast corner of the lot and go to beach level.

The bedrock exposed here consists of turbidites of the Miocene Hoh Formation (Fig. 10A). These rocks were deposited on the continental slope in at least 2 km of water and have been uplifted here to sea level. More importantly, the Hoh Formation was laid down 50–100 km west of its current position and has since followed a largely horizontal trajectory to the present Olympic coastline (Fig. 3B). Stewart and Brandon (2003) have shown, using fission-track grain ages from detrital zircons, that the Hoh Formation, which is also called the coastal unit of the Olympic structural complex, was deposited between ca. 24 and 16 Ma in water depths greater than 2000 m, which would correspond to the abyssal Cascadia basin, west of the modern Cascadia subduction zone. These rocks were accreted at the front of the Cascadia wedge, which is presently located 100 km west of our current position, and then slowly transported landward within the wedge (Fig. 3B), only reaching the coast at present. Figure 3C shows a reconstruction of this transport through the wedge, and Figure 3D shows the estimated history of landward translation.

The nearly vertically bedded bedrock is planed off ~2.7 m above mean sea level by a wave-cut unconformity. A thin boulder lag locally lies atop the unconformity, and it is superceded by gray pebbly beach sand texturally and structurally identical to the modern exposed shoreface. Exposures along the trail leading to the beach clearly show the unconformity continuing west under the 30-m coastal terrace (Fig. 9A). Cylindrical borings in the Hoh beneath the unconformity are interpreted as being shaped in part by pholad clams. The stratified sand, gravel, and peat overlying the beach deposits are part of the Brown's Point Formation (Heusser, 1972; Thackray, 1998; Fig. 10B). Gravel clasts in this deposit tend to be weathered and their provenance is consistent with a local source, most likely Kalaloch Ridge directly to the west, rather than the Hoh or Queets Rivers, whose deposits tend

Figure 10. A: Photograph of the bedrock and wave-cut unconformity exposed at Beach Trail 4. B: Annotated photograph of the Quaternary stratigraphy and numeric ages typical of the Brown's Point Formation in the vicinity of Beach Trail 4 (from Thackray, 1996). I.S.—isotope stage.

to be less weathered. The Brown's Point Formation seems to represent a long period of fluvial, glaciofluvial, and marsh-type sedimentation. The base of the unit directly above the beach sands is radiocarbon dead, but stratigraphically higher deposits have progressively younger radiocarbon ages (Heusser, 1972; Thackray, 1996). The youngest age comes from ~6 m below the terrace tread and is 16,700 ± 160 radiocarbon yr B.P.

The wave-cut unconformity here (Fig. 10A) is a key stratigraphic horizon, which can be traced for 80 km along the coast. It does not remain at the elevation viewed here. At Kalaloch, the unconformity is below sea level and not exposed. It rises to a maximum of 52 m above sea level south of the mouth of the Queets River. The average elevation between the mouth of the Hoh and Queets Rivers is 11 m. Deposits both above and below the unconformity are radiocarbon dead. At this stop we propose that the unconformity represents a wave-cut surface, produced during eastward migration of a shoreface during the last major interglacial eustatic highstand, at 122 ka (isotope stage 5e). We will develop the evidence at subsequent stops.

Cumulative Miles	(km)	Description
23.3	(37.6)	Cross Steamboat Creek.
25.2	(40.6)	Turn left. Stop 2-3.

Stop 2-3. Ruby Beach (lunch Stop)

The purposes of this stop will be to observe the Quaternary deposits above and below the wave-cut unconformity, key points for numeric ages of the deposits, and to observe the Hoh Formation. We follow the field observation of these units with a map-based correlation to glacial deposits in the Hoh and Queets River valleys. Park in the NPS lot and follow the trail to the beach. The exposures between Beach Trail 6 and Ruby Beach have changed significantly in recent years because of landsliding and coastal erosion.

The wave-cut unconformity is higher here, ~12 m above sea level, and is marked by a boulder lag (Fig. 11). The deposit below the unconformity was first named and described by Florer (1972) as the Steamboat Creek Formation. It is a complexly interbedded sequence of till, lacustrine deposits, glaciofluvial outwash, and sand dunes, which has yielded only infinite radiocarbon ages (below the detection limit for radiocarbon). Samples collected from lacustrine beds within this unit by Pazzaglia and Thackray, and analyzed by H. Rowe and J. Geissman at the University of New Mexico, show both normal and reversed polarities (Thackray, 1996). At this site in particular the polarity of the sample is reversed, indicating an age greater than 780 ka, the most recent reversal. Beach deposits and associated peat overlie the unconformity and are succeeded by predominantly glaciofluvial outwash sourced from the Hoh drainage. The outwash within 15 m of the unconformity is interbedded locally with peaty beds that have returned radiocarbon-dead ages of >33.7 and >48 ka (Florer, 1972). Closer to the terrace tread, in a bed locally separating two outwash units, finite radiocarbon ages of 36,760 ± 840 and 28,352 ± 504 ^{14}C yr B.P. have been determined on woody material by Thackray (1996). The interpretation of these ages suggested by Thackray (1996, 2001) is that the outwash and peat-rich sediments directly above the wave-cut unconformity are correlative to a marine isotope stage 4 (or possibly 5d, b) alpine glaciation (Lyman Rapids advance) and that the upper outwash was deposited by streams during isotope stage 3.

The new landslides here between Beach Trail 6 and Ruby Beach expose mud diapirs that are both onlapped by and that pierce the Steamboat Creek Formation (Rau and Grocock, 1974; Rau, 1975; Orange, 1990).

Return to the base of the trail and assemble for a view of Destruction Island. The broad topographic and structural low between the mouth of the Hoh and Queets Rivers filled with sediment from those two point sources, as well as from small streams draining Kalaloch Ridge (Fig. 9A). The center of the low near Kalaloch has more fine-grained sediment than do the regions proximal to the big river mouths. The general model is that the Queets and Hoh Rivers have periodically been point sources that built broad fans in front of the river mouths, spilling laterally

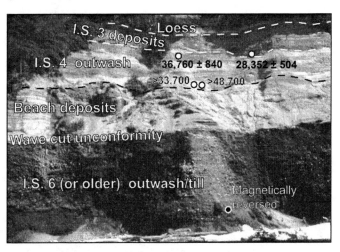

Figure 11. Annotated photograph of the Quaternary stratigraphy exposed between Beach Trail 6 and Ruby Beach (from Thackray, 1996). I.S.—isotope stage.

into the Kalaloch low. The fans formerly extended far west of the current coast. The top of Destruction Island 5 km offshore (20 m elevation) is correlative to the 30 m tread here at the coast, and the flanged base of Destruction Island that sits only 1 m above mean tide level is the westward continuation of the wave-cut unconformity. The level of the fan tread, represented by the 30 m terrace seen all morning of Day 2, must reflect a significant amount of vertical aggradation rather than tectonic uplift. The preservation of the isotope stage 5e unconformity and ca. 800 ka Pleistocene deposits in the outcrop below this viewpoint attest to the relative tectonic stability of the coast.

The most likely time for major periods of fan aggradation on a generally tectonically stable coastal setting is during a cycle of alpine deglaciation, as the river valleys are liberating a high sediment flux at the same time sea level is rising. It is difficult to imagine how the observed degree of aggradation could have occurred during a relative sea level low on a coast not undergoing rapid vertical uplift. Furthermore, map relationships show how the coastal terrace treads can be traced more or less continuously up the Hoh and Queets Valleys to heads of outwash (moraines). The general glacial stratigraphy of the Hoh and Queets Valleys (Thackray, 1996, 2001; Fig. 6) records a major isotope stage 4 alpine glaciation (55–125 ka, Lyman Rapids drift), which was responsible for the large body of outwash above the wave-cut unconformity and the construction of the 30 m terrace at the coast. Older alpine glacial periods, such as isotope stage 6 (150 ka, Whale Creek drift), are represented locally by the deposits below the wave-cut unconformity. However, there are clearly deposits older than isotope stage 6 below the unconformity, and these have a landward equivalent in various upland gravels, a Queets Valley moraine-outwash sequence called the Wolf Creek drift. Deposits of isotope stage 3 and 2 alpine glacial periods are underrepresented at the coast because of the relative small size of the glaciations. However, isotope stage 3 outwash is represented in the upper portion of the sequence south of Cedar Creek (Ruby Beach). Meltwater streams appear to have breached a divide at the head of Cedar Creek, permitting an outwash fill to be deposited in that valley and seaward of the last-interglacial seacliff.

Cumulative Miles	(km)	Description
25.3	(40.8)	Return to vehicles, exit parking lot, and turn left (north) onto U.S. Rt. 101.
27.0	(43.5)	U.S. Rt. 101 follows the Hoh River upstream. The road is built on a late Pleistocene outwash plain of valley glaciers with heads of outwash located several kilometers ahead in the direction of travel.
38.5	(62.1)	U.S. Rt. 101 turns to the north and ascends moraines of the Hoh Oxbow drift.
39.3	(63.4)	Turn right onto the Snahapish-Clearwater Road. The road climbs a bedrock-till high that represents the Hoh Oxbow 2 glacial limit (29,000–26,000 ^{14}C yr B.P.) (Thackray, 2001).
40.7	(65.6)	Cross Winfield Creek.
41.0	(66.1)	Slow, turn right on dirt road to ***Optional Stop 2-A. Department of Transportation gravel pit.*** The purpose of this stop is to observe the stratigraphy, sedimentology and soil development (Fig. 12) in the Hoh Oxbow (ca. 20 ka) outwash.
41.0	(66.1)	Return to the Snahapish-Clearwater Road, proceed east.
44.0	(71.0)	Dead-ice moraine on right. Heusser (1974) obtained a 14,480 ± 600 yr B.P. radiocarbon date on peat from a bog on the moraine, as well as a 15,600 ± 240 yr B.P. date from a bog lying 0.9 mi northwest.
46.0	(74.2)	Eocene-Oligocene bedrock on right (Tabor and Cady, 1978a), Hoh Oxbow 3 lateral moraine to left.
46.2	(74.5)	Turn left and begin to cross a broad, flat plain underlain by outwash and drift. Heusser (1974) reported an 18,800 ± 800 yr B.P. radiocarbon date on peat from a bog 1.6 mi north.
48.5	(78.2)	Turn right, cross Owl Creek.
48.8	(78.7)	Make a sharp left turn and ascend the nose of Huelsdonk Ridge. Note numerous debris flows deposits and stream channels affected by debris flows that originated in the short, steep tributaries draining this ridge to the north into the Hoh Valley.
53.5	(86.3)	Turn left, cross the South Fork of the Hoh, then turn right into the South Fork Hoh campground. Trail to the South Fork Hoh glacial exposures begins on the south side of the bridge and proceeds west (downstream).
53.8	(86.8)	Stop 2-4.

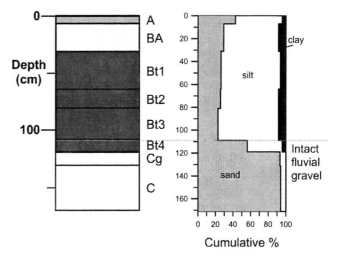

Figure 12. Soil described at Winfield gravel pit.

Stop 2-4. South Fork Hoh. Twin Creeks Drift Near Confluence of South Fork Hoh and Hoh Rivers

The purpose of this stop is to observe the well-exposed Twin Creeks drift and discuss the record of valley glaciation in the Olympic Mountains (Fig. 13). *Note: this exposure can be accessed via a similar road on the east side of the bridge, or by following the logging roads to the top of the terrace and descending the closed, terrace-edge logging road. Either access route can be rather treacherous, with the latter providing the best access to the entire exposure.* Following deposition of the Hoh Oxbow drift, the Hoh glacier retreated well upvalley, separating into two lobes at the confluence of the Hoh River and its South Fork. The two glaciers then readvanced to deposit the Twin Creeks drift. The maximum-phase Twin Creeks end moraine in the main valley lies ~2.5 km north of this exposure, adjacent the confluence area. Its South Fork counterpart lies ~3 km upvalley from this exposure, with a later-phase moraine ~2 km farther upvalley.

The stratigraphic sequence exposed here records events prior to and during the Twin Creeks 1 advance (Thackray, 2001). Three meters of grey, clast-rich till form the base of the exposure. This till correlates with the Hoh Oxbow 3 drift, the terminal moraine of which lies 12 km downvalley. Three meters of clay- and silt-rich lacustrine sediment overlie the till. Dropstones are common in the bottom meter. The lacustrine sediments yielded several wood samples, three of which yielded dates of 19,324 ± 165 (AA-18407), 19,274 ± 154 (AA-18408), and 19,169 ± 162 (AA-18406) ^{14}C yr B.P. Correlative, clast-rich glacial lacustrine sediments upvalley yielded a date on wood of 19,067 ± 329 ^{14}C yr B.P. (AA-18405). Delta sediments overlie the lacustrine beds. Ten meters of poorly exposed sand and gravel overlie the lacustrine sediments. The sediments appear to be horizontally bedded and may represent delta bottomset beds. Three meters of interbedded sand and silt with climbing ripples overlie the sand and gravel, and may also represent bottomset beds, perhaps in a

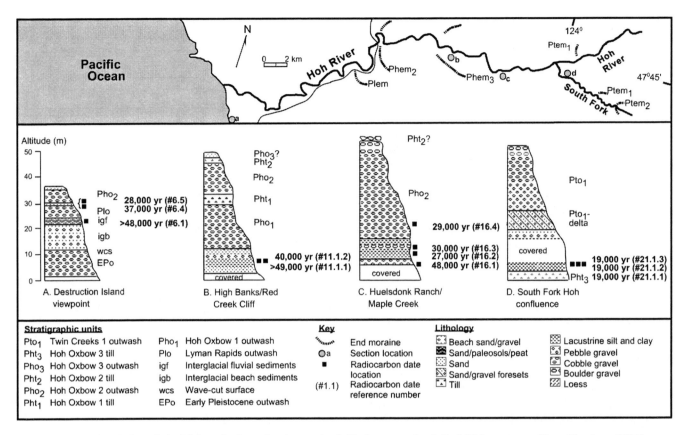

Figure 13. Correlation of glacial deposits from the coast to the field trip stop at South Fork Hoh exposures (from Thackray, 1996).

deepening lake. Northward-dipping gravel and sand delta foreset beds, 8 m thick, overlie the sand and silt. The foreset beds generally coarsen upward and are more gravel-rich in the top 2 m.

The delta sediments are overlain by ~25 m of outwash. The outwash is pebble and cobble dominated through most of its thickness, but coarsens upward to boulder-rich outwash. The top 5 m is very bouldery. The boulder-rich outwash can be observed especially well on the closed logging road ~700 m northwest of this exposure. The coarse outwash forms a prominent terrace that can be traced upvalley to the end moraine. The terrace merges with main-valley terraces in the confluence area. Till exposed in the moraine area yielded wood dated at 18,274 ± 195 ^{14}C yr B.P. (AA-16700), documenting the culmination of the Twin Creeks 1 advance at the time of the Northern Hemisphere ice-sheet maximum.

The stratigraphic sequence in this exposure reflects two advances of the South Fork glacier. The glacier first advanced past this location during the Hoh Oxbow 3 advance, depositing the till, and merged with the main Hoh glacier. It then retreated as a lake filled the valley. The valley was likely dammed by drift deposited during that advance. The lake appears to have extended into the main valley: thick sequences, clast-rich lacustrine, and deltaic sediments underlie Twin Creeks outwash on the north

side of the valley 3 km northwest of this exposure. As the South Fork glacier readvanced, an outwash delta was built into the lake. The lake was finally filled and/or the dam breached, and outwash was deposited fluvially at this location as the glacier approached its maximum position. Morphologic and stratigraphic evidence in the several kilometers upvalley of this exposure indicates that the glaciers subsequently readvanced to construct the younger Twin Creeks 2 end moraine and outwash terraces.

The Twin Creeks 1 drift was deposited at the time of the Northern Hemisphere ice-sheet maximum (ca. 18,000 ^{14}C yr B.P.). The glacier terminus was 12 km upvalley of the Hoh Oxbow 3 moraine, 17 km upvalley of the Hoh Oxbow 2 moraine (ca. 29,000–26,000 ^{14}C yr B.P.), and 20 km upvalley of the late Pleistocene-maximum Lyman Rapids moraine (55,000–130,000 yr B.P.). Thus, the alpine advance during the Last Glacial Maximum was far less extensive than advances during marine isotope stages 3 and 4. This pattern reflects the strength of westerly flow into the Olympic Mountains, which was diminished at the time of the ice-sheet maximum but sustained during earlier stadial events. The Twin Creeks 2 drift was likely deposited during the Vashon Stade (Puget Lowland maximum, ca. 14,000 ^{14}C yr B.P.), or during late-glacial time. Retrace route back to Snahapish-Clearwater Road.

Cumulative		
Miles	(km)	Description
61.4	(99.0)	Turn left. The road follows the Snahapish Valley south into the Clearwater Valley. The Snahapish Valley occupies a low divide between the Clearwater and Hoh Valleys. The Snahapish Valley is choked with outwash and drift. At the head of the valley, this drift is mapped as Lyman Rapids (isotope stage 4). Farther downstream, as the Snahapish begins to fall into the Clearwater Valley, there is another head of outwash that we map as Whale Creek drift (isotope stage 6). We view the Snahapish Valley as being the main conduit for both water and sediment entering the Clearwater Valley at least two times in the middle and late Pleistocene.
67.5	(108.9)	Turn left on dirt road and continue into gravel pit.
67.6	(109.0)	Stop 2-5.

Stop 2-5. Clearwater Corrections Pit

The purpose is to examine the coarse-grained, proximal portion of the Qt2 outwash in a gravel pit adjacent to the Qt2 head of outwash. (Fig. 4). The exposures at this stop are dominated by stratified sandy gravel that is particularly coarse. Large, subangular boulders in the deposit are unique to this site and indicate that it is proximal to a head of outwash (Whale Creek, isotope stage 6). A soil was described at this site. The profile is composed of a younger, yellowish-brown late Pleistocene soil ~1 m thick that overlies reddish-brown, clay-rich horizons (Fig. 14). The base of the buried soil is not exposed. The reddish-brown buried soil is poorly preserved in only a few localities on glacial-fluvial deposits in the Hoh, Clearwater, and Queets Valleys. In most

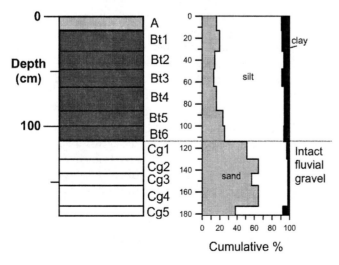

Figure 14. Soil described at the Clearwater corrections pit.

cases, the reddish-brown horizons are associated with a buried colluvial wedge; it is rarely if ever found on the deposit tread, despite the fact that the treads are typically flat. We will further develop the soil chronosequence and our interpretations of soil-landscape relationships and what they mean for landscape evolution and erosion at the stops for Day 3.

Cumulative		
Miles	(km)	Description
67.7	(109.2)	Return to the Snahapish-Clearwater Road.
70.8	(114.2)	Turn right and follow the main, paved Snahapish-Clearwater Road through the town of Clearwater.
84.3	(135.9)	Turn right onto Highway 101 and proceed north back to Kalaloch.
94.5	(152.4)	Turn left into Kalaloch Lodge, this is the end of Day 2.

Day 3. Clearwater Drainage, Terrace Stratigraphy, Age of Pleistocene Terraces, a Model for Holocene Terrace Genesis (Fig. 6)

Cumulative		
Miles	(km)	Description
0.0	(0.0)	**Start, Kalaloch Lodge.** Turn right on Highway 101 and continue on the 30 m tread.
10.2	(16.5)	Turn left on the Clearwater-Snahapish Road.
10.8	(17.4)	Cross the Queets River. The confluence of the Queets and Clearwater is ahead and to the left. Continue along the Clearwater River. As at the Highway 101 bridge, the engineering borings here demonstrate how the alluvium is thin and the stream is essentially on bedrock. The road is on a Holocene (Qt6) terrace. Older Holocene (Qt5) treads are exposed in clearcuts to the right and the treads of the two big Pleistocene fill terraces (Qt2 and Qt3) underlie the hills directly ahead.
11.5	(18.5)	Turn left onto dirt road for Clearwater Picnic bar and proceed out onto the point bar.
11.6	(18.7)	Stop 3-1.

Stop 3-1. Clearwater Picnic Bar

The purpose of this site is to observe the channel in the lower reaches of the Clearwater River and the stratigraphy of the river floodplain. The Clearwater River here is predominantly an alluvial stream, but the alluvium is typically less than 3 m thick. We envision most or all of this thickness of alluvium to be mobilized during large discharge events at which point it is in contact with the bedrock, driving incision. The floodplain stratigraphy is well exposed on the far bank. Depending on river stage conditions, it is possible to wade across to observe the cut bank. The cut bank exposes ~1 m of coarse sandy gravel that has a grain-size distribution similar to the bar we are parked on. In places, the tread of this deposit is above the mean flood level. In those places, we name the tread Qt6. Here the tread continues to be inundated by floods and is named Qt7. The coarse-grained facies locally contains wood or charcoal. C-14 dates from the coarse gravelly facies 5 km upstream from this site returned an age of 710 ± 48 yr B.P. Typically, this gravelly facies of Qt6 elsewhere in the watershed is dated ca. 1000 yr B.P. Stratified, locally cross-bedded sand and silt conformably overlie the gravel. The sand and silt facies is interpreted as overbank. Radiocarbon ages range from modern to several centuries before present and the virtual lack of any soil development attest to the continued aggradation of this unit. The Holocene terrace visible at this location is just one of several Holocene and late Pleistocene terraces that will be seen at the next stops (Fig. 15).

Return to the Snahapish-Clearwater Road.

Cumulative Miles	(km)	Description
11.7	(18.9)	Turn left.
14.4	(23.2)	Begin ascent of the Qt3 terrace.
14.9	(24.0)	To the left and down the bank is a 30 m high landslide headwall exposure of Qt3 gravels

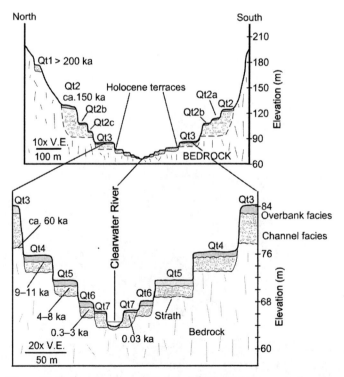

Figure 15. Composite cross section representing the terrace stratigraphy and general age ranges in the Clearwater drainage. We use the convention of naming terraces in order of increasing age where 1 is the oldest (highest) terrace in the landscape and assigning numbers to straths only. In this manner, terrace Qt2 may have more than one tread, which we designate with lower case alpha numeric subscripts (as Qt2a, Qt2b).

unconformably overlying lacustrine beds. The lacustrine beds have been dated at >47,000 radiocarbon yr B.P. and are likely correlative to Qt2. The overlying Qt3 alluvium has a reported finite age of 48,300 ± 3,300 radiocarbon yr B.P. (Thackray, 1996; likely also an infinite age).

16.4	(26.5)	Rise onto a degraded tread of Qt2 ~20 m above the Qt3 tread.
18.0	(29.0)	Cross Elkhorn Creek incised into the alluvium of Qt3. Slow for a left turn.
18.1	(29.2)	Turn left onto dirt road and park.

Stop 3-2. Qt3 at the Elkhorn Pit

The purpose of this stop is to show the sedimentology, stratigraphy, and weathering characteristics of the Qt3 terrace and to develop arguments that the underlying strath was buried at 60 ka, and the Qt3 terrace deposit and tread are correlative to the 30 m coastal terrace. Qt3 has a moderately developed yellow soil profile (Fig. 16) and an oxidation depth of 4–5 m, but locally 10 m in coarse alluvium. Post-depositional modification is minimal on the terrace treads, which retain a constructional morphology with well-preserved sandy overbank and silty loess deposits. The loess is more than 1 m near the Snahapish River and near the coast. Soil profiles consist of a 50–80-cm-thick B horizon composed of yellowish-brown (10YR) to brown (10YR–7.5YR) silt loam with numerous thin clay films preserved on soil ped faces. Soil profiles in fine-grained material have strong brown colors (7.5YR) and well-developed soil structure. Qt3 is best preserved below approximately km 24, where it locally has two treads, designated as Qt3a and Qt3b, with the Qt3b tread sitting ~4 m below the Qt3a tread. In this part of the valley, Qt3 straths maintain a gentle gradient, lying 6–10 m above the channel, and treads are 35 m above the channel. Above km 24, Qt3 has only one tread, and the straths take on a

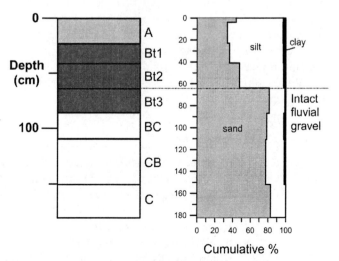

Figure 16. Soil described at Elkhorn Creek.

steeper gradient, climbing to a maximum height of 70 m above the channel. This gravel-pit outcrop of Qt3 exposes a highwall cut into the Qt3b tread. The Qt3 tread can be traced upstream through the Snahapish Valley to the Lyman Rapids (isotope stage 4, ca. 60 ka) head of outwash in the Hoh Valley (Fig. 3).

Return to vehicles and return to the Clearwater-Snahapish Road. Turn left.

Cumulative Miles	(km)	Description
18.2	(29.4)	Cross Shale Creek.
18.3	(29.5)	Stay left on the paved main road.
19.2	(31.0)	Slow and pull off the road to the left in the parking area just before the bridge. The outcrop is a short walk down the dirt road leading to the river. Stop 3-3.

Stop 3-3. Holocene Strath at the Grouse Bridge

The purpose of this stop is to introduce the concept of a strath, the terrace deposit, and some of the relative and numeric criteria in establishing strath age. Detailed discussions of these topics can be found in Wegmann (1999), Pazzaglia and Brandon (2001), and Wegmann and Pazzaglia (2002). Terraces are well preserved in the Clearwater drainage. There are two major flights of terraces: a higher, outer, older sequence that is underlain by thick alluvial-fill deposits, and a lower, inner, younger sequence underlain by thin alluvial deposits (Fig. 15). The terrace exposed here at the Grouse Bridge is a fine example of the lower, inner, younger sequence, and it contains all of the stratigraphic characteristics important to distinguishing and using terraces in tectonic interpretations (Figs. 17 and 18). The lower terraces like the one exposed here are composed of a basal, coarse-grained, 1–3-m-thick axial channel, sandy gravel facies, and overlying fine-grained 1–3-m-thick sandy silt overbank facies. The sandy gravel facies locally preserve sedimentary structures consistent with lateral accretion processes, as might be expected for point and transverse bars, which can be seen in the adjacent modern channel. So by analogy, we take the coarse-grained facies of the terrace deposit to represent the bedload being transported when the terrace strath was cut. In contrast, the fine-grained facies represent vertical accretion atop the floodplain, presumably related to deposition during floods.

Return to vehicles and continue across the Grouse Bridge on the Snahapish-Clearwater Road.

Cumulative Miles	(km)	Description
19.9	(32.1)	Cross the Clearwater River and ascend the Qt3 tread. As the road continues to climb, bedrock is exposed in the valley wall.
20.8	(33.5)	Cross Christmas Creek. Like the Snahapish River, alluvium from the Hoh River spilled into the Clearwater drainage through this valley.
21.2	(34.2)	Climb out of the Christmas Creek Valley and ascend onto the Qt2 tread.

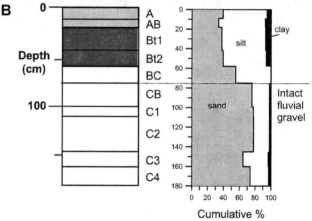

Figure 17. A: Annotated photograph of the Qt5 strath and strath terrace exposed at the Grouse Bridge. B: Soil described at the Grouse Bridge site in Qt5.

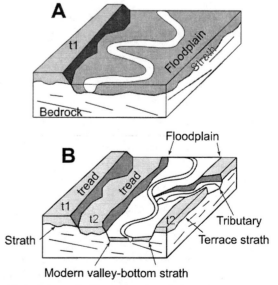

Figure 18. A: Schematic illustrating the relationship between a terrace (t1, t2), straths, the floodplain, and a valley bottom. B: The correspondence between the width of the floodplain and the width of the valley bottom strath is corroborated by exposure of the valley bottom strath in tributary channels, shown entering from the right part of the diagram.

22.0	(35.5)	Continue along a long, flat portion of the Qt2 tread.
22.8	(36.8)	Begin descending the Qt2 tread and stay right onto the dirt road leading to the Coppermine Bottom campground.
22.9	(36.9)	Proceed straight down to the campground. Descend Qt2, cross Crooks Creek slough, and a fill unit inset into the Qt3 tread.
23.5	(37.9)	Traverse the Qt3 tread and turn right.
24.0	(38.7)	Stop 3-4.

Stop 3-4 and Lunch. Coppermine Bottom

The purpose of this stop is to observe a reach of the Clearwater River channel near the middle part of the watershed and discuss the observed and modeled distribution and rates of incision of the Clearwater channel. Tomkin et al. (2003) uses the incision data for the Clearwater River to test a variety of models for incision of bedrock channels. We refer you there for details of this analysis. Figure 19 provides a simple summary of the main results. Models like the unit-shear stress model or stream-power model for incision assume a functional form of $I = kS^nQ_p^m$, where I is the long-term incision rate, S is the gradient of the channel, and Q is the discharge normalized for channel width. The parameters k, n, and m have different predicted values, depending on the model, but all models require that n and m are greater than zero. Figure 19 shows the best-fit solution for the model equation given. The transformation to logarithms allows the equation to be fitted by a plane, as indicated by $I = \ln k + n\ln S + m\ln Q_p$. The best-fit solution is shown in Figure 19. What should be obvious is that the slope of the plane is negative for both $\ln S$ and $\ln Q_p$, which means that the predicted values for m and n are less than zero. Thus, available models do not adequately fit the data for the Clearwater. We will summarize possible solutions for this paradox.

After lunch, loop through the campground and retrace route out to the Clearwater-Snahapish Road.

Cumulative		
Miles	(km)	Description
25.6	(41.3)	Stay left and then make a quick right into the Copper Pit (gravel pit).

Stop 3-5. Qt2 Terrace at the Copper Pit and Cosmogenic Surface Dating Profile

The purpose of this stop is to show the sedimentology, stratigraphy, and weathering characteristics of the Qt2 terrace and develop the arguments that the underlying strath is 150 ka, whereas the terrace deposit and tread are correlative to the 60 m coastal terrace. Cosmogenic results should be available at the time of the field trip. Qt2 is the thickest and most widespread fill terrace in the Clearwater Valley. Qt2 terrace deposits are made up of 5–40 m of coarse stratified sand and gravel that sit on straths 0–20 m above the level of the modern valley bottom (Figs. 13 and 20). Locally, the fill has buried not only the paleovalley bottom (equivalent to the strath) but also the side slopes of the river

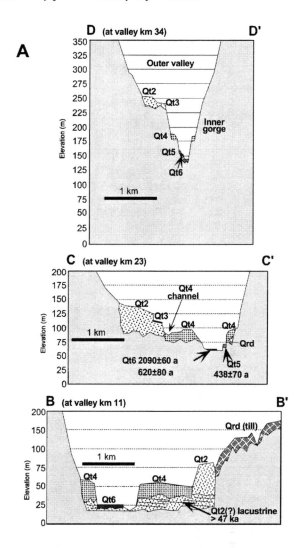

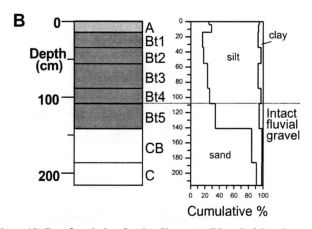

Figure 19. Best-fit solution for the Clearwater River incision data, assuming a stream-power type function. Dark circles are the Qt2 strath (ca. 140 ka) and open circles are the Qt3 strath (ca. 65 ka). I_{rate} is the incision rate, S is the channel gradient, and Q_p is the modern average discharge normalized for channel width. Each of these variables is plotted on logarithmic axes. The slope of the best-fit plane indicates that the parameters m and n are less than zero, which is not permitted by stream power model. See text for further discussion and also Tomkin et al. (2003).

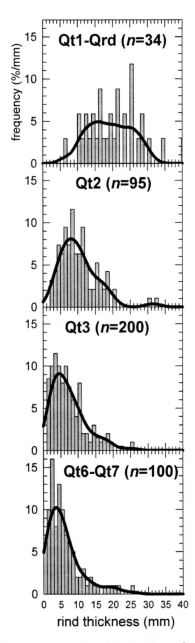

Figure 20. A: Cross-valley profiles of the lower (B–B′), middle (C–C′), and upper (D–D′) parts of the Clearwater Valley showing the relationship of the terrace units to the local valley geometry. Locations are shown in Figure 4A. B: Soil described in road cut, 100 m west of Copper Pit. V.E.—vertical exaggeration.

thin-bedded sand, locally laminated silt, and massive silt, which are interpreted as both overbank and loess deposits.

Soil profile development and clast weathering rinds (Figs. 20 and 21) in the terrace treads allow distinctions between deposits of different age and correlation between upstream and downstream remnants of the terraces. The Qt2 terrace alluvium represents a time of major Clearwater Valley aggradation, when the middle and lower portion of the drainage were hydrologically connected to the Hoh drainage (Fig. 4). Downstream, the Qt2 tread can be traced nearly unbroken to the 60 m coastal terrace, which we have already argued was likely deposited during isotope stage 6 or ca. 150 ka. So here, we have the upstream projection of that isotope stage 6 fill to its head of outwash in the Snahapish Valley. The timing of aggradation must be limited upstream by when the ice margin was stalled in the Snahapish Valley pumping out sediment and discharge, and downstream by when sea level was rising to produce the accommodation space leading to the high elevation of the Qt2 tread. This restricts the filling to between 150 and 125 ka. So the strath at the base of Qt2 is taken as 150 ka, and the tread is considered to be younger (Fig. 7).

Return to Snahapish-Clearwater Road, turn right and retrace route back to Grouse Bridge.

Cumulative Miles	(km)	Description
29.2	(47.1)	Cross Grouse Bridge.
30.0	(48.4)	Turn left onto the C1000 road. This road will ascend the Qt3 tread, here ~30 m above the valley bottom.
31.3	(50.5)	Ascend the degraded remnants of the Qt2 tread.
31.8	(51.3)	Stay right on the major C1000 road. The dirt road leading off to the left is the old approach to the former Goodyear Bridge, an old suspension bridge no longer suitable for vehicle traffic. C1000 continues to follow the Qt3 tread.
33.6	(54.2)	Cross Deception Creek and then begin ascending the bedrock ridge north of the Deception Creek drainage.
34.0	(54.8)	Find a safe place to park. You may have to take advantage of the C1200-C1000 road intersection 0.5 mile farther up the road. This is where we need to descend the bank down to the Clearwater River. At the river, find a gravel bar and follow it onto the big north-facing meander loop. Stop 3-6.

Stop 3-6. Crooks Creek Terraces

The objective here is to use the view to illustrate the magnitude of strath separation and resulting river incision rates. The maps and ages of Pleistocene and Holocene terraces will be used to develop models of terrace formation influenced by climatic and tectonic forcing (Meyer et al., 1995; Pazzaglia and Brandon, 2001; Wegmann and Pazzaglia, 2002; Figs. 22, 23, 24, 25, and 26).

Retrace route out the C1000 road.

valley itself, thus forming a buttress unconformity. Sedimentary structures within the terrace deposits include broad, shallow channel forms exhibiting 0.5–2-m-high tabular crossbeds and smaller-scale trough crossbeds of silty sand. These sedimentary structures are generally consistent with a braided channel form and mimic the features exhibited by glaciofluvial deposits that can be physically traced to heads of outwash in adjacent, glaciated drainages. The terrace alluvium is capped by ~1–2 m of

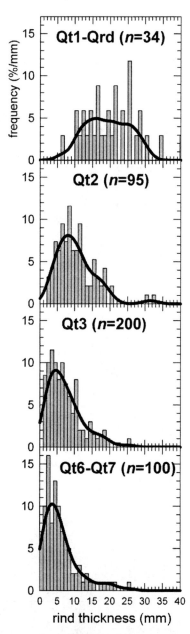

Figure 21. Thickness distributions for clast weathering rinds for terrace deposits of different stratigraphic age. The probability density curves were calculated using the Gaussian kernel method (Brandon, 1996), with the kernel size set to 2 mm. Qt1-Qrd is from an interfluve ~180 m elevation on the north divide of Shale Creek. Qt2 is from the Peterson Creek terrace ~130 m elevation. Qt3 is from the Quinault quarry pit ~30 m elevation. Qt6 is from an exposed gravel bar adjacent to the Clearwater River, near the Clearwater Picnic Area, ~2 km south of the town of Clearwater.

Cumulative Miles	(km)	Description
38.0	(61.3)	Proceed straight on the Snahapish-Clearwater Road, turn right on Highway 101 and return to Kalaloch.

Optional stops for Day 3, weather and interest permitting. Both routes are logged from the intersection of the Coppermine Bottom access road and the Snahapish-Clearwater Road.

Optional Log 3-A. Upper Basin and Terraces at Kunamakst Creek

Cumulative Miles	(km)	Description
0.0	(0.0)	Turn right onto the Snahapish-Clearwater Road.
0.5	(0.8)	Turn right directly before the Snahapish River Bridge (on the Clearwater-Snahapish Road) and begin driving up the Snahapish Valley.
4.5	(7.3)	Turn right at the major triangle intersection onto the C2000 road.
7.3	(11.8)	Loop around steep tributaries. The Qt2 tread is visible in the clearcuts on the right.
8.6	(13.9)	Slow and turn right into dirt road marked W-5 or C2017. Pull forward and park to the left in the opening.
8.7	(14.0)	Walk south on the overgrown dirt road leading out to an old clearcut.

Optional Stop 3-A. Terraces at Kunamakst Creek

This is an overview stop to illustrate the presence of terraces, straths, and their considerable separation from the modern channel in the upper Clearwater drainage.

Return to vehicles and retrace route out to the triangle intersection with the Clearwater-Snahapish Road, stay left back toward Coppermine Bottom.

Optional Log 3-B. Grouse Creek Landslide

Cumulative Miles	(km)	Description
0.0	(0.0)	Turn right onto the Snahapish-Clearwater Road.
0.5	(0.8)	Cross the Snahapish River and stay to the right on the paved road (toward Upper Clearwater campground).
1.4	(2.3)	Here, and at several other places, you will pass exposed gravels of the Qt2 terrace.
2.5	(4.0)	Cross the Qt2 tread.
3.9	(6.3)	Cross the Clearwater River at the Upper Clearwater campground. Qt5 terraces, like the one observed at the Grouse Bridge, are exposed both upstream and downstream of the bridge on the bank straight ahead.
4.5	(7.3)	Stay right on the C3100 road. Begin traversing the interfluve between the Solleks River and

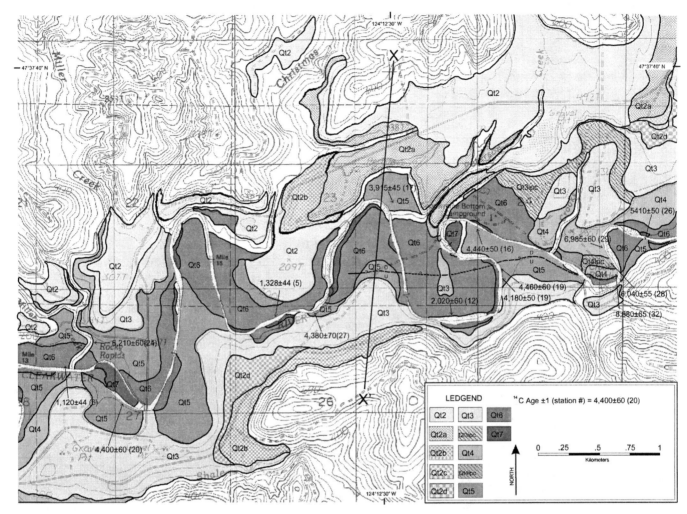

Figure 22. Map of river terraces in the medial portion of the Clearwater Valley (modified from Wegmann, 1999). Letters following terrace-name designations: a, b, c, d indicate treads that share a common strath; ipc indicates inset paleochannel.

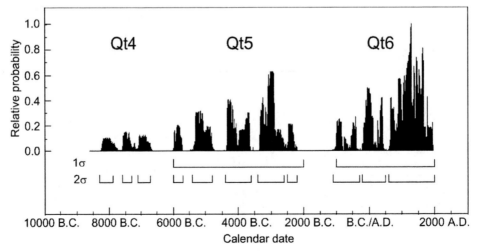

Figure 23. Probability density and frequency plots for 38 radiocarbon dates of the Holocene terraces. The plot was produced by calculating calendar ages, their accompanying 1σ errors, and then summing the probabilities using the program OxCal (Ramsay, 2000). Note the brackets beneath the frequency histograms. At the 1σ confidence level, the group of ages between 6000 and 2000 B.C. and 1000 B.C. and 2000 A.D. statistically coincide with terraces Qt5 and Qt6, respectively.

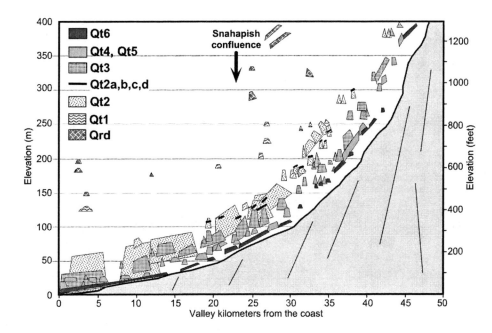

Figure 24. A long-profile section of the Clearwater Valley showing the vertical relationship of mapped terrace deposits (polygons) to the modern Clearwater River (continuous line). Terrace units were projected orthogonally into the valley profile from their mapped positions along the flanks of the Clearwater Valley (Fig. 4B). The bottom and top of each polygon corresponds to the strath and tread, respectively, for a mapped terrace deposit.

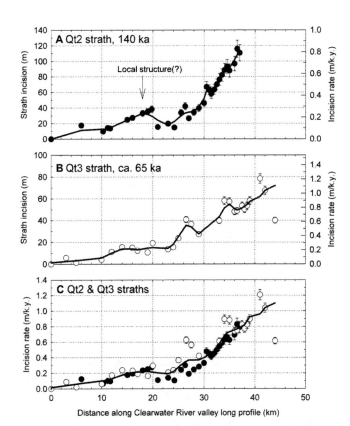

Figure 25. Incision of the (A) Qt2 and (B) Qt3 straths with respect to distance along the valley long profile (see Fig. 4B). C: Incision rates calculated from the Qt2 strath (solid circles) and Qt3 strath (open circles). Note the similarity in the estimated long-term incision rates for these two different-age straths. In general, incision and incision rates increase in an upstream direction. The small bump in the profiles at ~15–25 km suggests some localized uplift, such as a broad fold. Error bars show ±2 standard error uncertainties due to measurement errors for strath height. The thick lines are smoothed from the original data using a locally weighted regression method (Lowess algorithm of Cleveland, 1979, 1981, with the interval parameter set to 0.2).

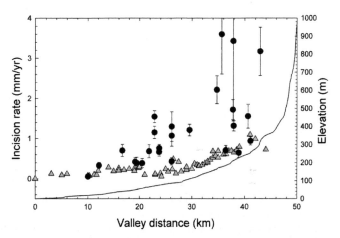

Figure 26. Incision rates (left vertical axis) for the Clearwater Basin determined from Holocene terraces (solid black circles) and Pleistocene terraces (gray triangles) plotted with respect to distance along the valley long profile (right vertical axis). Rates reflect calendar ages and error bars on the Holocene data are 1σ standard error. Note that the incision rates for both data sets increase upstream.

Stequaleho Creek, both major tributaries to the Clearwater River.

6.2	(10.0)	Turn left onto the C3140 road. Note that this road is gated by the Washington Department of Natural Resources just past the junction with the C3100 road. Vehicle traffic past this gate is allowed by permission of the department. To obtain permission and a key to the gate, contact the Olympic region office in Forks (360-374-6131). Foot and bike access is allowed without permission.
7.7	(12.4)	Stay on the main road.
10.0	(16.1)	Drop into the Solleks Valley bottom.
11.6	(18.7)	Cross the Solleks River.
12.1	(19.5)	To the south across the old gravel pit is an unobstructed view into the Grouse Creek drainage and the 1997 landslide.
13.0	(21.0)	Turn right onto the C3185 road. Cross the Solleks River again and begin ascending the ridge on the south side of the river.
14.0	(22.6)	Negotiate switchbacks.
15.1	(24.4)	Stay to the right, turning onto the C3100 road.
15.5	(25.0)	Stay to the right (straight on C3100 road).
15.7	(25.3)	Park vehicles near the short access roads connected to the main dirt road. Stop 3-7.

Stop 3-7. Grouse Creek Landslide

The objective here is to illustrate the magnitude of mass wastage and sediment delivery to the channel occurring during the Holocene. *Note: Due to road deconstruction, one has to walk ~1 km to the headscarp of the landslide on the unmaintained road.* On March 19th, 1997, 506,000 m³ of rock and regolith moved off this hillslope and into the Solleks River (Serdar, 1999; Gerstel, 1999). The landslide occurred during the waning phases of a major rain-on-snow flood event that started two days prior on March 17th (the St. Patrick's Day flood). Approximately 200,000 m³ of material was deposited on the upper slopes of the Grouse Creek channel, with the remainder moving as a debris flow that traveled down Grouse Creek, picking up additional material from side slopes, which were scoured up to 54 m high. When the debris flow reached the Solleks River, it deposited ~78,000 m³ of material. Landsliding here may have been exacerbated by both a road and associated logging activity; earlier slide scars and additional landslides cut adjacent hillslopes.

The debris delivered to the Solleks channel temporarily dammed the river and pushed the channel to the north. A plume of coarse sediment worked its way down the Solleks into the Clearwater channel by the summer of 1997. The net effect of such instantaneous, point introductions of sediment was to force the channel to enhance its lateral corrasion as the sediment was temporarily stored in floodplains and in channel bars. Downed trees recruited from floodplains trapped the sediment and locally made alluvial fills behind dams of woody debris. The abundance of woody debris dams increased in frequency

on the Clearwater River in the reach directly downstream of the Solleks confluence.

Return to vehicles and retrace route all the way back to the Upper Clearwater campground and then on the paved road back to the Snahapish River bridge.

Day 4. Kalaloch to Queets River, Lake Quinault, Hoquium, Seattle

Cumulative		
Miles	(km)	Description
0.0	(0.0)	Depart Kalaloch Lodge, turn right onto U.S. Rt. 101 south.
12.0	(19.4)	Turn left, entrance to Queets River valley.
24.0	(38.7)	Stop 4-1.

Stop 4-1. Queets River Valley

The purpose of this stop is to discuss cosmogenic erosion rates based on cosmogenic inventories of channel alluvium.

Return to U.S. Rt. 101.

Cumulative		
Miles	(km)	Description
36.0	(58.1)	Turn right onto U.S. Rt. 101.
53.5	(86.3)	Stop 4-2.

Stop 4-2. Lake Quinault

The purpose of this stop is to observe the Hurricane Ridge fault as well as the glacially dammed Lake Quinault. The Hurricane Ridge fault at this location is steeply dipping and juxtaposes the Crescent Formation against the Hoh Formation, which is largely buried by glaciofluvial deposits along the southeast flank of the Quinault Valley. Lake Quinault is dammed by late Pleistocene, presumably Hoh Oxbow-equivalent moraines.

Continue south on U.S. Rt. 101.

Cumulative		
Miles	(km)	Description
94.0	(151.6)	Stop 4-3.

Stop 4-3. Hoquium

The purpose of this stop is to observe a cut-bank exposure where a buried marsh marking rapid coastal emergence during the 1700 A.D. earthquake is exposed.

CONCLUDING REMARKS

Our study uses fluvial geomorphology and Quaternary stratigraphy to reconstruct the Quaternary landscape evolution and rock uplift across the Olympic sector of the Cascadia subduction zone. Glacial deposits and a well-preserved terrace stratigraphy in an unglaciated drainage allow us to use coastal exposures and the valley profile of the Clearwater as a crude geodetic datum. As such, we are able to quantify the effects of both the short-term earthquake cycle and permanent wedge deformation driving uplift across the interior of the Olympic Peninsula.

1. Uplift and formation of Olympic topography is primarily a result of northeast-directed shortening parallel to the direction of plate subduction. Thermochronology and other erosion rate data support removal of at least 12 km of rock from the Olympic Mountains core at a current modern rate of ~1 mm/yr. These results do not discredit the data which suggests north-south shortening along the Coast Ranges, but the north-south shortening simply cannot account for the degree of shortening and removal of rock that has been accomplished by growth and deformation of the Olympic accretionary wedge.

2. A buried sea cliff, probably formed at 122 ka, provides evidence of horizontal motion of rock relative to the modern shoreline. The rate of motion is 3.7 m/k.y. to the northeast, which is close to the 3 m/k.y. horizontal material velocity predicted for a frontally accreting steady-state wedge. These results are consistent with a kinematic model in which long-term horizontal velocity may account for 20%–35% of the geodetically measured horizontal velocity across the Olympics (Pazzaglia and Brandon, 2001). The remaining 65%–80% is presumably elastic deformation.

3. There are five recognized periods of alpine glaciation for the western Olympic Peninsula. The relative extent of these glaciations reflects a complex interaction between temperature and available moisture, controlling the equilibrium line altitude (ELA). The extent of alpine ice was limited when the Cordilleran Ice Sheet was farthest south, presumably because storm tracks bringing moisture to western North America were depressed southward.

4. Pleistocene terrace sequences in the Olympics seem to be closely tied to the glacial climate cycle through its influence on local climate, sediment supply from adjacent alpine-glaciated drainages, and eustasy. The sequence of terrace-forming events is consistent with the model of Bull (1991), but the timing of these events relative to the eustatic cycle is quite different. Bull (1991) proposed that strath formation occurred during rising sea level and aggradation during falling sea level. In the Olympics, strath formation seems to occur during peak glaciation (when sea level is low), and aggradation during late glacial and interglacial times, when sea level is rising. We suspect that this difference is a local effect, related to the strong influence that local deglaciation has had on sediment supply and the interaction of that enhanced sediment supply with rising sea level during interglacial times.

5. Holocene terraces represent local-scale processes of enhanced lateral incision, valley-bottom widening, and the carving of straths accomplished with the aid of the thin alluvial deposits preserved atop the straths. Valley bottom narrowing and rapid vertical incision into bedrock is accomplished during relatively brief (1000 yr) intervals between the carving of the major straths.

6. The upstream divergence of straths in the Clearwater drainages provides strong evidence that uplift is very slow at the coast and increases to a maximum in the center of the range. This conclusion runs counter to a commonly invoked assumption in fluvial geomorphology that long-term uplift rates can be taken as

uniform within a single drainage. The incision rates correspond closely with the pattern of long-term erosion rates indicated by apatite fission-track cooling ages. These observations indicate that, at long time scales (10–100 k.y.), the average form of the landscape remains close to steady state. This also implies that during each phase of strath cutting, the Clearwater Valley profile is able to return to the same steady-state form. Thus, bedrock incision rates seem to be a reasonable proxy for rock uplift rates in the Olympics.

7. The profile of incision and erosion rates across the Olympics indicates a close balance between the accretionary and erosional fluxes moving in and out of the wedge. This result supports the hypothesis of Brandon et al. (1998) that the Olympic sector of the Cascadia wedge is close to a flux steady state, but note that this conclusion does not require the topography to be steady as well.

ACKNOWLEDGMENTS

The authors would like to acknowledge long-standing working relationships with geoscientists that have worked with us, supported us, and inspired discussion on the topics presented in this paper. These include, but are not limited to Bill Lingley, Wendy Gerstel, Sean Willett, Joe Vance, Tony Garcia, John Garver, Mary Roden-Tice, and Brian Atwater. We thank the Quinault Nation, State of Washington, Olympic National Park, and Rayonier Inc. for access to their respective lands. Research supported by National Science Foundation grants EAR-8707442, -9302661, -9405659, and -9736748.

REFERENCES CITED

Adams, J., 1984, Active deformation of the Pacific Northwest continental margin: Tectonics, v. 3, p. 449–472.

Ahnert, F., 1970, Functional relationships between denudation, relief, and uplift in large mid-latitude drainage basins: American Journal of Science, v. 268, p. 243–263.

Armentrout, J.M., 1981, Correlation and ages of Cenozoic chronostratigraphic units in Oregon and Washington, *in* Armentrout, J.M., ed., Pacific Northwest Cenozoic biostratigraphy: Geological Society of America Special Paper 184, p. 137–148.

Atwater, B., 1996, Coastal evidence for great earthquakes in western Washington, *in* Rogers, A.M., Walsh, T.J., Kockelman, W.J., and Priest, G.R., eds., Assessing earthquake hazards and reducing risk in the Pacific Northwest: Volume 1: U.S. Geological Survey Professional Paper 1560: Reston, Virginia, U.S. Geological Survey, p. 77–90.

Atwater, B.F., 1987, Evidence for great Holocene earthquakes along the outer coast of Washington State: Science, v. 236, p. 942–944.

Atwater, B.F., Stuiver, M., and Yamaguchi, D.K., 1991, Radiocarbon test of earthquake magnitude at the Cascadia subduction zone: Nature, v. 353, p. 156–158.

Batt, G.E., Brandon, M.T., Farley, K.A., Roden-Tice, M., 2001, Tectonic synthesis of the Olympic Mountains segment of the Cascadia wedge, using two-dimensional thermal and kinematic modeling of thermochronological ages: Journal of Geophysical Research, v. 106, no. B11, p. 26,731–26,746.

Beaumont, C., Ellis, S., and Pfiffner, A., 1999, Dynamics of subduction-accretion at convergent margins: Short-term modes, long-term deformation, and tectonic implications: Journal of Geophysical Research, v. 104, p. 17,573–17,602.

Beck, M.E., Jr., and Engebretson, D.C., 1982, Paleomagnetism of small basalt exposures in the West Puget Sound area, Washington, and speculations on the accretionary origins of the Olympic Mountains: Journal of Geophysical Research, v. 87, p. 3755–3760.

Bigelow, P.K., 1987, The petrology, stratigraphy and drainage history of the Montesano Formation, southwestern Washington and southern Olympic Peninsula [M.S. Thesis]: Bellingham, Washington, Western Washington University, 263 p.

Bockheim, J.G., Kelsey, H.M., and Marshall, J.G., III., 1992, Soil development, relative dating, and correlation of late Quaternary marine terraces in southwestern Oregon: Quaternary Research, v. 37, p. 60–74.

Brandon, M.T., 1996, Probability density plots for fission-track grain age distributions: Radiation Measurements, v. 26, p. 663–676.

Brandon, M.T., and Calderwood, A.R., 1990, High-pressure metamorphism and uplift of the Olympic subduction complex: Geology, v. 18, p. 1252–1255.

Brandon, M.T., and Vance, J.A., 1992, Tectonic evolution of the Cenozoic Olympic subduction complex, Washington State, as deduced from fission track ages for detrital zircons: American Journal of Science, v. 292, p. 565–636.

Brandon, M.T., Roden-Tice, M.K., and Garver, J.I., 1998, Late Cenozoic exhumation of the Cascadia accretionary wedge in the Olympic Mountains, northwest Washington State: Geological Society of America Bulletin, v. 110, p. 985–1009.

Brozovic, N., Burbank, D., and Meigs, A., 1997, Climatic limits on landscape development in the northwestern Himalaya: Science, v. 276, p. 571–574.

Bucknam, R.C., Hemphill-Haley, E., and Leopold, E.B., 1992, Abrupt uplift within the past 1700 years at southern Puget Sound, Washington: Science, v. 258, p. 1611–1614.

Bull, W.B., 1991, Geomorphic response to climate change: Oxford University Press, New York, 326 p.

Campbell, K.A., and Nesbitt, E.A., 2000, High resolution architecture and paleoecology of an active margin, storm-flood influenced estuary, Quinault Formation (Pliocene), Washington: Palaios, v. 15, n. 6, p. 553–579.

Chappell, J., Omura, A., Esat, T., McCulloch, M., Pandolfi, J., Ota, Y., and Pillans, B., 1996, Reconciliation of late Quaternary sea levels derived from coral terraces at Huon Peninsula with deep sea oxygen isotope records: Earth and Planetary Science Letters, v. 141, p. 227–236.

Cleveland, W.S., 1979, Robust locally weighted regression and smoothing scatterplots: Journal of the American Statistics Association, v. 74, p. 829–836.

Cleveland, W.S., 1981, LOWESS: A program for smoothing scatterplots by robust locally weighted regression: American Statistician, v. 35, p. 54.

Clowes, R.M., Brandon, M.T., Green, A.C., Yorath, C.J., Sutherland Brown, A., Kanasewich, E.R., and Spencer, C., 1987, LITHOPROBE - southern Vancouver Island: Cenozoic subduction complex imaged by deep seismic reflections: Canadian Journal of Earth Sciences, v. 24, p. 31–51.

Crosson, R.S., and Owens, T.J., 1987, Slab geometry of the Cascadia subduction zone beneath Washington from earthquake hypocenters and teleseismic converted waves: Geophysical Research Letters, v. 14, p. 824–827.

Darenzio, M.E., and Peterson, C.D.,1990, Episodic tectonic subsidence of late Holocene salt marshes, northern Oregon coast, central Cascadia margin, U.S.A.: Tectonics, v. 9, p. 1–22.

DeMets, C., and Dixon, T., 1999, New kinematic models for Pacific-North America motion from 3 Ma to present; I, Evidence for steady motion and biases in the NUVEL-1A model: Geophysical Research Letters, v. 26, no. 13, p. 1921–1924.

DeMets, C., Gordon, R.G., Argus, D.F., and Stein, S., 1990, Current plate motions: Geophysical Journal International, v. 101, p. 425–478.

Dethier, D.P., 1986, Weathering rates and the chemical flux from catchments in the Pacific Northwest, U.S.A., *in* Colman, S., and Dethier, D.P., eds., Rates of chemical weathering of rocks and minerals: Academic Press, Orlando, Florida, p. 503–530.

Dragert, H., 1987, The fall (and rise) of central Vancouver Island: 1930–1985: Geological Survey of Canada Contribution 10586, p. 689–697.

Dragert, H., Hyndman, R.D., Rogers, G.C., and Wang, K., 1994, Current deformation and the width of the seismogenic zone of the northern Cascadia subduction thrust: Journal of Geophysical Research, ser. B, v. 99, p. 653–668.

Durham, J.W., 1944, Megafaunal zones of the Oligocene of northwestern Washington: University of California Publications in Geological Sciences, v. 27, p. 101–211.

Easterbrook, D.J., 1986, Stratigraphy and chronology of Quaternary deposits of the Puget lowland and Olympic Mountains of Washington and the Cascade Mountains of Washington and Oregon, *in* Sibrava, V., Bowen, D.Q., and Richmond, G.M., eds., Quaternary glaciations in the Northern Hemisphere: Quaternary Science Reviews, v. 5, p. 145–159.

Florer, L.E., 1972, Quaternary paleoecology and stratigraphy of the sea cliffs, western Olympic Peninsula, Washington: Quaternary Research, v. 2, p. 202–216.

Garcia, A.F., 1996, Active tectonic deformation and late Pleistocene and Holocene geomorphic and soil profile evolution in the Dosewallips River drainage basin, Olympic Mountains, western Washington State [M.S. thesis]: Albuquerque, The University of New Mexico, 152 p.

Gerstel, W.J., 1999, Deep-seated landslide inventory of the west-central Olympic Peninsula: Washington Division of Geology and Earth Resources Open File Report 99-2, 36 p.

Heusser, C.J., 1972, Palynology and phytogeographical significance of a late Pleistocene refugium near Kalaloch, Washington: Quaternary Research, v. 2, p. 189–201.

Heusser, C.J., 1974, Quaternary vegetation, climate, and glaciation of the Hoh River valley, Washington: Geological Society of America Bulletin, v. 85, p. 1547–1560.

Heusser, C.J., 1978, Palynology of the Quaternary deposits of the lower Bogachiel River area, Olympic Peninsula, Washington: Canadian Journal of Earth Sciences, v. 15, p. 1568–1578.

Holdahl, S.R., Martin, D.M., and Stoney, W.M., 1987, Methods for combination of water level and leveling measurements to determine vertical crustal motions, *in* Proceedings of Symposium on Height Determination and Recent Crustal Movement in Western Europe: Bonn, Germany, Dumler Verlag, p. 373–388.

Holdahl, S.R., Faucher, F., and Dragert, H., 1989, Contemporary vertical crustal motion in the Pacific northwest, *in* Cohen, S.C., and Vanicek, P., eds., Slow deformation and transmission of stress in the Earth: American Geophysical Union Geophysical Monograph 40, International Union of Geodesy and Geophysics Volume 4, p. 17–29.

Hyndman, R.D., and Wang, K., 1993, Thermal constraints on the zone of major thrust earthquake failure: The Cascadia subduction zone: Journal of Geophysical Research, v. 98, p. 2039–2060.

Irving, E., and Massey, N.W.D., 1990, Paleomagnetism of ocean layers 2 and 3: Evidence from the Metchosin Complex, Vancouver Island: Physics of Earth and Planetary Interiors, v. 64, p. 247–260.

Kelsey, H.M.,1990, Late Quaternary deformation of marine terraces on the Cascadia subduction zone near Cape Blanco, Oregon: Tectonics, v. 9, p. 983–1014.

Kelsey, H.M., and Bockheim, J.G., 1994, Coastal landscape evolution as a function of eustasy and surface uplift rate, Cascadia margin, southern Oregon: Geological Society of America Bulletin, v. 106, p. 840–854.

Koons, P., 1990, Two-sided orogen; collision and erosion from the sandbox to the Southern Alps, New Zealand: Geology, v. 18, p. 679–682.

McCrory, P.A., 1996, Tectonic model explaining divergent contraction directions along the Cascadia subduction margin, Washington: Geology, v. 24, p. 929–932.

McCrory, P.A., 1997, Evidence for Quaternary tectonism along the Washington coast: Washington Geology, v. 25, p. 14–19.

McNeill, L.C., Piper, K., Goldfinger, C., Kulm, L., and Yeats, R., 1997, Listric normal faulting on the Cascadia continental margin: Journal of Geophysical Research, ser. B, v. 102, p. 12,123–12,138.

McNeill, L.C., Goldfinger, C., Kulm, L.D., Yeats, R.S., 2000, Tectonics of the Neogene Cascadia forearc drainage; investigations of a deformed late Miocene unconformity: Geological Society of America Bulletin, v. 112, p. 1209–1224.

Meyer, G.A., Wells, S.G., and Jull, A.J.T., 1995, Fire and alluvial chronology in Yellowstone National Park: Climate and intrinsic controls on Holocene geomorphic processes: Geological Society of America Bulletin, v. 107, p. 1211–1230.

Mitchell, C.E., Vincent, P., Weldon R.J., and Richards, M., 1994, Present-day vertical deformation of the Cascadia margin, Pacific Northwest, U.S.A.: Journal of Geophysical Research, ser. B, v. 99, p. 12,257–12,277.

Montgomery, D.R., 2001, Slope distributions, threshold hillslopes, and steadystate topography: American Journal of Science, v. 301, p. 432–454.

Montgomery, D.R., 2002, Valley formation by fluvial and glacial erosion: Geology, v. 30, p. 1047–1050.

Montgomery, D.R., and Greenberg, H., 2000, Local relief and the height of Mount Olympus: Earth Surface Processes and Landforms, v. 25, p. 385–396.

Montgomery, D.R., and Brandon, M.T., 2002, Topographic controls on erosion rates in tectonically active mountain ranges: Earth and Planetary Science Letters, v. 201, p. 481–489.

Muller, J.E., Snavely, P.D., and Tabor, R.W., 1983, The Tertiary Olympic terrane, southwest Vancouver Island and northwest Washington: Geological Association of Canada, 1983 Annual Meeting, Guidebook for fieldtrip n. 12, 59 p.

Niem, W.A., Niem, A.R., and Snavely, P.D., Jr., 1992, Western Washington-Oregon coastal sequence, *in* Christiansen, R.L., and Yeats, R.S., eds., Post-Laramide geology of the U.S. Cordilleran region, The Geology of North America, v. G-3, The Cordilleran Orogen: Conterminous United States: Boulder, Colorado, Geological Society of America, p. 265–270.

Orange, D.L., 1990, Criteria helpful in recognizing shear-zone and diapiric melanges: Examples from the Hoh accretionary complex, Olympic Peninsula, Washington: Geological Society of America Bulletin, v. 102, p. 935–951.

Palmer, S.P., and Lingley, W.S., Jr., 1989, An assessment of the oil and gas potential of the Washington outer continental shelf: University of Washington, Washington Sea Grant Program, p. 83.

Pazzaglia, F.J., and Brandon, M.T., 2001 A fluvial record of rock uplift and shortening across the Cascadia forearc high: American Journal of Science, v. 301, p. 385–431.

Pillans, B., Chappell, J., and Naish, T.R., 1998, A review of the Milankovitch climatic beat: Template for Plio-Pleistocene sea-level changes and sequence stratigraphy: Sedimentary Geology, v. 122, p. 5–21.

Prothero, D.R., and Armentrout, J.M., 1985, Magnetostratigraphic correlation of the Lincoln Creek Formation, Washington: Implications for the age of the Eocene-Oligocene boundary: Geology, v. 13, p. 208–211.

Prothero, D.R., and Burns, C., 2001, Magnetic stratigraphy and tectonic rotation of the upper Oligocene-?lower Miocene (type Pillarian stage) Clallam Formation, Clallam County, Washington: Pacific Section Society for Sedimentary Geology Special Publication 91, p. 234–241.

Prothero, D.R., Streig, A., and Burns, C., 2001, Magnetic stratigraphy and tectonic rotation of the upper Oligocene Pysht Formation, Clallam County, Washington: Pacific Section Society for Sedimentary Geology Special Publication 91, p. 224–233.

Ramsay, B., 2000, OxCal v. 3.5: available from http://www.rlaha.ox.ac.uk/orau/index.htm.

Rau, W., 1973, Geology of the Washington coast between Point Grenville and the Hoh River: Washington Department of Natural Resources, Geology and Earth Resources Division Bulletin, v. 66, 58 p.

Rau, W., 1975, Geologic map of the Destruction Island and Taholah quadrangles, Washington: Washington Department of Natural Resources, Geology and Earth Resources Division Map GM-13, scale 1:62,500.

Rau, W., 1979, Geologic map in the vicinity of the lower Bogachiel and Hoh River valleys and the Washington coast: Department of Natural Resources, Geology and Earth Resources Division Map GM-24, scale 1:62,500.

Rau, W.W., 1970, Foraminifera, stratigraphy, and paleoecology of the Quinault Formation, Point Grenville-Raft River coastal area, Washington: Washington Department of Natural Resources Bulletin, v. 62, 34 p.

Rau, W.W., and Grocock, G., 1974, Piercement structure outcrops along the Washington coast: Washington, Department of Natural Resources, Division of Mines and Geology, Information Circular, v. 51, 7 p.

Reilinger, R., and Adams, J., 1982, Geodetic evidence for active landward tilting of the Oregon and Washington Coastal Ranges: Geophysical Research Letters, v. 9, p. 401–403.

Roering, J.J., Kirchner, J.W., Sklar, L.S., Dietrich, W.E., 2001, Hillslope evolution by nonlinear creep and landsliding: An experimental study: Geology, v. 29, p. 143–146.

Rogers, G.C., 1988, An assessment of megathrust earthquake potential of the Cascadia subduction zone: Canadian Journal of Earth Science, v. 25, p. 844–852.

Savage, J.C., Lisowski, M., and Prescott, W.H., 1981, Geodetic strain measurements in Washington: Journal of Geophysical Research, ser. B, v. 86, p. 4929–4940.

Savage, J.C., Lisowski, M., and Prescott, W.H., 1991, Strain accumulation in western Washington: Journal of Geophysical Research, ser. B, v. 96, p. 14,493–14,507.

Serdar, C.F., 1999, Description, analysis and impacts of the Grouse Creek landslide, Jefferson County, Washington, 1997–98: The Evergreen State College Master of Environmental Studies thesis, 171 p.

Snavely, P.D., Jr., Niem, A.R., and Pearl, J.E., 1978, Twin River Group (upper Eocene to lower Miocene) defined to include Hoko River, Makah, and Pysht Formations, Clallam County, Washington: U.S. Geological Survey Bulletin, v. 1457A, p. A111-A120.

Stewart, R.J., and Brandon, M.T., 2003, Detrital zircon fission-track ages for the "Hoh Formation": Implications for late Cenozoic evolution of the Cascadia subduction wedge: Geological Society of America Bulletin (in press).

Symons, D.T.A., 1973, Paleomagnetic zones in the Oligocene East Sooke Gabbro, Vancouver Island, British Columbia: Journal of Geophysical Research, v. 78, p. 5100–5109.

Tabor, R., 1975, Guide to the geology of Olympic National Park: Seattle, University of Washington Press, 144 p.

Tabor, R., and Cady, W., 1978b, The structure of the Olympic Mountains, Washington—Analysis of a subduction zone: U.S. Geological Survey Professional Paper 1033, 38 p.

Tabor, R.W., and Cady, W.M., 1978a, Geologic map of the Olympic Peninsula: U.S. Geological Survey Map I-994, 2 sheets, scale 1:125,000.

Thackray, G.D., and Pazzaglia, F.J., 1994, Quaternary stratigraphy, tectonic geomorphology, and fluvial evolution of the western Olympic Peninsula, Washington, *in* Swanson, D.A., and Haugerud, R.A., eds., Geologic field trips in the Pacific Northwest, 1994 Geological Society of America Annual Meeting, Seattle, Washington, p. 2A-1–2A-30.

Thackray, G.D., 1996, Glaciation and neotectonic deformation on the western Olympic Peninsula, Washington [Ph.D. dissertation]: Seattle, Washington, University of Washington, 139 p.

Thackray, G.D., 1998, Convergent-margin deformation of Pleistocene strata on the Olympic coast of Washington, USA, *in* Stewart, I.S., and Vita-Finzi, C., eds., Coastal tectonics: Geological Society [London] Special Publication 146, p. 199–211.

Thackray, G.D., 2001, Extensive Early and Middle Wisconsin glaciation on the western Olympic Peninsula, Washington, and the variability of Pacific moisture delivery to the northwestern United States: Quaternary Research, v. 55, p. 257–270.

Thatcher, W., and Rundle, J.B., 1984, A viscoelastic coupling model for the cyclic deformation due to periodically repeated earthquakes at subduction zones: Journal of Geophysical Research, ser. B, v. 89, p. 7631–7640.

Thorson, R.M., 1989, Glacio-isostatic response of the Puget Sound area, Washington: Geological Society of America Bulletin, v. 101, p. 1163–1174.

Tomkin, J.H., Brandon, M.T., Pazzaglia, F.J., Barbour, J.R., and Willett, S.D., 2003, Quantitative testing of bedrock incision models, Clearwater River, WA: Journal of Geophysical Research (in press).

Wang, K., 1996, Simplified analysis of horizontal stresses in a buttressed forearc sliver at an oblique subduction zone: Geophysical Research Letters, v. 23, p. 2021–2024.

Wegmann, K., 1999, Late Quaternary fluvial and tectonic evolution of the Clearwater River basin, western Olympic Mountains, Washington State [M.S. thesis]: Albuquerque, University of New Mexico, 217 p., 4 plates.

Wegmann, K., and Pazzaglia, F.J., 2002, Holocene strath terraces, climate change, and active tectonics: the Clearwater River basin, Olympic Peninsula, Washington State: Geological Society of America Bulletin, v. 114, no. 6, p. 731–744.

Wells, R., Weaver, C., and Blakely, R., 1998, Fore-arc migration in Cascadia and its neotectonic significance: Geology, v. 26, p. 759–762.

Wells, R.E., 1990, Paleomagnetic rotations and the Cenozoic tectonics of the Cascade Arc, Washington, Oregon, and California: Journal of Geophysical Research, v. 95, p. 19,409–19,417.

Wells, R.E., and Coe, R.S., 1985, Paleomagnetism and geology of Eocene volcanic rocks of southwest Washington: Implications for mechanisms of tectonic rotation: Journal of Geophysical Research, v. 90, p. 1925–1947.

West, D.O., and McCrumb, D.R., 1988, Coastline uplift in Oregon and Washington and the nature of Cascadia subduction zone tectonics: Geology, v. 16, p. 169–172.

Willett, S., 1999, Orogeny and orography: The effects of erosion on the structure of mountain belts: Journal of Geophysical Research, v. 104, p. 28,957–28,981.

Willett, S.D., and Brandon, M.T., 2002, On steady states in mountain belts: Geology, v. 30, p. 175–178.

Willett, S., Beaumont, C., and Fullsack, P., 1993, Mechanical models for the tectonics of doubly vergent compressional orogens: Geology, v. 21, p. 371–374.

Willett, S.D., Slingerland, R., and Hovius, N., 2001, Uplift, shortening, and steady state topography in active mountain belts: American Journal of Science, v. 301, p. 455–485.

Printed in the USA

Geological Society of America
Field Guide 4
2003

Wine and geology—The terroir of Washington State

Alan J. Busacca*

Department of Crop and Soil Sciences, Washington State University, Pullman, Washington 99164-6420, USA

Lawrence D. Meinert*

Department of Geology, Smith College, Northampton, Massachusetts 01063, USA

ABSTRACT

Washington State is second only to California in terms of wine produced in the United States, and some of its vineyards and wines are among the world's best. Most Washington vineyards are situated east of the Cascades on soils formed from Quaternary sediments that overlie Miocene basaltic rocks of the Columbia River Flood Basalt Province. Pleistocene fluvial sediments were deposited during cataclysmic glacial outburst floods that formed the spectacular Channeled Scabland. Late Pleistocene and Holocene sand sheets and loess form a variable mantle over outburst sediments. Rainfall for wine grape production ranges from ~6–18 in (150–450 mm) annually with a pronounced winter maximum and warm, dry summers. This field trip will examine the terroir of some of Washington's best vineyards. Terroir involves the complex interplay of climate, soil, geology, and other physical factors that influence the character and quality of wine. These factors underpin the substantial contribution of good viticultural practice and expert winemaking. We will travel by bus over the Cascade Mountains to the Yakima Valley appellation to see the effects of rain shadow, bedrock variation, sediment and soil characteristics, and air drainage on vineyard siting; we will visit the Red Mountain appellation to examine sites with warm mesoclimate and soils from back-eddy glacial flood and eolian sediments; the next stop will be the Walla Walla Valley appellation with excellent exposures of glacial slackwater sediments (which underlie the best vineyards) as well as the United States' largest wind energy facility. Finally, we will visit the very creatively sited Wallula Vineyard in the Columbia Valley appellation overlooking the Columbia River before returning to Seattle.

Keywords: wine, terroir, wine grapes, loess, soils, outburst floods.

INTRODUCTION

The purpose of this field trip is to examine the connection between wine and geology in Washington State, a connection that commonly is described by the word "terroir." Although the term originated in France, terroir increasingly is being used in other parts of the world to explore differences at the scale of appellations to individual vineyards to within-vineyard domains (Halliday, 1993, 1999; Wilson, 1998, 2001; Haynes, 1999, 2000).

But the word "terroir" is mysterious to many people; there is confusion about what it is, how it is documented, and even how it is pronounced (tehr-wahr). Terroir involves the complex interplay of climate, soil, geology, and other physical factors that influence the character and quality of wine. These factors are in addition to, or perhaps underlie, the substantial contribution of good viticultural practice and expert winemaking. One common illustration of the importance of terroir is the occurrence of adjacent or nearby vineyards that produce strikingly different wines even though many of the measurable aspects of climate, viticulture, and winemaking technique are very similar. It is also common,

*E-mail: Busacca—busacca@wsu.edu; Meinert—Lmeinert@smith.edu.

Busacca, A.J., and Meinert, L.D., 2003, Wine and geology—The terroir of Washington State, *in* Swanson, T.W., ed., Western Cordillera and adjacent areas: Boulder, Colorado, Geological Society of America Field Guide 4, p. 69–85. For permission to copy, contact editing@geosociety.org. © 2003 Geological Society of America.

although usually incorrect, to point to a single factor as the explanation: "it's the soil"; "it's the water"; "it's the limestone"; etc. Terroir is the integration of individual factors that contribute to wine quality, and to make matters even more complicated there is the complexity of year-to-year variation in climate. What may be good terroir in one year may be less so in another.

The Merriam-Webster's dictionary defines "appellation" as a geographical name (as of a region, village, or vineyard) under which a winegrower is authorized to identify and market wine (from http://www.m-w.com/cgi-bin/dictionary?book=Dictionary &va=appellation; accessed on September 4, 2003). Washington State has five wine appellations called American Viticultural Areas (AVAs) by the Alcohol and Tobacco Tax and Trade Bureau (TTB; formerly Bureau of Alcohol, Tobacco, and Firearms), the chief regulatory agency of the wine industry in the United States. The current Washington State appellations (AVAs) are Columbia Valley, Puget Sound, Red Mountain, Walla Walla Valley, and Yakima Valley (Fig. 1). Sub-appellations that may someday become AVAs include Alder Ridge, Canoe Ridge, Cold Creek, Columbia River Gorge, Horse Heaven Hills, Wahluke Slope, Zephyr Ridge (Peterson-Nedry, 2000), and the Okanogan Valley-Lake Chelan area.

As with most other wine growing regions, Washington AVAs can be nested such that the Columbia Valley appellation, which produces more than 90% of the state's wine grapes, includes the Yakima Valley, Walla Walla Valley, and Red Mountain appellations (Fig. 1). The area available for future planting is very large. In the 10.7-million-acre Columbia Valley appellation, only ~16,000 acres are planted with wine grapes. Even the smallest appellation, Red Mountain, has room for expansion with ~710 acres out of

the 4040 acres of the AVA planted with vines. In many cases the availability of water for irrigation is a larger limitation than the suitability of land for growing high-quality grapes.

Only ~18% of Washington's wine grapes are from vineyards more than 20 yr old, and of these older vineyards, white grapes (73%) predominate over red grapes (27%) (www.nass.usda.gov/wa/wine02.pdf). For example, Riesling was the most widely planted white wine grape prior to 1982 at 54% of its current (2002) acreage. In contrast, Cabernet Sauvignon, Merlot, and Syrah were the three most widely planted red grapes in 2002 and had only 12%, 5%, and 0%, respectively, of their current acreage planted prior to 1982.

Currently there are ~240 wineries in Washington State. Total wine grape production in 2001 was 100,000 tons from 24,000 acres of bearing vineyards. Wine grape production will continue to increase since there are an additional 6000 acres of wine grapes planted that were not yet bearing fruit in 2001. Most grape vines start producing commercial yields in their third year. Wine grape production in 2002 was 115,000 tons. Of the wine produced in Washington State in 2002, there was an equal split between white and red wine, down from a majority (62%) of white wine in 1998. For example, the production of Semillon and Chenin Blanc in this three-year period decreased 35%, whereas the production of Cabernet Sauvignon, Merlot, and Syrah increased 200%. This trend toward a predominance of red wine production in Washington State likely will continue in the future because of the increased plantings of red varietals and the higher prices realized from red grapes in general.

REGIONAL GEOLOGIC HISTORY

Most Washington vineyards lie in the geographic center of the Columbia Plateau, which is bordered on the north and east by the Rocky Mountains, on the south by the Blue Mountains, and on the west by the Cascade Mountains (Fig. 1). The area is underlain by the Columbia River Basalt Group, which covers an area of ~165,000 km². The Columbia River Basalt Group was erupted mostly between 17 and 11 Ma (early Miocene) from north-south fissures roughly paralleling the present-day Washington-Idaho border. The Columbia River Basalt Group has individual flows with estimated eruptive volumes of at least 3000 km³, making them the largest documented lava flows on Earth (Baksi, 1989; Landon and Long, 1989; Tolan et al., 1989). This dwarfs the erupted volumes of typical Cascade volcanoes: even the explosive eruption of Mount St. Helens in 1980 yielded only ~1 km³ of volcanic material (Pringle, 1993). The basalts are interstratified with volcaniclastic rocks of the Ellensburg Formation, mainly in the western part, including the Yakima fold belt through which we will be traveling.

The basalt bedrock is overlain by unconsolidated sediments deposited by glacial outburst floods and eolian processes described in some detail in Meinert and Busacca (2000). To briefly summarize: a lobe of the Cordilleran Ice Sheet blocked the Clark Fork River near the Canadian border in northern Idaho most recently ca. 18,000 ka and created glacial Lake Missoula (Fig. 2),

Figure 1. Location map of the Pacific Northwest showing wine appellations of Washington State and major geographical features described in the text.

which covered 7800 km² of western Montana (Pardee, 1910). At the ice dam the water was ~600 m deep (Weis and Newman, 1989). The ice dam failed repeatedly, releasing the largest floods documented on Earth (Baker and Nummedal, 1978). These floods overwhelmed the Columbia River drainage system and sent up to 2500 km³ of water across the Columbia Plateau with each outburst (called jökulhlaups in Iceland, where similar, though orders of magnitude smaller, events occur today). The floods eroded a spectacular complex of anastomosing channels, locally called "coulees," into southwest-dipping basalt surfaces. They also eroded huge cataracts in the basalt, now seen as dry falls, and "loess islands" that are erosional remnants of an early thick loess cover on the plateau. The floods deposited immense gravel bars and ice-rafted erratic boulders at high elevations. Collectively these features make up the Channeled Scabland as detailed in the early work by Bretz (1923, 1925, 1928a, b, and c, 1932).

In south-central Washington State, the many paths of the onrushing floods converged on the Pasco Basin, where floodwaters were slowed by the hydrologic constriction of Wallula Gap (Fig. 2) before draining out through the Columbia River Gorge to the Pacific Ocean. This constriction caused back flooding of local river valleys and basins, which resulted in deposition of relatively fine-grained slackwater sediments characterized by rhythmically graded bedding; these graded rhythmites, locally called touchet beds and multiple sets, have been recognized and are indicative of multiple floods during the Last Glacial Maximum (Flint, 1938; Waitt, 1980, 1985).

Loess, sand dunes, and sand sheets have been accumulating on the Columbia Plateau throughout much or all of the Quaternary Period (Busacca, 1989). The loess is thickest, up to 75 m, in a 10,000 km² area northeast of the Columbia Valley appellation in an area called the Palouse (Fig. 2; Baker et al., 1991). A major source of sediment for the dunes and loess has been slackwater and other glacial sediments from older episodes of outburst flooding (McDonald and Busacca, 1988; Sweeney et al., 2002). Most recently during the last stages of the Pleistocene (from ca. 20 ka to 14 ka) and continuing through the Holocene, prevailing southwesterly winds eroded slackwater and other glacial sediments and redeposited them into the present sand dunes, sand sheets, and loess that mantle much of the Columbia Plateau. Soils formed from these windblown sediments are the backbone of agriculture in all of eastern Washington (Boling et al., 1998).

Two major units of loess that span approximately the past 70,000 yr have been informally named L1 and L2 (McDonald and Busacca, 1992). Many layers of distal tephra have been described and sampled from loess exposures and fingerprinted by electron microprobe (Busacca et al., 1992). Distal tephra layers in L1 loess have been correlated to Glacier Peak layers G and B (ca. 13,300 cal. yr B.P.) and to Mount St. Helens set S distal tephra (MSH S; ca. 15,300 cal. yr B.P.), and those in L2 loess to Mount St. Helens set C distal tephra (MSH C; ca. 50,000–55,000 thermoluminescence [TL] yr B.P.).

The L1 and L2 loess units thin and fine away from major slackwater sediment areas in the Umatilla and Pasco Basins (Busacca and McDonald, 1994) and Eureka Flat (Sweeney et al., 2003). The patterns are consistent with evidence that two major episodes of scabland flooding, one ca. 70,000–60,000 yr B.P. and the other the classic Spokane Floods ca. 20,000–15,000 TL yr B.P., triggered the last two major cycles of loess deposition on the Columbia Plateau and that they accumulated with only temporary slowing of deposition for much of the succeeding interglacial intervals.

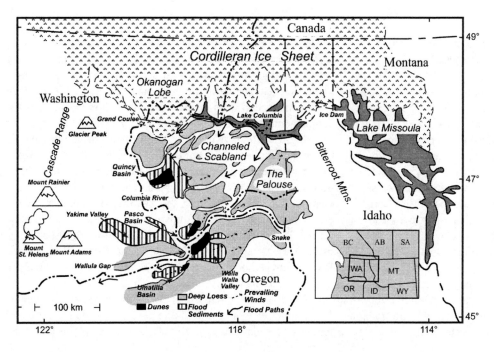

Figure 2. Schematic diagram showing the Pacific Northwest during the Last Glacial Maximum, the Lake Missoula–Channeled Scabland system, eolian sediments, and volcanoes of the Cascade Range.

SOILS, NATIVE VEGETATION, AND CLIMATE

Surface soils on the Columbia Plateau are dominantly Mollisols, Aridisols, and Entisols (Boling et al., 1998; Soil Survey Staff, 1999). In the central part of the Columbia Plateau where mean annual precipitation (MAP) is less than ~9 in (230 mm), soils developed in loess under sagebrush steppe are Aridisols, whereas soils developed in sand dunes under similar vegetation and precipitation are Entisols. Soils developed in loess under perennial bunchgrass vegetation where MAP is greater than or equal to ~9 in are Mollisols. Around the margins of the plateau loess soils formed under conifers are Alfisols. Some forest soils have a mantle of tephra-rich loess from Mount Mazama and are Andisols. Soils used for wine grapes commonly are Aridisols in which the upper horizons are formed in loess or sheet sands and lower horizons are formed in stratified silty to gravelly outburst flood sediments. Some have a lime-silica indurated pan at the interface between materials. Some wine grape soils are formed in loess or sand to 5 ft (1.5 m) or more. Thus, there are major differences in rooting depth, texture, and resulting water-holding capacity, which are key properties for inducing controlled water stress to improve grape quality.

Pre-agricultural vegetation in southeastern Washington ranged from sagebrush-steppe in the driest areas, to meadow steppe in areas of intermediate precipitation, to coniferous forest (Daubenmire, 1970). Xerophytic (drought tolerant) shrubs include several species of *Artemisia, Purshia,* and *Crysothamnus.* Perennial grasses include the major species bluebunch wheatgrass (*Agropyron spicatum*), Idaho fescue (*Festuca idahoensis*), and Sandberg bluegrass (*Poa sandbergii*) and a host of less common annual and perennial grasses and forbs. Mesophytic (moisture-loving) shrubs include *rosa* spp., Serviceberry (*Amelanchier alnifolia*), and Snowberry (*Symphoricarpos albus*). Several zones of conifer vegetation have been recognized with increasing effective moisture and decreasing temperature (Daubenmire and Daubenmire, 1984).

Climate is one of the more important components of terroir. In some ways it is the most difficult to evaluate because it varies in both space and time. There are many weather variables and these can be measured at three different scales. Macroclimate is on a continental to regional scale and controls the length of the growing season and other long-term trends and extremes. Mesoclimate is on a regional to vineyard scale and is affected by topography, elevation, slope, aspect, and proximity to bodies of water or other moderating influences. Microclimate ranges from the scale of a vineyard down to individual vines, grape clusters, and even smaller domains if measurement permits. Macroclimate changes on a geologic time scale (thousands to millions of years), but both mesoclimate and microclimate can vary seasonally, daily, or even hourly. Both mesoclimate and microclimate can be affected by human activities such as urban development, wind machines, irrigation, and canopy management.

Although many climatic variables can be measured, four of the more important are temperature, humidity, wind, and sunlight (solar radiation). These and others are collected systematically by a variety of meteorological services, but in the state of Washington we are fortunate to have the Washington State University (WSU) Public Agricultural Weather System (PAWS) that automatically and continuously collects climatic data (http://frost.prosser.wsu.edu/). Such data can be used for regional and worldwide comparisons, e.g., the excellent analyses of Gladstones (1992, 2001).

The climate of the Columbia Plateau is influenced to a great extent by prevailing westerly winds and by the Cascade and Rocky Mountains. The Cascade Mountains create a rain shadow, and as a result the climate of the Columbia Plateau is arid to sub-humid (15–100 cm of mean annual precipitation). The amount of precipitation is closely correlated with elevation, generally increasing from west to east and southeast. The Rocky Mountains protect this section of Washington from the coldest of the arctic storms that sweep down through Canada.

During the summer, high-pressure systems prevail, leading to dry, warm conditions and low relative humidity. Average afternoon temperatures in the summer range from 20 to over 35 °C. Most of the growing season is very dry and some vineyards experience no measurable precipitation during the summer months. The rainy season extends from October to late May or June, as frontal storms sweep across the area. In eastern Washington, most of the precipitation from mid-December to mid-February is in the form of snow.

As an example of climates of Washington appellations, Red Mountain is a warm vineyard site with 3409 degree days (50 °F) recorded in 1998 and an average of 3016 degree days for the years of record. For comparison, the Napa Valley in California and the Barossa Valley in Australia average 3280 and 3090 degree days, respectively (see the broader discussion by Meinert and Busacca [2000, 2002] of climatic measures in Washington and by Gladstones [1992, 2001] of general climatic measures relative to viticulture). Red Mountain also may be the driest viticultural area in Washington State, with an average annual precipitation of 17.8 cm and a low in 1999 of 8.4 cm. Typically, in most areas of the state, the time of year with lowest precipitation coincides with that of highest temperatures, and because of the low soil water-holding capacity and general absence of water tables, this creates a moisture deficit that requires irrigation in most vineyards. With the high evapotranspiration rates in such conditions, drip irrigation is the dominant method of supplying supplemental water.

Regional comparisons are possible using the above geologic history and climatic data. For example, more than 90% of Washington vineyards are located in areas affected by glacial outburst floods. In the Red Mountain appellation, these flood sediments were mostly deposited from the swirling back-eddies behind Red Mountain and include numerous lenses of relatively coarse gravel. In the Walla Walla Valley appellation, the flood sediments are generally finer grained due to deposition from ponded floodwaters, although there are some zones of coarse gravels in modern river channels.

The goal of this field trip is to examine the terroir of specific Washington vineyards and to attempt to correlate the observed features of soils, geology, climate, and other physical factors with variations in grape and wine quality. The specific field stops will include:

ROAD LOG

Day 1. Seattle to Walla Walla (Fig. 3A)

Cumulative		
Miles	*(km)*	
0.0	(0.0)	Check in for field trip at main entrance to the Washington State Convention and Trade Center at 7:00 am to enable 7:30 am departure. Start out going southwest on Union St. Go three blocks and turn left onto 5th Ave. Go three blocks and turn left onto Spring St. Take the I-5 south ramp toward Portland.
1.3	(2.1)	Junction of I-5 with I-90 east. Immediately merge onto I-90 east via the exit—on the left—toward Bellevue/Spokane.
50.3	(80.9)	Snoqualmie Pass, elevation 3022 ft, on I-90, is the lowest and most heavily traveled east-west highway crossing in Washington State. It is one of the state's two east-west highways with mountain passes open year-round. It is the drainage divide between generally cool, rainy lands west of the Cascades and warm, dry lands east of the Cascades.
109.3	(175.9)	Take offramp from I-90 east to I-82 east toward Yakima.
146.8	(236.2)	Take offramp from I-82 onto State Highway 97 south.
147.4	(237.2)	Here the Yakima River passes through Union Gap, a water gap in Ahtanum Ridge, part of the Yakima fold belt (see Reidel et al., 1984). The Yakima fold belt formed when basalt flows and their interstratified sediments were folded and faulted by north-south compression. The Yakima fold belt is composed of sharp anticlinal ridges separated by wide synclinal valleys. Most of the folding is younger than ca. 10.5 Ma (Reidel and Hooper, 1989), or after the end of the major outpourings of Columbia River basalt. The steep north sides of most anticline ridges are faults that consist of imbricated thrust zones (Reidel, 1984; West et al., 1996).
149.4	(240.4)	Exit from Highway 97 by turning right (south) onto Lateral A.
154.2	(248.1)	Turn right onto West Wapato Road and travel west. Here the road travels across the Holocene fan of the Yakima River created as flood-stage flows expanded into the upper Yakima Valley after passing through the constriction of Union Gap. The low-lying fan and floodplain soils are not well suited for vinifera grapes but support hops, field crops, concord grapes (Washington is the United States' largest producer of concord grapes for juice and jellies), and various fruit crops such as apples and cherries.
156.9	(252.4)	The road rises 5–8 m over a dissected remnant of glacial slackwater sediments deposited by outburst floods that backflooded more than 70 mi (112 km) up the Yakima Valley from the Columbia-Yakima Rivers' confluence. The Horse Heaven Hills, the largest anticline in the fold belt, are visible on the left (to the south).
164.8	(265.2)	Turn right onto Stephenson Road (at 164.2 mi) and continue to the end of the road at Mike Sauer's Red Willow Vineyard.

Stop 1. Red Willow Vineyard

Red Willow Vineyard is on the south slope of Ahtanum Ridge and is within the Yakama Indian Reservation. Unlike almost all other vineyards in Washington State, Red Willow is entirely above/outside the influence of the Missoula floods and thus occurs on much older soils developed on Miocene-age volcaniclastic sediments of the Ellensburg Formation (Waters, 1961; Bingham and Grolier, 1966; Smith, 1988). At this stop we will examine soils formed from loess over volcaniclastic sediments (Fig. 4A) and discuss vineyard siting by grape varietal with owner Mike Sauer.

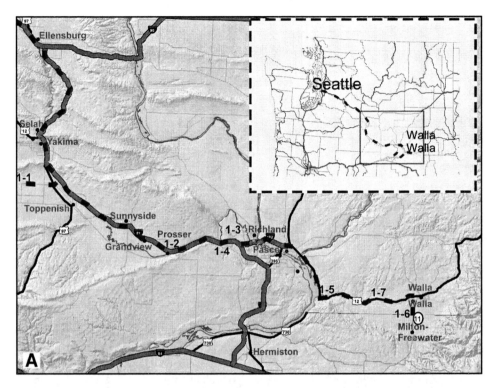

Figure 3. Maps of the routes and stops for Days 1 and 2 of the field trip.

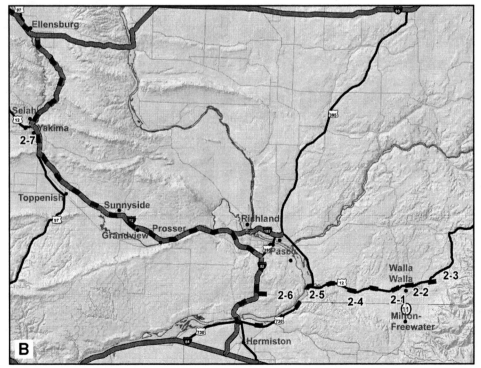

Figure 4. A: Outcrop of distinctive volcaniclastic sediments that form the soils at the Red Willow Vineyard in the Yakima Valley appellation. B: View to the south from Red Willow Vineyard showing Mount Adams (12276 ft/3472 m).

Known for being the inspiration for new varietals in the state, Red Willow has successfully pioneered varietals such as Syrah, Sangiovese, Malbec, Viognier and Cabernet Franc and has long produced award-winning Cabernet Sauvignon and Merlot.

At Red Willow Vineyard, the vines are planted on a peninsula of land jutting out from the south-facing Ahtanum Ridge. It is the only vinifera vineyard on the Ahtanum Ridge, and is the most westerly and the most northerly vineyard in the Yakima Valley appellation of Washington state. At 1300 feet above sea level, it is also the highest. To the west rise the foothills of the Cascades with Mt. Adams' snowcapped peak dominating the landscape.

While Red Willow itself is the highest vineyard in the Yakima Valley, it is also a relatively warm site—averaging 2700 degree days during the growing season. At 1300 feet above sea level, Red Willow stood above the cataclysmic Missoula floods at the end of the last ice age, floods that deposited silt and sand over the surrounding area. The ancient, well-drained, and nutritionally poor soil provides superb conditions for vinifera grapes.

Red Willow Vineyard first planted Cabernet Sauvignon in 1973 on the precipitous west-facing slope of the peninsula. In 1981, Columbia Winery released the first vineyard-designated wine from the vineyard with its Red Willow Cabernet Sauvignon. Cabernet Sauvignons from Red Willow are generally substantial, powerful wines with plum and blueberry fruit. Often the wine is blended with Cabernet Franc, which gives it an added finesse. Columbia Winery first produced Merlot from Red Willow grapes in 1987. This Merlot was the first in Washington to be blended with Cabernet Franc in the traditional Bordeaux manner.

With the release of Syrah from Red Willow Vineyard in 1988, Columbia Winery produced the first Syrah in the Pacific Northwest. In 1991, Columbia Winery released the first Red Willow Cabernet Franc and Columbia Winery's first vintage of Red Willow Sangiovese was 1995. (excerpted from http://www.columbiawinery.com/vineyards/redwillow.asp, accessed on August 5, 2003)

Cumulative		
Miles	(km)	
177.8	(286.1)	Retrace route on Stephenson Road, turning left and proceeding back to the east on West Wapato Road. Continue east on West Wapato Road across Lateral A to the intersection of State Highway 97. This is part of the Yakama Indian Nation. The reservation includes more than 1,300,000 acres, twice as large as Rhode Island. Its lands are used for agriculture, timber, range, and gathering of native plants.
178.3	(286.9)	Continue east across Highway 97 through the town of Wapato and angle left (NE) onto South Wapato (Donald Wapato) Road, crossing the railroad tracks.
179.3	(288.5)	Cross the Yakima River.
179.7	(289.1)	Enter onramp to I-82 east. There are good exposures of slackwater sediments on the left (north) side of the interstate.
186.7	(300.4)	On the left (north) side of the interstate are 10–20-m-high bluffs formed in slackwater sediments.
187.4	(301.5)	Slackwater sediment exposures on the left (north) side of the interstate have a distinctive white 2–5-cm-thick layer of 15 ka Mount St. Helens "S" ash near the top. This provides evidence for the timing of this emplacement of the slackwater sediments at the end of the Last Glacial Maximum.
198.3	(319.1)	On the right (south) side of the interstate is Snipes Mountain, a failed anticline.
217.6	(350.1)	Take exit 82 off of I-82 and turn right onto Wine Country Road.

218.5 (351.6) Continue east for ~0.9 mi on Wine Country Road, following signs to Hogue Cellars parking lot for lunch and tour.

Stop 2. The Hogue Cellars

Our stop here will provide a catered lunch, wine tasting, and tour of this state-of-the-art winery. The Prosser-based Hogue Cellars is Washington State's third largest winery, behind Stimson Lane (Château Ste. Michelle/Columbia Crest) and Constellation Brands (Columbia Winery/Covey Run), producing 450,000 cases of wine per year.

Mike and Gary Hogue's parents, Wayne and Shyla Hogue, began farming in the Yakima Valley in the 1940s, and eventually turned management of the business over to Mike. In 1974, he planted Hogue's first wine grapes, and in 1982, the first wine (2000 cases of Johannisberg Riesling) was produced by Hogue Cellars. Approximately 650 acres of the Hogue's 1600 acre farm is used to grow nine varieties of wine grapes. The additional acres produce hops, table grapes, apples, and vegetables, some of which are pickled and sold under the Hogue Farms label. In 2001, Vincor International bought Hogue Cellars, which continues to be run by Mike Hogue. Wade Wolfe is general manager and David Forsythe is director of winemaking.

| Cumulative | | |
Miles	(km)	
219.4	(353.0)	Return to onramp of I-82 and enter freeway heading east.
221.3	(356.1)	To the south on the steep slopes of the Horse Heaven Hills are numerous landslide scarps and zones of hummocky ground indicative of the numerous mass failures on these oversteepened slopes. Landslides may have been set off by relatively recent earthquakes.
224.8	(361.7)	To the left (north) is a spectacular scabland landscape with scoured basalt on the valley floor where flood waters rushing into the Yakima Valley were constricted by the narrow valley walls. Rattlesnake Mountain is visible on the skyline to the north.
228.8	(368.1)	To the north is a breached anticline on the north side of the Yakima River. The anticline was cut by the backflooding of Missoula floodwaters upvalley into the Yakima Valley. The valley narrows here to less than four miles wide between the Rattlesnake Hills to the north and the Horse Heaven Hills to the south, concentrating the floodwaters and their erosive power. Upstream (toward Prosser and Yakima), the valley widens to more than twenty miles.
230.0	(370.1)	Red Mountain is visible in the distance on the north side of the highway.

233.7	(376.0)	Take exit 96 off of I-82 to Benton City. At the bottom of the offramp, turn left and go under the freeway toward Benton City.
234.1	(376.7)	Turn right onto 224 east.
234.6	(377.5)	Turn left onto Sunset Road.
235.0	(378.1)	On the right (east) is the tower of the Public Agricultural Weather Station (PAWS). This is one of many automated data collection sites operated by Washington State University.
235.5	(378.9)	Turn right into the driveway of the Ciel du Cheval Vineyard.

Stop 3. Red Mountain AVA: Ciel du Cheval and Klipsun Vineyards

The Ciel du Cheval and Klipsun Vineyards are two of about fifteen vineyards located in Red Mountain, the newest appellation in Washington State. Growing conditions include ~3000 degree days, with a 210 day growing season, which are similar to other great wine growing areas. Our stops here will include discussions with Jim Holmes of Ciel and Fred Artz and the Gelles family of Klipsun regarding the pioneering spirit and good fortune that led them to plant wine grapes at Red Mountain, the challenges posed by soil variability, extreme temperatures during veraison (ripening), and drying winds on grape quality.

Our discussion of these topics will draw on information and figures presented in Meinert and Busacca (2002) on the terroir of Red Mountain. Copies of this paper will be handed out at the beginning of the field trip. Diverse soils such as the Scooteney, Warden, and Hezel form the backbone of wine production at Red Mountain (Fig. 5; also see Figs. 11 and 14 in Meinert and Busacca, 2002).

Wines made from Ciel grapes have been described in the wine press as being among the best in the world. No wine is made at the vineyard; however, Ciel supplies grapes to more than 20 wineries in Washington and Oregon, such as Andrew Will, Quilceda Creek, and McCrea Cellars. Plantings on the 120 acre ranch include most Bordeaux types (Cabernet Sauvignon, Cabernet Franc, Merlot), several Rhone varieties (Syrah, Grenache), and two Italian varieties. Recent plantings have emphasized use of clones selected for their exceptional wine quality in France and Italy.

Klipsun Vineyards was founded in 1982. In 2002, *Wine and Spirits* magazine acclaimed Red Mountain's Klipsun Vineyard as one of the world's top 25 vineyards. The first 40 acres was planted with Cabernet Sauvignon, Chardonnay, and Sauvignon Blanc in 1984 and has now been expanded to 120 acres, including Cabernet Sauvignon, Merlot, Syrah, Sauvignon Blanc, Semillon, and Nebbiolo. Klipsun sells to ~25 different wineries in the Pacific Northwest. Vineyard rows are numbered so each winery knows which rows will be theirs.

| Cumulative | | |
Miles	(km)	
238.3	(383.4)	Retrace route to and under the I-82 underpass.
238.4	(383.6)	Continue straight south, cross the railroad tracks, and curve left on Webber Canyon Road.

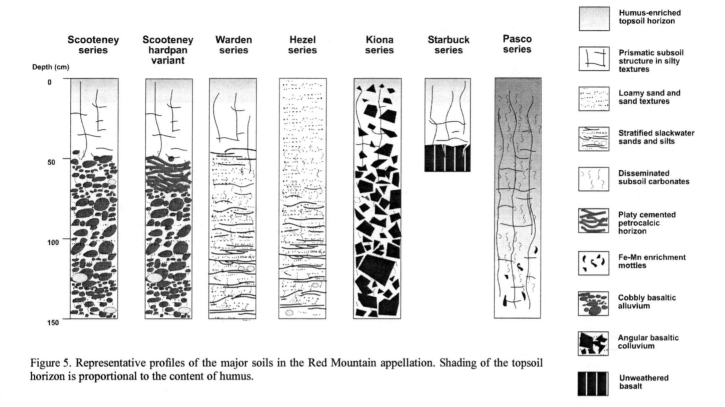

Figure 5. Representative profiles of the major soils in the Red Mountain appellation. Shading of the topsoil horizon is proportional to the content of humus.

239.0   (384.6)   Drive past the road cut and park in the wide section of Webber Canyon Road.

Stop 4. Exposure of Loess, Slackwater Sediments, and Tephra

On the left (north side) is a road cut that exposes slackwater sediments that contain a "doublet" of Mount St. Helens "S" ash that was deposited from an eruption that occurred simultaneously with an outburst flood slackwater ponding event. The Mount St. Helens S tephra layer is radiocarbon dated at 13,000 yr B.P. (ca. 15,300 cal. yr B.P.) at the volcano (Mullineaux, 1996). TL age dating of loess enclosing the same tephra (Busacca et al., 1992) yields a similar age estimate (Berger and Busacca, 1995). At this site, the slackwater sediments are covered by ~30 cm to 1 m of post-flood L1 loess (McDonald and Busacca, 1992).

Cumulative		
Miles	(km)	
239.7	(385.7)	Turn around and return to the onramp for I-82 and enter the interstate heading east.
244.7	(393.7)	Junction of I-82 and I-182. Take exit 12 on right onto I-182 toward Richland and Pasco.
257.6	(414.5)	At the intersection of State Highway 395 and State Highway 12, take Highway 12 east toward Walla Walla.
262.1	(421.7)	The Vaughan Hubbard Bridge crosses the Snake River, which flows in a channel cut

into mega gravel bars deposited by outburst flood flows into the Pasco Basin. The force of the largest of these floods was great enough to travel more than 100 mi (160 km) upstream on the Snake River past Lewiston, Idaho.

272.7   (438.8)   Turn left at the town of Wallula, then left again on the frontage road, and park in front of the post office.

Stop 5. Wallula Gap Overlook (Optional)

Wallula Gap, the water gap where the Columbia River passes through the Horse Heaven Hills today, was one of the major flow constrictions along the lower Columbia River that was responsible for hydraulic damming and ponding of outburst floods that then backflooded axial valleys upstream from the constrictions. Elevations of high divide crossings eroded in basalt above Wallula Gap indicate that the maximum flood stage was at least 1150 ft (350 m). The floor of the original river channel beneath Lake Wallula was ~240 ft (80 m). Flood-cut scarps in the deep loess cover on the hills at the entrance to the constriction allow that the maximum stage could have been as high as 1200 ft (365 m; O'Connor and Baker, 1992). This is close to the highest elevations at which ice-rafted granitic erratic are found around the Pasco Basin.

Recent calculations of maximum flood discharges based on this high-water evidence suggest that ~10 million m^3s^{-1} passed through Wallula Gap (O'Connor and Baker, 1992). This is ~300 times the maximum flows of the 1993 Mississippi River flood! We are standing on a gravel bar deposited by giant floods; notice the mixed lithologies of the gravel. Today the Columbia River is dammed by a series of 10 dams from Bonneville Dam, 65 km (40 mi) east of Portland, Oregon, to Grand Coulee Dam in north-central Washington. The flat water in front of us is Lake Wallula ponded behind McNary Dam.

Cumulative Miles	(km)	
272.8	(438.9)	Rejoin Highway 12 toward Walla Walla. Nearby is a historical marker denoting former site of Fort Walla Walla (now underwater), originally a fur trading post of Hudson's Bay Company dating from 1818.
274.8	(442.2)	Intersection of Highways 12 and 730. At stop sign, turn left, continuing east on Highway 12 toward Walla Walla.
282.5	(454.5)	FPL Energy's Stateline Wind Energy Facility, which will be a stop on Day 2, is visible on the Horse Heaven Hills to the right (south).
283.8	(456.6)	Beginning of the Walla Walla appellation. The Blue Mountains form the skyline ridge straight ahead (east).
286.6	(461.1)	On right (south) are incised remnants of the very thick fill of bedded slackwater deposits that today form terrace remnants throughout the Walla Walla Valley.
287.7	(462.9)	The relatively flat Holocene flood plain of the Walla Walla River, which is bounded by terrace remnants, can be seen here.
291.8	(469.5)	On the left (north) are the Woodward Canyon and L'Ecole No. 41 Wineries. After checking in at the hotel in Walla Walla, we will return for dinner at L'Ecole Winery.
298.0	(479.5)	On the right (south) is the Three Rivers Winery.
302.5	(486.7)	Turn right onto West Pine Road.
302.7	(487.0)	Turn right into the Holiday Inn Express parking lot.

Stop 6. Holiday Inn Express

Check in, rest, and dress for dinner, then re-board the bus to drive to L'Ecole Winery for dinner. Retrace steps out of the Holiday Inn Express parking lot, left onto West Pine Road, and left again onto Highway 12 west.

Cumulative Miles	(km)	
313.6	(504.6)	Turn right into the L'Ecole No. 41 parking lot.

Stop 7. L'Ecole No. 41 Winery

Dinner will be provided through the hospitality of owner/winemaker Marty Clubb and executive chef Cristiana Fagioli, who will prepare an extraordinary dinner expertly pairing L'Ecole's new wine releases with five innovative courses.

L'Ecole No. 41 has been producing premium handcrafted varietal wines since 1983 in the historic Frenchtown School in Lowden, Washington. L'Ecole No. 41 is a family-owned business. Founded by Jean and Baker Ferguson, the winery is now owned and operated by their daughter and son-in-law, Megan and Marty Clubb. Marty has been the general manager and winemaker since 1989.

Built in 1915, the schoolhouse is located in historic Frenchtown, a small community just west of Walla Walla, Washington. Frenchtown derived its name from the many French-Canadians who settled the valley during the early 1800s. Legend has it, these men of French descent were raising grapes and producing wine. By the 1860s, nurseries, vineyards and winemaking had become a part of the region's growing economy. The name—L'Ecole No. 41, French for "the school" located in district number 41—was chosen to salute these pioneer viticulture efforts.

The winery currently produces ~20,000 cases annually. Semillon, Chardonnay, Merlot, and Cabernet Sauvignon at L'Ecole are all barrel aged, creating quite a demand for barrel storage. Today, L'Ecole has over 1,000 French and American oak barrels. In recent years, the winery has produced more single-vineyard and Walla Walla Valley appellation designated wines to take advantage of the exceptional fruit from the Walla Walla AVA. (excerpted from http://www.lecole.com/, accessed on August 5, 2003)

After dinner, drive east again to the Holiday Inn Express.

Cumulative Miles	(km)	
324.5	(522.1)	Turn right onto West Pine Road and into the parking lot of the Holiday Inn Express. End of Day 1.

Day 2. Walla Walla to Seattle (Fig. 3B)

Cumulative Miles	(km)	
0.0	(0.0)	From parking lot of Holiday Inn Express, turn right (east) onto West Pine St. and veer left to stay on West Pine to N 9th Ave. (Highway 125).
0.9	(1.4)	Turn right (south) onto N 9th Ave. (Highway 125) and continue on Highway 125 south.
6.6	(10.6)	Oregon-Washington state line. Washington Highway 125 becomes Oregon Highway 11 at the state line. The Walla Walla Valley appellation is one of the few to cross state lines and is, appropriately, bounded by natural rather than political features. Along the route the highway is alternately sited on higher elevation remnants of slackwater terraces and on

lower alluvial surfaces of the Holocene and modern Walla Walla River and its tributaries.

9.3 (15.0) Turn right (west) onto Sunnyside Rd. This area around Milton-Freewater, Oregon, has been a highly productive center of production of orchard crops (cherries, peaches, apples, etc.) for more than 100 yr.

10.8 (17.4) On the right (north) side of the road is the Cailloux Vineyard, planted by vigneron Christophe Baron of Cayuse Vineyard on the cobbly former riverbed of the Walla Walla River.

Stop 1. Cailloux Vineyard

Discussion at Stops 1–3 today will draw heavily on information and figures presented in Meinert and Busacca (2000) on the terroirs of the Walla Walla Valley appellation. Copies of this paper will be handed out at the beginning of the field trip.

The soils of the Walla Walla Valley appellation have formed from four different types of surficial sediments or bedrock. Various combinations of soil parent materials and a strong gradient of mean annual precipitation across the appellation are key to determining vineyard potential performance. Soils formed from young alluvium vary tremendously in their properties, such as texture (cobbly to clayey), salt effects, and presence or absence of a water table within the rooting zone of vines. Soils formed from loess more than 150 cm deep are found around the margins of the Walla Walla Valley appellation and have dominantly silty, uniform soil profiles. Mean annual rainfall varies widely depending on location in the appellation, and this, along with slope steepness and aspect, determines suitability or potential for development of dryland or irrigated vineyards. Soils formed from thin to moderately thick loess overlying slackwater sediments (Fig. 6) have been the main focus of vineyard development up to the present time in the appellation. Soils located on steep slopes of the Blue Mountains that have bedrock at shallow depth have not been fully evaluated to determine their potential for vineyard development.

The objective of Stop 1 is to examine the Cailloux Vineyard, which is sited on one of the most unique agricultural soils in the Northwest, and to discuss the viticultural practices that have been tailored specifically to the stony, hot, droughty Freewater series soils (Figs. 6 and 7).

Christophe Baron was the first to envision the potential of these stony soils to produce grapes for fine wines. He planted the ten acres of this vineyard in 1996. Ten acres of En Cerise (French for cherry), and ten acres of Coccinelle (French for ladybug) followed in 1997. The ten acres of En Chamberlin were planted in spring of 2000. The majority of the vineyards are planted with Syrah alongside a few acres of Cabernet Sauvignon, Cabernet Franc, Merlot, Roussanne, Tempranillo, and Yiognier. A fifth vineyard, Armada, planted in 2001, contains 3 acres of Grenache, 3 acres of Syrah, and 1 acre of Mourvedre. Yields average 2–2.5 tons per acre, resulting in rich, highly-concentrated fruit.

Spacing is four feet between vines and five feet between rows, 2178 vines per acre. This is nearly double the standard vine quantity and easily marks it as the highest density vineyard in the Walla Walla Valley. The vines in all of the vineyards are trained

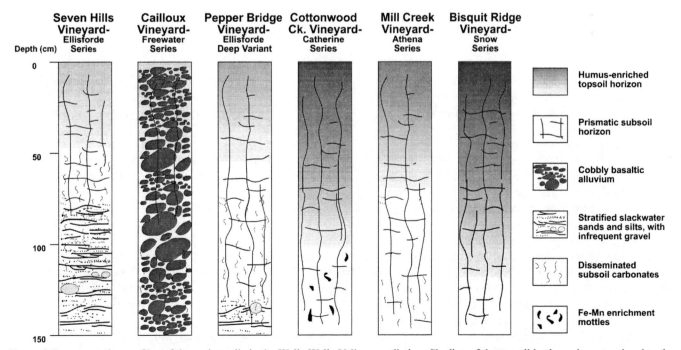

Figure 6. Representative profiles of the major soils in the Walla Walla Valley appellation. Shading of the topsoil horizons is proportional to the content of humus.

Figure 7. Cobbly surface horizon of the Freewater series soil in the Cailloux Vineyard.

Figure 8. View across a part of the floodplain of the Walla Walla River showing dissected remnants of outburst flood slackwater terraces that make excellent vineyard sites.

low to the ground in the belief that re-radiation of heat at night from the exposed cobbly surface aids in the development of the fruit. An extra cane is maintained on the trunk close to the ground and is buried in the fall to provide a new starter trunk for each vine against a killing freeze in this low-lying valley site. Irrigation includes the application of scant quantities of water by drip irrigation, inducing water stress that concentrates fruit flavors, and the vines are organic.

Cumulative Miles	(km)	
12.3	(19.8)	Turn around and go back to Highway 11. Turn left (north) toward Walla Walla.
15.0	(24.1)	Turn right onto Stateline Road.
15.2	(24.4)	Turn left onto Pepper Road.
15.6	(25.1)	Turn right onto J.B. George Road.
16.3	(26.2)	Turn left onto Larson Road.
16.7	(26.9)	Pepper Bridge Winery is on the left and the Northstar Winery is on the right.

Stop 2. Pepper Bridge Vineyard

This stop will highlight the concept of terroir or optimal vineyard siting to produce fine wines. The best vineyards in the Walla Walla Valley occupy the tops of flood slackwater terraces (Fig. 8) with optimal soil characteristics and air drainage. Soils at Pepper Bridge Vineyard are dominantly a deep variant of the Ellisforde series with ~100–120 cm of loess over stratified slackwater sediments (Fig. 6). At this stop, there will be discussion about viticultural practices for wine grape production and we will relate these practices to the Pleistocene and Holocene geology.

The original 10 acres were planted in 1991 and have expanded to a total of 180 acres of wine grapes. Pepper Bridge Vineyard has gained an outstanding reputation with winemakers throughout the state of Washington, and especially the Walla Walla Valley. Tom Waliser has been the vineyard manager at Pepper Bridge

Vineyard since its inception in 1991. All grapes are grown on split canopy trellises, in which the vines are trained both up and down off the cordon, or grape-bearing wire. With the exception of 5 acres of Merlot, which is on the Scott-Henry trellis system, all grapes are on the Smart-Dyson trellis system.

The vineyard uses cutting edge technology in its irrigation and weather systems. Weather data, temperature, humidity, wind, and sun energy units are recorded around the clock and the data are downloaded to computer by phone line. Over sixty moisture-measuring points are spread throughout the farm and moisture is data logged once an hour, 24 hours per day.

Cumulative Miles	(km)	
18.4	(29.6)	Return to the Pepper Road/J.B. George Road intersection, then turn left onto Pepper Road, right onto Stateline Road, and proceed to stoplight at Highway 125. Turn right onto Highway 125 and proceed toward Walla Walla (north).
23.6	(38.0)	Turn right onto W Poplar St.
24.1	(38.8)	Turn left onto S 2nd Ave.
24.7	(39.7)	Merge onto Highway12 east via the ramp on the left.
27.9	(44.9)	Take the offramp for Airport Road.
28.2	(45.4)	Turn right onto Airport Road, heading south, then left onto Isaacs Ave.
30.4	(48.9)	Isaacs Ave. becomes Mill Creek Road. Turn left at Walla Walla Vintners sign and drive up to Mill Creek Upland Vineyard of Leonetti Cellars and park opposite the vineyard.

Stop 3. Mill Creek Upland Vineyard of Leonetti Cellars

The purpose of this stop is to examine a state-of-the-art vineyard installation at Leonetti Cellars, one of the premier wineries in the Pacific Northwest. Chris Figgins, vineyard manager and assistant winemaker, will explain the use of the latest drip irrigation systems, virus-free rootstocks, new clones and varietals, and environmental monitors to take grape and wine quality to the next level. This vineyard receives ~16–18″ (400–460 mm) of annual precipitation, making it perhaps the vineyard with highest precipitation in eastern Washington. This and the very deep, organic matter–rich Athena soils in loess (Fig. 6) provide a new challenge for Leonetti Cellars to develop management strategies for this vineyard, which perhaps could be dry farmed because of the high rainfall and high water holding capacity of the silt-loam textured soils. This and the steep south-facing slope provide excellent sun exposure and air drainage.

Leonetti Cellars' owner and winemaker, Gary Figgins, honed his craft as a home winemaker and released his first commercial wine in 1978. Leonetti Cellars produces limited quantities of the highest quality Cabernet Sauvignon, Merlot, and Sangiovese.

Leonetti Cellars manages their vineyards very intensively, using a combination of the latest in technology, proven traditions, and sustainable agriculture. They draw their fruit increasingly only from vineyards in the Walla Walla Valley AVA. Yields are moderately low to very low, ranging from 1.5–4 tons per acre, depending on the vineyard and variety. Most vineyards are trellised to a vertically divided canopy, a method known as Smart-Dyson or Scott Henry. Deficit irrigation, monitored by neutron probes and a system called "enviroscan," is practiced in all of Leonetti's vineyards to control vegetative growth, reduce berry size, and intensify flavors in the berries. All grapes for Leonetti wines are handpicked at physiological maturity after a season of intensive hand pruning, hand leaf-plucking, shoot positioning, and cluster thinning.

Cumulative		
Miles	(km)	
32.9	(52.9)	Retrace route to Highway 12, enter Highway 12 west.
49.0	(78.8)	Continue on Highway 12 west through Walla Walla to the town of Lowden.

Stop 4. Burlingame Canyon, the "Little Grand Canyon"

The spectacular exposure of slackwater or touchet sediments from cataclysmic outburst flooding, which can be seen stopping at this site (Fig. 9), was created by a break in the irrigation canal a number of years ago. The purpose in stopping at this site is to discuss paleoflood dynamics, the "40-floods" hypothesis, and to show what underlies the terrace remnants in valleys like the Walla Walla and Yakima, forming the landscapes of some of the better vineyard sites in the Northwest.

Note: This exposure is on private property of the Gardena Farms Irrigation District and we are allowed to visit during this field trip only by special permission. Please do not return to this site at a later time on your own. Take any photos during this visit only. Please note the dangerous banks and stay clear of the edge. Our thanks to Stuart Durfee, manager.

Figure 9. Exposure of a loess-covered terrace formed of rhythmically bedded outburst flood sediments.

The following are guidelines set by the land owner for our safety during our visit:

1. If a representative of Gardena Farms Irrigation District No. 13 accompanies you during your visit, you will abide by all instruction or be asked to leave immediately.

2. You will park along the main irrigation canal at the top of the hill near the house and walk down the west side of the channel leading south to the Little Grand Canyon.

3. You will not travel outside of the fenced area on the east side of the canyon or travel down the west side of the canyon unless instructed to do so.

4. There will be no ascent into the canyon past the first ascending down slope at the northern entrance to the canyon.

5. Visitors to the canyon must act in an orderly fashion so as not to endanger any of the participants.

6. The visitation privileges for your organization or any organization associated with the visit will be revoked for future visits if you fail to follow directions or if there are unauthorized repeat visits by persons in attendance on this site visit.

Exposures of fine sand-, silt-, and clay-dominated slackwater rhythmites at sites such as Burlingame Canyon formed the inspiration for Richard Waitt's hypothesis for multiple outburst floods during the last glaciation (Waitt, 1980, 1984, 1985). It also has been argued that some rhythmites, rather than signaling individual outburst floods separated by decades-long intervals of subaerial exposure, instead reflect multiple flood surge events during a few floods (Baker, 1973; Bjornstad, 1982; Bunker, 1982; Baker and Bunker, 1985). This alternate argument is supported by the scarcity of definitive sedimentary evidence for subaerial exposure of rhythmites, an impression one gets while viewing the parallel, amalgamated nature of the multiple, stacked rhythmites exposed in the canyon. However, the paleoaridity and attendant sparse vegetation of this region both during and after slackwater flood accumulation, the isolation of broad reaches of slackwater deposits from flanking valleys (and hence colluvial deposits), as well as compelling ecologic and sedimentary evidence elsewhere on the plateau (e.g., Smith, 1993) tend to support Waitt's (1985) contention for multiple floods. Regardless of the explanation you prefer, sites such as Burlingame Canyon provide clear evidence that the glacial outburst floods provided an abundance of raw sedimentary material from which winds have created sand sheets, sand dunes, and loess-covered landscapes.

Cumulative		
Miles	(km)	
65.9	(106.0)	From Lowden, continue west on Highway 12 to Hatch Grade Road. Rendezvous with representative of FPL Energy at Hatch Grade Road turnoff. Follow representative to FLP Energy's Stateline Wind Energy Center.

Stop 5. Stateline Wind Energy Center

The purpose of this stop is to recognize the incredible wind energy that has partnered with the incredible energy of the cataclysmic floods to redistribute huge amounts of fluvial sediment into the eolian mantle of soils, which provide the basis for the agricultural wealth of the inland Pacific Northwest. In addition, this impressive facility is the largest wind energy facility in the United States.

Developed, owned, and operated by FPL Energy, the 300 megawatt Stateline Wind Energy Center provides clean, renewable energy to PacificCorp Power Marketing (PPM) for customers throughout the Pacific Northwest.

The Stateline Energy Center was fully operational in December 2001, just nine months after construction of the facility began. The first power from Stateline reached consumers in July 2001. The facility has the capacity to provide enough electricity to power ~60,000 homes and businesses. The Stateline Energy Center provides electricity for PPM customers including Bonneville Power Administration, Seattle City Light, the Eugene Water and Electric Board, and Avista Utilities.

FPL completed a 37 megawatt expansion of the Stateline Wind Energy Center in early December of 2002, making it the largest wind energy facility in the United States at 300 megawatts.

The Stateline Wind Project is the Northwest's largest commercial facility to generate electricity using wind. The project is located on Vansycle Ridge, an anticline ridge straddling the Washington-Oregon border, near Touchet, Washington. The ridge catches winds from the Columbia Gorge, which average 16–18 mph; this is considered excellent for wind farm development. The area around the project is used mostly for private farming, and this has continued beneath the completed wind project. The site is also close to preexisting transmission lines, reducing the need for new cables and minimizing the amount of power lost during transmission.

The Stateline Wind Project uses 660 kw Vestas wind turbines (Fig. 10), producing a maximum output of 300 megawatts (MW) of electricity. On average the project receives enough wind to deliver 30%–35% of its peak capacity year round: enough power for more than 21,600 Northwest homes. Electronic control systems point each turbine into the wind and adjust the pitch of the blades to make the best use of wind at any speed. The turbines can generate power at wind speeds of 7–56 mph. At higher speeds the turbines automatically shut down, a feature which allows them to withstand hurricane-force winds.

The Stateline Wind Project was planned carefully and underwent extensive review to minimize its environmental impact. Early biological studies indicated that the site receives little use by birds or other vulnerable species. The project uses tubular towers and buried cables in order to avoid adding new perching places for birds. Slower-moving blades and an upwind design further minimize any potential for avian fatality. As a clean power source, the project also eliminates some of the need for fossil fuel electric plants in the region. If natural gas or coal were used to generate the same amount of power, they would emit at least 310,000 tons of carbon dioxide per year, as well as air pollutants and acid rain precursors. Wind power produces no air emissions.

Return from the wind energy center to Highway 12. Turn left (south) onto Highway 730-395 south.

Cumulative		
Miles	(km)	
67.4	(108.4)	The highway and the Columbia River pass through Wallula Gap. Scabland topography can be seen up to ~1200 ft (366 m).
91.1	(146.6)	Take the exit onto Interstate 82 west, crossing the Columbia River from Oregon into Washington. Continue north, rising onto the Horse Heaven Hills. An application to the Bureau of Alcohol, Tobacco, and Firearms has recently been submitted to create a Horse Heaven Hills appellation.

Figure 10. Wind turbines seen during a dust storm at the Stateline Wind Energy facility near Walla Walla, Washington.

100.1	(161.1)	Get off of I-82 at exit 122, Coffin Road. At the stop sign at the bottom of the offramp, turn right and drive east on Coffin Road.
104.9	(168.9)	Turn right (south) onto 9 Canyon Road.
105.8	(170.2)	Turn left (east) onto Easterday Ranch private road.
110.3	(177.5)	Turn right (south) onto Finley Road.
112.3	(180.7)	Entrance to Wallula Vineyards.

Stop 6. Wallula Vineyards

The purpose of this stop is to view the Columbia River and Wallula Gap from the downstream side and to see and discuss a very interesting and potentially outstanding new vineyard with a diverse array of vineyard sites ranging from gentle, south-facing sites on deep loess soils to steeply sloping terraced sites, to microvineyards on shallow soils on small scabland buttes (Fig. 11A).

The family farm, which has expanded steadily to include Wallula Vineyards LLC today, was started by Andrew denHoed in 1954. Andy started out raising traditional row crops such as sugar beets, mint, potatoes, and dry edible beans. The original "home place" included 15 acres of concord grapes. Today his sons, Andy denHoed Jr. and Bill denHoed, are partners with Andy Sr. Today's operation includes 980 acres of vinifera, 125 acres of juice grapes, and 85 acres of orchard. The denHoeds first planted vinifera in 1979 at the Desert Hills Vineyards near Grandview in the Yakima Valley. Planting started at Wallula Vineyards in 1998. Desert Hills Vineyards includes 560 acres of wine grapes; Wallula Vineyards has 420 acres of wine grapes. The majority of the den-Hoed's grapes are delivered to the Stimson Lane group: Chateau Ste. Michelle, Columbia Crest, and Snoqualmie Wineries. A small portion go to the Hogue Cellars in Prosser, Washington.

Wallula Vineyards totals 1550 acres. Approximately 950 of that is plantable. Development has been divided into three phases. Phase I is currently in production. Phases II and III are in the planning stages. Before the irrigation system was installed and the first vines were planted, multiple earth-moving machines worked 5 days a week, 45–50 weeks a year, for 3 years to prepare the site. Unique aspects of this vineyard include plantings of vines that range in elevation from 350 ft (107 m) to 1200 ft (366 m), a difference which provides a greater range of meso-climates than any other vineyard in the state. Soil temperature is monitored in each vineyard with temperature probes to develop a database for management decisions and to inform future plantings. At the lower elevation sites, Lake Wallula provides a lake effect that moderates high temperatures in summer and low temperatures in winter.

Soils in deep loess at the upper elevations are classic Mollisols of the Ritzville series; those in deep loess below ~1000 ft elevation (305 m) actually receive enough less precipitation to be classed as Aridisols of the Shano series. Soils on margins of the scabland portion of the vineyard are formed in loess over stratified gravelly flood sediments (Fig. 11B) and are soils such as the Scooteney series (Fig. 5). Soils on the scabland buttes and

Figure 11. A: Vineyard terraces and scabland topography at the Wallula Vineyard in Columbia Valley appellation. B: Scooteney-type soil formed of loess over outburst flood slackwater sediments at the Wallula Vineyard. The loess mantle is ~75 cm thick.

benches are highly variable but include shallow and stony soils such as the Starbuck and Kiona series (Fig. 5).

The Phase I vineyard produces Cabernet Sauvignon, with clones 4, 6, 8, and 15. Clone 8 is the predominant clone, and is widespread throughout the state. Clones 6 and 15 of Merlot are also planted, as are the Piccolo and Grosso clones of Sangiovese. Other red wine grapes include Syrah, Mourvedre, Barbera, Cinsault and Dolcetto. White varieties include Chardonnay, Viognier, and Pinot Gris.

All of the water comes from the Columbia River, is pumped from the river to a settling pond, and is distributed from there through a drip irrigation system. Because of soil variability, the irrigation system is divided into many zones and subzones to provide the flexibility to tailor water delivery to the vines to maintain uniformity throughout the vineyard for crop and canopy.

Crop load ranges from 2.5 tons/yr (0.6 Mg/hectare) to 7 tons/yr (15.7 Mg/hectare) and is specified by the winery that takes each block of grapes.

124.5	(200.3)	Retrace steps to the intersection of Coffin Road and Interstate 82. Enter onramp for I-82 toward Yakima (west).
213.4	(343.4)	Continue on I-82 to Yakima. Take exit 33 (Yakima Ave., toward Terrace Hts.) to E Yakima Ave.
214.8	(345.6)	Drive west on E Yakima to N 1st St. Turn right and immediately look for the train station that houses Grant's Brew Pub. Dinner at Grant's Brewpub, 2 N Front St., 509-575-2922.
216.2	(347.9)	Retrace steps to I-82. Take onramp and enter I-82 west toward Seattle.
246.8	(397.1)	Take I-82 west toward Ellensburg to junction with I-90 west.
355.6	(572.2)	Take interchange from I-90 west to I-5 north.
356.4	(573.4)	Take the I-5 north exit, number 2C, toward Madison St./Convention Place.
357.7	(573.9)	Take the exit toward Madison St./Convention Place. Stay straight onto 7th Ave. Turn left onto Madison St.
357.9	(575.9)	Turn right onto 6th Ave.
358.0	(576.0)	Turn right onto 7th Ave.
358.1	(576.1)	Turn right onto Union St., then onto Convention Place. Stop in front of Seattle Convention and Trade Center. End of trip.

ACKNOWLEDGMENTS

We thank a number of vineyard owners and managers, winemakers, and other cooperators who made this field tour possible: Mike Sauer, Carlos Trevino, Jim Holmes, Tricia and David Gelles, Fred Artz, Marty Clubb, Cristiana Fagioli, Christophe Baron, Norm McKibben, Tom Waliser, Gary and Chris Figgins, Stuart Durfee, Ann Walsh, the den Hoed family, and Doug Gore. Thanks to Ron Bolton for preparing Figures 3A and B.

REFERENCES CITED

Baker, V.R., 1973, Paleohydrology and sedimentology of Lake Missoula flooding in eastern Washington: Geological Society of America Special Paper 144, 79 p.

Baker, V.R., and Bunker, R.C., 1985, Cataclysmic late Pleistocene flooding from glacial Lake Missoula—a review: Quaternary Science Reviews, v. 4, p. 1–44.

Baker, V.R., and Nummedal, D., eds., 1978, The Channeled Scabland (a guide to the geomorphology of the Columbia Basin, Washington): National Aeronautics and Space Administration, 186 p.

Baker, V.R., Bjornstad, B.N., Busacca, A.J., Fecht, K.R., Kiver, E.P., Moody, U.L., Rigby, J.G., Stradling, D.F., Tallman, A.M., 1991, Quaternary geology of the Columbia Plateau, in Morrison, R.B., ed., Quaternary nonglacial geology; conterminous U.S.: Geological Society of America, The geology of North America, v. K-2, p. 216–228.

Baksi, A.K., 1989, Reevaluation of the timing and duration of extrusion of the Imnaha, Picture Gorge, and Grande Ronde Basalts, Columbia River Basalt Group, in Reidel, S.P., and Hooper, P.R., eds., Volcanism and tectonism in the Columbia River flood-basalt province: Geological Society of America Special Paper 239, p. 105–112.

Berger, G.W., and Busacca, A.J., 1995, Thermoluminescence dating of late-Pleistocene loess and tephra from eastern Washington and southern Idaho, and implications for the eruptive history of Mount St. Helens: Journal of Geophysical Research, v. 100, no. B11, p. 22,361–22,374.

Bingham, J.W., and Grolier, M.J., 1966, The Yakima Basalt and Ellensburg Formation of south-central Washington: U.S. Geological Survey Bulletin 1224-G, p. 1–15.

Bjornstad, B.N., 1982, Catastrophic flood surging represented in the Touchet Beds, Walla Walla Valley, Washington: American Quaternary Association, Program and Abstracts of the Seventh Biennial Conference, Seattle Washington, 72 p.

Boling, M., Frazier, B., and Busacca, A., 1998, General soil map, Washington: Department of Crop and Soil Sciences, Washington State University, Pullman, and USDA Natural Resources Conservation Service, scale 1:750,000.

Bretz, J.H., 1923, The channeled scablands of the Columbia Plateau: Journal of Geology, v. 31, p. 617–649.

Bretz, J.H., 1925, The Spokane flood beyond the channeled scablands: Journal of Geology, v. 33, p. 97–115, 312–341.

Bretz, J.H., 1928a, Bars of channeled scabland: Geological Society of America Bulletin, v. 39, p. 643–702.

Bretz, J.H., 1928b, The channeled scabland of eastern Washington: Geographical Review, v. 18, p. 446–477.

Bretz, J.H., 1928c, Alternate hypothesis for channeled scabland: Journal of Geology, v. 36, p. 193–223, 312–341.

Bretz, J.H., 1932, The Grand Coulee: American Geographical Society Special Publication 15, 89 p.

Bunker, R.C., 1982, Evidence of late Wisconsin floods from Glacial Lake Missoula in Badger Coulee, Washington: Quaternary Research, v. 18, p. 17–31.

Busacca, A.J., 1989, Long Quaternary record in eastern Washington, U.S.A., interpreted from multiple buried paleosols in loess: Geoderma, v. 45, p. 105–122.

Busacca, A.J., Nelstead, K.T., McDonald, E.V., and Purser, M.D., 1992, Correlation of distal tephra layers in loess in the Channeled Scabland and Palouse of Washington State: Quaternary Research, v. 37, p. 281–303.

Busacca, A.J., and McDonald, E.V., 1994, Regional sedimentation of late Quaternary loess on the Columbia Plateau: sediment source areas and loess distribution patterns: Regional Geology of Washington State, Washington Division of Geology and Earth Resources Bulletin, v. 80, p. 181–190.

Daubenmire, R., 1970, Steppe vegetation of Washington: Washington Agricultural Experiment Station, Technical Bulletin 62, Washington State University, Pullman, 131 p.

Daubenmire, R., and Daubenmire, J.B., 1984, Forest vegetation of eastern Washington and northern Idaho: Cooperative extension: Washington State University, Pullman, 137 p.

Flint, R.F., 1938, Origin of the Cheney-Palouse Scabland Tract, Washington: Geological Society of America Bulletin, v. 49, p. 461–524.

Gladstones, J., 1992, Viticulture and environment: Winetitles, Underdale, Australia, 310 p.

Gladstones, J., 2001, Climatic indicators guide site selection: Practical Winery & Vineyard: v. 23, p. 9–18.

Halliday, J., 1993, Wine Atlas of California: Penguin Books, New York, 400 p.

Halliday, J., 1999, Wine Atlas of Australia & New Zealand: Harper Collins, Australia, 496 p.

Haynes, S.J., 1999, Geology and Wine 1. Concept of terroir and the role of geology: Geoscience Canada, v. 26, p. 190–194.

Haynes, S.J., 2000, Geology and Wine 2. A geological foundation for terroirs and potential sub-appellations of Niagara Peninsula wines, Ontario, Canada: Geoscience Canada, v. 27, p. 67–87.

Landon, R.D., and Long, P.E., 1989, Detailed stratigraphy of the N2 Grande Ronde Basalt, Columbia River Basalt Group, in the central Columbia Plateau, in Reidel, S.P., and Hooper, P.R., eds., Volcanism and tectonism in the Columbia River flood-basalt province: Geological Society of America Special Paper 239, p. 55–66.

McDonald, E.V., and Busacca, A.J., 1988, Record of pre-late Wisconsin giant floods in the Channeled Scabland interpreted from loess deposits: Geology, v. 16, p. 728–731.

McDonald, E.V., and Busacca, A.J., 1992, Late Quaternary stratigraphy of loess in the Channeled Scabland and Palouse regions of Washington State: Quaternary Research, v. 38, p. 141–156.

Meinert, L.D., and Busacca, A.J., 2000, Geology and Wine 3. Terroirs of the Walla Walla Valley appellation, southeastern Washington State, USA: Geoscience Canada, v. 27, p. 149–171.

Meinert, L.D., and Busacca, A.J., 2002, Geology and Wine 5: Terroir of the Red Mountain Appellation, Central Washington State, USA: Geoscience Canada, v. 29, p. 149–168.

Mullineaux, D.R., 1996, Pre-1980 tephra-fall deposits erupted from Mount St. Helens, Washington: U.S. Geological Survey Professional Paper 1563, 99 p.

O'Connor, J.E., and Baker, V.R., 1992, Magnitudes and implications of peak discharges from glacial Lake Missoula: Geological Society of America Bulletin, v. 104, p. 267–291.

Pardee, J.T., 1910, The Glacial Lake Missoula: Journal of Geology, v. 18, p. 376–386.

Peterson-Nedry, J., 2000, Washington wine country: Portland, Oregon, Graphics Art Center Publishing, 111 p.

Pringle, P.T., 1993, Roadside geology of Mt. St. Helens National Volcanic Monument and vicinity: Washington Department of Natural Resources Information Circular 88, 120 p.

Reidel, S.P., 1984, The Saddle Mountains: the evolution of an anticline in the Yakima fold belt: American Journal of Science, v. 284, p. 942–978.

Reidel, S.P., and Hooper, P.R., 1989, Volcanism and tectonism in the Columbia River flood-basalt province: Geological Society of America Special Paper 239, 386 p.

Smith, G.A., 1988, Sedimentology of proximal to distal volcaniclastics dispersed across an active foldbelt; Ellensburg Formation (late Miocene), central Washington: Sedimentology, v. 35, no. 6, p. 953–977.

Smith, G.A., 1993, Missoula flood dynamics and magnitudes inferred from sedimentology of slack-water deposits on the Columbia Plateau, Washington: Geological Society of America Bulletin, v. 105, p. 77–100.

Soil Survey Staff, 1999, Soil Taxonomy. 2nd edition, Agricultural Handbook Number 436: Washington, D.C., U.S. Department of Agriculture, Natural Resources Conservation Service, U.S. Government Printing Office, 869 p.

Sweeney, M.R., Gaylord, D.R., Busacca, A.J., and Halver, B.A., 2003, Eureka Flat, A long-term dust production engine of the Palouse loess, Pacific Northwest U.S.A. [abs]: XVI INQUA Congress, Reno, Nevada (in press).

Tolan, T.L., Reidel, S.P., Beeson, M.H., Anderson, J.L., Fecht, K.R., and Swanson, D.A., 1989, Revisions to the estimates of the areal extent and volume of the Columbia River Basalt Group, *in* Reidel, S.P., and Hooper, P.R., eds., Volcanism and tectonism in the Columbia River flood-basalt province: Geological Society of America Special Paper 239, p. 1–20.

Waitt, R.B., 1980, About forty last-glacial Lake Missoula jökulhlaups through southern Washington: Journal of Geology, v. 88, p. 653–679.

Waitt, R.B., 1984, Periodic jökulhlaups from Pleistocene glacial Lake Missoula—new evidence from varved sediment in northern Idaho and Washington: Quaternary Research, v. 22, p. 46–58.

Waitt, R.B., 1985, Case for periodic, colossal jökulhlaups from Pleistocene glacial Lake Missoula: Geological Society of America Bulletin, v. 96, p. 1271–1286.

Waters, A.C., 1961, Stratigraphy and lithologic variations in the Columbia River basalt: American Journal of Science, v. 259, p. 583–611.

Weis, P.L., and Newman, W.L., 1989, The Channeled Scablands of eastern Washington—The geologic story of the Spokane Flood: Cheney, Washington, Eastern Washington University Press, 25 p.

West, M.W., Ashland, F.X., Busacca, A.J., Berger, G.W., and Shaffer, M.E., 1996, Late Quaternary deformation, Saddle Mountains anticline, south-central Washington: Geology, v. 24, p. 1123–1126.

Wilson, J.E., 1998, Terroir: The role of geology, climate, and culture in the making of French wines: London, Mitchell Beazley, 336 p.

Wilson, J.E., 2001, Geology and Wine 4. The origin and odyssey of terroir: Geoscience Canada, v. 28, p. 139–142.

Geological Society of America
Field Guide 4
2003

The Columbia River flood basalts and the Yakima fold belt

Stephen P. Reidel

Pacific Northwest National Laboratory, MS-K6-81, P.O. Box 999, Richland, Washington 99352, USA, and Washington State University–TriCities Campus, Richland, Washington 99352, USA

Barton S. Martin

Department of Geology and Geography, Ohio Wesleyan University, Delaware, Ohio 43015, USA

Heather L. Petcovic

Department of Geosciences, Oregon State University, Corvallis, Oregon 97331, USA

ABSTRACT

This field trip guide covers a two-day trip to examine the characteristics of Columbia River Basalt Group flows and the Yakima fold belt. This field trip focuses on the main physical characteristics of the lavas, compositional variations, and evidence for their emplacement, and on the geometry of the anticlinal ridges and synclinal valleys of the fold belt and deformational features in the basalts.

BACKGROUND

Flood basalt volcanism occurred in the United States' Pacific Northwest between 17.5 and 6 Ma, when over 300 basaltic lavas of the Columbia River Basalt Group were erupted from fissures in eastern Washington, eastern Oregon, and western Idaho (Fig. 1) (Swanson et al., 1979a). These flood basalts cover over 200,000 km^2 of the Pacific Northwest and have an estimated volume of more than 234,000 km^3 (Camp et al., 2003). Concurrent with these massive basalt eruptions was the folding and faulting of the basalt in the western part of the Columbia Basin and development of generally east-west–trending anticlinal ridges and synclinal valleys collectively known as the Yakima fold belt.

Setting

The Columbia River Basalt Group covers much of eastern Washington, northern and eastern Oregon, and western Idaho. Recent studies (e.g., Cummings et al., 2000; Hooper et al., 2002; Camp et al., 2003) have shown the oldest flows occur in southern Oregon and that volcanism progressed northward to the Columbia Basin. The Columbia Basin is the northern part the Columbia River flood-basalt province (Reidel and Hooper, 1989) and forms an intermontane basin between the Cascade Range and the Rocky Mountains (Fig. 1).

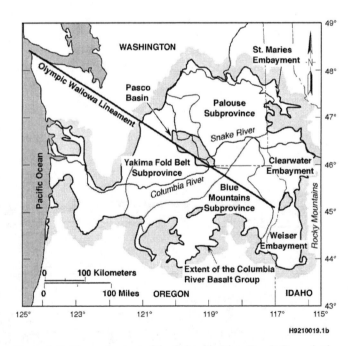

Figure 1. Setting and extent of the Columbia River Basalt Group in the northern portion of the Pacific Northwest Columbia Basin. This map shows the outline of northern portion of the Columbia River flood-basalt province. The southeast Oregon portion is not shown. See Hooper et al. (2002) and Camp et al. (2003) for that portion. The trace of the Olympic-Wallowa lineament is also shown.

Reidel, S.P., Martin, B.S., and Petcovic, H.L., 2003, The Columbia River flood basalts and the Yakima fold belt, *in* Swanson, T.W., ed., Western Cordillera and adjacent areas: Boulder, Colorado, Geological Society of America Field Guide 4, p. 87–105. For permission to copy, contact editing@geosociety.org. © 2003 Geological Society of America.

The Columbia Basin includes two structural subdivisions or subprovinces: the Yakima fold belt and the Palouse Subprovince (Fig. 1). The Yakima fold belt includes the western and central parts of the Columbia Basin and consists of a series of anticlinal ridges and synclinal valleys with northwest to southwest structural trends. The Palouse Slope is the eastern part of the basin and is the least deformed subprovince with only a few faults and low amplitude, long wavelength folds on an otherwise gently westward dipping paleoslope (Swanson et al., 1980).

The Blue Mountains subprovince (Fig. 1) of the Columbia River flood-basalt province forms the southeastern boundary of the Columbia Basin. The Blue Mountains is a northeast-trending anticlinorium that extends 250 km from the Oregon Cascades along the southeastern edge of the Columbia Basin. It overlies accreted terrane rock assemblages and Eocene and Oligocene volcaniclastic rocks.

In the Yakima fold belt portion of the Columbia Basin, 4–5 km of the Columbia River Basalt Group overlies > 6 km of Tertiary continental sedimentary rocks that, in turn, overlie accreted terranes of Mesozoic age (Fig. 2). These rocks are overlain by late Tertiary and Quaternary fluvial and glaciofluvial deposits (Campbell, 1989; Reidel et al., 1989a; Smith et al., 1988; Department of Energy, 1988). In the Palouse subprovince of the Columbia Basin, a thin (<100 m) sedimentary unit separates ~2 km or less of basalt from the crystalline basement, which consists of continental crustal rock typical of that underlying much of western North America (Reidel et al., 1994). The western edge of the late Precambrian-early Paleozoic continental margin and Precambrian North American craton occurs near the juncture of the Palouse and Yakima fold belt subprovinces.

STRATIGRAPHY OF THE COLUMBIA RIVER BASALT GROUP

The Columbia River Basalt Group consists of a thick sequence of ~300 continental tholeiitic flood-basalt flows that were erupted from ca. 17 to 5.5 Ma (Fig. 3; Swanson et al., 1979a). Although the eruption of Columbia River Basalt Group flows spans nearly a 12 m.y. period, the majority of the Columbia River Basalt Group (>96 vol%) was erupted over a period of ~2.5 m.y., between 17 and 14.5 Ma (Swanson et al., 1979a). Individual Columbia River Basalt Group flows typically extend over many tens of thousands of square kilometers. The source for these flows was a series of north-northwest–trending linear fissure systems located in eastern Washington, eastern Oregon, and western Idaho (generally within the Palouse subprovince of Fig. 1).

Detailed study and mapping of the Columbia River flood basalts have demonstrated that there are significant variations in lithological, geochemical, and paleomagnetic polarity properties between flows (and packets of flows), which has allowed for the establishment of stratigraphic units that can be reliably identified and correlated on a regional basis (e.g., Swanson et al., 1979a; Beeson et al., 1985; Reidel et al., 1989b). Based on the ability to recognize the flows, the Columbia River Basalt Group has been divided into five formations (Swanson et al., 1979a): Imnaha, Grande Ronde, Picture Gorge, Wanapum, and Saddle Mountains

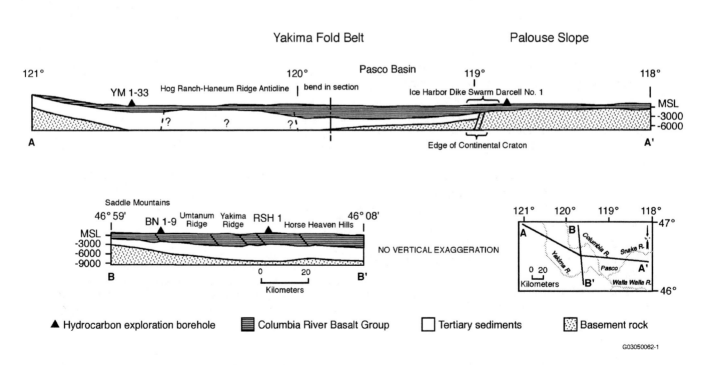

Figure 2. Simplified geologic sections across the Columbia Basin. This diagram shows the subsurface of the Columbia Basin, which is based on deep boreholes that penetrate the basalt and basement and geophysical surveys (see Department of Energy, 1988).

Series	Group	Formation	Member	Isotopic Age (m.y)	Magnetic Polarity
Miocene — Upper	Columbia River Basalt Group — Yakima Basalt Subgroup	Saddle Mountains Basalt	**Lower Monumental Member**	6	N
			Ice-Harbor Member	8.5	
			Basalt of Goose Island		N
			Basalt of Martindale		R
			Basalt of Basin City		N
			Buford Member		R
			Elephant Mountain Member	10.5	N, T
			Pomona Member	12	R
			Esquatzel Member	N	
			Weissenfels Ridge Member		
			Basalt of Slippery Creek		N
			Basalt of Tenmile Creek		N
			Basalt of Lewiston Orchards		N
			Basalt of Cloverland		N
			Asotin Member	13	
			Basalt of Huntzinger		N
			Wilbur Creek Member		
			Basalt of Lapwal		N
			Basalt of Wahluke		N
			Umatilla Member		
			Basalt of Sillusi		N
			Basalt of Umatilla		N
Miocene — Middle		Wanapum Basalt	**Priest Rapids Member**	14.5	
			Basalt of Lolo		R
			Basalt of Rosalia		R
			Roza Member		T, R
			Shumaker Creek Member		N
			Frenchman Springs Member		
			Basalt of Lyons Ferry		N
			Basalt of Sentinel Gap		N
			Basalt of Sand Hollow	15.3	N
			Basalt of Silver Falls		N, E
			Basalt of Ginkgo	15.6	E
			Basalt of Palouse Falls		E
			Eckler Mountain Member		
			Basalt of Dodge		N
			Basalt of Robinette Mountain		N
			Vantage Horizon		
Miocene — Lower		Grande Ronde Basalt	**Member of Sentinel Bluffs**	15.6	N_2
			Member of Slack Canyon		
			Member of Fields Spring		
			Member of Winter Water		
			Member of Umtanum		
			Member of Ortley		
			Member of Armstrong Canyon		
			Member of Meyer Ridge		R_2
			Member of Grouse Creek		
			Member of Wapshilla Ridge		
			Member of Mt. Horrible		
			Member of China Creek		N_1
			Member of Downy Gulch		
			Member of Center Creek		R_1
			Member of Rogersburg		
			Teepee Butte Member		
			Member of Buckhorn Springs	16.5	
		Imnaha Basalt			R_1
					T
					N_0
				17.5	R_0

(within Grande Ronde Basalt column: *Prineville Basalt*, *Picture Gorge Basalt*)

G02050100-1

Figure 3. Stratigraphic nomenclature of the Columbia River Basalt Group. The intercalated sediments of the Ellensburg Formation are not shown.

Basalt (Fig. 3). Flows of the Saddle Mountains Basalt are exposed at the surface, with underlying flows of the Wanapum and Grande Ronde Basalts mainly in the subsurface. Intercalated with, and in some places overlying, the Columbia River Basalt Group are epiclastic and volcaniclastic sedimentary rocks of the Ellensburg Formation (Waters, 1961; Swanson et al., 1979b; Smith et al., 1988). Most volcaniclastic material occurs in the western basin; in the central and eastern basin, epiclastic sediments of the ancestral Clearwater and Columbia Rivers form the dominant lithologies (Fecht et al., 1982, 1987).

Lava Features and Nomenclature

The immense size of Columbia River flood-basalt lavas makes it difficult to determine what constitutes a single eruption and what does not. Historically, Columbia River flood-basalt descriptions have used the term "cooling unit" to describe a lobe. A single cooling unit is defined as having a flow top and a base that show evidence of more rapid cooling or chilling compared to the interiors (Fig. 4). Although many of the surface features observed on recent eruptions of pahoehoe lavas have excellent counterparts on flood basalts (e.g., pahoehoe lobes, tumulus, vesicle sheets), some of the nomenclature that has been applied to those eruptions has proven difficult to adapt to flood basalts (Self et al., 1997). For example, in the terminology of Walker (1972), large single lobes as much as 30 m thick are called simple flows and many similar sized lobes totaling 100 m thick are called compound flows. Self et al. (1998) have attempted to resolve these differences by refining existing terms and introducing new ones. For example, they define a "lobe" as the smallest coherent package of lava, a "flow" as the product of a single outpouring of lava, and a "flow field" as lava covering a large area that has many separate outpourings. The term lobe is easy to apply, but problems still remain in recognizing what constitutes a flow or flow field.

The concept of inflated lava flows has also become important to models describing emplacement of flood-basalt lavas. Hon et al. (1994) described inflated flows from Hawaii, where they documented the growth and inflation of lobes of lava by the internal injection of more lava. With each pulse of new lava, the flow grows thicker. Flows advance by breakouts at the front of the flow. As more lava erupts it causes inflation of flows and breakouts of new lava at the flow front. This mechanism can insulate hot lava great distances from the vent (Self et al., 1997, 1998).

Inflated Columbia River flood-basalt lavas can be recognized by two criteria. The first is the recognition of lobes and breakouts as described by Self et al. (1998). These are equivalent to multiple lobes described above that may or may not be traced back to the main lobe. The second criterion is the recognition of vertical compositional zonation in the lavas. Reidel and Fecht (1987) and (Reidel, 1998) described lavas of the Saddle Mountains Basalt that were inflated lavas over 200 km from the vent areas. They recognized distinct compositions in an inflated lava flow at the Pasco Basin that matched several individual lava flows that occurred at the vent area. The oldest compositions were at the top and bottom of the lava and progressively younger lava compositions occurred toward the center of the flow.

Internal Basalt Flow Features

Intraflow structures are primary internal features or stratified portions of basalt flows exhibiting grossly uniform macroscopic characteristics. These features originate during the emplacement and solidification of each flow and result from variations in cooling rates, degassing, thermal contraction, and interaction with surface water. They are distinct from features formed by tectonic processes.

Columbia River Basalt Group flows typically consist of a flow top, a dense flow interior, and a flow bottom of variable thickness. Figure 4 shows the types of intraflow structures that

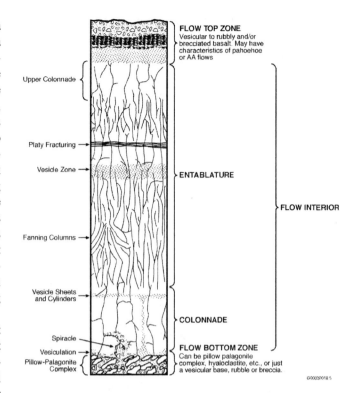

Figure 4. Generalized internal structure of a flood basalt lava flow.

are typically observed in a basalt flow; most flows do not show a complete set of these structures.

The flow top is the chilled, glassy upper crust of the flow. It may consist of vesicular to scoriaceous basalt, displaying either pahoehoe or a'a characteristics, or it may be rubbly to brecciated (Waters, 1961; Diery, 1967; Swanson and Wright, 1981). Typically, the flow top comprises ~10% of the thickness of a flow; however, it can be as thin as a few centimeters or occupy almost the entire flow thickness. Almost all Columbia River Basalt Group flows display pahoehoe features; some have rubbly to brecciated flow tops, but none is considered to represent a'a flows.

Flow top breccia occurs as a zone of angular to subrounded, broken volcanic rock fragments that may or may not be supported by a matrix and is located adjacent to the upper contact of the lava flow. A mixture of vesicular and nonvesicular clasts bound by basaltic glass often characterizes the breccia zone. The percentage of the breccia to rubbly surface is typically less that 30% but locally can be as much as 50% of the flow. This type of flow top usually forms from a cooled top that is broken up and carried along with the lava flow before it ceases movement.

The basal part of a Columbia River basalt lava flow is predominantly a glassy, chilled zone a few centimeters thick that may be vesicular. Where basalt flows encounter bodies of water or saturated sediments, the following features may occur:

Pillow-palagonite complexes. Discontinuous pillow-shaped structures of basalt formed as basalt flows into water. The space between the pillows is usually composed of hydrated basaltic glass (palagonite) and hyaloclastite.

Hyaloclastite complexes. Deposits resembling tuff that form when basalt shatters as it flows into water.

Foreset bedded breccias. These form as basalt flows into water and build out their own delta. Hyaloclastite and pillow-palagonite complexes usually compose the foreset beds.

Peperites. Breccia-like mixture of basalt (or hyaloclastite or palagonite) and sediment that form as basalt burrows into sediments, especially wet sediments.

Spiracles. A fumarolic vent-like feature that forms due to a gaseous explosion in fluid lava flowing over water-saturated soils or ground.

Internal fractures or joints in Columbia River Basalt Group lavas set them apart from smaller more recent lavas. Columbia River Basalt Group lavas have internal jointing that is classed as either entablature or colonnade. Entablature jointing consists of small columns of fine-grained basalt most commonly in the upper parts of the lavas that overlie larger columns of coarser basalt that form the colonnade. Entablature columns are defined by fractures with typical spacings of ~10 cm; fractures that define the colonnades typically have spacings of ~0.5–1 m. Long and Wood (1986) have shown that the entablatures have a quenched texture and represent rapid cooling compared to the colonnade, which developed under comparatively slow cooling. A typical entablature represents top down cooling and the colonnade represents comparatively slower bottom up cooling.

The colonnade consists of relatively well-formed polygonal columns of basalt, usually vertically oriented and typically 1 m in diameter or larger (some as large as 3 m have been observed). Colonnade, as defined by Tomkeieff (1940), occurs in the basal portion of flows. In Columbia River Basalt Group flows, the colonnade can make up the entire flow thickness, or there may be one or more colonnades present that are tiered with entablatures.

Zones or layers of vesicles occur in the interior portions of the lavas and are physically distinct from a vesicular flow top. These vesicle zones or sheets are nearly ubiquitous in the Columbia River basalt lavas. The vesicle zones can range from a few centimeters to as much as several meters in thickness. Typically, they are transitional between massive basalt above and below and are not physical boundaries of cooling units or lobes. The vesicle zones have been interpreted as gas trapped by an advancing solidification front (McMillan et al., 1989) and as distinct pulses of a continuing eruption (Walker et al., 1999).

Vesicle pipes and cylinders are cylindrical zones of gas bubbles that form as gas evolves from that lava and rises toward the top of the flow. Pipes and cylinders are distinguished by their size; cylinders are larger, but there is a continuum of sizes between the two. Vesicle cylinders, pipes, and sheets usually occur in relatively thin flows (5–30 m) composed mainly of colonnades and flow tops.

Vesicle sheets are horizontal to subhorizontal layers of vesicles. They typically are fed by vesicle cylinders and form below the solidification front. Vesicle zones within the interior of thicker flows can be thin (centimeters to meters thick) and can be laterally continuous, sometimes for kilometers.

Vesicle zones are usually thicker than vesicle sheets but probably form in much the same way. Vesicle zones can be up to several meters thick and are typically located in the dense interior of a lava flow.

Lava tubes have not been observed in Columbia River Basalt Group flows. This is because the flows were emplaced as sheets and were not tube fed as Hawaiian flows are. However, locally tube-like features have been observed but typically do not extend great distances.

THE YAKIMA FOLDS AND THE YAKIMA FOLD BELT

The Yakima Fold Belt Subprovince covers ~14,000 km^2 of the western Columbia Basin (Fig. 1) and formed as basalt flows and intercalated sediments were folded and faulted under north-south directed compression. Most of the present structural relief in the Columbia Basin has developed since ca. 10.5 Ma, when the last massive outpouring of lava, the Elephant Mountain Member, buried much of the central Columbia Basin. The main deformation is concentrated in the Yakima fold belt; there is only minor deformation on the Palouse Subprovince. Almost all the present structural relief exposed at the surface is post-Columbia River Basalt Group.

The Yakima fold belt consists of narrow anticlinal ridges separated by broad, synclinal valleys. The anticlines and synclines are typically segmented. Most have a north vergence; however, some anticlines such as the Columbia Hills, and a few segments of some other ridges, have a south vergence. Fold length ranges from 1 km to over 100 km; fold wavelengths range from several kilometers to as much as 20 km. The folds are segmented by crosscutting faults and folds (Reidel, 1984; Reidel et al. 1989b). Structural relief is typically less than 600 m but varies along the length of the fold. The greatest structural relief along the Frenchman Hills, the Saddle Mountains, Umtanum Ridge, and Yakima Ridge occurs where they intersect the north-trending Hog Ranch–Naneum Ridge anticline.

Anticlines in the southwest part of the Yakima fold belt, southwest of the Cle Elum–Wallula deformed zoned (Fig.5), generally have N50°E trends (Swanson et al., 1979b; Reidel et al., 1989b). Anticlines in the central part have east trends except along the Cle Elum–Wallula, where a N50°W trend predominates. The Rattlesnake Hills, Saddle Mountains, and Frenchman Hills have overall east trends, but Yakima Ridge and Umtanum Ridge change eastward from east to N50°W in the Cle Elum–Wallula. The Horse Heaven Hills, the N50°W trending Rattlesnake Hills, and the Columbia Hills abruptly terminate against the Cle Elum–Wallula.

Although rarely exposed, nearly all the steep forelimbs of the asymmetrical anticlines are faults. These frontal fault zones typically consist of imbricated thrusts (Bentley, 1977; Goff, 1981; Bentley, cited in Swanson et al., 1979b; Hagood, 1986; Reidel, 1984, 1988; Anderson, 1987) that are emergent at ground surface. Near the ground surface, the thrust faults merge into the shallow-dipping surface of the basalt (Reidel, 1984). Where erosion pro-

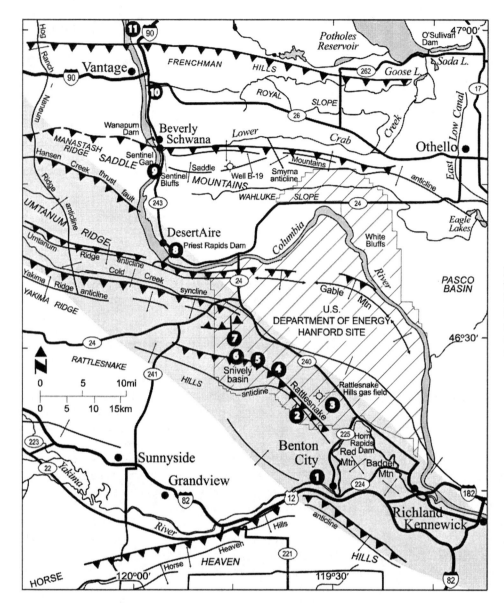

Figure 5. Simplified structure map of the Pasco Basin and eastern Yakima fold belt. The stops for the field trip are shown on the map. Shaded area is the area covered by the Cle Elum–Wallula deformed zoned and Rattlensake-Wallula portions of the Olympic-Wallowa lineament.

vides deeper exposures, these frontal faults are steep reverse faults (e.g., 45°S in the Frenchman Hills at the Columbia River water gap [Grolier and Bingham, 1971], and 50°–70° north in the Columbia Hills at Rock Creek, Washington [Swanson et al., 1979b]).

Hydrocarbon exploration boreholes provide direct evidence for the dips of these frontal faults. Reidel et al. (1989b) have shown that the Saddle Mountains fault must dip more than 60° where the Shell-ARCO 1-9 BN borehole was drilled. Drilling of the Umtanum fault near Priest Rapids Dam (PSPL, 1982) suggests that this fault dips southward under the ridge with a dip of at least 30°–40° (PSPL, 1982) but perhaps as high as 60° (Price, 1982; Price and Watkinson, 1989).

Although it is difficult to assess, total shortening increases from east to west across the Yakima fold belt. At ~120° longitude, it is estimated to be between 15 and 25 km (Reidel et al.,

1989b), or ~5%. Typically, shortening on an individual anticline as a result of folding is ~1–1.5 km. The amount of shortening on faults expressed at the surface is generally unknown. Estimates range from several hundreds of meters to as much as 3 km.

Synclines in the Yakima fold belt are structurally low areas formed between the gently dipping limb of one anticline and the steeply dipping limb of another where that limb was thrust up onto the gently dipping limb of the neighboring anticline. Few synclines within the Yakima fold belt were formed by synclinal folding of the basalt.

Tectonic Brecciation and Shearing

Tectonic breccia and shear zones are common in geologic structures in the Columbia River Basalt Group (Goff, 1981;

Reidel, 1984; Barsotti, 1986). Three types of breccias are recognized: shatter breccias, anastomosing breccias, and shear zone and fault breccias (Price, 1982).

Shatter breccias are simply shattered basalt in which the original primary features of the basalt are still preserved.

Anastomosing breccias are composed of lenticular basalt fragments with a submicroscopic, pulverized basalt matrix, and are nontabular basalt breccias of no apparent measurable orientation.

Shear zones and fault breccias are tabular breccia zones that have three stages of development. The first stage is the development of a set of parallel, sigmoidal, extension fractures superimposed on primary structures. The second stage involves rigid rotation of millimeter-scale basalt blocks, causing the initial granulation of basalt. The third stage involves development of discrete slip surfaces either within or bounding a tabular breccia.

Flow top breccias are distinguished from tectonic breccias by several characteristics. Tectonic breccia typically contains more angular clasts of smaller size, usually a few centimeters or less, than flow top breccia. Clasts in flow top breccia often are bound by original glass and are an admixture of vesicular and nonvesicular basalt, whereas clasts in tectonic breccia have a homogeneous texture. The presence of subparallel fracturing within a tectonic breccia zone results in clasts being arranged parallel to subparallel to each other, which also contrasts with the random, chaotic nature of clasts in flow top breccias. Slickensides are present on some surfaces in tectonic breccias and absent in flow top breccias without tectonic fracturing. Tectonic breccias typically display a crushed basalt matrix, while flow top breccias may be partially to fully filled with secondary minerals or palagonite between fragments, or the fragments may be welded together.

Major high-angle, reverse to thrust faults along anticlinal ridges are associated with very thick breccia zones. In the Saddle Mountains, these zones are very distinct and at Sentinel Gap consist of a several-hundred-meter–thick zone of shatter breccias (Reidel, 1984). Similar breccia zones have been found in Umtanum Ridge (Price, 1982; Barsotti, 1986) and at Wallula Gap.

ROAD LOG

Day 1. Benton City and Rattlesnake Mountain

The guidebook route begins at the Research and Operations Building parking lot at the Pacific Northwest National Laboratory on Battelle Blvd., Richland. All stops in this field trip guide are keyed to a field trip route map (Fig. 5). Mileage is given to the nearest tenth of a mile and kilometer and is cumulative. Because there are several optional side trips, mileage for each optional trip begins at 0.0 where the trip leaves the primary trip route.

Cumulative Miles	(km)	Description
0.0	(0.0)	Start at the Pacific Northwest National Laboratory Research and Operations Building parking lot. Turn right (west) on Battelle Blvd.

Cumulative Miles	(km)	Description
0.2	(0.3)	Turn left on Stevens Drive toward Richland.
2.6	(4.2)	Intersection of Stevens Drive and State Route (SR) 240. Bear right toward Pasco on SR 240.
6.6	(10.6)	Exit SR 240 at entrance Interstate 182 and head west toward Yakima.
9.5	(15.2)	Road cut on I-182. This road cut exposed the upper thrust fault along Badger Mountain anticline. The northeast part of the outcrop consists of vertical sedimentary beds of the Ellensburg Formation, probably of Mabton age (sediment between Wanapum Basalt and Saddle Mountains Basalt, Fig. 3). The south part of the road cut consists of south-dipping Saddle Mountains Basalt layers. A small thrust fault developed on the surface of the Pomona Member (12 Ma) that repeats the Esquatzel and Pomona Members. The thrust fault can be traced through the foundation of the house above and along the north side of Badger Mountain. The main thrust fault occurs at the base of the hill and is not exposed.
10.1	(16.2)	Intersection of I-182 and I-82. Bear west toward Yakima.
12.1	(19.4)	Optional Stop.

Optional Stop

View of Rattlesnake Mountain and Red Mountain and the Rattlensake-Wallula portion of the Olympic-Wallowa lineament. I-182 is paralleling the Rattlensake-Wallula section of the range. To the right are doubly-plunging anticlines of the "Rattles" that are aligned with Rattlesnake Mountain. To the left are the Horse Heaven Hills and the abrupt change from the northeast-trending segment to the northwest-trending segment. Also, at this locality one can see a profile of Red Mountain (closer) and Rattlesnake Mountain that show the overall geometry of the anticlines. Note the uniform south slope of Rattlesnake Mountain. This slope continues to within 2 km of the crest of the fold, where the dip increases abruptly. Long uniform slopes like this are typical of many of the Yakima folds, like the Horse Heaven Hills.

Cumulative Miles	(km)	Description
16.0	(25.6)	Benton City exit (exit 96). Turn off I-82.
16.4	(26.2)	Turn right on SR 225 (Horn Road) and follow signs to Benton City.
18.3	(29.3)	Turn left on Hazel (Old Inland Empire Highway). Follow Old Inland Empire Highway to Whan Road.
19.7	(31.5)	Pull off on the left side of the road in the broad spot just beyond Whan Road. Follow the trail down the hillside to the old Union Pacific Railroad grade. The stop consists of walking several hundred meters along the grade and examining the basalt flows exposed there.

Stop 1. Contact between Ward Gap and Yakima Bluffs Lava Flows of the Elephant Mountain Member

The Elephant Mountain Member is the last of the large-volume flood-basalt lavas, and is the youngest eruption in this area. It erupted from a vent system near the Washington-Idaho border and flowed down the ancestral Salmon-Clearwater River to the Pasco Basin, where it spread out. The member consists of two flows: the older Yakima Bluffs flow and the younger Ward Gap flow. Together they cover ~13,500 km^2 and have a total volume of ~440 km^3. Across the Yakima River (south) is the Horse Heaven Hills; the Elephant Mountain Member forms the top lava flow up to the base of the ridge and the top flow on the anticline nearly 2000 ft higher.

The exposure in the railroad grade provides an opportunity to examine the contact between two flood-basalt lavas, the Ward Gap flow (above) and the Yakima Bluffs flow (below). The starting point is the exposure just east of the trail down to the grade. Here the massive colonnade of the Ward Gap flow overlies a flow-top breccia of the Yakima Bluffs flow. As one follows the grade to the east, the flow top breccia becomes more vesicular and slabby. Continuing eastward, several tumuli up to 1 m high are found as well as several faulted tumuli. In places, the base of the Ward Gap flow has vesicle pipes that are bent in the direction of flow (to the west) and disappear into the flow above.

As you walk this exposure, look for evidence indicating the time between eruptions. This may be a key to the rates at which these lavas were emplaced. Although there are two lavas present, try to locate where the lower flow ends and the upper flow begins.

Cumulative Miles	(km)	Description
21.1	(33.8)	Return to Horn Road. Turn left on Horn Road. Between the road intersection and the turnoff to Rattlesnake Mountain, the Pomona Member is exposed at river level on the right; above the Yakima River on the opposite bank, the Ice Harbor Member overlies the Elephant Mountain Member, which, in turn, overlies the Pomona Member. These lavas have not been deformed where they cross the Rattlesnake-Rattles structural trend. Red Mountain is the next anticlinal ridge to the southeast.
28.9	(46.2)	Turn left onto Rattlesnake Mt. Road.
29.2	(46.7)	Go through locked gate (gate 106) onto the Hanford Reach National Monument. The road follows the surface of the Elephant Mountains Member, which forms a gentle dip slope into the Cold Creek syncline.
31.6	(50.6)	Note the landslide at one o'clock along the north side of Rattlesnake Mountain. This landslide occurred after the last ice-age flood (post 13 ka). The basis for this age determination is that there are no sediments from the last major flooding of the basin on the landslide.

This suggests that the landslide is younger than the ice-age floods, otherwise flood sediments would be expected to be found on the landslide. The landslide was probably triggered when the floodwaters removed basalt from along the base of Rattlesnake Mountain, allowing the basalt and sediment in the landslide area to slide. The slide plane is the Mabton interbed, which is ~50 ft thick in the area. The western edge of the landslide is marked by the point where the upper and lower faults along the higher parts of Rattlesnake Mountain merge into one fault zone.

32.4	(51.8)	The road forks; turn left.
33.9	(54.2)	Cross the approximate location of the Rattlesnake Mountain fault.
34.1	(54.6)	The road bends to the right; you are now climbing up the north side of Rattlesnake Mountain; the route is on basalt of the Pomona Member (12 Ma); some colluvium composed of basalt of the Elephant Mountains Member is also present.
35.3	(56.5)	Begin to climb up the plunging anticline that forms this part of Rattlesnake Mountain. The route follows the dip slope of the Pomona Member and is expressed by the topography. Basalts of the Ice Harbor and Elephant Mountain Members are farther down the south slope. The Ice Harbor Member pinches out farther down the flank of the ridge, but the Elephant Mountain Member covered the mountain when it erupted. The highest elevation of the Ice Harbor basalt is 1200 ft. This indicates that ~2400 ft of uplift has occurred on Rattlesnake Mountain since ca. 8.5 Ma, the age of the Ice Harbor Member. Mount Adams can be seen in the distance at approximately ten o'clock.
36.6	(58.6)	Cross the highest point on the landslide.
37.1	(59.4)	Radio telescope observatory.
38.0	(60.8)	Pull over by the radio towers and support buildings.

Stop 2. Overview of the Eastern Part of the Yakima Fold Belt from the Crest of Rattlesnake Mountain

This stop provides an overview of the central Columbia Basin and the Pasco Basin (Fig. 5) The view to the north allows one to see the main elements of the Palouse Subprovince (to the NE) and the Yakima fold belt (N and NW) and contrast their main features. The Palouse Subprovince is marked by very gentle dips (1°) to the east-southeast into the area near Ice Harbor Dam, where dikes that fed the basalts of the Ice Harbor Member 8.5 Ma are found. Structural dips increase to several degrees westward from the Ice Harbor dike swarm into the Pasco Basin (Fig. 2).

The Ice Harbor dikes generally mark the boundary between the Palouse Subprovince and the Yakima fold belt, which also marks the suture zone between the old continental craton (to the east) and the accreted terranes to the west (Fig. 2) (Reidel et al., 1989b, 1994). The Ice Harbor dikes are the westernmost known dikes of the Columbia River Basalt Group (excluding the Picture Gorge Basalt). The Pasco Basin lies within a graben that underlies almost the entire Yakima fold belt. This graben has been subsiding since Eocene time. The Hanford site lies within the part of the graben where the basalts and underlying sediments are thickest. The thick sequence of basalt and sediments indicates that subsidence has been occurring over a long period of time. The area continues to subside but at a very slow rate.

This vantage point also provides a good location to see the typical geometry of a Yakima fold. Throughout much of the Columbia Basin, the topography is a direct reflection of the geometry of geologic structures because very little erosion has occurred. As we discuss geologic structural features, much of what we describe at a distance can be seen in the topographic shape.

Note that the north side of Rattlesnake Mountain is very steep, while the south side has a very gentle slope. Also note the very abrupt change from a gentle south dip on the back slope to a steeper dip near the crest. This geometry is apparent on Rattlesnake Mountain and the Horse Heaven Hills (to the south) as well as other Yakima folds. From our viewpoint, we can see some of the doubly plunging anticlines that lie along the crest of the Horse Heaven Hills. These anticlines tend to be east-west oriented and en echelon to the overall northeast trend of the Horse Heaven Hills.

The view to the south and southwest allows us to see the Blue Mounatins anticlinorium. The Blue Mountains is one of the major mountain ranges of the Pacific Northwest that lies between the Cascade Range and the Rocky Mountains; it marks the southern boundary of the Palouse Subprovince and the Yakima fold belt.

Southeast of Rattlesnake Mountain, the Rattlesnake Mountain structural trend continues to Wallula Gap, where the Columbia River exits the Pasco Basin. This trend is blocked from our view at this locality but can easily be seen as we drive back down off Rattlesnake Mountain. The trend is marked by a series of doubly plunging anticlines that have been called the "rattles." The rattles and Rattlesnake Mountain form part of an alignment of topographic features called the Olympic-Wallowa lineament. This alignment of topographic features can be traced from the Olympic Mountains of Washington to the Wallowa Mountains of northeast Oregon. This portion of the alignment is called the Rattlesnake-Wallula alignment, which is a portion of the Cle Elum–Wallula deformed zone (Fig. 5).

The view to the southwest shows the continuation of the Horse Heaven Hills anticline as a topographic high on the horizon. The Horse Heaven Hills anticline can be traced as far as Portland, Oregon (Beeson et al., 1989). The highest elevations on the Horse Heaven Hills occur just north of Goldendale, Washington, and are probably related to the Simcoe volcanic field.

To the northwest, the Saddle Mountains and Sentinel Gap, a water gap cut through the mountain by the Columbia River, can be seen. Beyond Sentinel Gap is the Hog Ranch–Naneum Ridge anticline, which forms the western boundary of the Pasco Basin. This major anticlinal feature can be traced from the Mount Stuart area of the North Cascade Mountains onto the Columbia Basin. On clear spring days, Mount Stuart and Glacier Peak can be seen from our viewpoint.

Cumulative Miles	(km)	Description
38.5	(61.6)	View of doubly plunging anticlines along the Rattlesnake-Wallula, Badger Coulee, bend in the Horse Heaven Hills, Wallula Gap, and the Blue Mountains.
43.3	(69.3)	"Y" in road; bear left.
44.6	(71.4)	Intersection—go straight. Leave pavement. The road traverses north-dipping Elephant Mountain Member.
45.1	(72.2)	Stop 3.

Stop 3. View of Rattlesnake Mountain and Rattlesnake Mountain Fault

This locality provides an excellent view of the north flank of one of the largest of the Yakima anticlines, Rattlesnake Mountain (Fig. 6). All basalt flows encountered in boreholes at Hanford thin onto Rattlesnake Mountain; we interpret this to mean that Rattlesnake Mountain was growing during the eruption of the Columbia River basalt. The present relief developed in the past 10.5 m.y., but geologic and geophysical data show that even more structural relief on Rattlesnake Mountain is buried by the younger flows (Reidel et al., 1989b).

The prominent bench just below the crest marks the Wanapum–Saddle Mountains Basalt contact and is erosional. A sedimentary unit, the Mabton interbed, which has eroded back into the hillside, marks the contact.

The main fault runs along the base of the mountain, but sediments cover the fault. The fault dies out to the southeast before reaching the Yakima River. Basalt flows along the north side of Rattlesnake Mountain dip between 50° and 70° to the north. The second bench below the upper one marks an upper thrust fault, placing gently south-dipping basalt flows above the steeply north-dipping ones. The anticlinal axis has been thrust over and eroded from the present exposures. The upper thrust dies out to the southeast and northwest and is responsible for the greater structural relief along this part of Rattlesnake Mountain. This is typical of most Yakima folds where the ridge is segmented and each segment is defined by a distinct structural style. The structurally higher segments are usually the result of the development of a second thrust fault where the basalt ramps up onto steeply dipping basalt.

Cumulative Miles	(km)	Description
46.7	(74.7)	Leave pavement and gravel; follow dirt road. The old Rattlesnake gas field is to the left

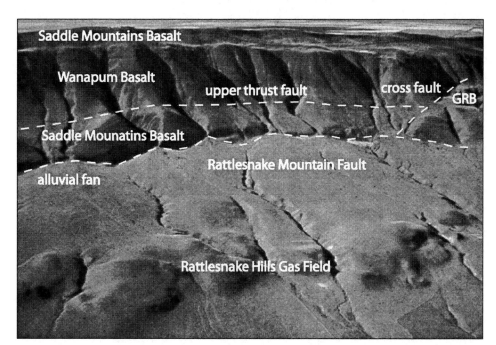

Figure 6. Geologic features of the northeastern face of Rattlesnake Mountain (see Fig. 5 for location).

at nine o'clock. The small hill 1 mi north of Rattlesnake Mountain is the location of the Rattlesnake gas field, which produced natural gas sold to local towns in the 1920s and 1930s. Although the Rattlesnake gas field was a very small gas field (the amount of natural gas produced would supply the present city of Richland for only one year), it played an important role in the history of oil and gas exploration in the Columbia Basin. The small hill is an anticline, like Rattlesnake Mountain only much smaller, and natural gas was found between the lava flows; probably most of the methane is at the top of the Priest Rapids Member. Natural gas occurred in a structural trap formed by the anticline.

The first deep hydrocarbon exploration well, RSH-1, was drilled to a depth of 10,660 ft in the 1950s on the Rattlesnake Hills south of here and was still in Columbia River basalt (Reidel et al., 1982). No significant quantity of gas was discovered in that borehole. Geophysical and basalt geochemical data indicate that there is at least another 1000 ft of basalt or more, making a total thickness of at least 12,000 ft.

Cumulative Miles	(km)	Description
47.7	(76.3)	Old gas well site at nine o'clock.
51.0	(81.6)	East end of the Snively Basin at nine o'clock.
52.5	(84.0)	Road intersection. Go straight off main road onto side road that goes up east side of alluvial fan.

52.9 (84.6) Stop at bend in road and walk to knoll that overlooks valley and alluvial fan, and then walk to the outcrops (Fig. 7).

Stop 4. Structure of Snively Basin Area at Bobcat Canyon

This portion of the Snively Basin is a complex geologic area that marks the intersection of the NW-trending Rattlesnake Mountain trend and the east-trending Rattlesnake Hills trend (Fig. 5). There are several thrust faults exposed in this area. The exposures of basalt above the alluvial fan are Saddle Mountains Basalt flows that strike to the northwest and dip ~50° to the northeast. Lying above these NE-dipping basalt flows are nearly horizontal Saddle Mountains Basalt flows that have been thrust up onto the lower flows.

This area shows the relationship between the northwest-striking structure of Rattlesnake Mountain (Olympic-Wallowa lineament [Cle Elum–Wallula deformed zone, Rattlesnake-Wallula] trend) and the E-W–striking structure of the Rattlesnake Hills. The structurally lower NW-striking basalt flows are some of the northwestern-most exposures of the Rattlesnake Mountain anticline. The structure has lost most of its structural relief by this point and appears to not extend beyond this exposure. The flows above the thrust fault are part of the E-W–striking Rattlesnake Hills structure. The E-W trend does not appear to extend any farther east than this area. This locality is interpreted to be where the E-W–striking Rattlesnake Hills have been thrust up on to the NW-trending Rattlesnake Mountain structure. The debris that lies in the north end of Snively Basin is interpreted to be "tectonic landslide" debris from the landsliding of basalt flows that were oversteepened as they were thrust over the top of the NW-trending Rattlesnake Mountain structure.

Figure 7. The geology at Bobcat Canyon, Snively Basin complex. Bobcat Canyon is an erosional valley near the intersection of the northwest Rattlesnake Mountain trend and the east-west Rattlesnake Hills trend. Tem—Elephant Mountain Member; Tp—Pomona Member; Telr—Rattlesnake Ridge sediment of the Ellensburg Formation.

There are several springs, including Benson Spring, that emanate from this area. These springs are coming from the top of the Priest Rapids Member flows. The several thrusted blocks in this area have repeated the exposures of the Priest Rapids Member, resulting in the creation of several springs flowing from the same repeated horizon. Up the creek to the east there is a tear fault that also has a small spring.

Return to the road intersection.

Cumulative Miles	(km)	Description
53.3	(85.3)	Turn left at the intersection and continue west on the road along the north side of the Snively Basin complex.
54.1	(86.6)	Road intersection; turn left at the road intersection. Continue following the road along the base of the ridge; this road follows approximate position of the frontal fault.
58.2	(93.1)	Park at wide spot where the road crosses gully bottom.

Stop 5. Snively Basin Frontal Fault Zone

At this locality, walk up the gully to the south of the road and examine faulted basalts, Ringold units (3.5 Ma), and Pliocene-Pleistocene deposits.

This locality is a good place to observe beds along the frontal fault zone; similar structures are found along many of the Yakima folds. Walking up the creek from the road, one first passes through caliche-covered colluvium and alluvial fan deposits of Ringold age (8.5–3.0 Ma). Thrust up on to these deposits is the basalt of the Elephant Mountain Member. The basalt is dipping ~50° to the south. Farther up the creek, resting on the Elephant Mountain Member, is a packet of sediments that is equivalent to the upper Ringold. These sediments are similar to other sediments in the area that are thought to be the upper part of the Ringold Formation. Thrust up on this sediment package is a repeat of the Elephant Mountain Member. The sediments in contact with the repeated Elephant Mountain Member at the upper end of the creek contain some ash, suggesting that the Rattlesnake Ridge interbed comprises some of these sediments. Farther up the creek and lying on the Elephant Mountain Member is Pomona Member.

In spite of the folding associated with the faulting, the basalt is surprisingly intact and unbrecciated. This is very typical of many fault zones although other frontal fault zones are brecciated and internal structure in the basalt flows is often unrecognizable. The basalt flows that remain intact are usually those that are in the Saddle Mountains Basalt and are interbedded with sediments. The Grande Ronde Basalt, which lacks interbedded sediments, is usually so brecciated that the internal features of the basalt cannot be recognized.

Cumulative Miles	(km)	Description
60.2	(96.3)	Intersection; go straight. The route now is over an alluvial fan complex. Springs fed from confined aquifers in Snively Basin flow into the fan and below ground. We believe that these springs make a significant contribution to the confined aquifer system at Hanford. The light-colored slope at nine o'clock is an exposure of Ringold muds similar to

those at previous stop. These muds are over-thrusted by the Columbia River basalt in the Snively Basin complex.

60.6	(97.0)	Gully on right. Colluvial debris flow deposits of the Snively fan are exposed on right. Ravine road is following exposures through core of anticline; basalts are dipping to north.
61.5	(98.4)	Stop 6.

Stop 6. Thrust faults of Snively Basin

Crossing the core of an anticline, basalts are of the Priest Rapids Member. At this locality, the basalts are tightly folded, and the road crosses the axis of a northwest trending anticline associated with the Snively Basin complex.

Cumulative Miles	(km)	Description
61.6	(98.6)	Cross creek. The water for this creek emanates from a series of springs farther up the creek. The first spring comes out of the flow top of the Priest Rapids Member exposed in the core of this anticline.
61.7	(98.7)	Optional Stop.

Optional Stop. Umatilla Member and Small Thrust Fault

Follow the road past the creek to the exposures of the Umatilla Member along the right side of the road. The Umatilla Member is a very fine-grained to glassy basalt flow. Throughout much of the area, a glassy entablature with many small cooling joints forms most of the flow. This can be observed on the east wall of the valley ~50 ft above the road. At road level, the flow is fine-to-medium grained and has columnar joins that are more widely spaced.

A small thrust fault cuts through the basalt of the Umatilla Member and is exposed on the right side of the road at road level. The faults dip ~15° to the south. Here, the thrust fault is marked by a small, several-centimeter-wide zone of basalt that has been fractured and brecciated. Many small clasts can be seen in the small fault zone. The relative sense of movement is for the upper block to have moved north with respect to the lower block.

Return to vehicles and turn around and retrace route to previous intersection.

Cumulative Miles	(km)	Description
63.2	(101.1)	Intersection; turn left.
64.1	(102.6)	At eleven o'clock see tear fault cutting through Yakima Ridge. We will see this at Stop 7.
65.3	(104.5)	Intersection; continue straight ahead.
66.0	(105.6)	Road to the right goes to the Benson Ranch site; follow road to the left.
66.1	(105.8)	Cross Dry Creek and turn left at the intersection immediately after crossing the creek. Just

past intersection is a fork in the road; take left fork. Dry Creek is filled with water year round. A U.S. Geological Survey gauging station is located where we cross the creek (upstream side). This station has shown that the water flow is relatively constant. The flow emanates ~2 km up the valley and disappears underground 3 km downstream to the east.

67.3	(107.7)	Stop 7.

Stop 7. Southern Block of Yakima Ridge

Yakima Ridge is an anticlinal ridge that forms the southern boundary of the Cold Creek syncline. It is one of the more complex anticlinal ridges in that it has a southern block that is separated from the main ridge trend. This southern block is a narrow (1 km), 5-km-long asymmetrical eroded anticline with a south vergence and a buried thrust fault on the south side.

This stop provides an opportunity to examine a small N-S–trending tear fault that cuts the southern block of Yakima Ridge (Fig. 5). The tear fault has ~500 m of sinistral movement. On the west side of the fault are simple south-dipping beds of the Saddle Mountains Basalt; here, the fold has been eroded away, exposing only the north flank. On the east side of the tear fault, an upper thrust fault is exposed in the Umatilla Member; north of the thrust fault are gently north-dipping units, but south of the thrust are steeply south-dipping units. The change in dip is controlled by the thrust fault and not an anticlinal axis. The west side is thrust 500 m farther south than the east side; the beds on the east side have been eroded away from the west side, giving an apparently different geometry when in fact, they are the same.

This ridge is similar to other anticlinal ridges such as in the Snively Basin area. The buried thrust fault along the southern block of Yakima Ridge probably places Umatilla or Priest Rapids basalt flows over basalt of the Elephant Mountain Member. Rattlesnake Springs occurs ~0.3 km east and downstream of the tear fault. One probable steady source of water for Rattlesnake Springs is the confined aquifer of the Priest Rapids Member where the thrust fault along the southern block of Yakima Ridge brought it near the surface. We suggest that where the tear fault and the thrust fault meet, a more permeable groundwater flow path was formed, accounting for the location of Rattlesnake Springs.

Quartzitic gravels containing quartzo-feldspathic matrix sands are exposed along the road and down the ravine to the north. These gravels directly overlie basalt of the Elephant Mountain Member and are similar to cores from the Ringold Formation taken from the Cold Creek syncline just a few kilometers to the east and the Snipes Mountain conglomerates.

Immediately prior to the beginning of, and during, the earliest stages of Ringold deposition, the Columbia River is interpreted to have followed a course across the easternmost end of Yakima Ridge (Fecht et al., 1987). During this period, the Columbia River exited the Pasco Basin in the area of Sunnyside Gap. The Ringold-like gravels exposed in the area of this stop are inferred to have been deposited by the Columbia River immediately prior to when it shifted position into the central part of the Pasco Basin.

Cumulative Miles	(km)	Description
67.7	(108.3)	Intersection at Dry Creek crossing. Continue straight ahead.
69.2	(110.7)	Cross the buried trace of the Yakima Ridge anticline. At this point, most of the structure has decreased, but an ancestral course of the Columbia River, ca. 13 Ma, probably eroded away much of the crest of the anticline (Fecht et al., 1987).
70.0	(112.0)	Gate 118 on SR 240; turn right and return to Richland. End of first day's road log.

Day 2. Richland to Vantage

Mileage begins on the north side of Richland at the stoplight on the corner of the Bypass Highway (SR 240) and Stevens Drive. This is the beginning of the portion of SR 240 that leads to Vantage, Washington.

Cumulative Miles	(km)	Description
0.0	(0.0)	Intersection of Bypass Highway (SR 240) and Stevens Drive. Turn west onto SR 240 toward Vantage.
8.0	(12.8)	The Horn Rapids Dam on the Yakima River is at nine o'clock. The river eroded a channel across the nose of a small north-plunging anticline and exposed the Ice Harbor, Elephant Mountain, and Pomona Members. This structure continues north across the road, where the gently northeast-dipping Elephant Mountain Member can be seen projecting above the ice-age flood gravels.
12.1	(19.4)	Milepost 17. For the next 1.2 mi, boulders left by the ice-age floods are visible on the right (east) side of the road. As floodwater that had been temporarily dammed at Wallula Gap began to drain, icebergs carrying boulders and smaller fragments of granite, basalt, and other rock types were left stranded.
21.9	(35.0)	At eleven o'clock, Snively Basin can be seen between the Rattlesnake Hills and Rattlesnake Mountain. Ahead you can see the southeast bend in the Yakima Ridge anticline. The fold plunges beneath basin fill (flood sediment and dune sand). Also ahead are several small basalt outcrops of the Elephant Mountain Member at road level along the south side of the road. These are at the crest of the buried Yakima Ridge anticline.

Cumulative Miles	(km)	Description
28.7	(45.9)	Climb onto the Cold Creek bar. Huge gravel bars lie on both sides of the Columbia River where the floodwater spilled out of Sentinel Gap, a narrow gorge in the Saddle Mountains to the north.
29.4	(47.0)	Junction with SR 24 (west) to Yakima and the "Yakima Barricade" (east); entrance to the Hanford site. Continue north on SR 24 east. The ridge at twelve o'clock is the Saddle Mountains anticline. A Shell Oil Company well site is at twelve-thirty. The well, drilled to a total depth of more than 17,000 ft, penetrated 11,500 ft of Columbia River basalt and passed into pre-basalt sedimentary rocks here (Reidel et al., 1989b).
31.2	(49.9)	At eleven o'clock is Sentinel Gap, a water gap cut by the Columbia River as the Saddle Mountains were uplifted.
33.0	(52.8)	At nine o'clock is a good view along the north limb of the Umtanum Ridge anticline. Here, flows of the Grande Ronde and Wanapum Basalts dip ~70° to the north. Farther west, the Grande Ronde Basalt is overturned and dips steeply to the south. To the east is Gable Butte, an en echelon anticline along Umtanum Ridge. Ice-age floodwaters eroded the crest of the anticline, producing the small valley visible in the distance.
34.4	(55.0)	Vernita Bridge over the Columbia River. This is the last free-flowing segment of the Columbia River in Washington. It is called the Hanford Reach and is part of the Hanford Reach National Monument.
43.6	(69.8)	Turn left at road to Priest Rapids Dam.
44.2	(70.7)	Stop 8.

Stop 8. Overview of Umtanum Ridge Anticline

Umtanum Ridge extends ~110 km from near the western margin of the Columbia Basin to the Palouse Slope (Fig. 5). The structural relief gradually decreases eastward, where it becomes a series of en echelon anticlines developed along the dying ridge. In the Priest Rapids Dam area, the north limb is overturned and dips 40° to the south. An upper thrust, the Buck thrust, and a lower thrust, the Umtanum thrust, define the overturned portion of the fold (Price and Watkinson, 1989). The Buck thrust merges with the Umtanum thrust fault to the east as the overturned portion becomes steeply dipping to the north. Drilling has constrained the fault to between 30° and 60° to the south.

To the west is a good view of the thrust faults of the Filey Road area. Many of the thrust faults are partially concealed by landslides, fanglomerates, and loess, but at least one is visible. The thrust faults place the Priest Rapids Member and older rocks onto flows of the Saddle Mountains and Wanapum basalts. Flood gravels near the river level mask the bedrock geology.

Cumulative Miles	(km)	Description
44.8	(71.7)	Return to SR 243 and continue north (left).
49.4	(79.0)	Road 26 SW. At twelve o'clock is an excellent view of Sentinel Gap. The antecedent Columbia River maintained its course as the Saddle Mountains were uplifted, cutting down through this ridge at Sentinel Gap. At nine o'clock, the Hansen Creek thrust fault extends to the west shore of the Columbia River. Ice-age flood bar gravels on this side of the river cover it. This thrust places Priest Rapids and Roza basalt flows over the Priest Rapids flows. The Asotin Member, which crops out at lower elevations of the Saddle Mountains anticline at ten o'clock, fills a former channel of the Columbia River through Sentinel Gap.
53.3	(85.3)	At one o'clock, along the right (east) side of Sentinel Gap, the Frenchman Springs, Roza, and Priest Rapids Members form the upper cliffs. The bench below the upper cliffs is the result of erosion along an interbed, the Vantage Member of the Ellensburg Formation. The lower cliffs here are flows of the Grande Ronde Basalt. At eleven o'clock, west of Sentinel Gap, the white patch high on the cliff face is the Vantage interbed. Frenchman Springs Member flows lie above it and Grande Ronde Basalt below.
54.0	(86.4)	The road to the right leads to a quarry exposing an interbed of the Ellensburg Formation. (The name "Beverly" was formerly applied to this interbed. The name is no longer used because this unit is a composite of several interbeds; the intervening flows are not present.) This quarry is a good place to see this 180-ft-thick interbed. The lower part contains conglomerate deposited by the ancestral Columbia River, and the upper part is made up of poorly indurated siltstone, sandstone, and tuff.
55.4	(88.6)	Milepost 18.

Stop 9. Member of Sentinel Bluffs, Grande Ronde Basalt

Along the walls of Sentinel Gap are excellent exposures of Grande Ronde Basalt, the Vantage interbed, and the upper part of the Frenchman Springs Member, as well as a good cross-sectional look at the Saddle Mountains. A large thrust fault at the base of the north flank of the Saddle Mountains places Grande Ronde Basalt on top of the Priest Rapids Member. Horizontal shortening is at least several kilometers. The fault cuts across the Columbia River somewhere between here and the town of Schwana. In addition, a north-trending right-lateral strike-slip fault cuts through the ridge at the gap (Reidel, 1984, 1988).

At this stop we will examine lava flows of the member of Sentinel Bluffs. The Grande Ronde Basalt has been divided into 17 units (Reidel et al., 1989a). The member of Sentinel Bluffs, the youngest of the Grande Ronde units, consists of flows that have high magnesium contents relative to other Grande Ronde units; the lavas also have normal polarity. The eruption of the member of Sentinel Bluffs lavas marked the end of Grande Ronde Basalt volcanism and the end of the greatest period of Columbia River Basalt Group volcanism. The member of Sentinel Bluffs lavas erupted from a northerly trending vent system in eastern Washington and northern Oregon and flowed westward down an ancestral paleoslope covering over 169,700 km^2 of the flood-basalt province and producing over 10,000 km^3 of lava. The member of Sentinel Bluffs is divided into six eruptions that are distinguished by their compositions. The first eruption was the most voluminous and reached the Pacific Ocean. The volume of basalt declined with later eruptions until the final eruption that produced the second largest volume of basalt.

Compositional variation in individual lavas is relatively small, both vertically through each basalt eruption and over their areal extents. This homogeneity allows the individual eruptions to be recognized throughout the province. One exception is in the central part of the province, where four of the eruptions combined to form one local lava, the "Cohassett flow." The Cohassett formed as one eruption after another inflated the first lava to form a new one that has a compositional zonation that reflects the sequence of eruptions. The compositions of the original lavas remained intact except for mixing where they came in contact. Thin zones of vesicles separate relatively uniform compositional zones. A thick vesicle zone, called the "interior vesicular zone," marks the boundary between the last two eruptions.

There are two compositional trends in the eruptions of the member of Sentinel Bluffs that are defined best by TiO_2 and P_2O_5. The six eruptions fall along one or the other trend, but neither trend is defined by timing of eruptions, nor by location along the vent system. The first and last eruptions follow one trend, and the eruptions between follow the other trend. Both trends represent decreasing TiO_2 and P_2O_5 with time.

Cumulative Miles	(km)	Description
55.9	(89.4)	Well-developed columns in the lowest flow of the member of Sentinel Bluffs are exposed here. This is the colonnade portion of the flow. Entablatures of the Grande Ronde flows are hackly or broken and were therefore easily removed by flood action.
57.4	(91.8)	Entering Schwana. Crab Creek floodway can be easily viewed to the east. Crab Creek acted as an overflow channel as floodwaters farther up the Columbia spilled out of the channel and spread across the Columbia Basin. Some of this water scoured out the Crab Creek drainage as it returned to the Columbia River

here. To the southeast, the thrust fault (or a high-angle reverse fault) has formed a very steep cliff along the north flank of the Saddle Mountains. To the west, fault breccia from the Saddle Mountains thrust fault is barely visible across the river.

58.6 (93.8) Town of Beverly and Crab Creek Road at three o'clock.

59.1 (94.6) At two o'clock, the flows are the Frenchman Springs (above), Roza, and Priest Rapids Members. At nine o'clock, across the river, is the same sequence.

62.0 (99.2) SR 243 lies on Frenchman Springs flows for the next 0.6 mi. The Roza Member forms cliffs above the highway level.

63.3 (101.3) Well-developed columns in the colonnade of the Roza Member can be seen at two o'clock. Ice-age floodwater scouring removed much of the entablature here. This is a good place to search for phenocrysts in basalt. The Roza has several hundred 2–5 mm phenocrysts per square meter of surface area; this is a distinctive characteristic of this flow. (In contrast, most Frenchman Springs flows have fewer phenocrysts, and Priest Rapids flows contain almost no phenocrysts.)

64.7 (103.5) Basalt of the Frenchman Springs Member (Basalt of Sand Hollow). Note the white "sand" clinging to the sides of the roadcut. This is tephra from the May 1980 Mount St. Helens eruption.

65.1 (104.2) View of Vantage, I-90 bridge, and the Columbia River. The town of Vantage rests on the Museum flow (member of Sentinel Bluffs). The Vantage interbed forms the bench along the river here; the Frenchman Springs and Roza flows lie above that level.

65.8 (105.3) Intersection of SRs 243 and 26; road sign for SR 26 to Royal City. Turn right onto SR 26.

66.0 (105.6) Park along the road to the west of the exposure.

Stop 10. Ginkgo Pillow Complex at Sand Hollow

Excellent exposures of the extensive pillow-palagonite complex at the base of the Basalt of Ginkgo occur throughout the Vantage region. The petrified wood shown in the museum at Ginkgo Petrified Forest State Park at Vantage comes from this pillow complex throughout the area.

The Basalt of Ginkgo, Frenchman Springs Member. The Basalt of Ginkgo is the second oldest and second largest unit in the Frenchman Springs Member (Beeson et al., 1985). Its type locality is along Schnebly Coulee at the Ginkgo Petrified Forest along the west side of the Columbia River at Vantage, Washington (Mackin, 1961). At this location, petrified logs occur in the basal pillow and palagonite breccias that rest on top of

clastic sediments of the Vantage Member. The Basalt of Ginkgo consists of 1–3 flows erupted from a north-northwest–trending linear vent system.

Basalt of Ginkgo may be readily distinguished from other Frenchman Spring flows by their distinct chemical composition, lithology, and excursional paleomagnetic polarity (Beeson et al., 1985). Hand samples of the Ginkgo are commonly characterized by the presence of abundant, medium- to coarse-grained plagioclase phenocrysts and glomerocrysts from 0.5 to > 3 cm in length. However, several other Frenchman Springs units (i.e., the Basalt of Silver Falls and Basalt of Sand Hollow) may also be plagioclase phyric; consequently, other criteria (i.e., geochemistry, paleomagnetism, stratigraphic position) should also be used to properly identify the Ginkgo (Beeson et al., 1985).

The vent system for the Basalt of Ginkgo is exposed along the Snake River near Sheffler, ~40 km northeast of Pasco, Washington. At this location, the dike is ~10 m wide; a second large (>3 m) dike and several smaller Ginkgo dikes also occur at this location. In addition, the dikes can be traced uphill along the walls of the Snake River canyon, where they widen out into a lava pond, merging with the Ginkgo flow. The Ginkgo vent system may continue southward and is again exposed along the Walla Walla River in Oregon. Several 2–6-m-wide abundantly plagioclase phyric dikes are exposed in the canyon walls of the South Fork of the Walla Walla River. Hand samples from these dikes bear a striking resemblance to those from the dikes at Sheffler. Geochemically, they are also similar to the Basalt of Ginkgo, with two major exceptions: the Walla Walla River dikes have significantly lower TiO_2 and higher Nb contents than the rest of the Ginkgo (TiO_2, 2.89 wt% vs. 3.10 wt%; Nb, 16.5–17 ppm vs. 14–15 ppm; B. Martin, unpublished data). There are exposures of plagioclase phyric lavas that geochemically resemble the Walla Walla River dikes near Milton-Freewater, Oregon. Consequently, the southern dikes may have fed a local variant of the Ginkgo (B. Martin, unpublished data).

At this location, the pillow-palagonite complex is ~15 m thick and constitutes one-third of the total flow thickness. The basal pillow complex grades upward into the colonnade that makes up the bulk of the flow. The presence of an extensive basal pillow complex suggests that the Ginkgo lavas flowed into the main valley of the ancestral Columbia River. Foreset beds within the pillow complex suggest that flow movement in this area was to the west or northwest, consistent with the location relative to the Ginkgo vent system.

Turn around and proceed west on SR 26 to its junction with I-90.

Cumulative		
Miles	(km)	Description
67.2	(107.5)	Enter I-90 headed east (toward Spokane). I-90 climbs from the Columbia River across the upper part of the member of Sentinel Bluffs (Grande Ronde Basalt), the Basalt of Ginkgo and the Basalt of Sand Hollow

(Frenchman Springs Member), and Roza Member into the lower part of the Priest Rapids Member before dropping back down onto the top of the Roza Member.

Cumulative Miles	(km)	Description
72.9	(116.6)	Take the Silica Road exit (exit 143) from I-90.
73.3	(117.3)	Turn left onto Silica Rd.

Optional Stop. Roza Peperite

At the foot of the exit, peperite in the Roza Member is well exposed in the roadcuts. Peperites are basalt-sediment breccias formed by the invasion of lava flows into sediment. Explosive interaction between the lava and the wet sediment produce a breccia of disintegrated basalt and sediment (Schmincke, 1967a). At this location, a Roza lava flow invaded and flowed over a 7-m-thick bed of soft diatomite in the Squaw Creek Member of the Ellensburg Formation. Rapid cooling along its base caused the lava to shatter, forming a peperite; continued advance of the lava buried the peperite beneath a pillow-palagonite complex.

Not only did the Roza lava invade the diatomite, but the resulting peperite was also apparently lifted and rafted along the top of the Roza flow. Invasion and rafting of sediments by Columbia River basalt flows was not an unusual occurrence, especially along the margins of the province where simultaneous clastic sedimentation was occurring (Swanson et al., 1989; Carson et al., 1987). The Basalt of Rosalia, Priest Rapids Member, overlies the Roza Member and peperite at this location.

Cumulative Miles	(km)	Description
74.1	(118.6)	Turn left onto Vantage Rd. at the first intersection. Vantage Rd. crosses the top of the Roza Member and then descends through to the Roza Member to the Roza–Sand Hollow contact.
75.4	(120.6)	Park in wide area on left.

Stop 11. The Roza Member at Frenchman Coulee

During the last Pleistocene glaciation, the two alcoves of Frenchman Coulee served as one of the outlets for the catastrophic floods from Glacial Lake Missoula that periodically crossed present-day eastern Washington. Floodwaters from the Quincy Basin to the east dropped more than 200 m through Frenchman Coulee to reach the Columbia River (Carson et al., 1987). The two alcoves of Frenchman Coulee were created by headward erosion as the floodwaters undercut the Wanapum Basalts at this location. Frenchman Coulee was one of several simultaneously operating spillways along the Columbia River between Quincy to the north and Wallula Gap to the south.

The Roza Member. The Roza Member is one of the most distinctive units within the Columbia River Basalt Group. It is characterized by the presence of single, clear to amber-colored, plagioclase phenocrysts between 0.3 and 1 cm in length that comprise 5–10 modal percent of the flow. This characteristic made the Roza Member an important marker bed during the development of the Columbia Basin Project (Grolier and Bing-

ham, 1978). In its type locality along the Yakima River, the Roza Member consists of one flow (Mackin, 1961); however, to the east and northeast in the Columbia Basin and Grand Coulee regions, two to three flow units can be recognized (Bingham and Walters, 1965; Bingham and Grolier, 1966; Lefevbre, 1966, 1970). Collectively, the Roza flows cover ~40,300 km^2 with ~1,300 km^3 of lava (Tolan et al., 1989).

Flow direction indicators (i.e., inclined pipe vesicles, spiracles, vesicle cylinders) within the Roza lavas of the northwestern Columbia Plateau generally yield northwesterly to westerly flow directions, suggesting a source to the southeast (Lefevbre, 1966, 1970; Schmincke, 1967b). Bingham and Grolier (1966) interpreted outcrops of phyric tephra interbedded with thin Roza flows near Winona, Washington, as Roza tephra cones. Subsequent reconnaissance mapping of the plateau delineated a narrow, 175-km-long linear vent system marked by dikes, spatter cones and ramparts, small volcanic shields, and near-vent flow facies in southeastern Washington and northeastern Oregon (Swanson et al., 1975, 1979b; Martin, 1989; Thordarson and Self, 1996).

Roza lavas are chemically similar to several units of the Frenchman Springs Member (i.e., the Basalt of Ginkgo and Basalt of Sentinel Gap); however, the combination of the Roza's plagioclase phyric nature and its higher MgO, CaO, and P_2O_5 contents permit it to be distinguished from the underlying Frenchman Springs lavas. The distinction between the Roza and overlying Priest Rapids lavas is unambiguous; the latter are aphyric, have lower SiO_2, higher P_2O_5, and very different trace element abundances.

Although multiple Roza cooling units were identified in the field during reconnaissance mapping of the Columbia Plateau, their major element chemistries were sufficiently uniform that geologists grouped them together (Wright et al., 1973; Swanson et al., 1979a; Wright et al., 1989). However, detailed stratigraphic study combined with lithologic variations and precise chemical analyses demonstrated that individual lavas within the Roza Member can be correlated across the Columbia Plateau. The integration of field studies of the lithologic variations and internal structures with major- and trace-element analyses suggested that the Roza Member is divisible into six units (Martin, 1989, 1991). Cr, Zr, Nb, TiO_2, P_2O_5, and CaO proved to be the most useful elements in subdividing the Roza Member. Within the Roza stratigraphic succession, the major-element chemistries of the lavas are relatively uniform; the bulk of the interflow chemical variation lies in the minor and trace elements. Subsequent work by Thordarson and Self (1996) verified the previous stratigraphic succession with one revision; they suggested that Roza I-A and I-B of Martin (1989) be combined into a single unit on the basis of the similarity of their chemical variations.

Along the western and southwestern margins of the Roza flow field examined on this field excursion, only the Roza II-A of Martin (1989) is present. Unit II-A is the largest volume unit within the Roza Member, covering ~21,000 km^2 with a volume of ~485 km^3. This flow was erupted from vents distributed along most of the 175 km length of the Roza vent system in southeastern Washington and northeastern Oregon. The lavas appear to have initially flowed northwestward from the vent system toward the Grand Coulee region, their westward progress apparently influenced by the constructional topography along the eastern and northeastern edges of the older Frenchman Springs flows. Roza II-A basalts spread out across the northern Columbia Basin and then flowed to the southwest, into the Columbia River Gorge region, where it terminated just west of Lyle, Washington. Throughout much of the western Columbia Basin, the base of the Roza is pillowed, contains spiracles, and overlies volcaniclastic and fluviolaccustrine sediments. Fecht et al. (1987) suggested that the Columbia River has occupied this region since the early to middle Miocene. It appears that the Roza lavas were captured by the drainage system of the ancestral Columbia River in the Grande Coulee region and channeled to the southwest. This helps to explain the predominance of southwest flow directions observed in the "Middle" Roza (II-A of Martin, 1989) by Lefevbre (1966) along the Grand Coulee and in the Frenchman Coulee by Schmincke (1967b).

The Roza Member in Frenchman Coulee consists of a single flow in excess of 30 m thick. A well-developed colonnade overlain by an irregularly jointed entablature is well exposed in the north wall of Frenchman Coulee. The colonnade represents just over 50% of the thickness of the flow at this location. Roza Member samples from Frenchman Coulee have the geochemical and petrographic characteristics of both units II-A and II-B of Martin (1989). The lower portion of the colonnade has the geochemical characteristics of Roza II-A. On the other hand, samples from the upper half of the exposure have slightly higher Sr and Cr concentrations and somewhat lower incompatible element abundances (TiO_2, K_2O, P_2O_5, Y, Nb, Zr) similar to the Roza II-B unit. The transition is also marked by increase in the abundance of plagioclase phenocrysts within the flow. At the location where Martin (1991) sampled the flow, the flow top had been removed, probably by erosion. A similar geochemical pattern is seen in the single, 50-m-thick Roza flow exposed at Potholes Coulee, ~15 km to the north. However, the II-B composition occurs within the colonnade at Potholes Coulee; as one proceeds upward through the section, the lava transitions back to the II-A composition.

Stop 11A. Roza-Sand Hollow Contact

Walking east from the parking area, you may observe the vesicular and brecciated flow top of the Basalt of Sand Hollow along with glass and flow structures along the base of the Roza Member.

Stop 11B. "Spiracle"/"Lava-Rise Suture" at the Base of the Roza Member

Spiracles are large, irregular, but crudely funnel-shaped chimneys that rise from the base of a lava flow. They may be several meters high and wide and may be partially filled with sediment, pillows, and palagonitized sideromelane breccia. Their tops commonly trail out in the flow direction (Schmincke, 1967b). They are believed to result from steam-induced fractur-

ing as the lava flowed across wet sediment. While sediment is generally absent along the contact between the Basalt of Sand Hollow and Roza Member, the spiracle appears to be localized above a low mound of sediment (peat and diatomite).

An alternate interpretation of this structure was provided by Thordarson and Self (1996). They interpret the structure as a suture between two lobes of a Roza sheet flow.

The sutures consist of interleaved subhorizontal lava plates, which thicken toward and connect to the coherent lava on either side. The plates are bound by glassy surfaces that are often striated. These striations are oriented perpendicular to the strike of the suture, showing that the plates were formed as a succession of lateral protrusions from the interior of the lava on either side....Subhorizontal cracks that originate at the junction of two plates extend into the coherent lava and show that the formation of these plates was associated with vertical extension of the lava lobe. (Thordarson and Self, 1996, p. 27,432)

Stop 11C. Flow Lobes in the Roza Member

Walk ~100 m to the east along Vantage Road. Along the north wall of the coulee, the Roza forms the uppermost flow; however, the flow does not appear to be continuous along the coulee wall. To the east of the prominent high-tension tower, you can see the upper surface of the flow dip downward before rising back to its previous elevation. In the swale created by the dip, there is an irregularly jointed knob of Roza that overlies a small pillow-palagonite complex that is above a small waterfall. Geochemical analysis of samples from the lower portion of the Roza Member on either side of the swale, as well as from the knob, indicate that these are all Roza II-A (B. Martin, unpublished data). Thordarson and Self (1996) suggest that these features represent the junction of two abutting flow lobes with the pillowpalagonite complex representing a rootless cone rampart overlain by the remnant of another flow lobe.

Stop 11D. Roza Colonnade

Return to the parking area. Along the south side of the parking area, you can observe the erosional remnant of a single row of Roza columns, pinch and swell structures in the Roza columns, and swirling platy jointing above the Roza colonnade. You may pass through the notch in the column wall onto the Basalt of Sand Hollow bench along the wall of the South Alcove of Frenchman Coulee in order to observe the 50-m-deep plunge pool at the base of its dry falls. The erosion that formed the plunge pool cut deeper into the section, exposing both the member of Sentinel Bluffs and the Basalt of Ginkgo.

Return east on Vantage Road.

Cumulative Miles	(km)	Description
76.7	(122.7)	Turn right onto Silica Rd.
77.5	(124.0)	Intersection of Silica Rd. and I-90. Turn right onto I-90 west toward Seattle. Take I-90 to Seattle.

Optional Stop. Saddle Mountains Anticline and Fault

Cumulative		
Miles	(km)	Description
0.0	(0.0)	Take Vantage exit 136.
0.2	(0.3)	Turn right onto Huntzinger Road.
6.4	(10.2)	Old Railroad crossing.
6.5	(10.4)	Gate to Yakima Training Center. Turn right onto gravel road.
7.4	(11.8)	Road to exposure of Saddle Mountains fault. Return to Vantage and I-90.

End of Road Log.

REFERENCES CITED

Anderson, J.L., 1987, Structural geology and ages of deformation of a portion of the southwest Columbia Plateau, Washington and Oregon [Ph.D. thesis]: Los Angeles, University of Southern California, 283 p.

Barsotti, A.T., 1986, Structural and paleomagnetic analysis of eastern Umtanum Ridge, south-central Washington [M.S., thesis]: Pullman, Washington State University, 204 p.

Beeson, M.H., Fecht, K.R., Reidel, S.P., and Tolan, T.L., 1985, Correlations within the Frenchman Springs Member of the Columbia River Basalt Group: New insights into the middle Miocene tectonics of northwest Oregon: Oregon Geology, v. 47, p. 87–96.

Beeson, M.H., Tolan, T.L., and Anderson, J.L., 1989, The Columbia River Basalt Group in western Oregon; Geologic structures and other factors that controlled flow emplacement patterns, in Reidel, S.P., and Hooper, P.R., eds., Volcanism and Tectonism in the Columbia River flood-basalt province: Boulder, Colorado, Geological Society of America Special Paper 239, p. 223–246.

Bentley, R.D., 1977, Stratigraphy of the Yakima Basalts and structural evolution of the Yakima Ridges in the western Columbia Plateau, in Brown, E.H., and Ellis, R.C., eds., Geology excursions in the Pacific Northwest: Bellingham, Washington, Western Washington University Press, p. 339–389.

Bingham, J.W., and Grolier, M.J., 1966, The Yakima Basalt and Ellensburg Formation of south-central Washington: U.S. Geological Survey Bulletin 1224-G, p. 1–15.

Bingham, J.W., and Walters, K.L., 1965, Stratigraphy of the upper part of the Yakima Basalt in Whitman and eastern Franklin Counties, Washington: U.S. Geological Survey Professional Paper 525-C, p. C87–C90.

Camp, V.E., Hanson, W.E., and Ross, M.E., 2003, Genesis of flood basalts and Basin and Range volcanic rocks from Steens Mountain to the Malheur River gorge, Oregon: Geological Society of America Bulletin, v. 115, p. 105–128.

Campbell, N.P., 1989, Structural and stratigraphic interpretation of the rocks under the Yakima fold belt based on recent surface mapping and well data, in Reidel, S.P., and Hooper, P.R., eds., Volcanism and Tectonism in the Columbia River flood-basalt province: Boulder, Colorado, Geological Society of America Special Paper 239, p. 209–222.

Carson, R.J., Tolan, T.L., and Reidel, S.P., 1987, Geology of the Vantage area, south-central Washington: An introduction to the Miocene flood basalts, Yakima fold belt, and the Channeled Scabland, in Hill, M.L., ed., Cordilleran Section of the Geological Society of America Centennial Field Guide, v. 1, p. 357–362.

Cummings, M.L., Evans, J.G., Ferns, M.L., and Lees, K.R., 2000, Stratigraphic and structural evolution of the middle Miocene synvolcanic Oregon-Idaho Graben: Geological Society of American Bulletin, v. 112, p. 668–682.

Diery, H., 1967, Stratigraphy and structure of Yakima canyon between Roza Gap and Kittitas Valley [Ph.D. thesis]: Seattle, University of Washington, 117 p.

Department of Energy, 1988, Site Characterization Plan, Consultation Draft Report DOE/RW-0164, 9 Volumes.

Fecht, K.R., Reidel, S.P., and Tallman, A.M., 1982, Evolution of the Columbia River system in the central Columbia Plateau of Washington from Miocene to present: Geological Society of America Abstracts with Program, v. 14, no. 4, p. 163.

Fecht, K.R., Reidel, S.P., and Tallman, A.M., 1987, Paleodrainage of the Columbia River system on the Columbia Plateau of Washington State: A summary, in Schuster, J.E., ed., Selected Papers on the geology of Washington, Division of Geology and Earth Resources, Bulletin 77, p. 219–248.

Goff, F.E., 1981, Preliminary geology of eastern Umtanum Ridge, south-central Washington: Richland, Washington, Rockwell Hanford Operations Report RHO-BWI-C-21, 100 p.

Grolier, M.J., and Bingham, J.W., 1971, Geologic map and sections of parts of Grant, Adams and Franklin Counties, Washington: U.S. Geological Survey Miscellaneous Geologic Investigations Series Map I-589, scale 1:62 500, 8 sheets.

Grolier, M.J., and Bingham, J.W., 1978, Geology of parts of Grant, Adams, and Franklin Counties, east-central Washington: Washington Division of Geology and Earth Resources Bulletin 71, 91 p.

Hagood, M.A., 1986, Structure and evolution of the Horse Heaven Hills, south-central Washington: Richland, Washington, Rockwell Hanford Operations Report RHO-BW-SA-344, 176 p., 1 plate.

Hon, K., Kauahikaua, J., Denlinger, R., and Mackay, K., 1994, Emplacement and inflation of pahoehoe sheet flows: Observations and measurements of active lava flows on Kilauea volcano, Hawaii: Geological Society of America Bulletin, v. 106, p. 351–370.

Hooper, P.R., Binger, G.B., and Lees, K.R., 2002, Ages of the Steens Mountain and Columbia River flood basalts and their relationship to extension-related calc-alkaline volcanism in eastern Oregon: Geological Society of America Bulletin, v. 114, p. 43–50.

Lefevbre, R.H., 1966, Variations of flood basalts of the Columbia River Plateau, central Washington [Ph.D. thesis]: Evanston, Illinois, Northwestern University, 211 p.

Lefevbre, R.H., 1970, Columbia River Basalt of the Grand Coulee area, in Gilmore, E.H., and Stradling, D., eds., Proceedings of the 2nd Columbia River Basalt Symposium: Cheney, Washington, Eastern Washington State College Press, p. 1–38.

Long, P.E., and Wood, B.J., 1986, Structures, textures, and cooling histories of Columbia River basalt flows: Geological Society of America Bulletin, v. 97, p. 1144–1155.

Mackin, J.H., 1961, A stratigraphic section in the Yakima Basalt and the Ellensburg Formation in south-central Washington: Washington Division of Mines and Geology Reports of Investigations 19, 45 p.

Martin, B.S., 1989, The Roza Member, Columbia River Basalt Group; chemical stratigraphy and flow distribution, in Reidel, S.P., and Hooper, P.R., eds., Volcanism and tectonism in the Columbia River flood-basalt province: Boulder, Colorado, Geological Society of America Special Paper 239, p. 85–104.

Martin, B.S., 1991, Geochemical variations within the Roza Member, Wanapum Basalt, Columbia River Basalt Group: Implications for the magmatic processes affecting continental flood basalts [Ph.D. thesis]: Amherst, University of Massachusetts, 513 p.

McMillan, K., Long, P.E., and Cross, R.W., 1989, Vesiculation in Columbia River Basalts, in Reidel, S.P., and Hooper, P.R., eds., Volcanism and Tectonism in the Columbia River flood-basalt province: Boulder, Colorado, Geological Society of America Special Paper 239, p. 157–167.

Price, E.H., 1982, Structural geology, strain distribution, and tectonic evolution of Umtanum Ridge at Priest Rapids Dam, and a comparison with other selected localities within Yakima fold belt structures, south-central Washington: Richland, Washington, Rockwell Hanford Operations Report RHO-BWI-SA-138, 197 p., 3 plates.

Price, E.H., and Watkinson, A.J., 1989, Structural geometry and strain distribution within east Umtanum Ridge, south-central Columbia Plateau, in Reidel, S.P., and Hooper, P.R., eds., Volcanism and Tectonism in the Columbia River flood-basalt province: Boulder, Colorado, Geological Society of America Special Paper 239, p. 265–282.

Puget Sound Power and Light Company (PSPL), 1982, Skagit/Hanford Nuclear Project, Preliminary Safety Analysis Report: Puget Sound Power and Light Co., Bellevue, Washington, v. 4, Ap. 20, Amendment 23.

Reidel, S.P., 1984, The Saddle Mountains: the evolution of an anticline in the Yakima fold belt: American Journal of Science, v. 284, p. 942–978.

Reidel, S.P., 1988, Geological map of the Saddle Mountains, south-central Washington: Washington Division of Geology and Earth Resources Geologic Map GM-38, scale 1:48 000, 5 sheets, 28 p.

Reidel, S.P., 1998, Emplacement of Columbia River flood basalt: Journal of Geophysical Research, v. 103, p. 27393–27410.

Reidel, S.P., and Fecht, K.R., 1987, The Huntzinger flow: Evidence of surface mixing of the Columbia River basalt and its petrogenetic implication: Geological Society of America Bulletin, v. 98, p. 664–677.

Reidel, S.P., and Hooper, P.R., 1989, Volcanism and Tectonism in the Columbia River flood-basalt Province: Boulder, Colorado, Geological Society of America Special Paper 239, 386 p.

Reidel, S.P., Long, P.E., Myers, C.W., and Mase, J., 1982, New evidence for greater than 3.2 km of Columbia River Basalt beneath the central Columbia Plateau [abs.]: Eos (Transactions, American Geophysical Union), v. 63, p. 173.

Reidel, S.P., Campbell, N.P., Fecht, K.R., and Lindsey, K.A., 1994, Late Cenozoic structure and stratigraphy of south central Washington: Washington Division of Geology and Earth resources Bulletin, v. 80, p. 159–180.

Reidel, S.P., Tolan, T.L., Hooper, P.R., Beeson, M.H., Fecht, K.R., Anderson, J.L., and Bentley, R.D., 1989a, The Grande Ronde Basalt, Columbia River Basalt Group; Stratigraphic descriptions and correlations in Washington, Oregon, and Idaho, *in* Reidel, S.P., and Hooper, P.R., eds., Volcanism and Tectonism in the Columbia River flood-basalt Province: Boulder, Colorado, Geological Society of America Special Paper 239, p. 21–54.

Reidel, S.P., Fecht, K.R., Hagood, M.C., and Tolan, T.L., 1989b, The geologic evolution of the central Columbia Plateau, *in* Reidel, S.P., and Hooper, P.R., eds., Volcanism and Tectonism in the Columbia River flood-basalt province: Boulder, Colorado, Geological Society of America Special Paper 239, p. 247–264.

Schmincke, H-U., 1967a, Fused tuff and peperites in south-central Washington: Geological Society of America Bulletin, v. 78, p. 319–330.

Schmincke, H-U., 1967b, Flow directions in Columbia River Basalt flows and paleocurrents of interbedded sedimentary rocks, south-central Washington: Geologische Rundschau, v. 56, p. 992–1020.

Self, S., Thordarson, T., and Keszthelyi, L., 1997, Emplacement of continental flood basalt lava flows, *in* Mahoney, J.J., and Coffin, M., eds., Large Igneous Provinces: American Geophysical Union Geophysical Monograph Series, v. 100, p. 381–400.

Self, S., Keszthelyi, L., and Thordarson, T., 1998, The importance of pahoehoe: Annual Reviews of Earth and Planetary Sciences, v. 26, p. 81–110.

Smith, G.A., Campbell, N.P., Deacon, M.W., and Shafiquallah, M., 1988, Eruptive style and location of volcanic centers in the Miocene Washington Cascade Range: Reconstruction from the sedimentary record: Geology, v. 16, p. 337–340.

Swanson, D.A., and Wright, T.L., 1981, Guide to the geologic field trip between Lewiston, Idaho and Kimberley, Oregon, *in* Johnston, D.A., and Donnelly-Nolan, J., eds., Guides to some volcanic terranes in Washington, Idaho, Oregon, and northern California: U.S. Geological Survey Circular 838, p. 1–28.

Swanson, D.A., Wright, T.L., and Helz, R.T., 1975, Linear vent systems and estimated rates of magma production and eruption for the Yakima Basalt on the Columbia Plateau: American Journal of Science, v. 275, p. 877–905.

Swanson, D.A., Wright, T.L., Hooper, P.R., and Bentley, R.D., 1979a, Revisions in stratigraphic nomenclature of the Columbia River Basalt Group: U.S. Geological Survey Bulletin 1457-G, 59 p.

Swanson, D.A., Anderson, J.L., Bentley, R.D., Camp, V.E., Gardner, J.N., Wright, T.L., 1979b, Reconnaissance geologic map of the Columbia River Basalt Group in Washington and adjacent Idaho: U.S. Geological Survey Open-file Report 79-1363, scale 1:250 000, 7 sheets.

Swanson, D.A., Wright, T.L., Camp, V.E., Gardner, J.N., Helz, R.T., Price, S.M., Reidel, S.P., and Ross, M.E., 1980, Reconnaissance geologic map of the Columbia River Basalt Group, Pullman and Walla quadrangles, southeast Washington and adjacent Idaho: U.S. Geological Survey Miscellaneous Investigations Map I-1139, scale 1:250 000, 2 sheets.

Swanson, D.A., Cameron, K.A., Evarts, R.C., Pringle, P.T., and Vance, J.A., 1989, Cenozoic volcanism in the Cascade Range and Columbia Plateau, southern Washington and northernmost Oregon: International Geological Congress, American Geophysical Union, 28th Field Trip Guidebook T106, 60 p.

Thordarson, T., and Self, S., 1996, Sulfur, chlorine, and fluorine degassing and atmospheric loading by the Roza eruption, Columbia River Basalt Group, Washington, USA: Journal of Volcanology and Geothermal Research, v. 74, p. 49–73.

Tolan, T.L., Reidel, S.P., Beeson, M.H., Anderson, J.L., Fecht, K.R., and Swanson, D.A., 1989, Revisions to the extent and volume of the Columbia River Basalt Group, *in* Reidel, S.P., and Hooper, P.R., eds., Volcanism and Tectonism in the Columbia River Flood-Basalt Province: Boulder, Colorado, Geological Society of America Special Paper 239, p. 1–20.

Tomkeieff, S.I., 1940, Basalt lavas of the Giants Causeway District of northern Ireland: Bulletin of Volcanology, v. 6, p. 90–143.

Walker, G.P.L., 1972, Compound and simple lava flows and flood basalts: Bulletin Volcanologique, v. 35, p. 579–590.

Walker, G.P.L., Canon-Tapia, E., and Herrero-Bervera, E., 1999, Origin of vesicle layering and double imbrication by endogenous growth in the Birkett basalt flow (Columbia River plateau): Journal of Volcanology and Geothermal Research, v. 88, p. 15–28.

Waters, A.C., 1961, Stratigraphy and lithologic variations in the Columbia River basalt: American Journal of Science, v. 259, p. 583–611.

Wright, T.L., Grolier, M.J., and Swanson, D.A., 1973, Chemical variation related to stratigraphy of the Columbia River Basalt: Geological Society of America Bulletin, v. 84, p. 371–386.

Wright, T.L., Mangan, M., and Swanson, D.A., 1989, Chemical data for flows and feeder dikes of the Yakima Basalt Subgroup, Columbia River Basalt Group, Washington, Oregon, and Idaho, and their bearing on a petrogenetic model: U.S. Geological Survey Bulletin 1821, 71 p.

Geological Society of America
Field Guide 4
2003

Cretaceous to Paleogene Cascades Arc:
Structure, metamorphism, and timescales of magmatism, burial, and exhumation of a crustal section

Robert B. Miller*

Department of Geology, San Jose State University, San Jose, California 95192-0102, USA

Jennifer P. Matzel

Department of Earth, Atmospheric, and Planetary Sciences, MIT, Cambridge, Massachusetts 02139, USA

Scott R. Paterson

Department of Earth Sciences, University of Southern California, Los Angeles, California 90089-0740, USA

Harold Stowell

Department of Geological Sciences, University of Alabama, Tuscaloosa, Alabama 35487-0338, USA

ABSTRACT

The crystalline core of the North Cascades (Cascades core) consists largely of oceanic and arc terranes that were metamorphosed to amphibolite facies and intruded by 96–45 Ma arc plutons. A crustal section recording paleodepths of ~5–40 km is preserved in the southern part of the core and facilitates evaluation of processes at different levels of the arc. After terrane juxtaposition, plutons were intruded during major arc-normal crustal shortening dominated by early recumbent folds and subsequent upright folds. Structural patterns emphasize the heterogeneous vertical partitioning of deformation and complex rheological stratification of arc crust at all scales. Metamorphic and geochronologic data indicate rapid burial of plutons and host rocks during the middle Cretaceous shortening. A subsequent major, cryptic event in the evolution of the Cascades core was the rapid underthrusting of Cretaceous sedimentary protoliths of the Swakane Gneiss to depths of ≥40 km between 73 and 68 Ma. The emplacement of the gneiss, which lacks arc magmas, may have removed the roots of many of the arc plutons. Plutonic rocks record both focused and unfocused, largely tonalitic magmatism that resulted in large plutons and abundant narrow sheets, respectively. Individual large-volume plutons were constructed over intervals of up to 5.5 m.y. Magmas probably ascended as visco-elastic diapirs, and emplacement was aided primarily by vertical material transfer, including ductile flow and stoping, and possibly regional folding. The magmatic, metamorphic, and structural processes recorded in the Cascades core exemplify the dynamic evolution of arcs and the large vertical and lateral displacements during arc construction.

Keywords: North Cascades, arcs, timescales, exhumation, pluton, rheology.

*E-mail: rmiller@geosun.sjsu.edu.

Miller, R.B., Matzel, J.P., Paterson, S.R., and Stowell, H., 2003, Cretaceous to Paleogene Cascades Arc: Structure, metamorphism, and timescales of magmatism, burial, and exhumation of a crustal section, *in* Swanson, T.W., ed., Western Cordillera and adjacent areas: Boulder, Colorado, Geological Society of America Field Guide 4, p. 107–135. For permission to copy, contact editing@geosociety.org. © 2003 Geological Society of America.

INTRODUCTION

We have been engaged in structural, metamorphic, igneous, and thermochronological study of the deep crustal evolution of a magmatic arc, the Cretaceous to Eocene crystalline core of the North Cascades (Cascades core). The large range in crustal levels exposed in the southern part of the Cascades core makes it an excellent natural laboratory to undertake such studies. In the following, we first briefly introduce some of the outstanding questions and problems for arcs that will be addressed on the field trip, using the Cascades core as a case study.

1. Are active arcs constructed during arc-perpendicular extension or arc-perpendicular shortening? Structural analysis of the plutonic roots of ancient arcs is one way to evaluate this question.

2. Is large vertical displacement of crust an important process in arcs, particularly contractional ones? What is the metamorphic signature (*P-T-t* path) of such motion? In the Cascades core, shallowly emplaced middle Cretaceous plutons and supracrustal rocks underwent major burial followed by rapid decompression (Whitney et al., 1999). What caused this burial: thickening during folding and thrusting (Paterson and Miller, 1998a; Whitney et al., 1999) or loading by plutons (Brown and Walker, 1993)?

3. What are the deformation patterns that result from vertical motion and how do they change with depth? In the Cascades core, much debate has focused on the significance of commonly gently plunging mineral lineations and the question of whether these lineations are parallel to the regional displacement direction (e.g., compare Brown and Walker [1993] with Miller and Paterson [1992] and Whitney et al. [1999]).

4. How do magmas ascend in arcs—in dikes, or as classical "hot-Stokes" diapirs? Are ascent mechanisms different at different crustal levels?

5. How are plutons emplaced in arcs and do emplacement mechanisms change with depth and time? Do horizontal material transfer processes such as extension marked by lateral displacement of host rock dominate, or is material largely transferred downward by ductile flow and stoping, which may result in large-scale exchange of crust (e.g., Paterson et al., 1996)?

6. What are the rates of tectonic and magmatic processes in this arc? Is the emplacement of a batholith better described as a long, continuous process or as short, episodic growth? What are reasonable rates of burial of supracrustal rocks, and can we infer tectonic processes from these rates?

7. Do simple models of brittle, downward-strengthening upper crust and ductile, downward-weakening lower crust fit the rheology of natural arcs? How important are magmatism, lithological heterogeneity, and mechanical anisotropy for crustal rheology (e.g., Miller and Paterson, 2001a; Klepeis et al., 2003)?

These issues are addressed in varying detail in the remainder of the guide. Much of the following is modified from our recent publications (Paterson and Miller, 1998a, 1998b; Whitney et al., 1999; Miller and Paterson, 1999, 2001a, 2001b; Valley et al., 2003; Stowell and Tinkham, 2003) and unpublished data. We also

rely on our previous field guides (Paterson et al., 1994; Miller et al., 2000), to which the reader is referred for more detail.

Overview of the Cascades Core

The crystalline core of the North Cascades (Cascades core) is the southernmost extension of the >1500-km-long Coast Belt of the northwest Cordillera (Fig. 1), and is characterized by Cretaceous and Paleogene arc plutons, amphibolite-facies metamorphism, and a major middle Cretaceous shortening event (e.g., Misch, 1966; Rubin et al., 1990; Journeay and Friedman, 1993). In the Cascades core, plutons ranging 96–45 Ma intrude oceanic, island arc, and clastic-dominated terranes, most of which were amalgamated before plutonism and middle Cretaceous orogenesis

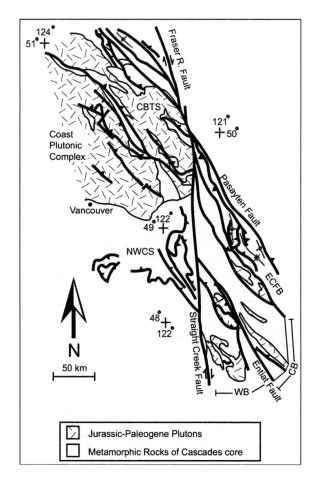

Figure 1. Sketch map emphasizing distribution of metamorphic rocks and plutons in the Cascades core and southern Coast Belt of Washington and southwest British Columbia. Cretaceous thrust faults in the Cascades core, Coast Belt thrust system (CBTS), and lower-grade rocks of the Eastern Cascades fold belt (ECFB), and Northwest Cascades system (NWCS) are also shown. The dextral Fraser-Straight Creek fault offsets the Cascades core from the main part of the Coast Belt. The Entiat fault and Pasayten fault are other major high-angle faults; the Entiat divides the Cascades core into the Wenatchee (WB) and Chelan blocks (CB).

(e.g., Tabor et al., 1989). In the middle Cretaceous, the orogen was deformed at shallow levels (<10 km) largely by SW-directed thrusting (e.g., Misch, 1966; Brandon et al., 1988), whereas at deeper levels (~10–40 km) complex ductile deformation during amphibolite-facies metamorphism included extensive folding and associated cleavage development, SW-directed shear in map-scale ductile shear zones, and subhorizontal, broadly orogen-parallel (NW-SE) stretching (e.g., Paterson and Miller, 1998a). Shortening continued until the latest Cretaceous in the deepest level of the arc.

The middle Cretaceous orogenic event records intra-arc shortening and final suturing of the insular superterrane to North America (e.g., Monger et al., 1982; McGroder, 1991). Subsequent to regional shortening, dextral strike slip was clearly active by 48 Ma (e.g., Miller and Bowring, 1990; Umhoefer and Schiarizza, 1996) and may have initiated much earlier. The magnitude of dextral displacement remains one of the most contentious issues in Cordilleran tectonics. Paleomagnetic data from the Mount Stuart batholith of the southern Cascades core and elsewhere in the Coast Belt and Insular superterrane imply ≥2000 km of dextral displacement relative to terranes to the east (Baja B.C. hypothesis), whereas mapped relationships across known faults appear to preclude such magnitudes (e.g., Cowan et al., 1997). In the middle and late Eocene, a transtensional tectonic regime was dominant (e.g., Johnson, 1985), resulting in several nonmarine basins with thick clastic sections and exhumation of parts of the Cascades core.

The post-metamorphic, high-angle Entiat fault divides the core into the Wenatchee and Chelan blocks (Figs. 1 and 2). In the southwestern part of the Wenatchee block, overlapping timing relationships between 96–88 Ma plutons and ductile structures and K-Ar and Ar/Ar cooling ages indicate that deformation largely occurred in the middle Cretaceous (ca. 96–85 Ma) (e.g., Tabor et al., 1989; Miller and Paterson, 1992; Paterson et al., 1994). The northeast Wenatchee block and Chelan block were deformed at this time but also record younger contractional deformation that overlapped with 73–65 Ma plutonism and still later deformation during Eocene plutonism and exhumation of the Chelan block (e.g., Tabor et al., 1989; Hurlow, 1992; Paterson and Miller, 1998b).

We have constructed a middle to Late Cretaceous crustal section for the Cascades core by qualitatively retrodeforming the Wenatchee block (Figs. 3 and 4) (Miller and Paterson, 2001a). On a regional scale, terrane boundaries, syn-metamorphic tight to isoclinal folds, compositional layers, and foliation have been folded by large, high-amplitude folds with steeply inclined axial planes. These large folds and possibly important NE-side-up tilt (e.g., Haugerud, 1987; Butler et al., 1989; Miller et al., 1990) have resulted in >15 km of structural relief in the Wenatchee block, as shown by the regional cross sections (Fig. 3). "Unfolding" these folds indicates that early structures, including terrane boundaries, were initially subhorizontal and allows us to construct the upper 25 km of the section.

The Wenatchee block is cut by the NE-dipping, reverse White River shear zone, which is marked in part by greenschist-

facies mylonites that postdate peak-metamorphism (Van Diver, 1967; Magloughlin, 1993). This shear zone divides the Wentachee block into two domains with somewhat different metamorphic histories (Brown and Walker, 1993; Miller et al., 1993a), which we informally refer to as the Mount Stuart (SW) and Tenpeak (NE) domains. The amount of slip on the White River shear zone is poorly constrained; however, we used thermobarometric data to restore the hanging wall to its appropriate position in the crustal section. The hanging-wall rocks (Napeequa unit) are also exposed NE of the Entiat fault in the Chelan block, and we use data from both the Wenatchee and Chelan blocks to characterize the lower (25–40 km) part of the section. The structurally lowest rocks in the Cascades core, the Swakane Gneiss, were underthrust beneath this Cretaceous section at a later time (see below).

Thermobarometric data from metamorphic and plutonic rocks in both blocks further constrain appropriate depths during the Late Cretaceous (Miller and Paterson, 2001a). We infer that the recorded pressures were those at the time the dominant mesoscopic ductile structures formed and that they at least in part predate late regional upright folds. This is a reasonable interpretation, because throughout much of the section, the dominant foliation is marked by syn-kinematic assemblages formed at or near the peak of metamorphism (e.g., Plummer, 1980; Paterson et al., 1994; Whitney et al., 1999). Similarly, all of the plutons are syn-tectonic (e.g., Plummer, 1980; Paterson and Miller, 1998a; Miller and Paterson, 1999), and barometric data are compatible with structural depths inferred from the cross sections. Thus, we suggest that this section gives reasonable positions and paleodepths for units during Late Cretaceous deformation.

In the following, we describe rock types and deformation patterns in the Cascades crustal section, beginning at the top of the exposed section.

WENATCHEE BLOCK

Ingalls Complex

The shallowest rocks involved in Cascades core metamorphism and deformation are in the ophiolitic Ingalls Complex (Miller, 1985). This Late Jurassic ophiolite structurally overlies the Chiwukum Schist along the middle Cretaceous Windy Pass thrust (Figs. 2 and 3). Metamorphic grade ranges from prehnite-pumpellyite facies in the southern part of the ophiolite to amphibolite facies in the structurally lower parts of the complex near the Windy Pass thrust.

The ophiolite is dominated by ultramafic tectonites, including a lherzolite belt typical of mantle associated with mid-ocean ridge basalt (MORB) and a harzburgite and dunite belt analogous to mantle in supra-subduction-zone settings (Miller and Mogk, 1987). Gabbros, diabase dikes, and pillowed basalts form the crustal section of the ophiolite. These rocks are dominantly MORB, transitional to island arc tholeiite (dated at 161 Ma) (Metzger et al., 2002a; Harper et al., this volume). Overlying Late Jurassic sediments include mudstone, chert, local gray-

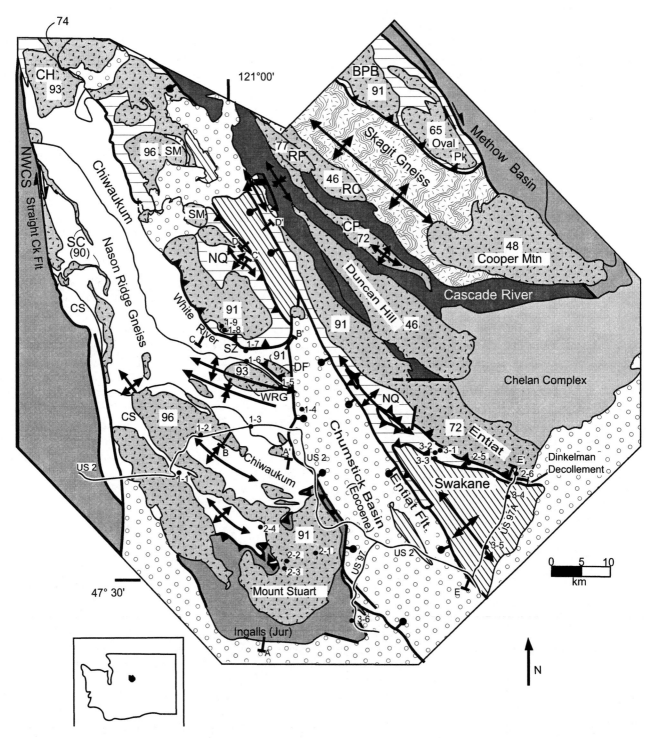

Figure 2. Simplified geologic map of the southern and central parts of the Cascades core. Plutons are shown by shaded, random dashes with ages; BPB—Black Peak batholith; CH—Chaval pluton; CP—Cardinal Peak pluton; CS—Chiwaukum Schist; DF—Dirtyface pluton; NQ—Napeequa Complex; NWCS—Northwest Cascades system; RC—Railroad Creek pluton; RP—Riddle Peaks pluton; SC—Sloan Creek plutons; SM—Sulphur Mountain pluton; SZ—shear zone; WRG—Wenatchee Ridge Gneiss. Open circle pattern—mid-Eocene and younger rocks. Also shown are the locations of field trip stops, major highways, and lines of cross sections in Figure 3.

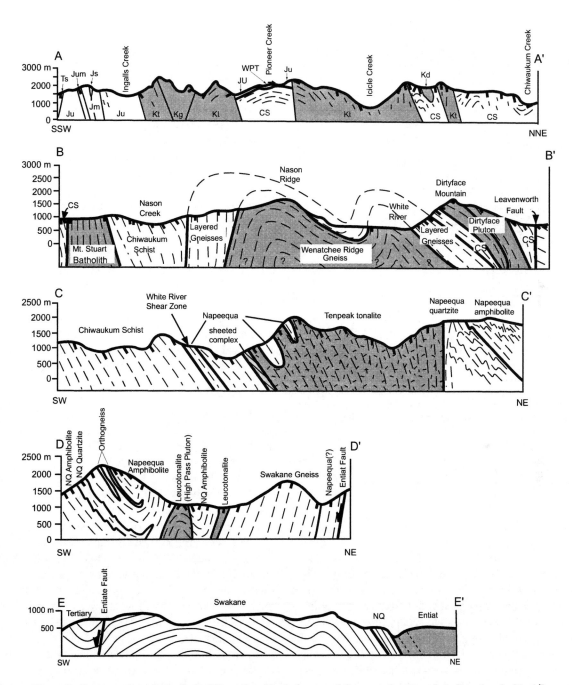

Figure 3. Cross sections through the Wenatchee block (arranged from southwest and shallow levels [A–A′] to northeast and deep levels [D–D′]) and southern part of Chelan block (E–E′). Tick marks at surface show dips of measured foliation, and dashes and thin lines (E–E′) are inferred foliation traces. Squiggles show foliation in sheared serpentinite. Note that foliation and some contacts define large upright, SW-vergent folds. CS—Chiwaukum Schist, Kd—Mount Stuart diorite, Kg—Mount Stuart granodiorite, Kt—Mount Stuart tonalite, Jm—Ingalls Complex mafic rocks, Js—Ingalls Complex sedimentary rocks, Ju—Ingalls Complex ultramafic rock, Jum—Ingalls Complex serpentinite mélange, NQ—Napeequa unit, Ts—Tertiary sandstone, WPT—Windy Pass thrust.

Cascades Crustal Section

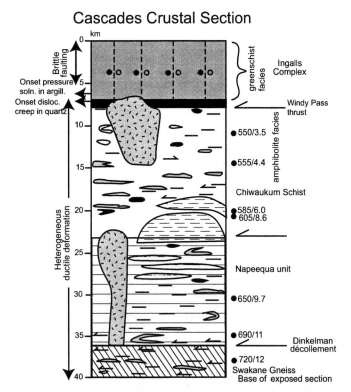

Figure 4. Diagram summarizing the Cascades crustal section. Random dashes note plutons containing mostly magmatic fabrics; dashed line pattern indicates plutonic rocks with considerable subsolidus deformation and/or intense magmatic foliation. Other lenses in Chiwaukum and Napeequa units are amphibolite, metaperidotite, and minor marble. Vertical dashed lines are faults in Ingalls Complex, and half-arrows show areas with non-coaxial shear. Note that the Dinkelman décollement marks the base of the crustal section, as the Swakane unit was tectonically juxtaposed at a younger time. Numbers on right side of section summarize representative temperatures (°C) and pressures (kbar). See text for references to sources of data.

wacke, and sedimentary breccia containing ophiolitic clasts. Within-plate basalts (WPB) with intercalated, locally oolitic limestone (MacDonald et al., 2002) are found in the southern part of the ophiolite, and are associated with Lower Jurassic radiolarian cherts (Miller et al., 1993b). Calc-alkaline dikes of uncertain age cut both the MORB and WPB rocks. The diversity of the ophiolite may reflect juxtaposition of different parts of a backarc basin across an oceanic fracture zone (Miller, 1985; Metzger et al., 2002a).

Deformation of the ophiolite is markedly heterogeneous. E-W-striking, steeply dipping serpentinite mélange that formed before emplacement on the Windy Pass thrust typifies lower grade parts of the ophiolite (Miller, 1985). The onset of ductile deformation is recorded by greenschist-facies argillaceous rocks containing slaty cleavage that is axial-planar to local folds of bedding (Paterson et al., 1994). Ductile deformation is much more widespread in the amphibolite-facies part of the ophiolite,

and foliation in metasedimentary rocks is tightly to isoclinally folded and transposed.

Foliation in the lower part of the Ingalls Complex is parallel to ductile imbricate thrusts associated with the Windy Pass thrust. In the footwall of Chiwaukum Schist, foliation is rotated into parallelism with the generally gently dipping, folded thrust (Miller, 1988; Taylor, 1994). Thrusting and folding were close in time (Paterson et al., 1994). The Windy Pass thrust is intruded by a ca. 94 Ma phase of the regionally discordant, 91–96 Ma Mount Stuart batholith (Matzel et al., 2002a). The ophiolite is also intruded by foliated sheets of tonalite and diorite that have been involved in thrusting. One sheet yields a U-Pb zircon age inferred to record crystallization ca. 94 Ma, suggesting that thrusting overlapped emplacement of early phases of the Mount Stuart batholith.

Chiwaukum Schist and Nason Ridge Migmatitic Gneiss (Nason Terrane)

The Chiwaukum Schist of the Nason terrane forms a significant proportion of the crustal section (Figs. 2, 3, and 4). It consists of interlayered metapelitic and metapsammitic schist, lesser lenses of amphibolite and ultramafite, and rare marble (Plummer, 1980; Tabor et al., 1987a). The metaclastic rocks were derived from an island arc (Anderson and Paterson, 1991; Magloughlin, 1993) and the protolith of the amphibolites has a spectrum of magma types, ranging from WPB to MORB, to local island arc tholeiite (Magloughlin, 1993; Metzger et al., 2002b). The volume of clastic rocks and association of intercalated rock types from oceanic and arc settings suggest the unit was assembled in an accretionary wedge before arc magmatism began (Paterson et al., 1994). The protolith age is only constrained as pre-96 Ma, but may be Late Jurassic based on comparisons with relations west of the Straight Creek–Fraser River fault in southwest British Columbia, where similar clastic rocks of the Cayoosh assemblage overlie Mississippian to Jurassic Bridge River Complex (= Napeequa unit, see below) (Monger and Journeay, 1994).

The southern, structurally higher part of the Chiwaukum Schist is intruded by the Mount Stuart batholith (Stop 1-2), and the lower half of the schist is intruded by numerous trondhjemitic sheets (Stops 1-3 and 1-6) that are mostly 10 cm to tens of meters thick. This extensively intruded region has been mapped separately as the Nason Ridge Migmatitic Gneiss (formerly banded gneiss) (Tabor et al., 2002). The internally sheeted, 91 Ma Dirtyface pluton intrudes the middle of this section (Hurlow, 1992; Miller et al., 2000). The lowermost exposed contact of the Chiwaukum Schist/Nason Ridge gneiss is a folded intrusive contact with the ca. 93 Ma Wenatchee Ridge Gneiss (Tabor et al., 1987a; Magloughlin, 1993; J.P. Matzel, unpublished data).

Metamorphic pressures of the Chiwaukum Schist are lowest at the south end of the schist near the Mount Stuart batholith where syn-kinematic andalusite- and cordierite-bearing assemblages define a 2-km-wide contact aureole. Over a distance of 10 km northeastward from the batholith, pressures for pelitic schists range 3–9 kbar, and temperatures 540–700 °C (Evans and Berti, 1986; Brown and Walker, 1993; Paterson et al., 1994; Whitney et al., 1999). Kyanite and staurolite replace andalusite in the Mount Stuart aureole and occur in a broad region northeast of the batholith. Sillimanite is present with kyanite in a relatively narrow swath in the center of the Nason terrane, and kyanite is the dominant aluminum-silicate polymorph in the northeastern part of the unit.

The replacement of andalusite by kyanite and thermobarometric data from samples containing zoned garnet indicate that the southern part of the schist underwent crustal loading of up to 5 kbar soon after (<7 m.y.) emplacement of the Mount Stuart batholith (Evans and Berti, 1986; Brown and Walker, 1993; Evans and Davidson, 1999; Whitney et al., 1999; Stowell and Tinkham, 2003). For example, rocks northeast of the Mount Stuart batholith contain pseudomorphs of andalusite and experienced a pressure increase of ~2 kbar before garnet growth and after emplacement of this part of the batholith (Fig. 5). It is uncertain if the Nason Ridge Gneiss and northeastern part of the Chiwaukum Schist experienced a similar history because of a paucity of geochronological data; however, *P-T-t* paths for the gneiss indicate that garnet grew at 7–9 kbar and loading was ≤3 kbar during garnet growth (Fig. 5) (Tinkham, 2002; Zuluaga and Stowell, 2002).

Deformation was broadly similar throughout the Chiwaukum Schist. The schist displays strong foliation and compositional layering, which in some places formed by metamorphic differentiation during folding but in others may be relict bedding. Folds are the dominant structures from the thin section to map scale, and a minimum of three cycles of folding is recognized throughout the schist (Paterson et al., 1994). Two cycles are defined by nearly coaxial, tight to isoclinal folds and new axial-planar foliation. Dips of foliation are typically moderate, reflecting the youngest cycle of larger, more upright folds. The early folds initially were recumbent or gently inclined. Late folds have WNW- to NW-striking, moderately to steeply NE-dipping axial planes (Figs. 2 and 3), are mostly open to tight, and commonly lack an axial-planar fabric. Axes of the various fold generations scatter considerably in orientation, but their maxima are gently WNW- or ESE-plunging (Paterson et al., 1994). Miller and Paterson (2001a) interpreted the folds to have formed by flexural mechanisms plus homogeneous shortening.

Mineral lineations have broadly similar orientations to the fold axes (Brown and Talbot, 1989; Paterson et al., 1994), but in some outcrops they are folded and lineation commonly lies at a higher angle to the axes in the limbs. Lebit et al. (1998) inferred that these lineation patterns reflect superposed folding. Lineation also steepens in pluton structural aureoles and shear zones next to plutons.

The structural style of plutonic rocks that intrude Chiwaukum Schist and Nason Ridge Migmatitic Gneiss depends in part upon the size of the body and structural position. The structurally higher trondhjemitic sheets include both concordant and discordant bodies that are commonly folded or boudinaged, but some only display weak subsolidus fabrics (Stop 1-3) (Getsinger,

1978; Magloughlin, 1993; Paterson et al., 1994). Deeper in the schist, sheets are more commonly concordant and possess a strong magmatic and/or subsolidus fabric continuous with that in the schist (Stop 1-6).

Kinematic indicators are rare in the Chiwaukum Schist. Excluding areas near large plutons, we have observed similar numbers of indicators in lineation-normal surfaces and lineation-parallel surfaces. The former primarily record SW-directed reverse shear in shear zones or flexural slip in fold limbs, whereas the lineation-parallel planes yield mixed shear sense. This kinematic pattern dominantly reflects flexural fold-dominated shear compatible with the mechanically active layering (Lebit et al., 1998).

Regional relationships differ in a 200-m-wide zone structurally below the 91 Ma Dirtyface pluton. In this domain, lineation swings to NE trends and steep pitches, and folding and non-coaxial shear increase in intensity. Strong magmatic and locally subsolidus foliations in the pluton are continuous with foliation in the underlying schist and both are associated with top-to-SW shear.

Down-dip lineation and reverse kinematics are also present in the moderately to steeply N- to NE-dipping White River shear zone. In this zone, the high-pressure (7–10 kbar) Tenpeak pluton and a thin (0–500-m-thick) rind of Napeequa unit structurally overlie the schist (Figs. 2 and 3) along a tectonic contact that locally is a modified intrusive contact. This shear zone is marked by a 100- to 500-m-wide zone of intense deformation and synkinematic retrogression to greenschist-facies assemblages in the Chiwaukum Schist (Stop 1-7) (Van Diver, 1967; Tabor et al., 1987a; Magloughlin, 1993) and by medium- and high-temperature S-C fabrics in the margin of the Tenpeak tonalite (Stop 1-8) (Miller and Paterson, 1999). Umhoefer and Miller (1996) postulated large displacements in this shear zone based in part on the presence of deeper-level plutons in the hanging wall relative to the footwall and the apparently different metamorphic history across the shear zone. They also suggested that if the Wenatchee block is viewed as a SW-vergent fold and thrust belt, then the White River shear zone is a late, out of sequence structure. Large post-Tenpeak displacements are unlikely, however, because of local intrusive relations between the pluton and Chiwaukum Schist, and petrologic similarities between the Dirtyface and Tenpeak plutons (Miller et al., 2000).

Major Intrusions in the Mount Stuart Domain

Mount Stuart Batholith

The composite Mount Stuart batholith, the largest intrusion in the Cascades core, consists of two major bodies of primarily biotite-hornblende tonalite: a larger (~1000 km²) northeastern body and smaller southwestern body (Fig. 6). It also includes two-pyroxene gabbro, diorite, granodiorite, and granite. The better-studied northeastern body has a mushroom-shaped southeastern end, the stem of which extends into a sheet-like central area, and a hook-shaped northwestern end (Stop 1-2). A mafic outlier (Big Jim Complex) lies close to the central area.

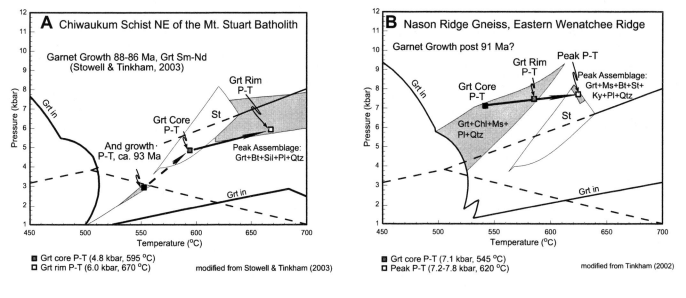

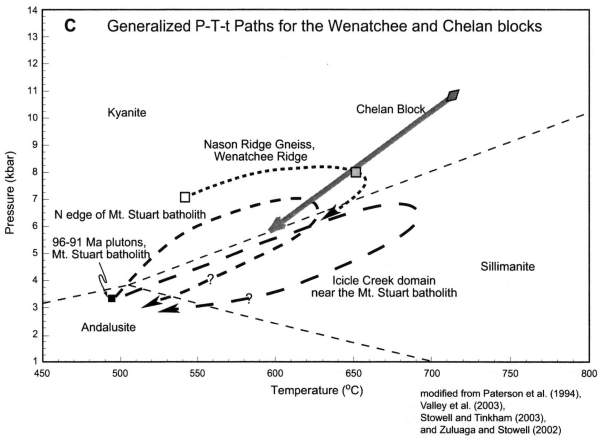

Figure 5. Pressure-temperature-time (*P-T-t*) paths for metamorphism in the Wenatchee and Chelan blocks of the North Cascades. All phase equilibria in (A) and (B) were calculated with THERMO-CALC v. 30 (Powell and Holland, 2001) and the data set of Holland and Powell (1998). A: *P-T-t* path for Chiwaukum Schist along the northeastern margin of the Mount Stuart batholith. The initial part of the path is constrained by andalusite and pseudomorphs after andalusite, both of which are widespread. Garnet growth pressures and temperatures are constrained by pseudosection phase equilibria modeling and garnet chemistry (Stowell and Tinkham, 2003). Garnet core chemistry provides a *P-T* estimate in the staurolite stability field, compatible with observation of staurolite inclusions in garnet cores. Peak *P-T* is constrained by rim thermobarometry and the peak mineral assemblage. The timing of garnet growth is constrained by core, rim, and whole grain garnet Sm-Nd ages of 88–86 Ma (Stowell and Tinkham, 2003). B: *P-T-t* path for Nason Ridge Migmatitic Gneiss along the eastern end of Wenatchee Ridge. The initial part of the path is not constrained. Garnet growth pressures and temperatures are constrained by pseudosection phase equilibria modeling and garnet chemistry (Tinkham, 2002). Garnet core chemistry provides a *P-T* estimate, and peak *P-T* is constrained by rim thermobarometry and peak mineral assemblage. No direct isotopic ages are available to directly date this *P-T-t* path; however, garnet growth is likely to postdate emplacement of the Mount Stuart batholith and may be synchronous with garnet growth discussed above for (A). C: Generalized *P-T-t* paths for the Wenatchee and Chelan blocks (see above, Paterson et al. [1994], and Valley et al. [2003]).

Figure 6. Map of the Mount Stuart batholith showing internal petrologic variations, regional foliation patterns, and age data. Foliations in the host rock represent average orientations of the most pervasively developed foliation in that region. Foliation dips are 0–29, filled squares; 30–59, filled triangles; and 60–90, no dip symbol. Solid circles—sample localities with geochronological data. Note that the regional foliation is only deflected within a short distance from the pluton. Geochronological data from Matzel et al. (2002a), Walker and Brown (1991), and Tabor et al. (1987a). H—K-Ar hornblende age; B—K-Ar or Ar/Ar biotite age; Z—U-Pb zircon age from Matzel et al. (2002a); Z*—U-Pb zircon age from Walker and Brown (1991).

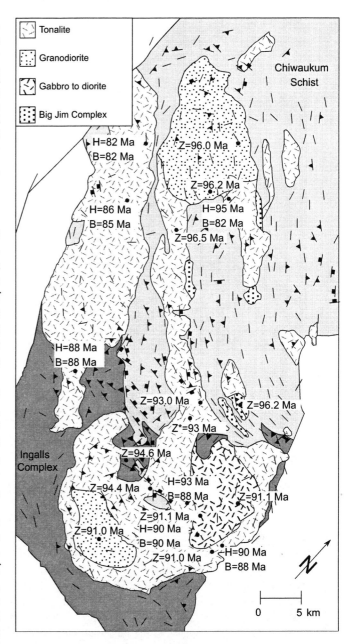

New U-Pb zircon geochronology from 10 samples indicates that the batholith was constructed over a ~5.5 m.y. period from ca. 96.5 Ma to 91.0 Ma (Fig. 6). Significant amounts of mafic magma were emplaced throughout this period, with crystallization dates from Big Jim Complex diorite ca. 96.2 Ma and two samples of gabbro from the mushroom-shaped, southeastern region of the batholith ca. 91.0 Ma. The oldest dated phases (ca. 96.5–96.2 Ma) are tonalite and granodiorite that crop out in the hook-shaped part of the batholith. The bulbous, SE part of the batholith comprises the youngest phases of gabbro, tonalite, and granodiorite.

The most mafic compositions within the batholith (51–55 wt% SiO_2) lie in the eastern part of the mushroom-shaped body, whereas the most evolved magmas (>68% SiO_2) are found near the center of the mushroom-shaped body and at the northwestern end of the hook-shaped area (Fig. 6) (Pongsapich, 1974; Erikson, 1977; Paterson et al., 1994). The batholith differs from other Cascades plutons in that it has unusually high MgO for a given SiO_2 content (Paterson et al., 1994) and contains orthopyroxene.

The origin of the petrological diversity of the batholith is controversial. Anderson and Paterson (1991) and Anderson (1992) attributed much of the diversity to fractional crystallization of a dioritic parental magma derived by partial melting of a mafic, MORB-like, garnet-bearing source (see Paterson et al., 1994). More mafic rocks are interpreted as cumulates. DeBari (*in* Miller et al., 2000) suggested that mixing of the dioritic parental magma with more mafic magmas may have been an important process based on relationships in the deeper plutons discussed below. Kelemen and Ghiorso (1986) proposed that the reverse zoning of the Big Jim Complex resulted from progressive interaction between ultramafic wall rocks in the Chiwaukum Schist and dioritic magma. Such a process could explain the high Mg content of the batholith.

Magmatic foliation in the mushroom-shaped region of the batholith dips steeply and defines margin-parallel, "onion-skin" patterns (Paterson et al., 1994; Paterson and Miller, 1998a). Weak mineral lineation plunges steeply. This pattern is interrupted by strong, gently dipping foliation near the Windy Pass thrust, which is semi-continuous with thrust-related ductile fabrics in the Ingalls Complex and Chiwaukum Schist. Miller and Paterson (1992, 1994) thus inferred that the batholith was deformed by thrusting while melt dominated, but upon reaching its solidus, locked up

displacement on the thrust. Subsolidus deformation along part of the northeast margin is associated with down-dip lineation in the moderately to steeply NE-dipping, reverse Tumwater Mountain shear zone, which carries amphibolite-facies Ingalls Complex up relative to the batholith. In the sill-like region, magmatic foliation defines complex patterns, including large and small magmatic folds with subhorizontal, NW-SE–trending axes (Paterson and Miller, 1998a). Magmatic foliation similarly is folded by decameter- to kilometer-scale magmatic folds in the hook-shaped region, and the hook-like shape itself may reflect a large such fold (Paterson and Miller, 1998a; Benn et al., 2001). These magmatic folds, syn-emplacement folds in the structural aureole, and overlap of emplacement with movement on the Windy Pass thrust, indicate that intrusion occurred during regional shortening (Miller and Paterson, 1992; Paterson and Miller, 1998a).

Paterson and Miller (1998a) speculated that the petrological variations, internal magmatic contacts, local magma mingling, and age data indicate that magma ascended in small to large (hundreds to thousands of cubic meters), variably shaped batches that were assembled at the site of emplacement. Paterson and Miller (1998a) demonstrated that material transfer processes accompanying emplacement include ductile flow in a narrow (1–2-km-wide) structural aureole (20%–50% of needed space), stoping (33%–63%), and regional crustal thickening by folding, foliation development, and movement along reverse shear zones, during which magma flowed into fold hinges and fractures (~15%).

Dirtyface and Related Plutons

Several discrete plutons intrude the Chiwaukum Schist and Nason Ridge Migmatitic Gneiss north of the Mount Stuart batholith, including the 91 Ma Dirtyface and 90 Ma Sloan Creek plutons (Fig. 2) (e.g., Vance, 1957; Van Diver, 1967; Tabor et al., 1987a; Miller et al., 2000). The Dirtyface pluton consists of 1–100-m-wide typically concordant sheets of quartz diorite-tonalite, and in its lower half some trondhjemitic sheets similar to those in the adjacent Chiwaukum Schist and Wenatchee Ridge Gneiss. The trondhjemites are petrologically unrelated to the dominant tonalite and quartz diorite sheets, but probably shared the same magma pathway (Miller et al., 2000). The tonalite sheets are much less heavy–rare earth element (REE) depleted than those of the Mount Stuart batholith and are probably fractionates from mantle-derived basaltic magma with little input from a garnet-bearing lower crust (Miller et al., 2000).

A contact aureole has not been recognized next to the Dirtyface pluton. Porphyroblasts of garnet, staurolite, and kyanite are common in the Chiwaukum Schist near the pluton, but were deformed during retrograde metamorphism in contrast to relations next to the Mount Stuart batholith. The continuity of magmatic and host rock foliation suggests that the main foliation in the schist near the Dirtyface pluton is ca. 91 Ma, but deformation of the pluton and porphyroblasts in the host rock indicates that deformation continued after 91 Ma.

Wenatchee Ridge Gneiss

The ca. 93 Ma Wenatchee Ridge Gneiss is the structurally lowest unit southwest of the White River shear zone. The intrusive contact between this trondhjemitic orthogneiss and Chiwaukum Schist is folded and relatively gently dipping, and is best described as an injection migmatite zone. Locally, this intrusive contact rolls over to a steep orientation (Getsinger, 1978), suggesting that orthogneiss exposures may represent the crest of a large pluton. The orthogneiss consists of numerous internal sheets that range from subparallel to crosscutting and vary from 10 cm to meters in thickness. Some contain metaperidotite xenoliths ranging up to hundreds of meters in length (Tabor et al., 1987a; Magloughlin, 1993).

All sheets contain high-temperature, subsolidus foliation. Local isoclinal folds of sheets and foliation are refolded by common upright structures that control the orientation of structures in most outcrops (Stop 1-5). These folds are continuous with folds in the Chiwaukum host rock.

Summary of Timing of Events in Mount Stuart Domain

The timing of metamorphism and ductile deformation is best constrained for the southern part of the Chiwaukum Schist. There, truncation of axial-planar fabrics by the Mount Stuart batholith (Paterson et al., 1994), syn-kinematic assemblages in its aureole, and subsequent loading of the aureole indicate pre-, syn-, and post-Mount Stuart dynamothermal metamorphism. Biotite cooling ages (Ar/Ar, K-Ar) of 90–81 Ma in the batholith and southern part of the schist (Engels et al., 1976; Tabor et al., 1982, 1987a; Evans and Davidson, 1999), and Sm-Nd garnet ages as young as 86 Ma in the schist indicate that metamorphism began before 96 Ma and ended by 81 Ma. Numerous >85 Ma biotite ages demonstrate that loading and exhumation of the schist was rapid.

Biotite K-Ar and Ar/Ar ages decrease by ~10 m.y. to the northeast of the batholith. U-Pb monazite ages of 88–91 Ma from three samples of schist may date prograde metamorphism (Brown and Walker, 1993). Some foliation and folding in the schist postdate the 93 Ma Wenatchee Ridge Gneiss and 91 Ma Dirtyface pluton.

Napeequa Schist (Napeequa Complex)

The base of the Wenatchee Ridge Gneiss and Chiwaukum Schist is not exposed, but we infer that the Napeequa Complex, which now structurally overlies the schist in the reverse-slip White River shear zone (Figs. 2 and 3), was beneath the schist in the middle Cretaceous (Fig. 4). This proposal is compatible with sparse barometric data (~9–11 kbar at 565–675 °C) (Brown and Walker, 1993; Valley et al., 2003) from the Napeequa Complex and middle Cretaceous plutons that intrude it, and with relationships west of the Straight Creek-Fraser River fault in British Columbia (Monger and Journeay, 1994).

The Napeequa Complex consists of amphibolite and quartzite (metachert), with some marble, metaperidotite, and biotite schist (Cater and Crowder, 1967; Tabor et al., 1987a, 1989). The inferred protoliths suggest that the complex was an oceanic unit assembled in a clastic-starved accretionary wedge (cf. Tabor et al., 1989; Haugerud et al., 1991; Miller et al., 1993c). These protoliths are similar to rock types in the Mississippian to Jurassic Bridge River Complex, which largely lies northeast of the Cascades core (Tabor et al., 1989; Miller et al., 1993c).

In the Wenatchee block, the Napeequa Complex was intruded by 96–87 Ma plutons that crystallized at 7–10 kbar (Al-in-hornblende barometry) including the 90–92 Ma Tenpeak pluton and 96 Ma Sulphur Mountain pluton (Brown and Walker, 1993; Dawes, 1993; Miller et al., 2000). It was also extensively intruded by concordant tonalite sheets ranging from 10 cm to 1 km in thickness, which make up >30% of the Napeequa outcrop belt.

The outcrop-scale structure of the Napeequa Complex is dominated by folds of generally NW- to WNW-striking foliation(s) (Miller and Paterson, 2001a). Early isoclinal folds and associated axial-planar foliations are widespread. One or more cycles of open to isoclinal, upright to overturned folds are superposed on the isoclinal folds, which originally were probably recumbent.

The major structure of the Napeequa Complex is an upright, asymmetric synform with a steeply NE-dipping axial plane and a minimum wavelength of 7 km (Figs. 2 and 3). This synform and the outcrop-scale upright to overturned folds generally plunge gently to moderately southeast. Mineral lineation is folded by the upright structures but has broadly similar trends and plunges as the folds. Kinematic indicators are sparse in all rock types in the Napeequa Complex.

Swakane Gneiss

The structurally lowest unit in the southern part of the Cascades core is the Swakane Gneiss. This gneiss underlies the Napeequa Complex in the northeast limb of the regional synform of the Wenatchee block and in the core of a regional antiform in the Chelan block (Figs. 2 and 3). The Swakane Gneiss consists almost entirely of quartzo-feldspathic biotite ± garnet and muscovite gneiss that displays subtle compositional layering. Minor pelitic schist and rare marble, amphibolite, quartzite, and metaperidotite form thin layers and lenses within the gneiss (Sawyko, 1994). No major plutons intrude the Swakane unit, but the gneiss is injected by thin (<10 m) leucocratic sheets, many of which are pegmatitic; some contain garnet and muscovite, and probably formed by partial melting of crustal source rocks (Boysun and Paterson, 2002). Syn-kinematic garnet-kyanite ± sillimanite-bearing gneisses in the Swakane Gneiss in the Wenatchee block yield peak temperatures of 650–700 °C at ~8.5–10 kbar (Whitney, 1992; Brown and Walker, 1993; Sawyko, 1994; Valley et al., 2003).

U-Pb analyses of detrital zircons from the Swakane Gneiss yield dates from 73 Ma to 1610 Ma with a dominant Late Cretaceous population (Matzel et al., 2002b). Cathodoluminescence images show oscillatory-zoned (igneous) cores with chaotically zoned (metamorphic) overgrowths. Several zircons were plucked from a grain mount after imaging, abraded to remove the rim, and then analyzed. These grains exhibited a wide range of $^{206}Pb/^{207}Pb$ crystallization dates, from 1609.8 ± 1.4 Ma to 73.5 ± 0.8 Ma. The observation that even the youngest zircon populations have oscillatory-zoned cores is consistent with this population being derived from an igneous rock that was subsequently eroded and deposited to form the Swakane protolith. Thus, the maximum age of deposition is 72.6 ± 0.6 Ma, the $^{206}Pb/^{238}U$ date of the youngest detrital grain. The fact that all of the imaged zircons displayed chaotically zoned rims regardless of the age of their core suggests that the rims grew during metamorphism.

The timing of metamorphism has been difficult to determine, and our efforts have focused on determining the crystallization ages of leucocratic dikes that we infer are locally derived melts of the gneiss. U-Pb zircon analyses from a peraluminous dike that postdates initiation of foliation in the gneiss yield a weighted mean $^{206}Pb/^{238}U$ date of 68.47 ± 0.04 Ma (mean square of weighted deviates = 0.01), which is interpreted to date crystallization of this dike (Matzel et al., 2002b). Additional zircon analyses have $^{206}Pb/^{238}U$ dates that range from 72.9 to 87.1 Ma and are interpreted as inherited from the protolith.

The geochronologic data also indicate that sources of a variety of ages contributed to the Swakane protolith, which bears on the question of whether the protolith was a silicic volcanic rock (e.g., Cater, 1982; C. Hopson *in* Mattison, 1972) or a sedimentary rock (e.g., Miller et al., 2000). The main piece of evidence used to support a predominantly silicic volcanic protolith is the "remarkable homogeneity" of the Swakane Gneiss (C. Hopson *in* Mattinson, 1972; Cater, 1982); however, if the protolith was volcanic, then the geochronologic data require that it inherited zircons from a highly diverse crustal column. These observations indicate that the Swakane Gneiss almost certainly had a predominantly clastic protolith.

In the Wenatchee block, the gneiss displays NW-striking, moderately to steeply SW-dipping foliation. Mineral lineation plunges gently NW-SE. Outcrop-scale folds are much less common than in the Chiwaukum Schist and Napeequa unit. Early recumbent folds and transposition foliation were folded by nearly coaxial, more open upright folds; mineral lineations lie close to the axes, but are folded by both types of folds (Miller and Paterson, 2001a). Leucocratic intrusive rocks have well-developed subsolidus foliation and lineation but are less deformed than the host gneisses. They commonly are boudinaged, and some define tight, originally recumbent folds.

Plutons in the Tenpeak Domain—Tenpeak Pluton

The 90–92 Ma Tenpeak pluton was emplaced at ~25–35 km depth (7–10 kbar [Dawes, 1993; S.M. DeBari, 1999, personal commun.]) into the Napeequa unit and locally, the Chiwaukum Schist (Fig. 7). The margin of the pluton is marked by a mafic complex of sheeted and mingled gabbro and tonalite (e.g., Cater and Crowder, 1967; Miller and Paterson, 1999). Sheeted tonalite is also found in internal zones. Two larger phases of tonalite (Schaeffer Lake and Indian Creek phases) make up most of the pluton interior, and intrude out much of the sheeted zones. Widely scattered inclusions of Napeequa amphibolite range up to 1 km in width and are particularly common as meter-scale rafts in sheeted zones.

U-Pb zircon dates from six samples indicate that the Tenpeak pluton was intruded over ~2.5 m.y., from ca. 90.0–92.5 Ma. A sample of the Schaeffer Lake phase near the southern margin of the pluton and a nearby sample from a sheeted, internal zone (Stop 1-8) represent the oldest dated phases (Fig. 7). The Indian Creek phase is the youngest dated unit at ca. 90.0 Ma.

Petrological studies by Dawes (1993) and DeBari (DeBari et al., 1998; Miller et al., 2000) demonstrate that the volumetrically dominant tonalites formed by mixing of mafic (high-Al basalt), mantle-derived magmas with trondhjemitic, crustally derived

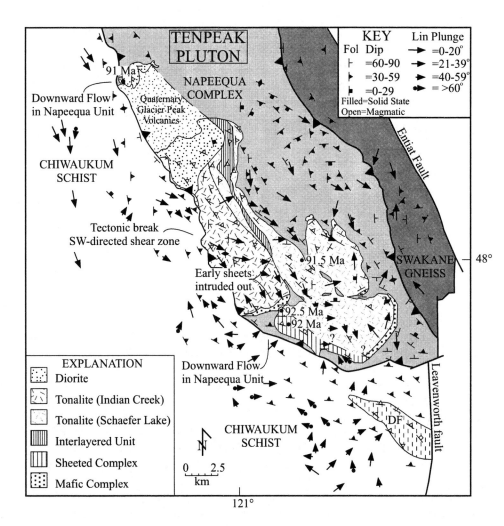

Figure 7. Map of Tenpeak pluton showing rock types, U-Pb zircon localities, foliation, and lineation. Important points for emplacement mechanisms are also shown (after Miller and Paterson, 1999). DF—Dirtyface pluton.

magmas; the latter are only locally exposed in the pluton. The mantle-derived magmas are typically basaltic and occur as gabbros in the mafic complex, mafic sheets in the main phase tonalite, and as abundant enclaves in the tonalite. Very strong enrichment in light REE and strong depletion in heavy REE indicate that the crustal melts were derived from a garnet-bearing, mafic source (Miller et al., 2000), and this melting probably occurred at 15–16 kbar (DeBari et al., 1998).

Moderate to strong magmatic foliation and lineation are well developed in the pluton and are variably overprinted by a largely margin-parallel, medium- to high-T subsolidus fabric. Foliation mainly strikes NW-SE, subparallel to the regional trend. Mineral lineation has moderate to high pitches throughout the pluton.

A 500-m- to 1-km-wide (0.09–0.18 body radius) structural aureole is defined by deflections of regional host rock structures (Fig. 7). Foliation and contacts between lithologic units steepen and mostly bend downward in the aureole, and fold axes and subhorizontal mineral lineation swing to a down-dip orientation (Miller and Paterson, 1999). The south and southwest margins of the pluton are modified by the White River shear zone. The reverse-shear (pluton-side-up) indicators associated with folia-

tion in the pluton and deflection of host rock markers suggest downward flow of host rock in the aureole during rise of the pluton (Miller and Paterson, 1999). Downward movement in the area now occupied by the pluton is indicated by the stoping out of sheeted zones by larger tonalite bodies and abundant host rock inclusions. Downward ductile flow, stoped blocks, moderate to steep lineation in the pluton and aureole, cross sectional shape of the pluton, and orientation of the pluton at a high angle to the regional shortening direction are more compatible with diapiric than dike ascent (Miller and Paterson, 1999).

Summary of Timing of Events in Tenpeak Domain

The congruity of 8–10 kbar pressures in the Napeequa unit and plutons that intrude it are compatible with peak metamorphism ca. 91–96 Ma. The high pressures of the plutons and lack of evidence for significant pressure increase in the metamorphic rocks after plutonism contrast with relations near the Mount Stuart batholith. This suggests that initiation of metamorphism was earlier in the Tenpeak domain, or at least that the Napeequa unit was buried by 96 Ma, when the Sulphur Mountain pluton was emplaced.

Brown and Walker (1993) noted that the Sulphur Mountain and Tenpeak plutons crosscut fabric in the Napeequa Complex and thus at least some deformation is pre-96 Ma. Substantial ductile deformation also postdates these plutons, as intense, gently dipping solid-state foliation in the southeastern end of the Sulphur Mountain pluton is continuous with the fabric in the Napeequa Complex, and the northeast margin of the Tenpeak pluton displays strong subsolidus foliation. The intrusive sheets in the Napeequa Complex display solid-state fabrics, and some sheets are tightly folded and boudinaged. These sheets are poorly dated, but include the 84 Ma Buck Creek pluton (Hurlow, 1992). Foliation and lineation swing from WNW to NNW orientations as the contact with the Swakane Gneiss is approached, implying that ductile deformation also coincided with that in the post-73 Ma gneiss (see above). Sparse K-Ar biotite ages indicate significant cooling and cessation of ductile deformation in the Napeequa Complex by 60–70 Ma.

RELATIONS IN THE SOUTHERN PART OF THE CHELAN BLOCK

Our studies in the southern part of the Chelan block (NE of Entiat fault) have focused on the Entiat pluton and relationships across the tectonic contact between Napeequa Complex and Swakane Gneiss. Deformation and lithological patterns in these units are well displayed in transects along the Columbia River, and Roaring and Tamarack Creeks (Fig. 8) (Paterson and Miller, 1998c; Miller and Paterson, 2001a). These transects cross the NE limb of a regional upright, gently plunging antiform, and expose vertical structural thicknesses of ~4 km (Fig. 8). Deformation in this transect involves the 73 Ma Entiat pluton and thus in part postdates middle Cretaceous deformation in the Wenatchee block.

The Roaring Creek transect starts at its NE end in the Entiat pluton, which is a >70-km-long sheeted body of tonalite and lesser diorite and gabbro emplaced at 20–25 km depths (Dawes, 1993; Paterson and Miller, 1998b). The sheets are centimeters to hundreds of meters wide, and dip steeply to moderately NE. Sheets are particularly common in the NW end and along the SW margin of the pluton, and give way in the SE to a few broad sheets up to 2.5 km wide, which possess numerous schlieren and layers that may reflect sheets with more diffuse boundaries. Our (Matzel) new U-Pb zircon analyses indicate that sheets in the NW end of the pluton are significantly older (ca. 91 Ma) than the thick tonalite sheets of the SW end (ca. 72 Ma [Mattinson, 1972; Hurlow, 1992]).

The widely variable compositions in the Entiat pluton cannot all be related by closed-system fractional crystallization (Miller et al., 2000). Mafic rocks have compositions of typical high-alumina basalts from island arcs (Dawes, 1993; S.M. DeBari, 1999, personal commun.; Miller et al., 2000). REE patterns are similar to those of the Tenpeak pluton, but the Entiat pluton lacks the direct evidence for a garnet-bearing, crustal source (Miller et al., 2000). The geochemical similarity of the tonalites to those in the Tenpeak pluton, however, suggests that the Entiat tonalites were better mixed at deeper levels, such that the felsic endmember is not preserved.

Well-developed magmatic and locally subsolidus foliation and lineation in the Entiat pluton define complex patterns that overprint sheets and internal layers (Paterson and Miller, 1998b; Miller and Paterson, 2001b). Foliation and local sheets define meso- and map-scale zones of syn-emplacement, upright, NW-SE–trending magmatic folds and margin-parallel shear similar to and/or continuous with those in the adjacent host rock (Paterson and Miller, 1998b; Paterson et al., 1998). Magmatic foliation and lineation intensify toward the southwest contact with the Napeequa Complex and are progressively overprinted by lower amphibolite- to greenschist-facies subsolidus fabric. The contact with the Napeequa Complex is sharp and concordant to host rock foliation, and no obvious thermal aureole is present.

The structurally highest rocks in the Napeequa Complex in the Roaring Creek transect form a weak, thinly interlayered unit of marble, metachert, hornblende- and biotite-rich gneisses, and amphibolite, bounded below by a large orthogneiss sheet (Fig. 8) (Miller and Paterson, 2001a). Layers are commonly transposed and disrupted by folding and boudinage. A thick section of dominantly thinly layered, folded and boudinaged biotite gneiss, schist, quartzite, and local calc-silicate rock, marble, and metaperidotite lies below the orthogneiss. Structurally downward, biotite gneiss is the most common rock type, and the lowest mappable Napeequa unit in the transect is dominated by thicker (2–75 m), rheologically strong layers of garnet amphibolite.

All Napeequa units display at least two generations of folds. Foliation axial-planar to tight to isoclinal folds is folded by open to tight, upright to moderately plunging folds. Kinematics are complex with early top-to-SW, younger top-to-N, and fold-related (i.e., local reversals across axial planes) shear preserved (Paterson and Miller, 1998c). Foliation and folding intensity increase in several zones tens of meters wide, commonly in thinly layered units next to more massive units; we interpret these as ductile shear zones.

Mapping by Paterson indicates that the metasupracrustal rocks were intruded by internally sheeted, tonalitic to granodioritic orthogneisses that range up to 700 m in thickness and extend laterally for at least several kilometers. These orthogneisses make up ~50% of the Napeequa Complex. Internal sheets, centimeter-thick petrographic layers, and host-rock rafts are folded and boudinaged, although not as commonly as host rock units. Orthogneiss sheets display high-T subsolidus foliation and lineation, which typically obliterated magmatic foliation.

The Dinkelman décollement separates the Napeequa Complex from the structurally underlying Swakane Gneiss (Paterson and Miller, 1998c; Alsleben, 2000; Miller et al., 2000). This fault contact is commonly abrupt and can be mapped at the meter scale, whereas in other places it is an ~20–60-m-wide zone of interleaved slices of Napeequa Complex and Swakane Gneiss. Some slices are truncated by the décollement, but the regional foliation continues across the contact.

In the Roaring Creek and Columbia River transects, thin granitoid sheets and irregularly shaped intrusive bodies ranging

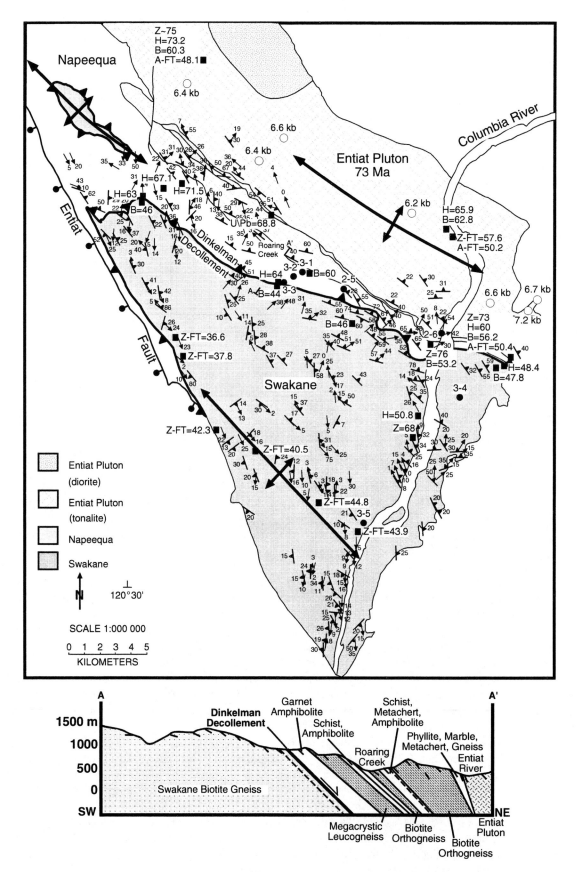

Figure 8. Geologic map of southern part of the Chelan block emphasizing structural data, geochronology (filled circles), and field trip stops. A-FT—apatite fission track date; B—biotite Ar/Ar or K-Ar date; H—hornblende Ar/Ar or K-Ar date; Z—U-Pb zircon date; Z-FT—zircon fission track date. Numbers in Entiat pluton are pressures from Al-in-hornblende barometry. Sources for dates are Tabor et al. (1987a) and our unpublished data. A–A′ is a cross section through the Roaring Creek area. Tick marks along topographic surface show dip of measured foliation. Note the intrusive sheets in the structurally higher Napeequa Complex (between Dinkelman décollement and Entiat pluton) and their absence in the Swakane Gneiss.

from meters to tens of meters in diameter intrude the deepest exposed levels of the Swakane Gneiss. Large intrusive sheets such as those seen in the overlying Napeequa Complex are conspicuously absent.

The Swakane Gneiss displays local tight to isoclinal recumbent folds followed by local gentle to tight upright folds. The dominant foliation and early folds are overprinted by pervasive top-to-N non-coaxial shear on originally sub-horizontal surfaces, as best shown by extensional crenulation cleavages (shear bands). The top-to-N shear began at medium or possibly high temperatures, as biotite in shear bands is optically similar to that in the deflected foliation. Continuation under lower-T conditions, presumably during exhumation, is indicated by synkinematic(?) retrograde muscovite and chlorite in some shear bands, local pseudotachylite, and micaceous lineations that are less pervasive and are typically oriented 10–30° counterclockwise from the dominant mineral lineation (Alsleben, 2000). The involvement of pegmatites (poorly dated at 68 Ma; Mattinson, 1972) provides some control on the timing of shearing. The strong development of the top-to-N shear in the deepest exposures of the Swakane suggests that the gneiss behaved as a thick (minimum structural thickness of 1.1 km) shear zone.

Metamorphic barometry records similar *P-T* conditions for the Napeequa and Swakane units in the southern Chelan block. Valley et al. (2003) calculated peak conditions of 640–740 °C, 10–11 kbar, for the Napeequa unit and 660 to >730 °C, 11–12 kbar, for the Swakane Gneiss. In the gneiss, replacement of ilmenite by rutile, both in matrix grains and in inclusions in garnet, implies that garnet grew during burial (Valley et al., 2003). Modification of Ca zoning in garnet, particularly in the vicinity of plagioclase inclusions, and plagioclase coronas after garnet (Stein and Stowell, 2002) indicate high-T decompression following burial (Valley et al., 2003). A clockwise *P-T* path is thus inferred for the Swakane Gneiss (Fig. 5).

K-Ar and Ar/Ar cooling ages allow evaluation of potential crustal excision across the Dinkelman décollement. In the footwall of Swakane Gneiss, biotite ages are <50 Ma, and a single hornblende age is 50.8 Ma, whereas in the Napeequa Complex, a single biotite date is 59.8 Ma and hornblende ages range from 64 to 71 Ma (Tabor et al., 1987a). These sparse age data support excision by the décollement.

This differences in cooling ages, the regional extent of the décollement, pervasiveness of non-coaxial shear in the Swakane Gneiss, and contrasts in style, scale, and abundance of intrusive material across the Dinkelman décollement suggest large displacements. Moreover, the recent recognition that the Swakane Gneiss was deposited ca. 73–68 Ma indicates that the gneiss was rapidly underthrust to depths of ≥40 km at a time when parts of the Cascades Arc were already substantially exhumed (Matzel et al., 2002b) and may have removed the underpinnings of the older arc plutons. This underthrusting also broadly coincided with tectonic emplacement of meta-clastic rocks beneath the southern part of the Sierra Nevada batholith (e.g., Jacobsen et al., 2000; Saleeby, 2003).

DISCUSSION OF PROCESSES IN THE CRETACEOUS CASCADES ARC

We raised numerous questions about magmatic arcs in the introduction. In the following, we take the observations and interpretations presented in the preceding sections and synthesize them to evaluate processes operating at different crustal levels in the Cascades Arc.

Summary of Magma Ascent and Emplacement

A variety of controversial topics regarding magma ascent and emplacement mechanisms can be addressed using the Cascades crustal section. Below, we briefly list these topics and our main conclusions, without much supporting data, and hope that this will generate discussion during the field trip.

- Are plutons more sheet-like with depth? Sheet-like bodies are present at all crustal levels, but the percentage increases with depth.
- Are plutons more compositionally heterogeneous with depth? Heterogeneous bodies are found at all depths, but more sheet-like and heterogeneous bodies are gradually replaced by more homogeneous, typically tonalitic, elliptical bodies at higher levels.
- Do plutons become more concordant with depth? Concordant and discordant bodies occur at all levels; as depth increases, smaller, sheet-like bodies are more concordant with each other and host rock structures, but larger elliptical bodies remain discordant.
- Do plutons become more deformed with depth? All of the plutons have both magmatic and subsolidus fabrics. Magmatic fabrics are more strongly developed with depth and better coupled with host rock fabrics. Most larger elliptical bodies display subsolidus deformation of their margins at all crustal levels; this deformation is more pervasive with depth in smaller sheet-like bodies.
- Does the volume of plutonic material change with depth? There is no significant increase in plutonic rock with depth preserved in large bodies, but the amount of rock in smaller, sheet-like bodies increases with depth.
- Do plutons ascend as dikes or diapirs, or by other mechanisms? Miller and Paterson (1999) have proposed that many magmas ascended as "visco-elastic diapirs," which consist of one or multiple batches of magma rising together that have length-to-width ratios of <100, ascend through host rocks deforming by both ductile and brittle processes, and ascend by buoyancy and regional stress.

  Many of the thinner, sheet-like bodies, however, probably reflect a process transitional between elastic dike and visco-elastic diapir-type mechanisms.
- Are faults and pluton ascent and emplacement genetically linked? Regional faults did not localize magmatism, although faults, fault motion and/or stress fields associated with nearby faults may have influenced internal characteristics of the Cascades plutons.

- Do emplacement mechanisms change with depth? There is strong evidence for multiple material transfer processes during emplacement at all crustal levels, including ductile flow, variable but mostly downward deflection of host rock markers in narrow aureoles (return flow), stoping, and possibly regional folding.

In summary, two types of large-volume, magmatic systems are displayed in the Cascades core. The first systems are injection complexes of sheeted bodies of all sizes and orientations, reflecting domains of poorly focused magma ascent and emplacement (e.g., Stops 1-3, 1-6, 3-2). The second are discrete sheet-like and elliptical plutons representing regions of more focused magmatism (e.g., Mount Stuart, Tenpeak, Entiat plutons). We consider both the focused and unfocused systems as frozen conduits that may have fed larger magma chambers at shallower levels.

For the focused flow case, we envision the model described in Miller and Paterson (2001b) in which several processes act as a "filter" between lower crustal zones of magma generation and upper crustal zones of large, relatively homogenous plutons. This is illustrated in part by differences between deep bodies like the Entiat and Tenpeak plutons and shallower plutons such as the Mount Stuart batholith. For example, early, predominantly mantle-derived, mafic sheets may have risen from deep magma ridges to an emplacement site, solidifying along the sides of the system but also forming heated and, at times, molten pathways. These mafic magmas may have been modified chemically during ascent by fractionation and/or interaction with lower arc crust. Space for thin, mafic bodies probably was made partly by folding and ductile flow during magma wedging, and partly by local stoping (Miller and Paterson, 2001b). Felsic magmas could not reach shallow levels until heated pathways formed because they lack sufficient heat to rise as diapirs for long distances through the crust (e.g., Marsh, 1982).

With time, continued influx of mafic magmas from the mantle may have increased melting in the lower crust. Mixing of mantle- and crustally derived magmas formed tonalite at the site of crustal melting, during ascent, and/or at the emplacement level. The earliest tonalites were likely emplaced as thin sheets that wedged apart, ductilely strained, and locally stoped parts of the mafic complex and host rocks. These sheets were then detached during wedging by younger and probably more voluminous tonalite sheets and dispersed by ductile flow and stoping as sheets coalesced and a larger magma chamber formed. Regional shortening deformation continued throughout pluton construction.

Development of heated pathways allowed larger, and thus potentially more homogenized, magma batches to reach higher crustal levels. As subsequent magma passed through these systems, it may have preferentially mixed with more evolved magmas in the centers of chambers but not interacted with the less evolved magmas frozen in sheets along the chamber margins (Miller and Paterson, 2001b). Mafic magmas were progressively overwhelmed and/or intruded out by these mixed tonalitic melts in some, but not all (e.g., Mount Stuart), plutons. Finally, larger chambers may have formed with a larger percentage of crustal

melt. All the while, host rock was displaced by multiple processes in aureoles and/or inside and through the chamber, and regional deformation persisted throughout the life of this conduit system.

Metamorphism

Many of the rocks in the Coast Mountains–North Cascades orogen display evidence for large-scale vertical motion during construction and exhumation of the magmatic arc (Whitney et al., 1999; Stowell and Crawford, 2000). Supracrustal rocks contain evidence for burial to depths ≥ 40 km, followed by rapid cooling (based on Ar/Ar mineral ages) and probable decompression to depths of <10 km. In some cases, this exhumation occurred within a few million years.

Middle Cretaceous plutons in the Nason terrane, such as the Mount Stuart batholith, have low-pressure contact metamorphic minerals in their aureoles that were partially overprinted by high-pressure regional metamorphic minerals, as evidenced by kyanite ± staurolite pseudomorphs after andalusite in metapelitic rocks. In many rocks, andalusite relics remain despite burial of the rocks to kyanite zone conditions. U-Pb zircon ages from the Mount Stuart batholith and Sm-Nd garnet ages from the aureole indicate that the high-pressure metamorphism postdated intrusion by ~5 m.y. (Stowell and Tinkham, 2003).

Rocks near middle Cretaceous plutons record the loading history with kyanite after andalusite and in some cases kyanite after sillimanite that apparently replaced andalusite (Fig. 5). Rocks that have been regionally metamorphosed (no contact effects) only preserve evidence for high-pressure conditions and/or high-temperature decompression, as indicated, for example, by sillimanite and cordierite after kyanite and garnet, respectively (e.g., Whitney et al., 1999). Evidence for earlier, prograde metamorphism has been obliterated by high-temperature reactions, and only the peak conditions and part of the decompression path are recorded (kyanit $\rightarrow$ sillimanite) (Fig. 5).

The principal driving forces for deep burial are controversial. The common clockwise shape of P-T paths that cross from the kyanite into the sillimanite field, relative timing and regional extent of the high-pressure metamorphism, and comparison of observations with thermal models are consistent with thrusting and/or pure shear thickening as primary loading mechanisms throughout the orogen (Whitney et al., 1999), rather than magma-dominated loading (Brown and Walker, 1993).

Thermal models that most successfully predict the inferred P-T paths for Cascades metamorphic rocks are those that invoke a high initial geothermal gradient (35–40 °C/km), such as would be expected in a magmatic arc (Whitney et al., 1999). Magmatism was therefore important in the tectonic-thermal history of the orogen, although it was not the primary loading mechanism for deep burial of supracrustal rocks.

We conclude that major vertical displacement of rocks in arcs during magmatism (arc construction) is typical of magmatic arcs and an important element of the tectonic evolution of this type of orogen.

Timescales of Magmatic and Tectonic Processes

The Cascades core provides an excellent opportunity to determine rates of magmatic and tectonic processes at a range of structural levels. Geochronologic results from the shallowly emplaced Mount Stuart batholith indicate that it was constructed over an ~5.5 m.y. time period. Magmas of a wide range of composition, from gabbro to granodiorite, were emplaced over the entire interval, and mafic magma was volumetrically significant throughout. Magmatic foliations are continuous across pulses of different age and composition, suggesting that parts of the batholith remained a crystalline mush while new phases were emplaced, and/or that the strain field remained constant.

In contrast, the deep-seated Tenpeak pluton was emplaced during a shorter, ~2.5 m.y., time period. Older mafic complexes and interlayered mafic and tonalitic sheets were intruded out by voluminous tonalite pulses.

We have also determined rates of tectonic burial from contrasting parts of the Cascades core using a combination of geochronologic and *P-T* studies. Rates of vertical motion can be calculated utilizing zircon U-Pb ages for the northernmost Mount Stuart batholith (93.5 ± 1.5 Ma; Tinkham, 2002) and garnet Sm-Nd ages from adjacent host rock (86.1 ± 1 Ma; Stowell and Tinkham, 2003). The northwest edge of the batholith was emplaced at ≤4 kbar, based on the presence of andalusite in the contact metamorphic aureole, and *P-T-t* paths for garnet growth indicate peak pressures of ~6 kbar (Stowell and Tinkham, 2002). Taking the age difference as 7.4 m.y. and the pressure difference as 2.2 kbar, the rate of pressure increase would have been 0.3 kbar/m.y. This is equivalent to a loading rate of ~1.1 km/m.y.

In contrast, rates of burial of the 9–12 kbar Swakane Gneiss must have been much faster. U-Pb analyses of detrital zircons from the gneiss indicate that the maximum age of deposition of its protolith is 73 Ma (Matzel et al., 2002b). After burial, the gneiss was intruded by several generations of peraluminous dikes that are thought to have been derived from partial melting of the gneiss (Boysun and Paterson, 2003). One such peraluminous dike yielded a zircon crystallization age of 68.5 Ma (Matzel et al., 2002b). If we assume that this dike was intruded close to the time that the gneiss reached its peak pressure, then burial rates are on the order of 6–9 km/m.y.

Structure, Kinematics, and Rheology of the Cascades Crustal Section

In the introduction, we pointed out the controversy over whether arc magmatism occurs during regional shortening or extension. We have presented evidence that plutons emplaced at various crustal levels were intruded during arc-perpendicular shortening. This shortening is best illustrated by syn-magmatic folds in plutons and syn-emplacement folds and reverse shear zones in host rocks.

We also introduced the debate over whether Cretaceous structures formed in response to pervasive dextral strike slip or regional shortening. Folds of strong foliation dominate the struc-

ture from the outcrop to map scale over most structural levels of the arc (Miller and Paterson, 2001a). Layering was mechanically active during folding, and shear on fold limbs dominates the kinematic pattern of the Chiwaukum Schist and Napeequa unit. This pattern is interrupted by syn- to post-emplacement, SW-vergent reverse shear zones next to several large plutons where lineation is down-dip, and kinematic indicators that are well formed in lineation-parallel surfaces. Thus, the structures indicate the Cretaceous tectonic regime was one of regional shortening.

Crustal strength profiles constructed for the Cascades core emphasize the heterogeneous vertical partitioning of deformation and complex rheological stratification of arc crust at all scales (Miller and Paterson, 2001a). Heterogeneous brittle deformation dominates the shallow part of the section (≤10 km), and strength presumably increases with depth. The brittle-ductile transition is inferred to lie at relatively shallow levels, as implied by ductile deformation in deeper parts of the Ingalls Complex, and is compatible with the high upper-crustal geothermal gradients in arcs.

Heterogeneous ductile deformation characterizes the rest (10–35 km depth) of the exposed part of the Cascades section. The overall pattern of decreasing strength with depth fits idealized crustal strength profiles, but in more detail, relative strengths differ markedly. The importance of competence contrasts is shown by the mechanically active role of layering during folding, cleavage refraction, boudinage, and focusing of shearing in weaker layers at all crustal levels (Miller and Paterson, 2001a). Analogous patterns result from small compositional differences at the outcrop to thin section scale, indicating that rocks in the mid to deep crust of this arc were able to accumulate large ductile strains but had distinct relative strengths.

Magmatic bodies strongly influenced the rheology of the Cascades crustal profile. The clearest effect of the Cretaceous plutons is the intensification of ductile deformation in their narrow structural aureoles and the kinematic partitioning of strain adjacent to several plutons. In contrast, displacement on the relatively gently dipping Windy Pass thrust was probably terminated by solidification of the Mount Stuart batholith, and shortening was transferred elsewhere in the Chiwaukum Schist (Miller and Paterson, 1994; Paterson and Miller, 1998a). Thus, once plutons solidified, many acted as highly competent bodies, and the ensuing deformation was either focused in adjacent host rocks, which may have also been thermally weakened, or the plutons created strain shadows, partitioning strain to other domains.

We cannot demonstrate that deformation was localized in the pluton magma chambers before they reached the solidus (e.g., Davidson et al., 1992; Neves et al., 1996; Pavlis, 1996), because of the poor strain memory of these bodies. All of the plutons display magmatic foliations that are continuous with regional foliations and at least in part reflect regional strains (see above) (Paterson and Miller, 1998a; Miller and Paterson, 2001b). Deeper plutons may have accumulated much more regional strain than shallower ones, as they should have taken longer to crystallize.

Relationships between thinner magma sheets and host rock deformation are more variable. Strain may have been focused in the sheets while they were melt-rich systems, but their crystallization durations were much shorter than the larger plutons at similar depths. After solidification, the sheets behaved as competent bodies during folding and boudinage. Shear zones formed in weak host rocks adjacent to some sheets, but the sheets generally did not cause marked strain gradients in their hosts (Miller and Paterson, 2001a). This relationship probably reflects their concordance, quartz-bearing mineralogy, and relatively limited size, which would have caused little thermal perturbation of the host rock.

ROAD LOG

The road log begins northeast of Seattle at Skykomish, on U.S. Highway 2. A few miles east of Skykomish, the route along U.S. 2 crosses the trace of the dextral Eocene Straight Creek fault, which forms the western boundary of the Cascades core. From there to Stop 1, the road crosses through the Tonga Formation, a low-grade equivalent of the Chiwaukum Schist; the Beckler Peak stock, a 92–94 Ma satellite of the Mount Stuart batholith; the Evergreen fault, a mostly dip-slip Tertiary fault; a thin strip of the Chiwaukum Schist; and the southwest body of the Mount Stuart batholith. Yeats et al. (1977), Haugerud et al. (1994), and Paterson et al. (1994) provide road logs discussing the bedrock geology between Everett and Stevens Pass.

Day 1

The focus of Day 1 is the middle Cretaceous structure, plutonism, and metamorphism of the Wenatchee block. The general sequence of stops is from structurally higher to lower levels, although logistics will require a few stops to deviate from this pattern. Universal Transverse Mercator (UTM) coordinates (all zone 10, North America datum 1927) are given for all stops.

Miles	Description
0.0	Junction of U.S. Highway 2 and road that leads across Skykomish River into the town of Skykomish.
7.7	Deception Falls parking area.
12.5	Turn right onto gravel Forest Service Road 6095.
12.8	Stay right at junction.
13.2	Stay straight at junction.
13.4	At junction, pull off to the right and park.

Stop 1-1. Chiwaukum Schist in the Aureole of the Mount Stuart Batholith (UTM 0641689E; 5286346N)

This stop is in a narrow septum of Chiwaukum Schist between the southwest and northeast bodies of the Mount Stuart batholith. We will introduce the regional geology and then observe andalusite-bearing mineral assemblages and syn-emplacement structures in the contact metamorphic and structural aureole of the batholith. Roadcuts contain folded andalusite + garnet + biotite schist typical of the Chiwaukum Schist exposed adjacent to the northern contact of the hook-shaped part of the batholith. Most rocks display steep foliation. Fold axes and mineral lineations are variably

oriented in the septum and regionally have gentle northwest or southeast plunges, but a steep andalusite lineation prevails here. Felsic dikes that are thought to be related to the batholith show significant deformation and recrystallization. Based on oxygen isotopes, plutonic rocks on the northeast side of the screen exhibit variable degrees of host rock interaction. Engels and Crowder (1971) reported K-Ar ages of 85 Ma from both hornblende and biotite from nearby, suggesting rapid cooling.

This screen of Chiwaukum Schist is less than 1 km wide and lies entirely within overlapping contact aureoles of the southwest and northeast bodies of the Mount Stuart batholith. Spectacular andalusite porphyroblasts, which are typically several centimeters in length and display chiastolite crosses, can be found with smaller porphyroblasts of staurolite and garnet. The andalusite is relatively pristine, but some crystals are replaced by sillimanite or staurolite. Fibrolite also replaced biotite as aligned crystals in the necks of boudinaged andalusite crystals and in bundles overprinting the foliation (Evans and Berti, 1986). Euhedral staurolite crystals locally replaced andalusite or grew across the matrix foliation. Much of the andalusite in this part of the Wenatchee block has been replaced by sillimanite, kyanite, and/or staurolite that grew during regional metamorphism that postdates the Mount Stuart batholith.

Thermobarometry for sillimanite + garnet + biotite schist from nearby host rock indicates a *P-T* path originating at andalusite conditions similar to those of pluton emplacement and ending at higher pressures near the kyanite-sillimanite transition (Paterson et al., 1994). Stowell and Tinkham (2003) report a quantitative *P-T-t* path and a Sm-Nd garnet age of 86.1 ± 1 Ma for metamorphism of similar rocks adjacent to and in the aureole north of the Mount Stuart batholith (Fig. 5). Rocks in this area also experienced andalusite zone metamorphism synchronous with batholith emplacement. Garnet growth initiated during subsequent, regional metamorphism at ~4.8 kbar and 595 °C and culminated at ~6.0 kbar and 670 °C.

Continue SE on gravel road toward trailhead.

Miles	Description
13.7	Turn around and retrace route to Highway 2.
14.9	Turn right (E) onto Highway 2.
19.1	Stevens Pass. We are driving through the hook-shaped part of the 91–96 Ma Mount Stuart batholith.
23.2	Turn left onto Smith Brook Road (FS 6700) and park. We will examine roadcuts between fast moving west- and east-directed traffic on Highway 2.

Stop 1-2. Mount Stuart Batholith (UTM 0646792E; 5294096N)

These roadcuts are typical of the hook-shaped region of the northeast body of the batholith. Hornblende-biotite tonalite (this outcrop) contains microgranitoid enclaves and has moderately strong magmatic foliation and weak lineation defined best by hornblende and plagioclase. In this area, the foliation is folded by decameter- to kilometer-scale magmatic folds (Paterson and

Miller, 1998a; Benn et al., 2001). The outer 1–2 km of the batho-lith to the north of this stop commonly contains peraluminous garnet- and muscovite-bearing biotite tonalite and granodiorite, unlike the hornblende-bearing tonalite in most of the batholith. Whole rock samples from near the margin exhibit elemental variations off the main batholith trends and have elevated delta ^{18}O, indicating variable but significant wall rock interaction (Paterson et al., 1994).

Several zircon grains from this outcrop, analyzed by conventional U-Pb geochronology, contained slightly older inherited cores, but the tip of one grain yielded a concordant analysis with a ^{206}Pb/^{238}U date of 96.2 Ma (Matzel et al., 2002a). Zircon grains from nearby outcrops of granodiorite also contain inherited cores, but concordant analyses yield an age that overlaps within error of the age obtained from this outcrop (Matzel et al., 2002a). Engels and Crowder (1971) reported K-Ar ages of 95 Ma from hornblende and 82 Ma from biotite in the batholith near this locality.

Miles	Description
26.7	Contact between the Mount Stuart batholith and Chiwaukum Schist.
29.0	Outcrops of folded kyanite, staurolite, and garnet-bearing Chiwaukum Schist lie on both sides of the highway, and we are now outside of the aureole of the Mount Stuart batholith. See Paterson et al. (1994), their Stop 3-4, for detailed discussion.
31.0	Carefully pull off highway onto dirt road on right side just past roadcut beneath powerlines. Cross highway and examine large roadcuts on north side of road.

Stop 1-3. Nason Ridge Migmatitic Gneiss (UTM 0657900E; 5294372N)

Biotite schist and amphibolite of the Chiwaukum Schist are intruded by numerous dikes and sills in this outcrop, which is part of a regional injection zone named the Nason Ridge Migmatitic Gneiss (Tabor et al., 2002). The outcrop lies ~3 km northeast of the Mount Stuart batholith. It is dominated by dikes and sheets; amphibolite and biotite schist are most abundant at the west end of the outcrop. The dominant assemblage in the pelitic schists is garnet-kyanite-biotite-quartz-plagioclase-muscovite, and staurolite is found locally (Magloughlin, 1994). Magloughlin (1994) calculated *P-T* conditions of 6.5 kbar and 572 °C from this outcrop. The Chiwaukum Schist has been multiply folded; a prominent hinge of folded foliation in amphibolite is particularly well displayed on the south side of the highway.

The intrusive sheets are texturally variable, range from diorite to granite, and are dominantly tonalitic (cf. Getsinger [1978] and Magloughlin [1994] for detailed descriptions). Rb-Sr and Sm-Nd analyses suggest that some of the compositional variability results from contamination by the biotite schist (Magloughlin, 1994). The sheets also display variable orientations and crosscutting relations. Most lack subsolidus foliation, but some are folded or boudinaged and have strong, gently plunging magmatic and locally subsolidus

mineral lineation. Numerous small faults and a steep, low-temperature shear zone are also present. Overall, the sheets record less subsolidus deformation and recrystallization than to the northeast, in the Nason terrane (e.g., Stop 1-6) (Haugerud et al., 1994). Paterson et al. (1994) inferred that the sheets were emplaced during the latest stages of NE-SW shortening and NW-SE extension in the southern part of the Wenatchee block.

U-Pb zircon ages of 89–90 Ma (Walker and Brown, 1991) from northwest of this outcrop in the Nason Ridge Gneiss suggest that the tonalite sheets are younger than the Mount Stuart batholith. Age constraints for this outcrop are a biotite Ar/Ar date of 83 ± 1 Ma (S.A. Bowring, *in* Paterson et al. [1994]) from a tonalitic sheet, Rb-Sr whole-rock muscovite dates of 73 ± 1 Ma and 83 ± 1 Ma from two pegmatites (Magloughlin, 1994), and a Rb-Sr whole-rock biotite date of 75 ± 5 Ma from a schist (Magloughlin, 1994).

The significance of the sheets remains enigmatic. The multiple magmatic pulses may mark unfocused magma pathways or represent the plumbing system for a larger pluton, but this is unlikely to be the Mount Stuart batholith, given the apparently different ages. Brown and Walker (1993) postulated that the sheets were intruded at the margin of a now eroded batholith, which was emplaced above, and caused the loading recorded by the metamorphism of the Chiwaukum Schist.

Continue east on U.S. 2.

Miles	Description
36.4	Rest area on left. A short distance west of here we crossed the Eocene Leavenworth fault and are now in the middle Eocene Chumstick basin.
39.2	Junction with State Route 207. Turn left (north) on Route 207 toward Lake Wenatchee State Park.
42.7	Turn left into Lake Wenatchee State Park.
43.6	Lake Wenatchee. Park in parking lot by the lake.

Stop 1-4. Overview of Geology in Lake Wenatchee Region (UTM 0670200E; 5297513N)

We are looking west across the Eocene Leavenworth fault to get an overview of the regional structure and rock types examined for the remainder of Day 1. Across the lake, the Little Wenatchee River separates Nason Ridge to the southwest of Lake Wenatchee from Wenatchee Ridge, immediately to the west. The White River drainage and Dirtyface Mountain lie to the northwest. Regional upright, gently plunging folds occur throughout this region, with the hinge line of a major antiform passing through the lake and along Wenatchee Ridge. The axis of a smaller synform occurs along the Little Wenatchee River and another antiformal axis lies along Nason Ridge. The Dirtyface pluton is exposed in the upper part of Dirtyface Mountain in the northeast limb of the Wenatchee Ridge antiform. We have qualitatively unfolded these regional folds to construct our crustal section (Figs. 3 and 4).

Much of the eastern end of Nason and Wenatchee Ridges consists of the Wenatchee Ridge Gneiss (Stop 1-5) and overlying gneisses. Structurally higher areas (e.g., Dirtyface Mountain and northwest) are dominated by Chiwaukum Schist intruded by tonalitic orthogneiss sheets. The Tenpeak pluton (Stops 1-8 and 1-9) is located in a separate structural domain behind Dirtyface Mountain.

Retrace route to Route 207.

Miles	Description
44.5	Turn left on Route 207.
44.7	Junction with road to Plain—continue straight on Route 207.
45.2	Junction with Chiwawa Loop Road—bear left on Route 207.
46.1	Views of Dirtyface pluton to the northwest. The Leavenworth fault passes through the break in slope near the low saddle at the east end of the high ridge where it truncates this pluton.
49.9	Lake Wenatchee U.S. Forest Service Ranger Station.
51.4	Turn right onto White River Road.
52.0	Pull off road on right next to roadcut.

Stop 1-5. Boudinage of Wenatchee Ridge Gneiss (UTM 0661547E; 5302422N)

This outcrop consists of leucocratic sheets of the ca. 93 Ma Wenatchee Ridge Gneiss in the hinge region of an open, upright, gently plunging fold (oriented 7/320). Spectacular boudinaged layering and a strong mineral lineation oriented parallel to the fold axis are well displayed. We do not see consistent kinematic indicators here, and interpret the strong lineation as reflecting strain resulting from superposition of the upright folds on earlier recumbent isoclinal folds. Microstructures typically are annealed.

Continue northwest on White River Road.

Miles	Description
53.9	Turn left across White River toward Sears Creek (outcrop of Wenatchee Ridge Gneiss on right). We are now on the northeast limb of the major antiform cored by the Wenatchee Ridge Gneiss.
56.1	Pull off road on right next to a pond, across from large roadcut.

Stop 1-6. Chiwaukum Schist/Nason Ridge Migmatitic Gneiss Contact and Overview of Dirtyface and Tenpeak Plutons (UTM 0658024; 5306927N)

This stop contains rocks that display complex folding and shearing in schist and gneiss, orthogneiss sills, and mylonitic fabrics. Rocks display characteristics of both the Chiwaukum Schist and adjacent Nason Ridge Migmatitic Gneiss, and these rocks may mark a transitional zone between the units. The outcrop contains proportions of concordant orthogneiss that vary from ~15% on the north end to ~35% on the south. Approximately 100 m south of the large outcrop, a small exposure of gneiss displays mylonitic foliation with a distinctly different orientation from foliation in the nearby gneiss.

The large outcrop contains parasitic folds on the limbs (270/60N) of the major folds that show variable geometries, but typically plunge moderately to the east (53–60°). Some of the leucocratic tonalite sills have been deformed into spectacular asymmetric boudins, indicating top-to-S motion. The concordant tonalitic sheets contain well-developed solid-state fabrics and are significantly more strained than similar rocks at Stop 1-2. The gneiss is cut by subhorizontal (Magloughlin, 1994) and foliation-parallel cataclastic zones and pseudotachylite. The mylonite outcrop has a foliation at 320/70NE and mineral lineations and fold axes that plunge 20–65° toward 345°.

The structural setting and significance of these features are uncertain. These outcrops are in the northeast-dipping limb of the Wenatchee Ridge antiform and the southwest margin of the Dirtyface domain, which is the zone of complex folding and top-to-SW shearing immediately below (southwest of) the Dirtyface pluton. The northwest tip of the Dirtyface pluton is directly east-southeast across the river. Finally, the White River shear zone is located a short distance northeast of this outcrop. All of these structural settings are compatible with top-to-SW shearing. The complexity of the structures in the outcrops may result from strain partitioning due to rheological contrasts between pluton and host rock.

Rocks in the adjacent Nason Ridge Gneiss and in rafts within the Dirtyface pluton have porphyroblasts of garnet and kyanite ± staurolite. Thermobarometric results for these rocks are some of the highest pressures (7–9 kbar) determined in the Chiwaukum Schist (Magloughlin, 1993; Brown and Walker, 1993; Miller et al., 2000; Tinkham, 2002). Tinkham (2002) constructed a quantitative P-T-t path for garnet + biotite + kyanite schist from the ridge 5.5 km west of the outcrop. Garnet growth was near isobaric beginning at 545 °C and ~7.1 kbar and ending at 620 °C and 7.6 kbar (Fig. 5). No data directly constrain the timing of this metamorphism, but it is inferred to correlate with the 88–86 Ma post-Mount Stuart metamorphism northeast of the batholith.

Turn around and retrace route to White River Road.

Miles	Description
58.3	Turn left (north) on White River Road.
61.8	Pull off road at Napeequa Crossing Campground and Twin Lakes Trailhead. Walk ~20 m along road to outcrop just before paved road ends.

Stop 1-7. (Optional) Retrograded Chiwaukum Schist in White River Shear Zone (UTM 0653879E; 5312837N)

We are in the southern part of the White River shear zone, which carries the Tenpeak pluton and Napeequa Complex above the Chiwaukum Schist. In the footwall schist, mineral lineation swings from its general subhorizontal, NW-SE trend to more steeply pitching NE orientations and the NE-dipping foliation intensifies. Relict garnet, staurolite, and kyanite porphyroblasts exist in nearby outcrops. The schist also shows syn-kinematic greenschist-facies retrogression in the shear zone associated with porphyroblasts of plagioclase that contain crenulated inclusion trails.

Return to vehicles and continue on White River Road, crossing the Napeequa River.

Miles	Description
62.1	Stay left at junction with road to Tall Timbers Ranch.
62.7	Ultramafic slice of Napeequa Complex on right. The Napeequa forms a thin rind between the 90–92.5 Ma Tenpeak pluton and the Chiwaukum Schist. As we continue up river, cliffs in the Tenpeak pluton rise above us to the right.
64.9	Turn left into White River Falls Campground.
65.1	Park at end of road.

Stop 1-8. White River Shear Zone—Tenpeak Pluton (UTM 0653879E; 5312837N)

In outcrops next to the White River, tonalites of the Tenpeak pluton are strongly deformed in the White River shear zone. Foliation dips ≤30° to the east, less steeply than in much of the shear zone, and mineral lineation has high pitches. S-C fabrics and asymmetric plagioclase, biotite, and hornblende porphyroclasts indicate reverse (top-to-WSW) shear. C-surfaces in the Tenpeak tonalites are generally marked by biotite, finely recrystallized quartz, and locally, amphibole. These microstructures and the presence of recrystallized plagioclase mosaics are compatible with deformation at moderate temperatures. Darker, fine-grained, hornblende-bearing mylonites (best seen at low water levels) may be strongly deformed dikes and/or xenoliths of the Napeequa unit. They reach 80 cm in thickness.

Return to White River Road.

Miles	Description
65.3	Turn left onto White River Road.
65.8	Pull off road where small stream and avalanche chute come down to the road on the right and open views occur to the left toward Mount David. Walk up the stream to first outcrops.

Stop 1-9. Sheeted Tonalite in Tenpeak Pluton (UTM 0653848E; 5313541N)

This streamcut provides a spectacular view of the sheeted part of the Tenpeak pluton, where evidence of magma mingling and mixing processes is easily observed. As one walks up the drainage, stream-polished outcrop exposes light and dark sheets of mafic and felsic tonalite, which at this locality cannot be related to each other by crystal fractionation (Miller et al., 2000). However, the range of observed compositions can be reproduced by mixing the most mafic end-member with the most felsic end-member at this outcrop (S.M. DeBari, 2000, personal commun.). Physical evidence for this mixing can be seen where large phenocrysts of plagioclase from the felsic tonalites have been intermixed into the finer-grained mafic tonalites. The felsic tonalites have the same REE signature as Archean tonalite-trondhjemite-granodiorite (TTG) suites, suggesting that they were derived from a mafic, garnet-bearing source. This source is most likely the mafic lower crust (S.M. DeBari *in* Miller et al., 2000).

The ca. 92 Ma sheeted zone has been mapped as a NW-trending unit for ~4 km along strike (Fig. 7). It is truncated to the northwest by the more homogeneous, ca. 90 Ma Indian Creek phase of the batholith, implying that the sheeted zone was stoped out by this comparatively voluminous body. The sheeted zone is also cut by dikes, some of which resemble the fine-grained mylonites at Stop 1-8. The dikes are strongly deformed, and in many places it is difficult to distinguish them from xenoliths of the Napeequa unit. The sheets display strong NE-dipping (335/35NE) magmatic and parallel high-T subsolidus foliations that are concordant to sheet contacts. Subsolidus lineation plunges nearly down-dip. Weak C-surfaces deflect the foliation and indicate reverse shear (pluton interior up) compatible with that in the White River shear zone.

From this outcrop, we have a nice view of Mount David to the west. High on the east flank of the peak is the moderately north-dipping contact between the Chiwaukum Schist and Napeequa Complex in the White River shear zone.

Continue north on White River Road.

Miles	Description
66.2	Turn vehicles around at the White River trailhead, and retrace route on White River Road.
76.1	Turn left on Route 207.
86.5	Junction with U.S. 2. Turn left (east) for ~15 mi (24 km) to Leavenworth, where we will spend the night. En route to Leavenworth, we will drive through the Mount Stuart batholith and Ingalls Complex, crossing the syn-emplacement, reverse-slip, Tumwater Mountain shear zone (see Paterson et al. [1994], Stop 3-1).

Day 2

We begin the day viewing the Mount Stuart batholith and Chiwaukum Schist. We then drive east into the Chelan block and examine outcrops of the Napeequa Complex and Entiat pluton.

Leave Leavenworth heading west on U.S. 2.

Miles	Description
0.0	Turn left (west) on Icicle Road. Mount Stuart batholith is to right.
0.8	Cross Wenatchee River.
3.1	First exposures of Mount Stuart batholith.
3.6	Outcrop of Mount Stuart tonalite dated ca. 91.0 Ma (Matzel et al., 2002a). Tonalite here is slightly coarser-grained and contains a higher proportion of hornblende to biotite than at Stop 1-2 in the hook region of the batholith.
4.2	Snow Creek Trailhead to left.
5.1	Turn into pullouts on left.

Stop 2-1. Geochronology, Petrology, and Structure of the Mount Stuart Batholith (UTM 0671264E; 5267634N)

This outcrop is in the mafic part of the batholith and exhibits the scale of heterogeneity that can be found elsewhere. The mafic rocks have plagioclase, orthopyroxene, hornblende ± clinopyroxene, and biotite as major phases. Enclaves at this outcrop contain more plagioclase than the host diorite, but the more common case elsewhere is that the enclaves are more enriched in mafic phases. Besides the enclaves and host rock, note the feathery net-veining, which we interpret to mean that relatively viscous diorite magma cracked and was reintruded by more evolved melts.

Zircon from the eastern end of this outcrop yields a ca. 91.0 Ma crystallization age. This age agrees, within uncertainty, with zircon ages obtained from gabbro along U.S. 2, granodiorite in the southern mushroom-shaped region, and tonalite along Icicle Canyon Road (see mile 3.6 above).

Magmatic foliation is well developed in this roadcut and is defined by aligned plagioclase and mafic minerals, particularly hornblende. The foliation dips moderately to steeply west, but changes to east-dipping orientations to the west as we cross the "hinge" of an internal, magmatic structure. This structure cuts internal compositional boundaries, implying that the foliation formed after different magma pulses had arrived at their present position in the chamber. We thus argue that this foliation only records strain during final crystallization (cf. Paterson et al., 1998).

Also note the prominent joint sets, some sealed by quartz-feldspar–rich dikelets and some with slickensided surfaces. In the sill-like part of the batholith, dike-sealed joints are at high angles to the WNW-ESE direction of maximum stretch in the host rock (Miller and Paterson, 1994).

Continue west on Icicle Road.

Miles	Description
5.5	Magmatic foliation has changed to an east dip.
6.3	This exposure lies near the southwest edge of the mafic part of the batholith. Engels and Crowder (1971) reported K-Ar ages of 90 Ma from hornblende and 88 Ma from biotite from near this locality. See Paterson et al. (1994) for a more complete discussion.
8.4	Turn left onto Forest Service Road 7601, heading toward Eightmile Lake and Stuart Lake trailheads.
8.6	Gravel road, stay right. Cross bridge over Icicle Creek.
9.9	Outcrop of Mount Stuart tonalite with strong, steeply dipping foliation.
11.2	Park on left across from brown U.S. Forest Service post labeled 7601. Walk back up the undrivable logging road ~50 m.

Stop 2-2. Mount Stuart Fabrics, Geochronology, and the Windy Pass Thrust (UTM 0665016E; 5267279N)

This outcrop illustrates the strong, gently dipping magmatic foliation with weak subsolidus overprint that characterizes the Mount Stuart batholith near the Windy Pass thrust and a large (2.2 km²), strongly deformed inclusion of Chiwaukum Schist. We infer that movement on the thrust formed the intense foliation (see introductory text) while melt was present, and solidification of the batholith may have led to cessation of thrust movement and partitioned shortening elsewhere into the weaker schist (Miller and Paterson, 1994).

This stop lies at the northern contact of the schist inclusion with the structurally overlying batholith. In the inclusion, the schist is commonly migmatitic (Stop 2-3). Foliation in the schist is axial-planar to folds of compositional layers, and weak lineation trends approximately E-W. Lineation is also weak in the tonalite, compatible with the flattening fabric in this domain. Foliation in the tonalite is discordant to the contact but subparallel to foliation in the schist, supporting evidence throughout the batholith that magmatic foliation formed late during crystallization and records strain, not flow parallel to the margin of the body.

A tonalite similar to that at this outcrop collected 500 m to the west yielded a U-Pb zircon age of ca. 94.4 Ma (Matzel et al., 2002a).

Continue south up Eightmile Road.

Miles	Description
12.2	Stuart Lake trailhead. Park.

Stop 2-3. Windy Pass Thrust-Related Structures and Migmatitic Chiwaukum Schist (UTM 0664136E; 5265883N)

This locality is in the southwest part of the large inclusion of Chiwaukum Schist mentioned at Stop 2-2. Outcrops along the fire road that switches back to the northeast provide excellent examples of migmatitic schist and intense, flat-lying structures that we attribute to Windy Pass thrusting. Foliation is parallel to boudinaged quartz-rich layers and leucosomes, and is axial-planar to rootless fold hinges. Tourmaline-bearing, leucocratic dikes cut the foliation at nearly right angles.

The dominant metamorphic assemblage in the schists is garnet-biotite-plagioclase-quartz-cordierite-opaque phase. Whitney et al. (1999) reported temperatures of ~700 °C from these rocks, which are higher than those recorded in much of the thermal aureole of the Mount Stuart batholith. This is compatible with the development of the stromatic migmatites and may have led to thermal weakening of the schists during Windy Pass thrusting. Leucosomes have thin selvages of cordierite and biotite; the cordierite replaces biotite, is presumably peritectic, and locally contains inclusions of sillimanite and hercynite. In the mesosome, garnet inclusion trails are crenulated, with axial planes parallel to matrix foliation. This foliation wraps the garnet porphyroblasts, and strain shadows are well developed. These relations imply that garnet grew before or early during formation of thrust-related foliation.

Retrace route to Icicle Creek Road.

Miles	Description
15.8	Turn left (west) and continue up Icicle Canyon.
17.9	Grindstone Mountain (at 12 o'clock) preserves host rock in the roof of the batholith (cf. Paterson and Miller, 1998a).
18.6	At base of mountain on right (north), red-weathering rocks are part of a klippe of the Ingalls Complex. Engels and Crowder (1971) reported K-Ar ages of 93 Ma from hornblende and 88 Ma from biotite near here.

18.8 End of paved road.

20.6 Ida Creek Campground. Walker and Brown (1991) obtained a 93 Ma U-Pb zircon age from a sample of the Mount Stuart batholith near here, and Matzel et al. (2002a) determined a ca. 93 Ma U-Pb zircon age for a tonalite at the head of Ida Creek in the sill region.

21.8 Contact between the Chiwaukum Schist and Mount Stuart batholith.

22.4 Chatter Creek Guard Station. Park on left at Icicle Gorge Trailhead 30 m beyond guard station.

Stop 2-4. Structures, Metamorphism, and Ultramafic Blocks in Chiwaukum Schist (UTM 0658946E; 5274579N)

We are in the Chiwaukum Schist, a short distance southwest of the central sill-like region of the Mount Stuart batholith and well below the Windy Pass thrust. We start by examining stream-polished surfaces (be careful!) directly west of the footbridge on the north bank of Icicle Creek. These surfaces illustrate styles of folding in multiply deformed quartz-mica-garnet-fibrolite schists. Overprinting relations (e.g., cross-cutting cleavages and refolded folds or cleavages) preserve evidence for at least two transposition cycles, which we relate to pre- to syn-emplacement regional NE-SW shortening. Note the subhorizontal, NW-SE–trending mineral lineation and fold axes, asymmetry of folds, and north-dipping axial planes that shallow with decreasing inter-limb angle. Most of the youngest folds and some of the older folds have asymmetries indicating top-to-SW motion. Older quartz-rich layers are folded, whereas quartz-feldspar-rich "leucosomes" and garnet-bearing dikes cut most of the older structures, but are locally deformed. Paterson et al. (1994) reported a temperature of 685 °C from biotite-garnet schist at this locality, and a sample near this stop yielded a K-Ar biotite age of 85 Ma (Engels and Crowder, 1971).

Return to the main road and walk up this road (northwest) 25 m to the first roadcut where garnet-bearing dikes cut the multiply deformed Chiwaukum Schist. Similar tonalite dikes found throughout the southern Nason terrane have plagioclase coronas around single or multiple garnet grains (Stein and Stowell, 2002). Garnet zoning, and plagioclase zoning in the coronas and matrix, are compatible with garnet growth, consumption of early high-An plagioclase, and growth of lower-An plagioclase during prograde metamorphism. This is in marked contrast to earlier interpretations of garnet breakdown to plagioclase due to high-T decompression.

A NE-SW cross section through this region (Taylor, 1994) shows that foliation is steep within a few hundred meters of the Mount Stuart contact, dips moderately northeast in the central Icicle Canyon area (e.g., this locality), and gradually flattens to near horizontal dips near the Windy Pass thrust, as seen at Stops 2-2 and 2-3. Also note that from the road we can see orange-weathering metaperidotite in the hanging wall of the Windy Pass thrust on the distant, high ridge to the southwest.

Walk back down (southeast) the road and climb east up the embankment of the turnoff to Chatter Creek Campground. On the knoll a few tens of meters north of the road (UTM 0659014; 5274605) are blocks of metaperidotite, encased within schist, which are part of a ~300-m-wide, E-W–trending belt that passes through the campground and has been traced for ~1.1 km to the east. Typical mineral assemblages include forsterite-tremolite-talc (amphibolite facies). Amphibolites are also common in and near this belt and may be related to the ultramafites.

The relationship of the ultramafites in the southern part of the Chiwaukum Schist to the pelitic schists can be explained by at least three hypotheses, which are discussed in detail in Paterson et al. (1994): (1) The ultramafites are imbricate slices and/or infolded klippen of the Ingalls Complex, presumably associated with the Windy Pass thrust; (2) The ultramafites represent serpentinite blocks that slid from an uplifted mass of the Ingalls Complex into the sedimentary protoliths of the schist (Tabor et al., 1987b), implying that the protoliths are Late Jurassic or Early Cretaceous; (3) The ultramafites are not part of the Ingalls Complex, but are slices of oceanic mantle imbricated with the Chiwaukum clastic protolith, perhaps in an accretionary wedge before or during metamorphism.

Return to vehicles, and retrace route to U.S. 2.

Miles	Description
37.5	Turn right (east) onto U.S. 2 and pass through Leavenworth.
39.4	For the next 18 mi (29 km), we will pass by intermittent outcrops of a thick sequence of fluvial and lacustrine sedimentary rocks of the middle Eocene Chumstick basin. This basin is considered by Evans and Johnson (1989) to have formed during dextral transtension and slip on the bounding Entiat and Leavenworth faults. The internal structure of the basin is dominated by fault-parallel folds that record modest late Eocene(?) NE-SW shortening.
43.1	Junction with U.S. 97; stay straight on U.S. 2 toward Wenatchee.
57.3	Exit right, staying on U.S. 2 and U.S. 97; do not go into Wenatchee.
58.3	Exit right on U.S. 97A toward Entiat (do not cross the Columbia River). Note flows of Miocene Columbia River Basalt on the opposite (east) side of the river.
58.7	Begin outcrops of Swakane Gneiss, the deepest exposed unit in the southern part of the Cascades core. We have crossed the Entiat fault near the interchanges and are now in the Chelan block. The Entiat fault dips steeply southwest; it experienced Eocene normal slip and has been postulated to have substantial dextral slip (~30 km, Tabor et al., 1987a). We will drive for the next 12 mi (19.4 km) through the Swakane Gneiss in the canyon of the Columbia River. The river provides an excellent cross section through the regional, open, gently NW-SE–plunging antiform, which controls the outcrop pattern of the gneiss (Figs. 2, 3, and 8).
62.5	Rocky Reach Dam.

63.9   Swakane Canyon Road. We are in the hinge of the regional antiform in the deepest part of the Cascades core.

66.5   Note the overall moderate northeast dips of foliation as we cross onto the northeast limb of the antiform.

68.3   We cross the unexposed trace of the Dinkelman décollement, which separates the Swakane Gneiss from overlying Napeequa Complex.

68.5   Unexposed trace of moderately to steeply NE-dipping contact between the Napeequa Complex and Entiat pluton, which has probably been detached from its roots by the Dinkelman décollement.

73.1   Cross the Entiat River and immediately turn left (west) on Entiat River Road. We will pass through outcrops of the Entiat pluton for the next few miles, then cross back into the Napeequa Complex.

76.2   Continue past Mills Canyon Road.

77.4   Napeequa Complex in road cut on right; pull off at best parking on left.

Stop 2-5. Napeequa Complex—Structure and Rock Types (UTM 0702139E; 5283517N)

We have moved structurally downward into the central part of the Napeequa Complex. This part of the complex is dominated by quartzite (metachert), siliceous schist, and hornblende ± biotite gneiss, and contains less common amphibolite, ultramafic rock, and marble, and rare pelitic layers. These rocks are intruded by orthogneiss bodies; a thin body in this outcrop has a U-Pb zircon age of 68 Ma (H. Hurlow, 1996, personal commun.). All units were pervasively foliated and lineated, and these fabrics were folded by open to isoclinal folds. In this roadcut, folds tend to predominate in certain zones parallel to foliation, suggesting strain partitioning at the outcrop scale. Mineral lineation and fold axes are statistically subparallel and here plunge moderately northeast. The roadcut also displays a set of late conjugate normal faults indicating brittle N-S extension. We have used meter-scale grids for detailed structural analysis at this and other outcrops, which will be discussed at this stop.

Turn around and drive southeast on the Entiat River Road.

Miles	Description
81.6	Turn right onto U.S. 97A.
81.9	Carefully pull off highway to your right to view prominent cliffs.

Stop 2-6. Southwest Margin of Entiat Pluton (UTM 0708433E; 5282023N)

These outcrops illustrate features in the relatively homogeneous and wide southern end of the highly elongate, 72–73 Ma Entiat pluton. In the cliffs, the dominant biotite-hornblende tonalite (62%–63% SiO$_2$ here) contains microgranitoid enclaves, schlieren layers, and local diffuse contacts reflecting the important role of mixing and mingling in the evolution of this pluton, which crystallized at ~6 kbar (Dawes, 1993; Miller et al., 2000).

These igneous features are overprinted by generally strong magmatic foliation, which typically dips gently in the southern end of the pluton, as seen near the bridge, but steepens toward the west where it dips 50–70° NE, subparallel to the contact with the Napeequa Complex (Alsleben, 2000). The relatively steep contacts of the pluton are discordant to regional structures, except in a narrow, discontinuous structural aureole, which in the host rocks next to the southwest margin of the pluton is marked by high flattening strains (Hurlow, 1992; Miller and Paterson, 2001b). The generally gently dipping foliation in the pluton is subparallel to regional foliation. This structural coupling between the pluton and host rock, plus widespread magmatic folds, suggests that the magmatic foliation largely reflects regional strains (Paterson et al., 1998; Miller and Paterson, 2001b).

Note the Eocene dikes, brittle deformation, and associated veining and chloritization in these outcrops, features that are more common here than elsewhere in the pluton. These types of dikes also cut the Swakane Gneiss and thus intrude both plates of the Dinkelman décollement, but we have not traced individual dikes across the décollement.

Continue southwest on U.S. 97A and re-trace route to Leavenworth.

Day 3

We begin the day by looking at the deepest rocks in the southern part of the Cascades core and finish by examining the Ingalls Complex, the shallowest rocks in the Cascades section.

Follow same route as on Day 2, heading east on U.S. 2 toward Wenatchee.

Miles	Description
0.0	Reset odometer at the junction with U.S. 97. Continue straight on U.S. 2.
14.2	Exit right, staying on U.S. 2 and U.S. 97.
15.2	Exit right on U.S. 97A toward Entiat.
30.0	Cross the Entiat River and immediately turn left (west) on Entiat River Road.
34.3	Drive past Napeequa Complex seen at Stop 2-5.
36.0	Turn left onto Roaring Creek Road at sign for Entiat River fish hatchery. Follow road bend to left past hatchery.
36.7	Pavement ends—continue straight on middle dirt road through Forest Service gate.
36.9	Pass over cattle guard and through another Forest Service gate.
37.0	We are passing outcrops in the thickest orthogneiss sheet intruding this part of the Napeequa Complex. It extends for ≥25 km (Tabor et al., 1987a) and has a discordant U-Pb zircon age of 78–74 Ma. Here, it is ~1.5–2 km thick, dips 50–65° to the northeast, and is largely hornblende-biotite tonalite, displaying a high-T subsolidus foliation and weaker lineation.
37.1	Stay to the left on main dirt road.
37.4	Pull off the road to the right and park.

Stop 3-1. Metacherts, Gneisses, and Pelites in Central Napeequa Complex (UTM 0699896E; 5284839N)

We will walk past intermittent outcrops of amphibolite, quartzite (metachert), and biotite schist along the road. Fine-grained quartzite (metachert) displays relict bedding, strong foliation, and mineral lineation. Units surrounding the quartzite include amphibolite, hornblende ± biotite gneiss, local marble, and rare pelitic layers. Small-scale folds are common in the gneiss and pelitic layers. In one pelitic unit northeast of this outcrop, leucosomes, interpreted to represent largely in situ melting in the contact aureole of a large orthogneiss sheet, are more common in fold hinges, suggesting melting during folding. A biotite Ar/Ar cooling age of 59.8 Ma comes from a sample a short distance to the east.

Continue driving west.

Miles	Description
38.0	Pass through Forest Service gate.
38.1	Exposed at the corner of the road is a tectonic slice of metaperidotite (~50 m wide and 200 m long) surrounded by strongly deformed gneisses in the middle part of the Napeequa Complex. Foliation in the ultramafite is continuous with that in the other rocks.

Stop 3-2. Megacrystic Orthogneiss Body and Unfocused Magmatism in the Napeequa Complex (UTM 0698795E; 5284474N)

The southern tip of another orthogneiss body crops out along the road here. Mapping by Paterson shows that the body has an unusual sigmoidal shape, the asymmetry of which is compatible with top-to-N shear. This locally megacrystic orthogneiss is distinguished by large biotite books and strong constrictional fabrics particularly well displayed at the crest of the ridge above us. The orthogneiss is only one of many orthogneiss sheets in the Napeequa Complex, which represent widespread unfocused magmatism. These bodies are not well dated, but Hurlow's (1992) and our work indicate that they probably range from at least 68–86 Ma.

Directly east of this outcrop is hornblende ± biotite gneiss displaying two generations of open to tight folds, both of which fold a strong foliation and have variably oriented axes, although statistical maxima are parallel to a mineral lineation. These rock types, structures, and structural position are similar to those at Stop 2-5 and probably define a zone of more complex folding and shear at this level in the Napeequa Complex.

Continue driving west.

Miles	Description
38.7	Park in flat area to left.

Stop 3-3. Dinkelman Décollement and Uppermost Swakane Gneiss (UTM 0697990E; 5284151N)

We have crossed the Dinkelman décollement and can compare structures in the Swakane Gneiss to those in the Napeequa unit. Foliation appears continuous across the contact, but folding is much less common in the Swakane Gneiss, and the large intrusive sheets observed so commonly in the Napeeqa Complex a few meters to the east are absent. Top-to-N shearing is much more pervasive in the Swakane (see introductory text). From these outcrops, we have obtained an Ar/Ar age of 44 Ma from biotite in the gneiss and a few meters to the east an Ar/Ar age of 64 Ma from hornblende in an amphibolite that we interpret to be a thin imbricate sliver of the Napeequa Complex. Note the much younger biotite age at this stop relative to that in the Napeequa Complex at Stop 3-1 (ca. 44 vs. 59.8 Ma), compatible with major excision across the décollement.

If time permits, we will cross Roaring Creek to several prominent outcrops directly beneath the décollement on the west side of a small stream drainage. Excellent examples of lower-temperature and more discrete shear zones occur in this outcrop (see also Stop 3-5). These shear surfaces typically are marked by an increase in biotite and chlorite and have a slickenside-striae–like mineral lineation oriented counterclockwise from the moderate-temperature lineation.

Farther up on the spur, west of the small drainage, the Swakane Gneiss near the décollement coarsens and contains garnets that reach ~1 cm in diameter and kyanite crystals up to 3 cm long. These assemblages provide consistent estimates of 650–700 °C and 10–12 kbar (Valley et al., 2003). Temperature estimates of retrogressed samples containing chlorite and muscovite, grown synchronously with top-to-N shearing, are ~550 °C. Metamorphic reaction textures involving rutile and ilmenite in these samples, and garnet with zoning compatible with retrograde equilibration and surrounded by narrow plagioclase coronas (Stein and Stowell, 2002), record a burial event and high-T decompression (Valley et al., 2003). Thus, here the Swakane Gneiss records a clockwise *P-T* path, similar to other high-grade Cascades core rocks.

Turn vans around and retrace route to Entiat River Road.

Miles	Description
41.4	Turn right on the Entiat River Road.
47.4	Turn right, southward on U.S. 97A.
50.5	Pull off to the right on dirt turnout near small pond.

Stop 3-4. Semi-Pelitic Swakane Gneiss (UTM 0707518E; 5277672N)

These Swakane outcrops give us the opportunity to examine local garnet-staurolite schist and amphibolite lenses within the biotite gneiss. Folds are also well developed, in contrast to the dominant biotite gneiss, presumably reflecting the rheology of these thinly layered schists. The sequence of early recumbent isoclinal folds followed by upright folds is similar to that of the other Cascades core units, but the Swakane structures formed at a younger time than most of the folds in the older units (particularly the Chiwaukum Schist).

Continue south on U.S. 97A.

Miles	Description
56.2	Exit on left to parking area at Vista Point (signed).

Stop 3-5. Structure of Deepest Exposed Levels of Swakane Gneiss, Detrital Zircons, and Tectonic Speculation (UTM 0704308E; 5269487N)

We are in the hinge of the regional, gently NW-SE–plunging antiform that controls the map pattern of the Swakane Gneiss in the Chelan block. We first walk southward, examining roadcuts on the east side of the highway. The gneiss displays the assemblage quartz-plagioclase-biotite-clinozoisite ± muscovite ± garnet and is cut by numerous pegmatite sheets and quartz veins, which decrease in abundance at higher structural levels. The composition of the pegmatites (muscovite- and garnet-bearing) suggests they did not rise far from their source.

We have carried out detailed structural studies near this stop, including producing grid maps of two areas (individual grid squares of 25 × 25 cm and 1 × 1 m). The gneiss displays its typical strong foliation and mineral lineation and widespread but subtle compositional layering. This foliation wraps garnet and plagioclase porphyroblasts, which contain crenulated inclusion trails that record an early cycle of folding (Alsleben, 2000). Foliation and compositional layering are in turn folded by tight to isoclinal recumbent folds; a few examples of these folds are present on both sides of the highway. They typically trend SSW, ~25° clockwise of the mineral lineation (Alsleben, 2000).

These features are overprinted by structures formed during top-to-N shear on subhorizontal surfaces. The latter structures include extensional crenulation cleavage (shear bands) developed best in more micaceous layers, asymmetric boudinage (photogenic examples in roadcuts) and pinch-and-swell of pegmatite sheets and quartz veins oriented close to the foliation (those at high angles are ptygmatically folded), and local shear surfaces with associated pseudotachylite. The pervasiveness of non-coaxial shear is illustrated by the grid mapping, as top-to-N shear was recognized at 71% of the grid points (top-to-S <1%).

The top-to-N shear probably began at medium or possibly high temperatures (see introductory text). Continuation under lower-T conditions, presumably during exhumation, is indicated by synkinematic(?) retrograde muscovite and chlorite in some shear bands, the local pseudotachylite, and micacous lineations that are less pervasive and are typically oriented 10–30° counterclockwise from the dominant mineral lineation (Alsleben, 2000). The involvement of the pegmatites, one of which has been dated elsewhere at 68 Ma (Mattinson, 1972), provides some control on the timing of the shearing.

Detrital zircons separated from biotite gneiss from this outcrop yield a large range of dates from 1610–73 Ma. The youngest detrital zircons bracket the maximum age for the deposition of the deepest structural levels of this unit. These ages imply that the protolith of the Swakane Gneiss is younger than the other terrranes of the Cascades core; we will discuss the tectonic implications of this observation (see introductory text).

After viewing the roadcuts south of the vista point, we will retrace our steps to the turnout, carefully cross the highway (to the west side of road), and walk northward. Near the north end of the roadcuts, pseudotachylite is associated with a top-to-N

shear zone that displaces a granitic dike. Isoclinal recumbent folds and a possible sheath fold are visible a few meters south of the pseudotachylite.

Continue south on U.S. 97A through the Swakane Gneiss; note the moderate increase in southwest dips.

Miles	Description
62.1	Near Wenatchee, U.S. 97A rejoins U.S. 2 and U.S. 97. Continue west on U.S. 2 toward Leavenworth.
76.9	Turn left (south) on U.S. 97 toward Ellensburg and Cle Elum.
88.9	King Creek Road.
89.8	Pull off at large area on left.

Stop 3-6. Ingalls Complex and Deformation at Shallow Crustal Levels (UTM 0676780E, 5252714N)

Outcrops on both sides of Peshastin Creek illustrate the structural style of the eastern, low-grade part of the Jurassic Ingalls Complex. In the large stream exposures on the west side of the road is a N-dipping section of pillow lava (within-plate basalt affinity), breccia, limestone, and jasper. These rocks make up the Early Jurassic rifted(?) basement to the ophiolite, and lie on the south limb of a kilometer-scale syncline. At the south end of the outcrop, sheared serpentinite structurally overlies the basalt on a south-dipping fault with steeply pitching slickenside striae. Drag of bedding and a small fault indicate reverse motion (see Harper et al., this volume, their Stop 2B, for more detail). On the east side of the road, next to a marshy area, is sheared argillite with chert lenses. These rocks are cut by a plagioclase-phyric body that is similar to a nearby porphyry, dated ca. 91 Ma by Ar/Ar hornblende (Harper et al., this volume). These rocks presumably represent the shallow level equivalent of the coeval Mount Stuart batholith to the north. They also constrain the timing of folding and reverse faulting in the eastern part of the Ingalls Complex, which we interpret as the upper-crustal manifestation of the middle Cretaceous shortening in the Cascades core.

Continue south on U.S. 97 and then to Seattle.

ACKNOWLEDGMENTS

We thank Helge Alsleben and Scott McPeek for help preparing the manuscript, Sue DeBari and Donna Whitney for their long-term collaboration, and Mike Brown for his insights on migmatites in the Chiwaukum Schist. This research was supported by National Science Foundation grants to Sam Bowring (EAR-9980623), Miller (EAR-9614521, EAR-9628280, EAR-9980662, and EAR-0087829), Paterson (EAR-8916325, EAR-921874, EAR-9614758, and EAR-9627986), and Stowell (EAR-0207777).

REFERENCES CITED

Alsleben, H., 2000, Structural analysis of the Swakane terrane, North Cascades core, Washington [M.S. thesis]: San Jose, San Jose State University, 168 p.

Anderson, J.L., 1992, Compositional variation within the high-Mg, tonalitic Mount Stuart Batholith, North Cascades, Washington: Geological Society of America Abstracts with Programs, v. 24, no. 5, p. 3.

Anderson, J.L., and Paterson, S.R., 1991, Emplacement of the Cretaceous Mt. Stuart batholith, central Cascades, Washington: Geological Society of America Abstracts with Programs, v. 23, no. 5, p. A387.

Benn, K., Paterson, S.R., Lund, S.P., Pignotta, G.S., and Kruse, S., 2001, Magmatic fabrics in batholiths as markers of regional strains and plate kinematics: Example of the Cretaceous Mt. Stuart Batholith: Physics and Chemistry of the Earth, v. 26, p. 343–354.

Boysun, M.A., and Paterson, S.R., 2002, Melt injection in the Swakane Biotite Gneiss, North Cascades core: Implications for melting and dike emplacement in deep crust: Geological Society of America Abstracts with Programs, v. 34, no. 6, p. 374.

Boysun, M.A., and Paterson, S.R., 2003, Partial melting and melt collection and transport in the Swakane terrane, North Cascades crystalline core, WA: Geological Society of America Abstracts with Programs, v. 35 (in press).

Brandon, M.T., Cowan, D.S., and Vance, J.A., 1988, The Late Cretaceous San Juan thrust system, San Juan Islands, Washington: Boulder, Colorado, Geological Society of America Special Paper 221, 81 p.

Brown, E.H., and Talbot, J.L., 1989, Orogen-parallel extension in the North Cascades crystalline core, Washington: Tectonics, v. 8, p. 1105–1114.

Brown, E.H., and Walker, N.W., 1993, A magma-loading model for Barrovian metamorphism in the Southeast Coast Plutonic Complex, British Columbia and Washington: Geological Society of America Bulletin, v. 105, p. 479–500.

Butler, R.F., Gehrels, G.E., McClelland, W.C., May, S.R., and Klepacki, D., 1989, Discordant paleomagnetic poles from the Canadian Coast Plutonic Complex—Regional tilt rather than large-scale displacement: Geology, v. 17, p. 691–694.

Cater, F.W., 1982, Intrusive rocks of the Holden and Lucerne quadrangles, Washington; the relation of depth zones, composition, textures, and emplacement of plutons: U.S. Geological Survey Professional Paper 1220, 108 p.

Cater, F.W., and Crowder, D.F., 1967, Geologic map of the Holden Quadrangle, Snohomish and Chelan Counties, Washington: U.S. Geological Survey Map GQ-646, scale 1:62,500.

Cowan, D.S., Brandon, M.T., and Garver, J.I., 1997, Geologic tests of hypotheses for large coastwise displacements—A critique illustrated by the Baja British Columbia controversy: American Journal of Science, v. 297, p. 117–173.

Davidson, C., Hollister, L.S., and Schmid, S.M., 1992, Role of melt in the formation of a deep-crustal compressive shear zone—the Maclaren Glacier metamorphic belt, South Central Alaska: Tectonics, v. 11, p. 348–359.

Dawes, R.L., 1993, Mid-crustal, Late Cretaceous plutons of the North Cascades; petrogenesis and implications for the growth of continental crust [Ph.D. thesis]: Seattle, University of Washington, 273 p.

DeBari, S.M., Miller, R.B., and Paterson, S.R., 1998, Genesis of tonalitic plutons in the Cretaceous magmatic arc of the North Cascades; mixing of mantle-derived mafic magmas and melts of a garnet-bearing lower crust: Geological Society of America Abstracts with Programs, v. 30, no. 7, p. 257–258.

Engels, J.C., and Crowder, D.F., 1971, Late Cretaceous fission-track and potassium-argon ages of the Mount Stuart granodiorite and Beckler Peak stock, North Cascades, Washington: U.S. Geological Survey Professional Paper 750-D, p. D39-D43.

Engels, J.C., Tabor, R.W., Miller, F.K., and Obradovich, J.D., 1976, Summary of K-Ar, Rb-Sr, U-Pb, Pb-alpha, and fission-track ages of rocks from Washington State prior to 1975 (exclusive of Columbia Plateau basalts): U.S. Geological Survey Miscellaneous Field Studies Map MF-710.

Erikson, E.H., 1977, Petrology and petrogenesis of the Mount Stuart batholith - Plutonic equivalent of high-alumina basalt association?: Contributions to Mineralogy and Petrology, v. 60, p. 183–207.

Evans, B.W., and Berti, J.W., 1986, Revised metamorphic history for the Chiwaukum Schist, North Cascades, Washington: Geology, v. 14, p. 695–698.

Evans, B.W., and Davidson, G.F., 1999, Kinetic control of metamorphic imprint during synplutonic loading of batholiths: An example from Mount Stuart, Washington: Geology, v. 27, p. 415–418.

Evans, J.E., and Johnson, S.Y., 1989, Paleogene strike-slip basins of central Washington, in Joseph, N.L., ed., Geologic Guidebook for Washington and Adjacent Areas: Washington Division of Geology and Earth Resources Circular 86, p. 213–237.

Getsinger, J.S., 1978, A structural and petrologic study of the Chiwaukum Schist on Nason Ridge, northeast of Stevens Pass, North Cascades, Washington [M.S. thesis]: Seattle, University of Washington, 151 p.

Haugerud, R.A., 1987, Argon geochronology of the Tenpeak Pluton and untilting of the Wenatchee Block, North Cascades Range, Washington: Eos (Transactions, American Geophysical Union), v. 68, p. 1814.

Haugerud, R.A., Brown, E.H., Tabor, R.W., Kriens, B.J., and McGroder, M.F., 1994, Late Cretaceous and Early Tertiary orogeny in the North Cascades, in Swanson, D.A., and Haugerud, R.A., eds., Geologic field trips in the Pacific Northwest: 1994 Geological Society of America Annual Meeting, Geological Society of America, p. 2E1–2E51.

Haugerud, R.A., van der Heyden, P., Tabor, R.W., Stacey, J.S., and Zartman, R.E., 1991, Late Cretaceous and Early Tertiary plutonism and deformation in the Skagit Gneiss Complex, North Cascade Range, Washington and British Columbia: Geological Society of America Bulletin, v. 103, p. 1297–1307.

Holland, T.J.B., and Powell, R., 1998, An internally consistent thermodynamic data set for phases of petrological interest: Journal of Metamorphic Geology, v. 16, p. 309–343.

Hurlow, H.A., 1992, Structural and U/Pb geochronologic studies of the Pasayten Fault, Okanogan Range Batholith, and southeastern Cascades crystalline core, Washington [Ph.D. thesis]: Seattle, University of Washington, 180 p.

Jacobson, C.E., Barth, A.P., and Grove, M., 2000, Late Cretaceous protolith age and provenance of the Pelona and Orocopia Schists, southern California: Implications for evolution of the Cordilleran margin: Geology, v. 28, p. 219–222.

Johnson, S.Y., 1985, Eocene strike-slip faulting and non-marine basin formation in Washington, in Biddle, K.T., and Christie-Blick, N., eds., Strike-slip deformation, basin formation, and sedimentation: Society of Economic Paleontologists and Mineralogists, Special Publication 37, p. 283–302.

Journeay, J.M., and Friedman, R.M., 1993, The Coast Belt thrust system - Evidence of Late Cretaceous shortening in southwest British Columbia: Tectonics, v. 12, p. 1301–1302.

Kelemen, P.B., and Ghiorso, M.S., 1986, Assimilation of peridotite in zoned calc-alkaline plutonic complexes—Evidence from the Big Jim Complex, Washington Cascades: Contributions to Mineralogy and Petrology, v. 94, p. 12–28.

Klepeis, K.A., Clarke, G.L., and Rushmer, T., 2003, Magma transport and coupling between deformation and magmatism in the continental lithosphere: GSA Today, v. 13, p. 4–11.

Lebit, H., Lueneburg, C.M., Paterson, S.R., and Miller, R.B., 1998, The geometry of folds and mineral lineations; examples from the Cascades crystalline core, Washington: Geological Association of Canada, NUNA Conference, Canadian Tectonic Studies Group, 18th Annual Meeting, Abstracts with Program.

MacDonald, J.H., Harper, G.D., Miller, R.B., and Miller, J.S., 2002, Within-plate magmatic affinities of a lower pillow unit in the Ingalls ophiolite complex, Northwest Cascades, Washington: Geological Society of America Abstracts with Programs, v. 34, no. 5, p. A-22.

Magloughlin, J.F., 1993, A Nason Terrane trilogy; I, Nature and significance of pseudotachylyte; II, Summary of the structural and tectonic history; III, Major and trace element geochemistry and strontium and neodymium isotope geochemistry of the Chiwaukum Schist, amphibolite, and metatonalite gneiss of the Nason Terrane [Ph.D. thesis]: Minneapolis, University of Minnesota, 325 p.

Magloughlin, J.F., 1994, Migmatite to fault gouge: fault rocks and the structural and tectonic evolution of the Nason terrane, North Cascades Mountains, Washington, in Swanson, D.A., and Haugerud, R.A., eds., Geologic Field Trips in the Pacific Northwest: 1994 Geological Society of America Annual Meeting, Geological Society of America, p. 2B1–2B17.

Marsh, B.D., 1982, On the mechanics of igneous diapirism, stoping, and zone-melting: American Journal of Science, v. 282, p. 808–855.

Mattinson, J.M., 1972, Ages of zircons from the Northern Cascade Mountains, Washington: Geological Society of America Bulletin, v. 83, p. 3769–3783.

Matzel, J., Bowring, S.A., and Miller, R.B., 2002a, Timescales of pluton construction and the crystallization history of the Mt Stuart Batholith, North Cascades, WA: Eos (Transactions of the American Geophysical Union), Fall Meeting Supplemental, v. 83, no. 47, p. T11E-09.

Matzel, J., Bowring, S.A., and Miller, R.B., 2002b, Geochronologic evidence of a Late Cretaceous protolith age for the Swakane Gneiss, North Cascades, WA: Geological Society of America Abstracts with Programs, v. 34, no. 6, p. 510–511.

McGroder, M.F., 1991, Reconciliation of two-sided thrusting, burial metamorphism, and diachronous uplift in the Cascades of Washington and British Columbia: Geological Society of America Bulletin, v. 103, p. 189–209.

Metzger, E.P., Miller, R.B., and Harper, G.D., 2002a, Geochemistry and tectonic setting of the ophiolitic Ingalls Complex, North Cascades, Washington: Implications for correlations of Jurassic Cordilleran Ophiolites: Journal of Geology, v. 110, p. 543–560.

Metzger, E.P., Miller, R.B., and Harper, G.D., 2002b, Geochemistry of mafic rocks in the Chiwaukum Schist of the Cascades core and possible correlatives; tectonic implications for the North Cascades: Geological Society of America Abstracts with Programs, v. 34, no. 5, p. A-17.

Miller, R.B., 1985, The ophiolitic Ingalls Complex, North-Central Cascade Mountains, Washington: Geological Society of America Bulletin, v. 96, p. 27–42.

Miller, R.B., 1988, Fluid-flow, metasomatism and amphibole deformation in an imbricated ophiolite, North Cascades, Washington: Journal of Structural Geology, v. 10, p. 283–296.

Miller, R.B., and Bowring, S.A., 1990, Structure and chronology of the Oval Peak batholith and adjacent rocks: Implications for the Ross Lake fault zone, North Cascades, Washington: Geological Society of America Bulletin, v. 102, p. 1361–1377.

Miller, R.B., and Mogk, D.W., 1987, Ultramafic rocks of a fracture-zone ophiolite, North Cascades, Washington: Tectonophysics, v. 142, p. 261–289.

Miller, R.B., and Paterson, S.R., 1992, Tectonic implications of syn-emplacement and post-emplacement deformation of the Mount Stuart batholith for middle Cretaceous orogenesis in the North Cascades: Canadian Journal of Earth Sciences, v. 29, p. 479–485.

Miller, R.B., and Paterson, S.R., 1994, The transition from magmatic to high-temperature solid-state deformation: Implications from the Mount Stuart batholith, Washington: Journal of Structural Geology, v. 16, p. 853–865.

Miller, R.B., and Paterson, S.R., 1999, In defense of magmatic diapirs: Journal of Structural Geology, v. 21, p. 1161–1173.

Miller, R.B., and Paterson, S.R., 2001a, Influence of lithological heterogeneity, mechanical anisotropy, and magmatism on the rheology of an arc, North Cascades, Washington: Tectonophysics, v. 342, p. 351–370.

Miller, R.B., and Paterson, S.R., 2001b, Construction of mid-crustal sheeted plutons: Examples from the North Cascades, Washington: Geological Society of America Bulletin, v. 113, p. 1423–1442.

Miller, R.B., Johnson, S.Y., and McDougall, J.W., 1990, Discordant paleomagnetic poles from the Canadian Coast Plutonic Complex: Regional tilt rather than large displacement: Geology, v. 18, p. 1164–1165.

Miller, R.B., Brown, E.H., McShane, D.P., and Whitney, D.L., 1993a, Intra-arc crustal loading and its tectonic implications, North Cascades crystalline core, Washington and British Columbia: Geology, v. 21, p. 255–258.

Miller, R.B., Mattinson, J.M., Funk, S.A., Hopson, C.A., and Treat, C.L., 1993b, Tectonic evolution of Mesozoic rocks in the southern and central Washington Cascades, *in* Dunne, G., and McDougall, K., eds., Mesozoic Paleogeography of the Western United States—II: Pacific Section of the Society of Economic Mineralogists and Paleontologists, Book 71, p. 81–98.

Miller, R.B., Whitney, D.L., and Geary, E.E., 1993c, Tectonostratigraphic terranes and the metamorphic history of the northeastern part of the crystalline core of the North Cascades—Evidence from the Twisp Valley Schist: Canadian Journal of Earth Sciences, v. 30, p. 1306–1323.

Miller, R.B., Paterson, S.R., DeBari, S.M., and Whitney, D.L., 2000, North Cascades Cretaceous crustal section: changing kinematics, rheology, metamorphism, pluton emplacement and petrogenesis from 0 to 40 km depth, *in* Woodsworth, G.J., Jackson, L.E., Nelson, J.L., and Ward, B.C., eds., Guidebook for geological field trips in southwestern British Columbia and northern Washington: Vancouver, Geological Association of Canada, p. 229–278.

Misch, P., 1966, Tectonic evolution of the Northern Cascades of Washington State; a west-Cordilleran case history: Canadian Institute of Mining and Metallurgy, Special Volume 8, p. 101–148.

Monger, J.W.H., and Journeay, J.M., 1994, Guide to the geology and tectonic evolution of the southern Coast Mountains: Geological Survey of Canada, 97 p.

Monger, J.W.H., Price, R.A., and Tempelman-Kluit, D.J., 1982, Tectonic accretion and the origin of the two major metamorphic and plutonic belts in the Canadian Cordillera: Geology, v. 10, p. 70–75.

Neves, S.P., Vauchez, A., and Archanjo, C.J., 1996, Shear zone-controlled magma emplacement or magma-assisted nucleation of shear zones? Insights from northeast Brazil: Tectonophysics, v. 262, p. 349–364.

Paterson, S.R., and Miller, R.B., 1998a, Magma emplacement during arc-perpendicular shortening: An example from the Cascades crystalline core, Washington: Tectonics, v. 17, p. 571–586.

Paterson, S.R., and Miller, R.B., 1998b, Mid-crustal magmatic sheets in the Cascades Mountains, Washington: implications for magma ascent: Journal of Structural Geology, v. 20, p. 1345–1363.

Paterson, S.R., and Miller, R.B., 1998c, Rheological controls on meso-scale structures and kinematics near a mid-crustal detachment: an example from the Napeequa Complex, Cascades core, Washington: Geological Association of Canada NUNA conference, Canadian Tectonic Studies Group, 18th Annual Meeting, Abstracts with Programs.

Paterson, S.R., Miller, R.B., Anderson, J.L., Lund, S.P., Bendixen, J., Taylor, N., and Fink, T., 1994, Emplacement and evolution of the Mt. Stuart batholith, *in* Swanson, D.A., and Haugerud, R.A., eds., Geologic field trips in the Pacific Northwest: 1994 Geological Society of American Annual Meeting, Geological Society of America, p. 2F1–2F47.

Paterson, S.R., Fowler, T.K., and Miller, R.B., 1996, Pluton emplacement in arcs: A crustal-scale exchange process: Transactions of the Royal Society of Edinburgh–Earth Sciences, v. 87, p. 115–123.

Paterson, S.R., Fowler, T.K., Schmidt, K.L., Yoshinobu, A.S., Yuan, E.S., and Miller, R.B., 1998, Interpreting magmatic fabric patterns in plutons: Lithos, v. 44, p. 53–82.

Pavlis, T.L., 1996, Fabric development in syn-tectonic intrusive sheets as a consequence of melt-dominated flow and thermal softening of the crust: Tectonophysics, v. 253, p. 1–31.

Plummer, C.C., 1980, Dynamothermal contact-metamorphism superposed on regional metamorphism in the pelitic rocks of the Chiwaukum Mountains area, Washington Cascades: Geological Society of America Bulletin, v. 91, p. 386–388.

Pongsapich, W., 1974, The geology of the eastern part of the Mount Stuart Batholith, central Cascades, Washington [Ph.D. thesis]: Seattle, University of Washington, 170 p.

Powell, R., and Holland, T.J.B., 2001, Course notes for "THERMOCALC Workshop 2001: Calculating metamorphis phase equilibria" (on CD-ROM).

Rubin, C.M., Saleeby, J.B., Cowan, D.S., Brandon, M.T., and McGroder, M.F., 1990, Regionally extensive middle Cretaceous west-vergent thrust system in the northwestern Cordillera: Implications for continent-margin tectonism: Geology, v. 18, p. 276–280.

Saleeby, J.B., 2003, Segmentation of the Laramide Slab—evidence from the southern Sierra Nevada region: Geological Society of America Bulletin, v. 115, p. 655–668.

Sawyko, L.T., III, 1994, The geology and petrology of the Swakane Biotite Gneiss, North Cascades, Washington [M.S. thesis]: Seattle, University of Washington, 134 p.

Stein, E., and Stowell, H.H., 2002, Plagioclase coronas around garnet in the Washington North Cascades; preliminary interpretation for coronas from the Dinkelman Detachment: Geological Society of America Abstracts with Programs, v. 34, no. 5, p. A-95.

Stowell, H.H., and Crawford, M.L., 2000, Metamorphic history of the western Coast Mountains orogen, western British Columbia and southeastern Alaska, *in* Stowell, H.H., and McClelland, W.C., eds., Tectonics of the Coast Mountains, southeastern Alaska and British Columbia: Boulder, Colorado, Geological Society of America Special Paper 343, p. 257–283.

Stowell, H.H., and Tinkham, D.K., 2002, Integration of phase equilibria modelling and garnet Sm-Nd geochronology for construction of PT-t paths: Examples from the Cordilleran Coast Plutonic Complex, USA: Abstracts of the 12th Annual V.M. Goldschmidt Conference, Davos, Switzerland, Geochimica et Cosmochimica Acta, v. 66, p. A746.

Stowell, H.H., and Tinkham, D.K., 2003, Integration of phase equilibria modelling and garnet Sm-Nd chronology for construction of *P-T-t* paths: Examples from the Cordilleran Coast Plutonic Complex, USA, *in* Vance, D., Muller, W., and Villa, I., eds., Geochronology: Linking the Isotopic Record with Petrology and Textures, Geological Society [London] Special Publication 220, p. 119–145.

Tabor, R.W., Waitt, R.B., Jr., Frizzell, V.A., Jr., Swanson, D.A., Byerly, G.R., and Bentley, R.D., 1982, Geologic map of the Wenatchee 1:100,000 Quadrangle, central Washington: U.S. Geological Survey Map I-1311, scale 1:100,000.

Tabor, R.W., Frizzell, V.A., Jr., Whetten, J.T., Waitt, R.B., Jr., Swanson, D.A., Byerly, G.R., Booth, D.B., Hetherington, M.J., and Zartman, R.E., 1987a, Geologic map of the Chelan 30' by 60' Quadrangle, Washington: U.S. Geological Survey Map I-1661, scale 1:100,000.

Tabor, R.W., Zartman, R.E., and Frizzell, V.A., Jr., 1987b, Possible tectonostratigraphic terranes in the North Cascades crystalline core, Washington, *in* Schuster, J.E., ed., Selected Papers on the Geology of Washington: Washington Division of Geology and Earth Resources, Bulletin 77, p. 107–127.

Tabor, R.W., Haugerud, R.A., and Miller, R.B., 1989, Overview of the geology of the North Cascades, International Geologic Congress Trip T307: American Geophysical Union, 62 p.

Tabor, R.W., Booth, D.B., Vance, J.A., and Ford, A.B., 2002, Geologic map of the Sauk River 30- by 60-minute quadrangle, Washington: U.S. Geological Survey Map I-2592, scale 1:100,000.

Taylor, N., 1994, Structural geology of the Chiwaukum Schist in the Mount Stuart region, central Cascades, Washington [M.S. thesis]: Los Angeles, University of Southern California.

Tinkham, D.K., 2002, MnNCKFMASH phase equilibria, garnet activity modeling, and garnet samarium-neodymium chronology: applications to the Waterville Formation, Maine, and south-central Nason terrane, Washington [Ph.D. thesis]: Tuscaloosa, The University of Alabama, 234 p.

Umhoefer, P.J., and Miller, R.B., 1996, Middle Cretaceous thrusting in the southern Coast Belt, British Columbia and Washington, after strike-slip fault reconstruction: Tectonics, v. 15, p. 545–565.

Umhoefer, P.J., and Schiarizza, P., 1996, Latest Cretaceous to early Tertiary dextral strike-slip faulting on the southeastern Yalakom fault system, southeastern Coast belt, British Columbia: Geological Society of America Bulletin, v. 108, p. 768–785.

Valley, P.M., Whitney, D.L., Paterson, S.R., Miller, R.B., and Alsleben, H., 2003, Metamorphism of the deepest exposed arc rocks in the Cretaceous to Paleogene Cascades belt, Washington: evidence for large-scale vertical motion in a continental arc: Journal of Metamorphic Geology, v. 21, p. 203–220.

Van Diver, B.B., 1967, Contemporaneous faulting-metamorphism in Wenatchee Ridge area, Northern Cascades, Washington: American Journal of Science, v. 265, p. 132–150.

Vance, J.A., 1957, The geology of the Sauk River area in the northern Cascades of Washington [Ph.D. thesis]: Seattle, University of Washington, 312 p.

Walker, N.W., and Brown, E.H., 1991, Is the southeast Coast Plutonic Complex the consequence of accretion of the Insular Superterrane? Evidence from U-Pb zircon geochronometry in the Northern Washington Cascades: Geology, v. 19, p. 714–717.

Whitney, D.L., 1992, High-pressure metamorphism in the western Cordillera of North America: An example from the Skagit Gneiss, North Cascades: Journal of Metamorphic Geology, v. 10, p. 71–85.

Whitney, D.L., Miller, R.B., and Paterson, S.R., 1999, *P-T-t* evidence for mechanisms of vertical tectonic motion in a contractional orogen: Northwestern U.S. and Canadian Cordillera: Journal of Metamorphic Geology, v. 17, p. 75–90.

Yeats, R.S., Erikson, E.H., Frost, B.R., Hammond, P.E., and Miller, R.B., 1977, Structure, stratigraphy, plutonism, and volcanism of the Central Cascades, Washington, *in* Brown, E.H., and Ellis, R.C., eds., Geological Excursions in the Pacific Northwest, Geological Society of America, p. 265–308.

Zuluaga, C.A., and Stowell, H.H., 2002, Anatectic versus non-anatectic origins for migmatite in the banded gneiss, Cascades Core, Washington: Geological Society of America Abstracts with Programs, v. 34, no. 5, p. 95–96.

Geological Society of America
Field Guide 4
2003

Cordilleran Ice Sheet glaciation of the Puget Lowland and Columbia Plateau and alpine glaciation of the North Cascade Range, Washington

Don J. Easterbrook

Department of Geology, Western Washington University, Bellingham, Washington 98225, USA

ABSTRACT

The advance of the Cordilleran Ice Sheet (CIS) during the Vashon Stade is limited by ^{14}C dates from sediments beneath Vashon till, which indicate that ice advanced southward across the Canadian border sometime after ca. 18 ka ^{14}C yr B.P. and reached the Seattle area soon after 14.5 ka ^{14}C yr B.P. The Puget lobe underwent sudden, large-scale terminus recession and downwasting not long after 14.5 ka ^{14}C yr B.P., and backwasted northward from its southern terminus past the Seattle area by ca. 14 ka ^{14}C yr B.P. Rapid thinning of Vashon ice after the terminus had receded north of Seattle allowed marine water from the Strait of Juan de Fuca to flood the lowland, floating the remaining ice and disintegrating the remaining CIS northward all the way to Canada, except for a narrow band along the eastern margin of the lowland.

Everson glaciomarine drift (gmd), consisting mostly of poorly sorted stony clay deposited from floating ice, was deposited essentially contemporaneously over the central and northern Puget Lowland. Unbroken, articulated, marine shells, some in growth positions, indicate that the gmd represents in situ deposition. More than 150 ^{14}C dates from Washington and British Columbia fix the age of the Everson gmd at 11,500 to ca. 12,500 ^{14}C yr B.P., making it a valuable stratigraphic marker over the central and northern Puget Lowland.

Ice-contact marine deltas and shorelines were produced on Whidbey Island as the CIS thinned and disintegrated in the central Puget Lowland, allowing marine water from the Strait of Juan de Fuca to penetrate beneath the ice. During this time, the CIS had disintegrated in the deeper water of the inland waterways, but grounded ice remained along the eastern side of the mainland, changing the ice flow direction from N-S to NE-SW, from the grounded ice on the mainland toward the open deep water to the west at the Strait of Juan de Fuca. A well-defined, marine ice-margin existed along the south and west sides of Penn Cove and isostatically raised shorelines and marine deltas were formed at elevations up to ~33 m on southern Whidbey Island and up to ~88 m on northern Whidbey. The shorelines are best preserved along the sides of marine embayments on the island.

Following the deposition of Everson gmd and the emergence of the northern Puget Lowland, the CIS readvanced several times, defining four phases of the Sumas Stade: Sumas I represents grounding of the CIS and deposition of till in the western Fraser Lowland. Sumas II consists of a well-defined moraine and meltwater channels deeply incised into Everson gmd. A series of Sumas III moraines that occur in British Columbia shed meltwater that built a broad outwash plain behind the Sumas II moraine. A Sumas IV moraine occurs across a Sumas III meltwater channel at the eastern margin of the Fraser Lowland.

Keywords: Pleistocene, glaciation, Younger Dryas, Puget Lowland, Cascade Range, Columbia Plateau.

Easterbrook, D.J., 2003, Cordilleran Ice Sheet glaciation of the Puget Lowland and Columbia Plateau and alpine glaciation of the North Cascade Range, Washington, *in* Swanson, T.W., ed., Western Cordillera and adjacent areas: Boulder, Colorado, Geological Society of America Field Guide 4, p. 137–157. For permission to copy, contact editing@geosociety.org. © 2003 Geological Society of America.

THE CORDILLERAN ICE SHEET

The Puget Lowland

The Puget Lowland (Fig. 1) is mantled with deposits of Pleistocene glaciations recording repeated advances and retreats of the Cordilleran Ice Sheet from British Columbia. Deposits of till, outwash sand and gravel, glaciomarine drift, and related deposits are superbly exposed in sea cliffs along hundreds of kilometers of shorelines. The morphology of glacial deposits is well displayed and plays an important role in interpretation of the glaciation of the region.

The stratigraphic record exposed in the sea cliffs of Whidbey Island provides a definitive basis for much of what we know about glaciation of the Puget Lowland. The first Quaternary studies of Whidbey Island were made by Bretz (1913), focusing on the pre-Vashon sections (his Admiralty Glaciation). No other Quaternary studies were published for a half century, until Easterbrook (1962a, 1962b, 1963a, 1963b). Since then more than a dozen papers on the Quaternary geology of Whidbey have been published by Easterbrook.

Fraser Glaciation

The late Wisconsin Fraser Glaciation of the lowland consists of several stades: (1) the Vashon Stade, the maximum advance of the Cordilleran Ice Sheet (CIS); (2) the Everson Interstade, an interval of glaciomarine deposition during deglaciation of the lowland; and (3) the Sumas Stade, several short readvances of the CIS during deglaciation (Easterbrook, 1963b, 1992; Armstrong et al., 1965; Kovanen and Easterbrook, 2002a; Kovanen, 2002).

The Vashon Stade

The Vashon Stade was the last major climatic episode, during which drift was deposited by the CIS in British Columbia and Washington (Armstrong et al., 1965). It began with

advance of the CIS into the lowland and ended with the beginning of marine and glaciomarine conditions of the Everson Interstade. Meltwater streams deposited an apron of outwash sand and gravel in front of the advancing glacier that was later overridden by Vashon ice and covered with till. Timing of the advance is limited by ^{14}C dates from sediments beneath Vashon till. These dates indicate that Vashon ice advanced southward across the Canadian border sometime after ca. 18 ka ^{14}C yr B.P. and reached the Seattle area soon after 14.5 ka ^{14}C yr B.P. (Easterbrook, 1992; Porter and Swanson, 1998).

Elongate drumloidal hills and flutes, as well as grooves, and striations carved in bedrock indicate directions of movement of Vashon ice in the central and northern Puget Lowland (Easterbrook, 1968, 1979; Kovanen and Easterbrook, 2002a; Kovanen and Slaymaker, 2003a; Haugerud et al., 2003). At its maximum, the Vashon ice extended to ~25 km (15 mi) south of Olympia. Ice was >1830 m (6000 ft) thick near Bellingham, ~915 m (3000 ft) thick near Seattle, and ~550 m (1800 ft) thick southeast of Tacoma (Fig. 2).

The Everson Interstade

The Everson Interstade represents the final stages in the disintegration of the CIS in central and northern Puget Lowland. The CIS experienced sudden, rapid, large-scale recession and downwasting not long after 14.5 ka ^{14}C yr B.P. The Puget lobe retreated northward from its southern terminus by backwasting past the Seattle area by ca. 14 ka ^{14}C yr B.P. (Fig. 2B) (Leopold et al., 1982; Porter and Swanson, 1998).

Retreat of the Juan de Fuca lobe from its terminal position is limited by ^{14}C dates of 14,460 ± 200 yr from a bog on Vashon Drift and dates of 13,380 ± 250, 13,100 ± 180, 13,080 ± 260, 13,010 ± 240, and 12,020 ± 210 from wood in till at five localities in the terminal zone of the Juan de Fuca lobe (Heusser, 1973a, 1973b, 1982). Bottom sediments in the Strait of Juan de Fuca, dated at 14,400 ± 400 and 13,150 ± 400 ^{14}C yr B.P., were depos-

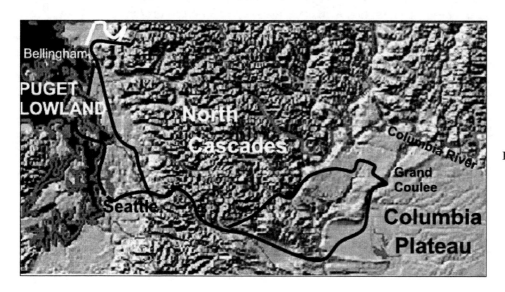

Figure 1. Field trip route map.

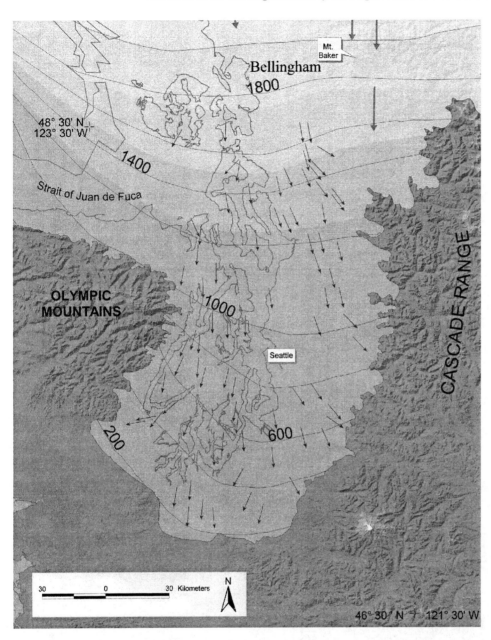

Figure 2. The Cordilleran Ice Sheet at its late Wisconsin maximum. The ice flowed due south in the North Cascades near Mount Baker and was continuous across the range at this latitude. (From Kovanen and Easterbrook, 2002a.)

ited during recession of the lobe (Hewitt and Mosher, 2001). The Juan de Fuca lobe retreated essentially contemporaneously with the Puget lobe (Easterbrook, 1992).

Rapid thinning of Vashon ice after the terminus had receded north of Seattle allowed marine water from the Strait of Juan de Fuca to flood the lowland, floating the remaining ice and rapidly disintegrating the remaining CIS northward all the way to Canada (Easterbrook, 1969, 1992) (Fig. 3C) except for a narrow band along the eastern margin of the lowland

During the Everson, glaciomarine drift (gmd) was deposited contemporaneously over a large area (18,000 km²) in the central and northern Puget Lowland (Easterbrook, 1963b, 1966a, 1966b, 1968, 1969, 1971, 1976a, 1976b, 1979, 1986, 1992, 2003a, 2003b;

Blunt et al., 1987; Dethier et al., 1995; Kovanen and Easterbrook, 2002a; Kovanen, 2002) and British Columbia (Armstrong and Brown, 1954; Armstrong, 1981, 1984; Armstrong and Hicock, 1980a, 1980b).

Everson gmd consists largely of poorly sorted diamictons deposited when melting of floating ice released rock debris that accumulated on the sea floor. Unbroken, articulated marine shells, some in growth positions, indicate that the gmd represents in situ deposition. The origin of the gmd from floating ice in seawater was first recognized in Washington by Easterbrook (1963b, 1966a, 1966b, 1968) and in British Columbia by Armstrong (Armstrong and Brown, 1954; Armstrong 1957, 1960, 1981, 1984). More than 150 ¹⁴C dates from Washington and British Columbia fix the age

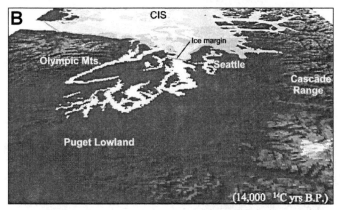

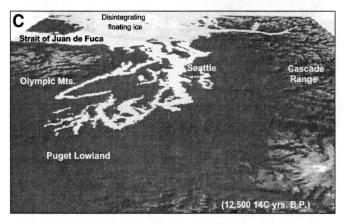

Figure 3. Deglaciation of the Puget Lowland. (A) Ice margin and the Vashon maximum, (B) backwasting and downwasting to the central lowland, (C) floating and disintegration of the remaining CIS.

of the Everson gmd at 11,500 to ca. 12,500 ^{14}C yr B.P. (Easterbrook, 1992, 2003a, 2003b), making it a valuable stratigraphic marker over the central and northern Puget Lowland.

Deglaciation Marine Deltas and Shorelines

Deglaciation marine deltas and shorelines were produced as the CIS thinned and disintegrated, allowing marine water from the Strait of Juan de Fuca to penetrate beneath the ice (Fig. 3C).

Marine deltas on Whidbey Island were first recognized in 1962 (Easterbrook, 1962a, 1962b, 1968), but images recently produced by D.J. Kovanen from high-resolution, LIDAR (light detection and ranging) digital data have revealed new, striking, geologic evidence of multiple former sea levels and ice-flow directions (Kovanen and Slaymaker, 2003a).

The Kovanen and Slaymaker (2003a) images show superimposed drumlins, flutes, and grooves that record two quite different flow directions on northern Whidbey Island, suggesting two ice-flow events: (1) an older N-S ice-flow direction (main Vashon), and (2) a younger NE-SW ice-flow direction during deglaciation (Fig. 4). The NE-SW ice-flow indicators overlap the N-S indicators and therefore did not form contemporaneously. A well-defined marine ice margin along the south and west sides of Penn Cove (Easterbrook, 1968, 1994; Domack, 1983; Kovanen and Slaymaker, 2003a) demonstrates the presence of ice filling Penn Cove during building of ice-marginal deltas to the west and southwest. The surface of the marine delta adjacent to the west end of Penn Cove is deeply pitted with kettles, indicating that it was built on stagnating ice. The marine delta south of Penn Cove is kettled at the former ice margin and the LIDAR imagery shows that the delta still retains relict braided channels on its surface. The surface of both deltas is 200 ft (60 m).

Raised shorelines and marine deltas on southern Whidbey Island occur at elevations up to ~33 m and up to ~88 m above present sea level on northern Whidbey Island (Kovanen and Slaymaker, 2003a). The shorelines are best preserved along the sides of marine embayments on the island (Fig. 5). Changes in relative sea level occurred rapidly following deglaciation. The shorelines shown in Figures 4 and 5 have not yet been directly dated and are the subject of ongoing work by Kovanen and Slaymaker.

DAY 1. VASHON ADVANCE AND RETREAT; DISINTEGRATION OF THE CORDILLERAN ICE SHEET AND RELATIVE SEA LEVEL CHANGES ON WHIDBEY ISLAND

Drive from Seattle to Mukilteo to catch the ferry to Clinton on Whidbey Island. Drive 7 mi west to Stop 1-1 (Figs. 5, 6).

The drumlins, flutes, and grooves on Whidbey Island, shown in remarkable clarity on LIDAR images, are developed on Vashon till that mantles much of the Puget Lowland and records the main advance of the CIS to its southern limit near Olympia. Radiocarbon dates beneath Vashon till limit the timing of the CIS advance. A broad outwash plain built in front of the advancing ice in the central Puget Lowland contains interbedded peat, yielding dates of 18,265, 18,000, 17,350, 17,250, 17,000, and 16,510 ^{14}C yr B.P. (Easterbrook, 1992; Deeter, 1979). The peat consists of thin, in situ layers interbedded with the outwash and is clearly not allochthonous. The ^{14}C dates from coarse-grained, Canadian-provenance outwash indicate that Vashon ice had reached 115 km south of the Canadian border by the time of deposition of the peat. However, mixed wood dates from deposits just north of the Canadian border, ranging from 16,000 to 20,000 ^{14}C yr B.P., led Clague et al. (1988)

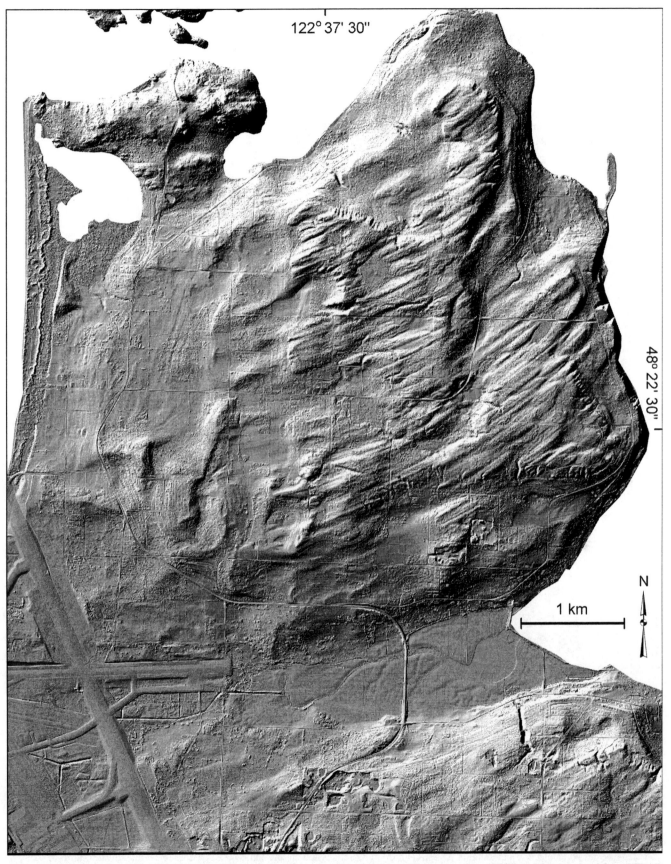

Figure 4. Light detection and ranging (LIDAR) image of ice-flow directions and shorelines on northern Whidbey Island derived from LIDAR topography base maps produced from data obtained from the Puget Sound Lidar Consortium. (Unpublished information from Kovanen and Slaymaker.)

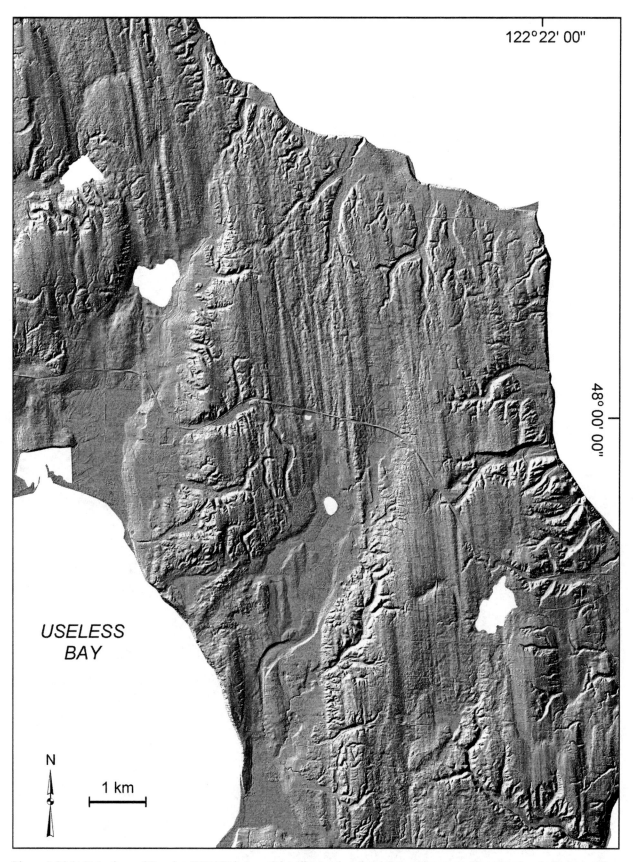

122° 22' 00"

48° 00' 00"

USELESS
BAY

N

1 km

Figure 5. Light Detection and Ranging (LIDAR) image of shorelines and marine deltas cutting across linear ice-flow indicators at Useless Bay, southern Whidbey Island. The shorelines extend to 40 m above present sea level near the lower lake at the upper left. Derived from LIDAR topography base maps produced from data obtained from the Puget Sound Lidar Consortium. (Unpublished information from Kovanen and Slaymaker.)

and Porter and Swanson (1998) to conclude that Vashon ice didn't cross the border until *after* 16 ka ^{14}C yr B.P. These dates are anomalous because allochthonous wood with dates ranging from 16 to 20 ka occur together. The reason for the incongruous dates remains uncertain, but the variability is almost certainly due to reworking of wood of mixed ages. No Vashon till occurs in this section so the deposit may not even be pre-Vashon. Their conclusion that Vashon ice didn't cross the border until after 16,000 ^{14}C yr B.P. would require throwing out all six 16,000–18,000 ^{14}C yr B.P. dates on peat in outwash far to the south. The most critical questions are: (1) was the height of the ice at 16 ka below the level at which trees could have been growing at that latitude, (2) how did 20 ka wood become mixed with 16 ka wood, and (3) are the sediments in Canada really beneath Vashon till or could they be younger?

The higher parts of the upland between Clinton and Stop 1 consist of Vashon till in well-defined drumlins, ridges, and grooves (Kovanen and Slaymaker, 2003a) (Fig. 5) produced during the Vashon maximum. The CIS initially retreated from the southern Puget Lowland by backwasting and downwasting (Fig. 3B), but when the ice margin reached the Strait of Juan de Fuca, marine water entered the lowland, floating the greatly thinned ice and causing widespread disintegration over the entire lowland from Whidbey Island to the Canadian border (Fig. 3C) (Easterbrook, 1992), and Everson gmd was deposited contemporaneously on the sea floor beneath the floating, disintegrating ice. Wood in subaerial, postglacial peat bogs (Kovanen and Easterbrook, 2002a), ^{14}C-dated ca. 13 ka, demonstrates that the CIS had retreated from northern Puget Lowland near the Canadian border by then, proving that the proposed model of a transgressively calving, retreating ice margin (Domack, 1983) is not possible. The distribution of 150 ^{14}C dates confirms this conclusion because the age of Everson gmd is the same on southern Whidbey Island and Vancouver, British Columbia. The thinning ice must have disintegrated almost simultaneously from the southern and northern Puget Lowland as it floated and broke up. As the lowland began to rise isostatically in the late Pleistocene, multiple shorelines were cut into the uplands and marine deltas were deposited.

Stop 1-1. Useless Bay Marine Delta and Shorelines

At Stop 1-1, a marine delta is banked against the upland at ~140 ft, fed by two streams that flowed off the adjacent upland (Figs. 5, 7). Streams in many gullies were graded to, and truncated at, the highest, raised shoreline (Kovanen and Slaymaker, 2003a). The shorelines shown in Figures 5, 7, and 8 have not yet been directly dated and are the subject of ongoing work by Kovanen and Slaymaker.

The shorelines and marine deltas appear to slightly postdate Everson gmd at Double Bluff along the SW margin of Useless Bay. Marine shells there gave a date of ca. 12,670 ^{14}C yr B.P., uncorrected for the marine reservoir effect (K. Stoffel, 1984, personal commun.). The corrected age is 11,570 ^{14}C yr B.P.

The upper shoreline can be traced laterally from the surface of the marine delta around the entire Useless Bay embayment (Fig. 8).

Figure 6. Whidbey Island showing field trip stops. (Modified from shaded digital topographic model by PRISM, University of Washington.)

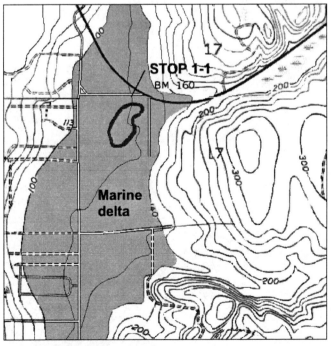

Figure 7. Marine delta at Stop 1-1.

At the northern end of the embayment, about a dozen lower shore-lines can be seen (Fig. 5).

The surfaces of the marine deltas and shorelines here occur at elevations up to ~42 m. On northern Whidbey Island, upper shoreline limits that extend to ~88 m above present sea level (Kovanen and Slaymaker, 2003a) are believed to be equivalent to the shorelines here but extend to higher levels there because of greater isostatic uplift.

Drive 29 mi on Highway 525 to Stop 1-2 at the intersection with the Parker Road at the north end of an abandoned airstrip.

Stop 1-2 Coupeville Marine Kame Delta

As the Vashon glacier retreated northward from its maximum stand past the latitude of the Strait of Juan de Fuca, it had thinned enough for marine water to float the ice in the deeper

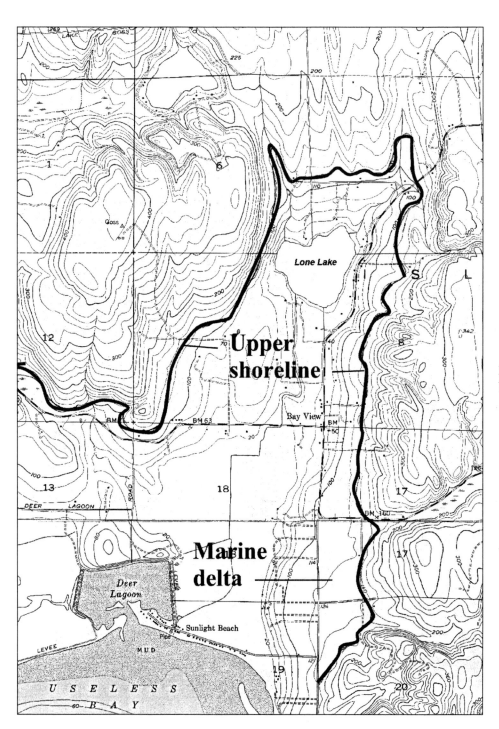

Figure 8. Upper shoreline limit around the Useless Bay embayment.

water and caused widespread disintegration of the glacier all the way to the Canadian border. However, grounded ice remained on parts of the eastern mainland and the disintegration of ice to the west produced a sharp change in the flow direction of the grounded ice from N-S to NE-SW. South of Penn Cove (Fig. 9), ice-flow indicators are all N-S, but to the north, a strong NE-SW trend is superimposed on the N-S lineations (Kovanen and Slaymaker, 2003a). The glacier margin at this time was along the south shore of Penn Cove, where meltwater deposited a large marine kame delta ~220 ft (67 m) above present sea level (Fig. 9). Braided channels on this surface are readily apparent on LIDAR imagery (Kovanen and Slaymaker, 2003a). Another large, deeply kettled, marine, kame delta was built over stagnant ice west of the glacier margin at Penn Cove (Fig. 9). Long, westward-sloping foreset beds are exposed in sea cliffs and gravel pits cut into these deltas (Easterbrook, 1968). West of the distal margin of the Coupeville delta, distinct marine shorelines are present along the sides of the adjacent uplands (Kovanen and Slaymaker, 2003a).

Stop 1-3. Deeply Kettled Ice-Contact Marine Delta and Everson Glaciomarine Drift, Partridge Point

The surface topography in the Partridge Point area is deeply indented with kettles (Fig. 9), many >30 m deep. Just south of West Beach Park, wave erosion has breached a 30-m-deep kettle, exposing the Partridge Point gravel that makes up the kettled topography. Collapse structures are common around the margins of the kettles. The gravel typically has strong foreset bedding with dips up to 20–25° and contains pumice clasts and small shell fragments (Easterbrook, 1966a, 1968; Carlstad, 1992). The top of the Partridge Point gravel, where not indented by kettles, consists of a flat surface ~220 ft (67 m) above present sea level.

Sea cliffs just north of the beach access consist of Everson gmd containing marine shells dated at 12,535 ± 300 ^{14}C yr B.P. (uncorrected) (Easterbrook, 1968). Applying the marine reservoir correction for this region (Kovanen and Easterbrook, 2002b) gives a corrected age of 11,435 ^{14}C yr B.P. The gmd caps the Partridge Point gravel in the bluffs just south of the beach access.

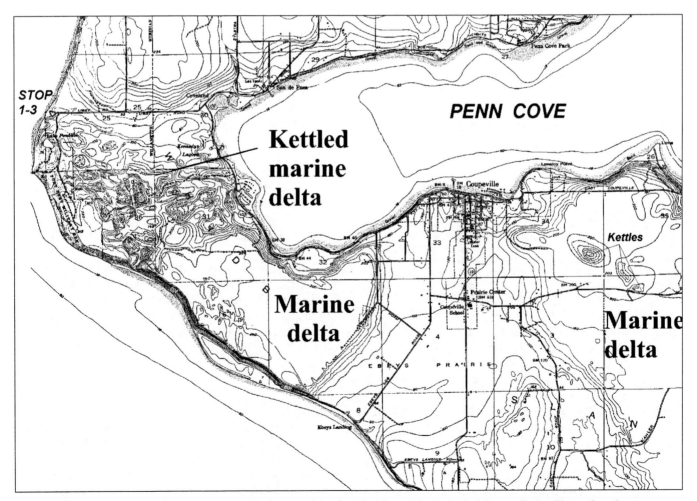

Figure 9. Marine deltas, Coupeville and Partridge Point Modified from U.S. Geological Survey Coupeville quadrangle.

North of Penn Cove on the upland south of Deception Pass, strong, linear, ice-flow indicators are truncated by multiple shorelines (Fig. 5) (Kovanen and Slaymaker, 2003a), the highest at an elevation similar to the surface of the Penn Cove marine deltas. The shorelines wrap around the east side of the linear, NE-SW, glacial features, indicating that the grounded ice had left this area by then.

The morphologic relationships between the linear glacial features, marine deltas, gmd, and shorelines are clearly shown on the LIDAR images (Kovanen and Slaymaker, 2003a). Grounded ice clearly changed flow direction from N-S to NE-SW when marine invasion caused disintegration of the CIS in the lowland to the west. At that time, ice filled Penn Cove and deposited a large, deeply kettled ice-contact marine delta upon stagnating ice to the west, between Coupeville and Partridge Point, and a large marine delta with braided channels to the south (Fig. 9) (Easterbrook, 1968, 1994). A line of kettles marks the thick, former ice margin along the northern end of the marine delta (Kovanen and Slaymaker, 2003a). The only date so far obtained for the deltas is 13,650 ± 350 ^{14}C yr B.P. (uncorrected), 12,550 ± 350 ^{14}C yr B.P. corrected for the marine reservoir effect (Kovanen and Easterbrook, 2002b), from shells in deltaic gravel near the Naval Air Station north of Oak Harbor (Dethier et al., 1995). Everson gmd overlies the Partridge Point deltaic gravel at West Beach, so it is slightly younger. Although the shorelines north of Oak Harbor lie at about the same elevation as the surface of the marine deltas, the shorelines wrap around the east side of the linear glacial features, indicating that the ice must have retreated from the marine deltas by then, so the shorelines must be slightly younger than the marine deltas. Radiocarbon ages are summarized below:

Locality	Uncorrected age (^{14}C yr B.P.)	Corrected age (^{14}C yr B.P.)
Marine delta		
Oak Harbor	13,650 ± 350	12,550 ± 350
Everson glaciomarine drift		
Double Bluff	12,670 ± 90	11,570 ± 90 (K. Stoffel, 1984, personal commun.)
West Beach	12,535 ± 300	11,435 ± 300 (Easterbrook, 1968, 1969)
Penn Cove	13,010 ± 170	11,910 ± 170 (Easterbrook, 1968, 1969)
Penn Cove	11,850 ± 240	10,750 ± 240 (Easterbrook, 1968, 1969)
Penn Cove	12,300 ± 180	11,200 ± 180 (Easterbrook, 1968, 1969)
Hope Island	12,400 ± 190	11,300 ± 190 (Easterbrook, 1969)
Kiket Island	12,865 ± 110	11,765 ± 110 (Easterbrook, 1969)

These dates are consistent with the 150 ^{14}C dates from Everson gmd throughout the Puget Lowland. Basal bog dates indicate that gmd deposition ceased in the Puget Lowland by 11,500 ^{14}C yr B.P. Swanson and Caffee (2001), however, contended that deposition of Everson gmd ended 13,500 ^{14}C yr B.P., fully 2000 yr earlier. This could be true only by ignoring the 150 ^{14}C dates. Unfortunately, Swanson and Caffee use their erroneous 13.5 ka age as the basis for calculating their [36]Cl production rate for [36]Cl, so consequently their [36]Cl ages are off by 20% (2000 yr) (Easterbrook, 2003a).

Return to main highway and drive 19.4 mi to Deception Pass.

Deception Pass is a narrow gorge in bedrock separating Whidbey Island from Fidalgo Island. Tidal currents in the gorge are unusually high as seawater pours in and out of the inlet to the east.

Cross the bridge and continue 6.2 mi to the intersection with the Mount Vernon/Anacortes Highway. Turn right, drive to Interstate 5, turn north on I-5, and continue to the Bow Hill exit (18 mi).

Stop 1-4. Sumas Outwash Channel Incised across Everson gmd and Shorelines

At the Samish River (Fig. 10), gmd extends to elevations of 65 m, about the same as the highest shorelines in the area, and is deeply incised by the underfit Samish River. The Samish River presently has a meander radius of ~0.15 km (0.1 mi) on a valley floor having meander radii >1.7 km (1 mi). The Samish Valley clearly was made by a much larger stream than the present Samish River. Meanders are incised into Everson gmd and cut across beach ridges at the marine limit of 65 m (215 ft) before terminating at a marine delta ~30 m (100 ft) above present sea level (Easterbrook, 1979, 1994). The outwash was graded to a sea level ~35 m lower than that of the gmd. The outwash channel and terraces belong to the Sumas Stade, ^{14}C dated elsewhere ca. 10,000–11,500 ^{14}C yr B.P. A limiting age for similar strandlines nearby is given by a date of 11,700 ± 110 ^{14}C yr B.P. from basal peat at 47 m (Siegfried, 1978).

Return to I-5 and drive to Bellingham. Continue to the Sunset Drive exit; turn right on Sunset Drive. Turn left on Hannegan Road.

Sumas I Squalicum Outwash Channel

Immediately upon turning onto Hannegan Road, it drops down into the Squalicum outwash channel, which originated just to the east from a brief stand of the earliest phase of the Sumas Glacier (Sumas I). The channel incises Bellingham gmd, so it is therefore post-11,600 ^{14}C yr B.P. and is older than Sumas II (11,400 ^{14}C yr B.P.). No well-defined moraine occurs at the head of the channel. It merely pinches out near the former glacier terminus.

Just north of the Squalicum channel, turn right on Bakerview Road and drive 1.1 mi. Turn left on Dewey Road and continue.

Stop 1-5. Sumas II Tenmile Creek Outwash Channel

Following deposition of Everson gmd and emergence of the region, meltwater from the Sumas II phase carved deep outwash channels into the gmd, indicating that emergence closely followed deposition of the gmd and providing a lower limiting age for the Sumas II phase. The Sumas outwash channel is incised 33 m (100 ft), exposing Bellingham gmd overlying Deming sand in the valley walls (Fig. 11). Basal peat from a 10 m (33 ft) bog that floors the channel has been dated at 11,080 ± 100 and 11,113 ± 88 ^{14}C yr B.P. (Kovanen and Easterbrook, 2002a). These dates correlate well with ^{14}C dates from a prominent Sumas II moraine in southwestern British Columbia that we will see at the next stop. A basal bog date of 10,400 ± 85 ^{14}C yr B.P. was obtained from Fazon Lake, a large kettle just to the north.

Drive west on Axton Road to the Guide Meridian; turn right and drive across peat-filled Sumas IV outwash channels. Continue north through Lynden for 14 mi to the Canadian border.

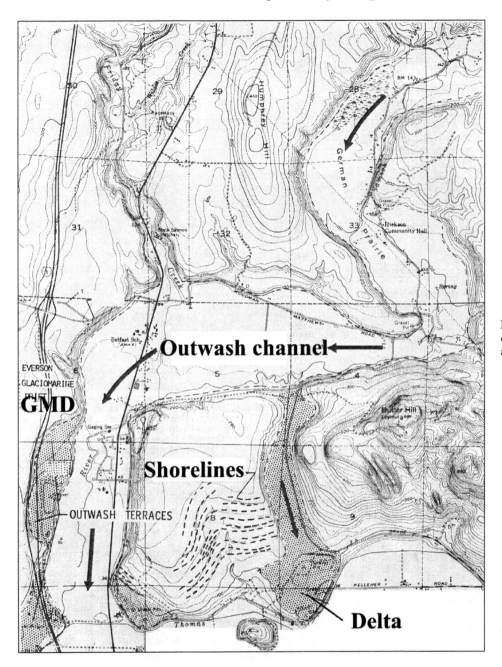

Figure 10. Sumas outwash channel incised into Everson glaciomarine drift and across late Everson shorelines.

Stop 1-6. Sumas II Moraine, Sumas III Outwash Plain

Just across the border is an elongate, tapering, Sumas II moraine (Fig. 12) that rests on Everson gmd, from which many ^{14}C dates on wood and shells have been obtained, clearly limiting the Sumas II advance to younger than 11,600 ^{14}C yr B.P. (Easterbrook and Kovanen, 1998; Kovanen and Easterbrook 2002a; Kovanen, 2002). A ^{14}C date of 11,413 ± 75 ^{14}C yr B.P. from basal peat in an outwash channel from the moraine (Fig. 12) indicates that the glacier had retreated back from the end moraine by 11.4 ka, cutting off meltwater flow in this channel. A ^{14}C date of 10,980 ± 250 ^{14}C yr B.P. from basal peat in a channel *behind* the moraine indicates that the glacier had receded to the Sumas III moraines to the north by this time. We are presently inside the Sumas II moraine on the broad Abbotsford outwash plain that heads at Sumas III moraines farther north (Figs. 13, 14).

Turn right on Border Rd., then left on Visser Rd., left on Bender Rd., and left on Halverstick Rd. Descend into a Sumas III outwash channel and turn right on Swanson Rd. Stop briefly at the distal end of the channel to see remnants of the Sumas IV moraine blocking the lower end of the channel. Turn left on

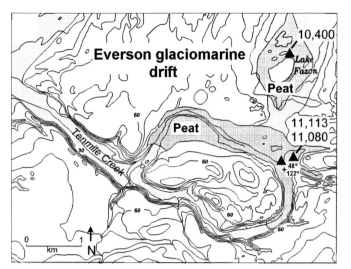

Figure 11. Sumas II Tenmile outwash channel deeply incised in glacio-marine drift. Basal peat dates indicate that meltwater ceased to flow in the channel by 11.1 ka.

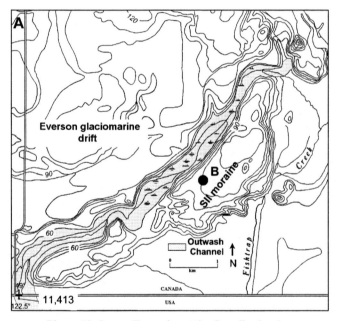

Figure 12. Sumas II moraine at the Canadian border.

Clearbrook Rd., cross the Everson-Sumas Highway, turn right on Hill Rd. then right on Telegraph Rd., and left on North Pass Rd. Continue 2.4 mi to Stop 1-7.

Stop 1-7. Sumas IV Moraine and Outwash Terrace

The Sumas IV moraine consists of a hummocky, linear ridge that extends in an E-W direction away from the mountains for ~3.5 km (2.2 mi) at the head of an outwash plain that slopes southward away from the moraine (Fig. 15) (Easterbrook, 1963a, 1976b). Erosion has removed the moraine from the central part of the Sumas Valley, but a remnant occurs on the west side of the valley, blocking the distal end of a Sumas III meltwater channel, showing that it represents a readvance of the glacier.

Return to Bellingham via Everson Goshen Rd.

DAY 2. GLACIATION OF THE NORTH CASCADES

In the North Cascade Range adjacent to the Fraser Lowland, long alpine glaciers (Fig. 16) occupied the Nooksack drainage following the disappearance of the CIS (Kovanen and Easterbrook, 2001). During the first two stops today, we will look at moraines of these alpine glaciers.

Drive east on the Mt. Baker Highway through the town of Kendall to Stop 2-1 at a large roadcut.

Stop 2-1. Kendall Moraine and Outwash Terrace

The Kendall moraine provides evidence for extension of a post-CIS, Cascade, alpine glacier ~45 km downvalley from its sources on Mount Shuksan and Mount Baker. It forms a distinct ridge several hundred meters long, whose axis is transverse to the valley. It is composed of ~12 m (40 ft) of till, containing abundant, glacially faceted, striated boulders and cobbles derived from the upper North Fork valley. The upper North Fork has a distinct U-shaped cross-profile at right angles to the direction of flow of the CIS.

The number of well-faceted and striated cobbles and boulders in the till is striking, probably because the valley glacier was so long. The lithology of cobble and boulders in the Kendall moraine is dominated by Chuckanut sandstone (54%), Mount Baker andesite (12%), and graywacke/greenstone (14%) (presumably from the Jurassic Nooksack Group) (Kovanen and

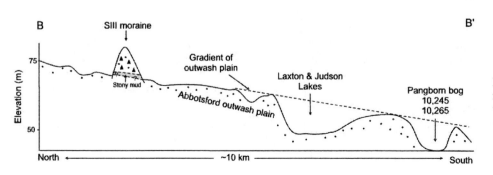

Figure 13. N-S profile from Sumas SIII moraines across the Abbotsford out-wash plain.

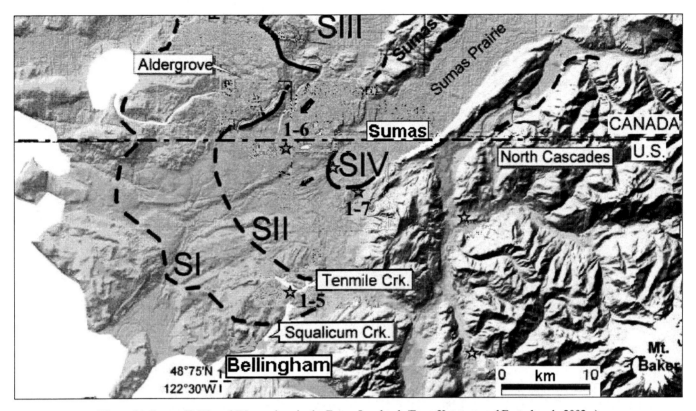

Figure 14. Sumas II, III, and IV moraines in the Fraser Lowland. (From Kovanen and Easterbrook, 2002a.)

Easterbrook, 1996, 2001). These lithologies demonstrate that the source of ice was from the North Fork valley to the east, rather than from the CIS to the north. The Mount Baker andesite was contributed from Mount Baker *northward* via Bagley Creek, Glacier Creek, and Wells Creek into the North Fork and carried westward to the Kendall moraine. The many pebbles and cobbles of Mount Baker andesite could not have been carried here by the CIS because the andesite had to be transported northward from Mount Baker, which would have been upglacier for the CIS, clearly impossible. Striations in bedrock on ridge crests in the North Cascades demonstrate that the CIS flowed N-S with no westward component (Easterbrook, 1963b, 1992; Kovanen and Easterbrook 2002a). Chuckanut sandstone was derived from the valley sides immediately upvalley, and the graywacke/greenstone came from the valley sides near the town of Glacier.

Not far to the west, outwash from the Kendall moraine and the Maple Falls moraine just upvalley merge with the outwash terrace in the Columbia Valley. A few kilometers downvalley, the outwash terrace rests on gmd dated at 11,910 ± 80 ^{14}C yr B.P. Distinct charcoal layers interbedded with sandy outwash terrace gravel gave dates of 10,788 ± 77 and 10,603 ± 69 ^{14}C yr B.P. (Kovanen and Easterbrook, 2001) but their relationship to the moraines is not entirely clear. ^{14}C-dated Younger Dryas moraines in the Nooksack Middle Fork are much farther upvalley, so the Kendall and Maple Falls moraines are most likely pre–Younger Dryas.

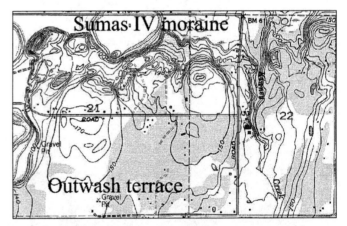

Figure 15. Topographic map of the Sumas IV moraine and outwash terrace at Stop 1-6.

Return to Welcome and turn left on Mosquito Lake Rd. Drive across the Nooksack Middle Fork bridge to Deep Kettle Bog.

Stop 2-2. Deep Kettle Bog, Mosquito Lake Kettle-Kame Complex and Lateral Moraine, Younger Dryas Moraines

In recent years, the lateral moraine that is draped across a kettle-kame complex at this stop has become badly overgrown

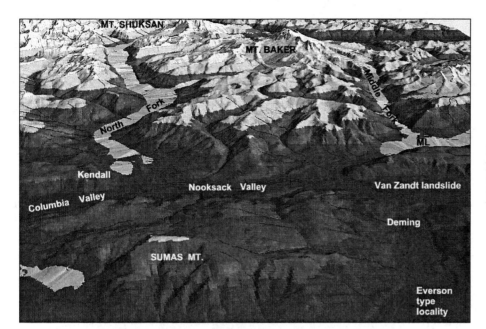

Figure 16. Post–Cordilleran Ice Sheet alpine glaciers in the North and Middle Forks of the Nooksack River, North Cascades. (Image by D.J. Kovanen from digital elevation model data.)

Figure 17. Pre–Younger Dryas (YD) alpine lateral moraine draped over a kettle-kame complex near Mosquito Lake, Nooksack Middle Fork. YD moraines occur upvalley (above the word "Nooksack" in the photo).

with vegetation and is now difficult to observe (see Fig. 17 for an aerial view). The moraine had been freshly exposed by logging when Kovanen and Easterbrook (2001) gathered evidence for the alpine ice origin of the post-CIS lateral moraine and kettle-kame complex. The moraine is composed of till containing many glacially faceted and striated boulders and cobles, dominated by Mount Baker andesite (17.4%), dunite from the Twin Sisters Range (16.6%), and Chuckanut sandstone (45.2%) from upvalley, demonstrating that the glacier that transported them must have originated in the headwaters of the Nooksack Middle Fork (Kovanen and Easterbrook, 1996, 2001). A landslide, now also mostly overgrown with vegetation, exposed collapsed sand and gravel of the kettle-kame complex.

Rootlets and wood in gravel at the base of the 20-m-deep bog were ^{14}C dated at 12,356 ± 115 ^{14}C yr B.P. The rootlets had to be growing in rock debris on stagnant ice before the ice melted out to make the kettle because the vegetation could not have grown at the bottom of such a deepwater lake. A date on wood lying on the basal gravel was 12,165 ± 115 ^{14}C yr B.P. (Kovanen and Easterbrook, 2001).

The age of the lateral moraine must be younger than the 12,000-yr-old kettle-kame complex because it overlies it, both morphologically and stratigraphically.

The strong dominance of glacially faceted boulders and cobbles from the Middle Fork headwaters 23 km (14.3 mi) upvalley demonstrates that the kettle-kame complex and the lateral moraines were deposited by true alpine glaciers, not by remnants of the CIS, and must therefore postdate the CIS. These Nooksack alpine glaciers could not have been fed by Cordilleran ice because of the lithology of the drift and, as shown by gmd in the Nooksack drainage, the area to the north was free of Cordilleran ice at that time.

Well upvalley, logs in a lateral moraine have been dated at 10,680 ± 70 and 10,500 ± 70 ^{14}C yr B.P. (Kovanen and Easterbrook, 2001). Figure 18 shows the position of the moraines.

Continue westward along Mosquito Lake Rd. to its intersection with Highway 9 at the town of Acme. Turn left and drive down Samish River Rd. to I-5. Turn left on I-5 and drive to Ever-

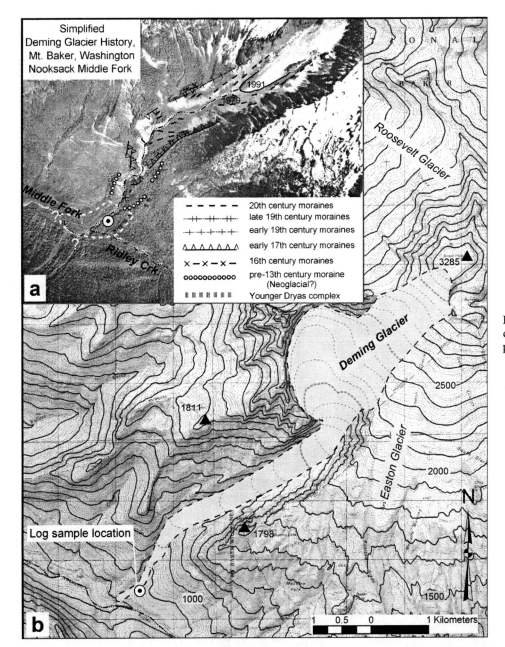

Figure 18. Deming glacier history. (Image courtesy of Kovanen and Slaymaker, unpublished data.)

ett. Exit I-5 to Highway 2; drive to Monroe. Turn right on the road to Carnation/Fall City and drive to Snoqualmie Falls.

Ice-Marginal Features of the Cordilleran Ice Sheet

The eastern margin of the CIS was banked against the Cascade Range as ice flowed southward down the Puget Lowland. The ice sheet extended up major Cascade valleys, blocking drainage and impounding ice-marginal lakes into which large deltas were built (Cary and Carlston, 1937; Mackin, 1941). An ice-marginal embankment was constructed across the Snoqualmie Valley east of North Bend. The embankment consists of a high, flat-topped ridge that extends from the north valley wall almost all the way across the valley of the Snoqualmie River. Deltaic sediments,

consisting mostly of coarse sand and gravel, grading upvalley to sand, then to silt and clay, were deposited in an ice-dammed lake upvalley. Wood from rhythmically bedded (varved?) lake sediments upvalley was ^{14}C dated at 13,570 ± 130 (UW-35) yr B.P. As the ice retreated to the northwest, the newly opened, ice-free valleys held ice-marginal lakes that became successively lower as new outlets were uncovered (Mackin, 1941; Booth, 1987).

Stop 2-3. Tokul Creek Delta and Snoqualmie Falls

As the Puget lobe retreated northward, an ice-marginal lake formed in the Snoqualmie Valley south of the ice margin. Tokul Creek, a tributary of the Snoqualmie River, brought large quantities of sand and gravel into the lake, building a delta that extended across the valley. As the ice front retreated northward, new, lower outlets were uncovered and the lake eventually drained. When the Snoqualmie River re-established its postglacial course, the river established its course on the west side of the valley at the distal edge of delta. As the Snoqualmie cut down through these sediments, it was superposed upon the bedrock of the west valley wall. Downstream, the river quickly eroded the valley-filling sediments, but where the river was notched into the bedrock, it could not erode its bed as rapidly, and Snoqualmie Falls was born. As the falls retreat upvalley, they will eventually erode through the bedrock spur and into the valley fill upstream. When this happens, the river will quickly cut down through the sediments and bring an end to the falls.

Continue through North Bend, turn left on I-90, and drive 22 mi (37 km) to Snoqualmie Pass. Exit I-90 to the ski area.

Stop 2-4. Late Pleistocene Alpine Moraines

Multiple, as-yet-undated, post–Last Glacial Maximum (LGM) alpine moraines occur at Snoqualmie Pass (Porter, 1978). Half a dozen closely spaced moraines at Snoqualmie Pass, well upvalley from the LGM, record several periods of glacial retreat and stillstand. No numerical ages have been published for any of these moraines, although a limiting date of 11,050 ± 50 ^{14}C yr

B.P. was obtained from wood in gravel on till between two of the moraines (Porter, 1978).

Rejoin I-90 and drive to exit 85 to Blewett Pass. Exit and drive north to Highway 2. Turn left on Highway 2 and drive to Leavenworth.

Stop 2-5. Leavenworth Moraines

During the LGM, a long alpine glacier extended down Icicle Creek and terminated at Leavenworth (Page, 1939). Two end moraines occur within the town and connect to lateral high moraines high along the sides of Icicle Creek (Fig. 19). At Stop 2-4, we will examine the outermost of the two moraines.

Stop 2-6. Younger Dryas (?) Moraine at Rat Creek

Upvalley from Leavenworth, Rat Creek, a tributary of Icicle Creek, has two moraines at its mouth (Fig. 20). The moraines occur on the opposite side of Icicle Creek from the road and are not easily accessible, so we will observe them from across the creek.

Porter (2003, personal commun.) has suggested that [36]Cl ages of boulders on the moraines place them within the Younger Dryas.

Return to Leavenworth and drive to Wenatchee for the night.

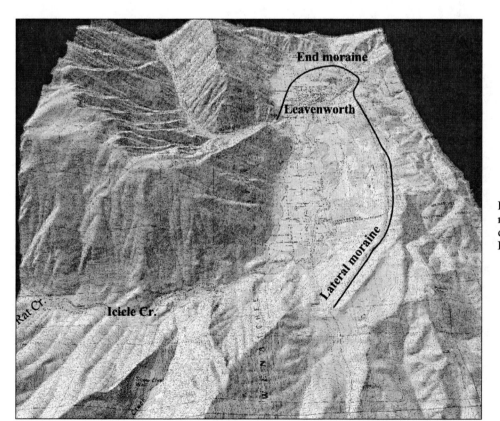

Figure 19. Icicle Creek lateral and end moraines at Leavenworth. Image from digital elevation model data by D.J. Kovanen.

Figure 20. Double moraines at Rat Creek. Image by D.J. Kovanen from digital elevation model data.

DAY 3. GLACIAL LANDFORMS OF THE OKANOGAN LOBE OF THE CORDILLERAN ICE SHEET; MISSOULA FLOOD FEATURES

During the LGM, the Cordilleran Ice Sheet flowed down both sides of the North Cascades, overwhelming the crest of the range and connecting the Okanogan lobe with Puget Lowland ice across the North Cascades. The Okanogan lobe reached ~50 km south of the Columbia River on the Waterville Plateau, where it built a massive end moraine and left a museum of glacial landforms magnificently preserved in the arid climate (Fig. 21). Drumlins and flutes (Fig. 22) show a radial flow of ice during the maximum stand, and eskers, kames, and recessional moraines were deposited during ice recession. The Okanogan lobe played an important role in directing giant floods from the bursting of ice dams and draining of Glacial Lake Missoula. Glacial landforms are intimately associated with Missoula flood features on the plateau (Bretz, 1959, 1969; Baker, 1978).

Drive through Waterville on Highway 2 to Moses Coulee.

Stop 3-1. Withrow Moraine Draped Across a Giant Missoula Flood Bar; Outwash Terrace, Moses Coulee

Moses Coulee was scoured out by a giant flood from Lake Missoula prior to the LGM, leaving a large, dry gorge over 1.7 km (1 mi) wide and 180 km (600 ft) deep. The Withrow

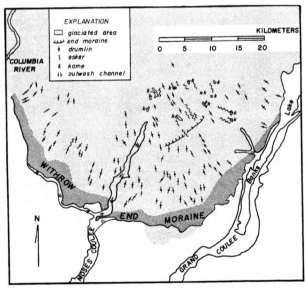

Figure 21. Configuration of the Okanogan lobe of the Cordilleran Ice Sheet showing the relationship between the Withrow moraine and Moses and Grand Coulees. (From Easterbrook, 1979.)

moraine extends for many kilometers across the plateau and drops into Moses Coulee across an earlier giant flood bar (Fig. 23). Moses Coulee is thus older than the Withrow moraine and must have been formed prior to the Okanogan lobe LGM, and no floods could have come down Moses Coulee since then

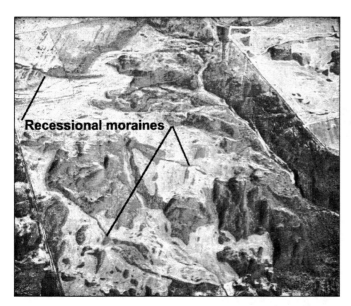

Figure 22. Drumlins and recessional moraines north of the Withrow moraine. Photo by D.A. Rahm.

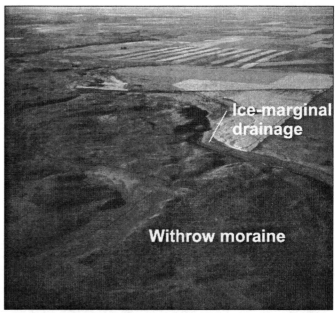

Figure 24. The Withrow moraine north of Farmer. Note the ice-marginal drainage between the moraine and the north-sloping topography beyond the moraine. Looking SE near Stop 3-2. Photo by D.A. Rahm.

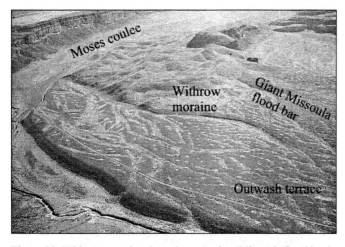

Figure 23. Withrow moraine draped over a giant Missoula flood bar in Moses Coulee. The outwash terrace from the moraine still shows the braided pattern of the outwash stream. Photo by D.A. Rahm.

without destroying the moraine and outwash terrace (Easterbrook, 1979).

Return to Highway 2; turn right and return to Farmer. Turn right at Farmer and drive 9 mi to the Withrow moraine.

Stop 3-2. The Withrow Moraine

The Withrow moraine (Fig. 24) is a hummocky moraine, 1–2 km wide and 30–70 m high, that extends across the Waterville Plateau. Large basalt erratics commonly dot the surface. The Columbia Plateau here slopes gently northward so meltwa-

ter streams could not flow directly away from the ice terminus. They flowed parallel to the ice margin, then down Moses to the Columbia River (Easterbrook, 1979). Dutch Henry Draw, which we crossed just before this stop, is one such channel (Fig. 24).

East of Stop 3-2, what appears to be a buried moraine, overridden by the Okanogan lobe, makes an arc across the plateau (Kovanen and Slaymaker, 2003b). Drumlins and other linear features are perched atop it, confirming that it was indeed overridden by actively advancing ice. The age of the buried moraine is not known. It could be a moraine from a pre-Wisconsin glaciation or perhaps just an earlier late Wisconsin advance. The position of the buried moraine relative to Moses Coulee invites consideration of the possibility that a pre–Grand Coulee Missoula flood was diverted by ice in a manner similar to the diversion of flood water down Grand Coulee by the Vashon CIS.

Drive to Mansfield, crossings several recessional moraines and passing several huge basalt erratics.

Stop 3-3. Eskers, Recessional Moraine, and Kame Fans

The Waterville Plateau north of the Withrow moraine contains many eskers (Fig. 25), kames, kettles, and related ice-contact deposits formed during deglaciation of the Okanogan lobe. Eskers that terminate at ice-marginal fans and low moraine ridges mark temporary halts in the recession of the glacier terminus (Fig. 26). Some eskers end abruptly at ice-marginal fans that still retain traces of abandoned braided channels on their surfaces. The proximal side of the fans mark the former ice margin (Easterbrook, 1979).

Figure 25. Esker on the Waterville Plateau. Photo by D.A. Rahm.

Stop 3-4. Grand Coulee, Dry Falls, Missoula Flood

Grand Coulee was described by Bretz (1969) as the "deepest, longest, and steepest in gradient. Its system possesses the largest number of abandoned cataracts, the widest one, a record of the highest, the greatest cascade, and the largest number of distributary canyons. It is unique in that, at its initiation, it found no preglacial valley head but crossed 20 miles or so of the northern divide of the plateau before encountering such a valley, and it trenched that divide nearly 1000 feet deep."

Dry Falls (Fig. 27) extends ~3 mi across, and is one of several falls excavated by spectacular quarrying of jointed basalt during catastrophic floods spilling down Grand Coulee when the Okanogan lobe blocked flow down the Columbia River near the site of Grand Coulee Dam and forced it to spill over into Grand Coulee. Much of the large size and depth of Grand Coulee is due to waterfall recession.

Drive 14 mi to the junction with Highway 2 and Dry Falls. Drive down Grand Coulee to Elphrata and return to Seattle.

ACKNOWLEDGMENTS

I would like to thank D.J. Kovanen for providing images derived from LIDAR topography base maps produced from data obtained from the Puget Sound Lidar Consortium. Ralph Haugerud has played a significant role in making LIDAR data available through the Puget Sound Lidar Consortium. The images here represent unpublished information from Kovanen and Slaymaker. Kovanen also produced images for several figures from digital elevation model data, which have allowed new interpretations of the geology of many areas.

REFERENCES CITED

Armstrong, J.E., 1957, Surficial geology of New Westminster map-area, British Columbia: Geological Survey of Canada, Map 57-5, 25 p.

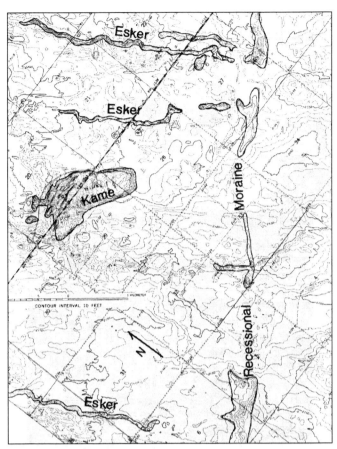

Figure 26. Esker terminating at a recessional moraine near Sims Corner.

Figure 27. Dry Falls. Photo by D.A. Rahm.

Armstrong, J.E., 1960, Surficial geology of the Sumas map-area, British Columbia: Geological Survey of Canada, Map 92G/1, 27 p.

Armstrong, J.E., 1981, Post-Vashon Wisconsin glaciation, Fraser Lowland, British Columbia: Geological Survey of Canada, Bulletin, v. 322, 34 p.

Armstrong, J.E., 1984, Environmental and engineering application of the surficial geology of the Fraser Lowland, British Columbia: Geological Survey of Canada, Paper 83-23, 54 p.

Armstrong, J.E., and Brown, W.L., 1954, Late Wisconsin marine drift and associated sediments of the lower Fraser Valley, British Columbia, Canada: Geological Society of America Bulletin, v. 65, p. 349–364.

Armstrong, J.E., Crandell, D.R., Easterbrook, D.J., and Noble, J.B., 1965, Late Pleistocene stratigraphy and chronology in southwestern British Columbia and northwestern Washington: Geological Society of America Bulletin, v. 76, p. 321–330.

Armstrong, J.E., and Hicock, S.R., 1980a, Surficial geology of New Westminster map-area, British Columbia: Geological Survey of Canada, Map 1484A.

Armstrong, J.E., and Hicock, S.R., 1980b, Surficial geology of Vancouver map-area, British Columbia: Geological Survey of Canada, Map 1486A.

Baker, V.R., 1978, Large-scale erosional and depositional features of the Channeled Scabland, in Baker, V.R., and Nummedal, D., eds., The Channeled Scabland: Washington, D.C., NASA, p. 81–115.

Blunt, D., Easterbrook, D.J., and Rutter, N.A., 1987, Chronology of Pleistocene sediments in the Puget Lowland, Washington: Washington Division of Geology and Earth Resources Bulletin, v. 77, p. 321–353.

Booth, D.B., 1987, Timing and processes of deglaciation along the southern margin of the Cordilleran ice sheet, in Ruddiman, W.F., and Wright, H.E., Jr., eds., North America and adjacent oceans during the last deglaciation: Boulder, Colorado, Geological Society of America, The Geology of North America, v. K-3, p. 71–90.

Bretz, JH., 1913, Glaciation of the Puget Sound region: Washington Geological Survey Bulletin, no. 8, 244 p.

Bretz, JH., 1959, Washington's Channeled Scabland: Washington Department of Conservation, Division of Mines and Geology Bulletin, v. 45, 57 p.

Bretz, JH., 1969, The Lake Missoula floods and the Channeled Scabland: Journal of Geology, v. 77, p. 505–543.

Carlstad, C.A., 1992, Late Pleistocene deglaciation history at Point Partridge, central Whidbey Island, Washington [M.S. thesis]: Bellingham, Washington, Western Washington University, 140 p.

Cary, A.S., and Carlston, C.W., 1937, Notes on Vashon stage glaciation of the South Fork of the Skykomish River valley, Washington: Northwest Science, v. 11, p. 61–62.

Clague, J.J., Saunders, I.R., and Roberts, M.C., 1988, Ice-free conditions in SW British Columbia at 16,000 years BP.: Canadian Journal of Earth Sciences, v. 25, p. 938–941.

Deeter, J.D., 1979, Quaternary geology and stratigraphy of Kitsap County, Washington [M.S. thesis]: Bellingham, Washington, Western Washington University, 175 p.

Dethier, D.P., Pessl, F., Keuler, R.F., Balzarina, M.A., 1995, Late Wisconsinan glaciomarine deposition and isostatic rebound, northern Puget Lowland, Washington: Geological Society of America Bulletin, v. 107, p. 1288–1303.

Domack, E.W., 1983, Facies of late Pleistocene glacial-marine sediments on Whidbey Island, Washington: An isostatic glacial-marine sequence, in Molnia, B.F., ed., Glacial-Marine Sedimentation: New York, Plenum, p. 535–570.

Easterbrook, D.J., 1962a, Pleistocene sequence on Whidbey Island: Bellingham, Washington, Northwest Science Annual Meeting Abstracts.

Easterbrook, D.J., 1962b, Vashon glaciation and Late Wisconsin relative sea-level changes in the northern part of the Puget Lowland, Washington: Los Angeles, California, Geological Society of America Abstracts.

Easterbrook, D.J., 1963a, Late Pleistocene glaciation of Whidbey Island, Washington: Berkeley, California, Geological Society of America Abstracts.

Easterbrook, D.J., 1963b, Late Pleistocene glacial events and relative sea-level changes in the northern Puget Lowland, Washington: Geological Society of America Bulletin, v. 74, p. 1465–1483.

Easterbrook, D.J., 1966a, Radiocarbon chronology of late Pleistocene deposits in northwest Washington: Science, v. 152, p. 764–767.

Easterbrook, D.J., 1966b, Glaciomarine environments and the Fraser Glaciation in northwest Washington: Guidebook for First Pacific Coast Friends of the Pleistocene Field Conference, 52 p.

Easterbrook, D.J., 1968, Pleistocene stratigraphy of Island County, Washington: Washington Division of Water Resources Bulletin, v. 25, p. 1–34.

Easterbrook, D.J., 1969, Pleistocene chronology of the Puget Lowland and San Juan Islands, Washington: Geological Society of America Bulletin, v. 80, p. 2273–2286.

Easterbrook, D.J., 1971, Geology and geomorphology of western Whatcom County, Washington: Western Washington University Press, 68 p.

Easterbrook, D.J., 1976a, Quaternary geology of the Pacific Northwest, in Mahaney, W.C., ed., Quaternary Stratigraphy of North America: Stroudsburg, Pennsylvania, Dowden, Hutchinson, and Ross, p. 441–462.

Easterbrook, D.J., 1976b, Geologic map of western Whatcom County, Washington: U.S. Geological Survey Miscellaneous Investigations, Map I-854B.

Easterbrook, D.J., 1979, The last glaciation of northwest Washington, in Armentrout, J.M., Cole, M.R., and Terbest, H., eds., Cenozoic Paleogeography of the Western United States. Pacific Coast Paleogeography Symposium 3: Los Angeles, Pacific Coast Section, Society of Economic Paleontologists and Mineralogists, p. 177–189.

Easterbrook, D.J., 1986, Stratigraphy and chronology of Quaternary deposits of the Puget Lowland and Olympic Mountains of Washington and the Cascade Mountains of Washington and Oregon, in Sibrava, V., Bowen, D.Q., and Richmond, G.M., eds., Quaternary Glaciations in the Northern Hemisphere: Quaternary Science Reviews, v. 5, p. 145–169.

Easterbrook, D.J., 1992, Advance and retreat of Cordilleran ice sheets in Washington, USA: Geographie Physique et Quaternaire, v. 46, p. 51–68.

Easterbrook, D.J., 1994, Stratigraphy and chronology of early to late Pleistocene glacial and interglacial sediments in the Puget Lowland, Washington, in Swanson, D.A., and Haugerud, R.A., eds., Geologic Field Trips in the Pacific Northwest: Boulder, Colorado, Geological Society of America, p. 1J23-38.

Easterbrook, D.J., 2003a, Comment on the paper "Determination of ^{36}Cl production rates from the well-dated deglaciation surfaces of Whidbey and Fidalgo Islands, Washington" by T.W. Swanson and M.C. Caffee: Quaternary Research, v. 59, p.132–134.

Easterbrook, D.J., 2003b, Cordilleran Ice Sheet glaciation of the Puget Lowland and Columbia Plateau and alpine glaciation of the North Cascade Range, Washington, in Easterbrook, D.J., ed., Quaternary Geology of the United States, INQUA 2003 Field Guide Volume: Reno, Nevada, Desert Research Institute, p. 265–286.

Easterbrook, D.J., and Kovanen, D.J., 1998, Pre–Younger Dryas resurgence of the southwestern margin of the Cordilleran Ice Sheet, British Columbia, Canada: Comments: Boreas, v. 27, p. 225–230.

Haugerud, R.A, Harding, D.J., Johnson, S.Y., Harles, J.L., Weaver, C.S., and Sherrod, B.L., 2003, High-resolution Lidar topography of the Puget Lowland: GSA Today, v. 13, no. 6, p. 4–10.

Heusser, C.T., 1973a, Environmental sequence following the Fraser advance of the Juan de Fuca Lobe, Washington: Quaternary Research, v. 3, p. 284–306.

Heusser, C.T., 1973b, Age and environment of peat clasts from the Bogachiel River Valley, Washington: Geological Society of America Bulletin, v. 84, p. 797–804.

Heusser, C.T., 1982, Quaternary vegetation and environmental record of the western Olympic Peninsula: American Quaternary Association, 7th Biennial Conference, Seattle, Guide for Day 2 of Field Trip G, 23 p.

Hewitt, A.T., and Mosher, D.C., 2001, Late Quaternary stratigraphy and seafloor geology of eastern Juan de Fuca Strait, British Columbia and Washington: Marine Geology, v. 177, p. 295–316.

Kovanen, D.J., 2002, Morphologic and stratigraphic evidence for Allerød and Younger Dryas age glacier fluctuations of the Cordilleran Ice Sheet, British Columbia, Canada and Northwest Washington, U.S.A.: Boreas, v.31, p. 163–184.

Kovanen, D.J., and Easterbrook, D.J., 1996, Extensive readvance of late Pleistocene (Y.D.?) alpine glaciers in the Nooksack River Valley, 10,000 to 12,000 years ago, following retreat of the Cordilleran Ice Sheet, North Cascades, Washington: Friends of the Pleistocene, Pacific Coast Cell Field Trip Guidebook, 74 p.

Kovanen, D.J., and Easterbrook, D.J., 2001, Late Pleistocene, post-Vashon, alpine glaciation of the Nooksack drainage, North Cascades, Washington: Geological Society of America Bulletin, v. 113, p. 274–288.

Kovanen, D.J., and Easterbrook, D.J., 2002a, Timing and extent of Allerod and Younger Dryas age (ca. 12,500?10,000 ^{14}C yr B.P.) oscillation of the Cordilleran Ice Sheet in the Fraser Lowland, western North America: Quaternary Research, v. 57, p. 208–224.

Kovanen, D.J., and Easterbrook, D.J., 2002b, Paleodeviations of radiocarbon marine reservoir values for the northeast Pacific: Geology, v. 30, p. 243–246.

Kovanen, D.J., and Slaymaker, 2003a, Sea-level changes, isostatic movement, and ice-flow indicators on Whidbey and Camano Islands, Puget Lowland:

Geologic data from LIDAR-aided observations, *in* Easterbrook, D.J., ed., Quaternary Geology of the United States, INQUA 2003 Field Guide Volume: Reno, Nevada, Desert Research Institute, p. 268–270.

Kovanen, D.J., and Slaymaker, 2003b, Glacial geomorphology and ice-flow indicators of the Okanogan lobe of the Cordilleran Ice Sheet: an archive of glacial features, *in* Easterbrook, D.J., ed., Quaternary Geology of the United States, INQUA 2003 Field Guide Volume: Reno, Nevada, Desert Research Institute, p. 281–283.

Leopold, L.B., Nickmann, R., Hedges, J.L., and Ertel, J.R., 1982, Pollen and lignin records of late Quaternary vegetation, Lake Washington: Science, v. 218, p. 1305–1307.

Mackin, J.H., 1941, Glacial geology of the Snoqualmie-Cedar area, Washington: Journal of Geology, v. 49, p. 449–481.

Page, B.M., 1939, Multiple glaciation in the Leavenworth area, Washington: Journal of Geology, v. 47, p. 785–815.

Porter, S.C., 1978, Glacier Peak tephra in the North Cascade Range, Washington: Stratigraphy, distribution, and relationship to late-glacial events: Quaternary Research, v. 10, p. 30–41.

Porter, S.C., and Swanson, T.W., 1998, Radiocarbon age constraints on rates of advance and retreat of the Puget Lobe of the Cordilleran Ice Sheet, during the last glaciation: Quaternary Research, v. 50, p. 205–213.

Siegfried, R.T., 1978, Stratigraphy and chronology of raised marine terraces, Bay View Ridge, Skagit County, Washington [M.S. thesis]: Bellingham, Washington, Western Washington University, 52 p.

Swanson, T.W., and Caffee, M.L., 2001, Determination of ^{36}Cl production rates derived from the well-dated deglaciation surfaces of Whidbey Island, Washington: Quaternary Research, v. 56, p. 366–382.

Geological Society of America
Field Guide 4
2003

Engineering geology in the central Columbia River valley

Thomas C. Badger*

Washington State Dept. of Transportation, P.O. Box 47365, Olympia, Washington 98504-7365, USA

Richard W. Galster*

Consultant, P.O. Box 908, Edmonds, Washington 98020-0908, USA

ABSTRACT

The deeply incised central Columbia River valley of Washington State and its tributaries expose mid to late Tertiary basalt flows and clastic sedimentary rocks, pre-Tertiary crystalline bedrock outcrops where the river flows along the eastern slope of the Cascade Mountains between Wenatchee and Chelan. River incision has primarily been driven by the uplift of the Cascades, deposition of the voluminous Columbia River basalts, and the formation of the Yakima fold belt. Glaciation during the Pleistocene, the terminus of which reached Chelan and the northern Waterville Plateau, infused large quantities of sediment into the valley. Concurrently, catastrophic glacial outburst floods, unprecedented in size, repeatedly swept down the river from the north and over the Quincy Basin in the south. Trip stops include some of the early engineering works, principally the dams, where much of the regional stratigraphy was developed and challenging engineering solutions were required for difficult geologic conditions. Stops also exemplify the pervasive large-scale landsliding, common where basalts overlie weak sedimentary rocks. Due to the steep topography, transportation corridors and other developments are widely threatened by rockfall and debris flow hazards. Seismicity is also a regional hazard; the largest historic earthquake in eastern Washington, moment magnitude 6.5–7.0, was sited near Chelan.

Keywords: geologic hazards, engineering geology, dams, landslides, Columbia River basin.

GEOLOGIC SETTING

The central Columbia River is situated between the rugged Cascade Mountains on the west and the eastward-tilted Columbia Plateau to the east (Fig. 1). The Columbia River flows southward through the valley and has deeply incised into this uplifted plateau, resulting in a relatively steep-sided valley with several thousand feet of relief.

The pre-Tertiary metamorphic rocks that are exposed in the northern portion of the central valley have been subdivided into five tectono-stratigraphic terranes (Tabor et al., 1987). Four terranes are relevant to the field trip: (1) the Ingalls, (2) the Swakane, (3) the Mad River, and (4) the Chelan Mountains. The Ingalls Tectonic Complex, which outcrops in the western portion of the field trip near Leavenworth, is composed of Late Jurassic to Early Cretaceous ophiolite mélange. The Swakane terrane, which is found north of the confluence of the Columbia and Wenatchee Rivers, is composed entirely of Precambrian(?) Swakane Biotite Gneiss. The Swakane is tectonically overlain by the Mad River terrane to the north that is composed of schist, quartzite, marble, and orthogneiss. Protolith age of the Mad River terrane is probably Paleozoic or older. The Late Cretaceous Entiat Pluton separates the highly migmatitic Chelan Complex and the Mad River terrane. While these terranes have unique histories, all were last metamorphosed during the Late Cretaceous.

An extensional episode started in the Eocene, producing large depositional basins accompanied by volcanism to the west.

*E-mail: Badger—badgert@wsdot.wa.gov;
Galster—georichgal@connectexpress.com.

Badger, T.C., and Galster, R.W., 2003, Engineering geology in the central Columbia River valley, *in* Swanson, T.W., ed., Western Cordillera and adjacent areas: Boulder, Colorado, Geological Society of America Field Guide 4, p. 159–176. For permission to copy, contact editing@geosociety.org. © 2003 Geological Society of America.

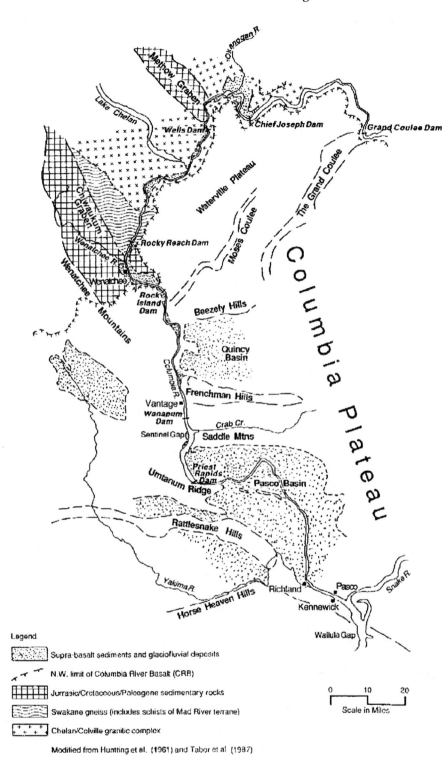

Figure 1. Physiographic map of the middle Columbia River valley, including generalized geologic units.

Legend

- Supra-basalt sediments and glaciofluvial deposits
- N.W. limit of Columbia River Basalt (CRB)
- Jurrasic/Cretaceous/Paleogene sedimentary rocks
- Swakane gneiss (includes schists of Mad River terrane)
- Chelan/Colville granitic complex

0   10   20
Scale in Miles

Modified from Huntting et al. (1961) and Tabor et al (1987)

Extremely thick sedimentary sequences are preserved around and west of Wenatchee, recording this time of rapid, uninterrupted sedimentation (Tabor et al., 1983). Unconformably overlying the pre-Tertiary crystalline basement, these sedimentary rocks are predominantly composed of fluvial arkosic sandstones with lesser conglomerates (Swauk Formation) and tuff and sandstone beds and shale (Chumstick Formation) (Gresens, 1987). Deposition of the Chumstick had, until recently, been interpreted to form within a tectonic graben, the Chiwaukum Graben, bound on the west by the Leavenworth fault and by the Entiat fault on the east.

Cheney (2000) proposed that these and other similar-age formations were deposited over a large area, that these formations are erosionally preserved in structural lows, and that these faults were active after, not concurrent with, deposition. Coeval dike swarms related to this extensional episode occur southwest of the Chiwaukum Graben, including the Teanaway Basalt, and north of Wenatchee in Corbaley Canyon, predominantly composed of rhyolite (Gresens, 1987).

The region was tectonically and magmatically quiet from the late Eocene to early Oligocene, during which time an extensive erosional surface developed. The Wenatchee Formation was deposited on this surface.

Extensive basalt flows covered central Washington during the Miocene. North of Wenatchee, the western extent of these flows reached the present location of Columbia River, constraining the river against the eastern slope of the Cascades (Frecht et al., 1987). South of Wenatchee, basalts extend ~50 mi west of the river. Sediment deposition within the ancestral Columbia River basin produced the Ellensburg Formation, which is intercalated with the basalts (Reidel and Campbell, 1989). Regional uplift along with northward compression and formation of the Yakima fold belt accompany and mostly postdate the basalts.

Multiple advances of Cordilleran ice occurred during the Pleistocene. During the late Pleistocene, the Okanogan lobe extended south of the Okanogan and Columbia River confluence onto the Waterville Plateau and blocked the Columbia River. The depositional "great terrace" of Russell (1893) extends nearly 20 mi south of the glacier terminus located near Chelan. Also during this time, glacial outburst floods repeatedly inundated the Columbia Basin, leaving behind a dramatic erosional landscape and equally dramatic constructional landforms and associated deposits. Repeated damming of the Columbia River by outwash deposits at Moses Coulee, and possibly by landsliding upstream of Rock Island, are evidenced by thick sequences of lacustrine deposits and hanging deltas as far upstream as Rocky Reach Dam. These intermittent lakes reached elevations of 850–1050 ft (260–320 m) (Tabor et al., 1983).

Landslides are abundant and voluminous along the Columbia River, where basalts overlie weak sedimentary rocks of the Ellensburg Formation. The Chumstick Formation is also a source of large landslides within the basin.

This field trip highlights human interaction with the geologic environment, primarily focusing on transportation, electrical power generation, and development. The rugged physiography and problematic geologic conditions create classic engineering geology challenges—creating complex design and construction of large hydroelectric dams, mitigating slope hazards from huge landslides, rockfall, and debris flows, and maintaining and expanding infrastructure to support growth in the region.

FIELD TRIP LOG

Our first day's route will follow Interstate 90 into central Washington, beginning with a passage through the world's largest soft-ground tunnel (Mount Baker Ridge Tunnel) and crossing Lake Washington via the Lacey V. Murrow (floating) Bridge. We will enter the Cascade Range through the Snoqualmie embayment, characterized by late Pleistocene delta morainal embankments. We cross the Cascade drainage divide through Snoqualmie Pass (elev. 3022 ft [921 m]) at mile 50 (km 80), and follow the Yakima River valley as far as Ellensburg (mile 106 [km 170]). From here, the route crosses the Saddle Mountains uplift (one of the folds of the Yakima fold belt) over Ryegrass Summit, then drops to the Columbia River at Vantage, mile 136 (km 219). Here we exit Interstate 90 and drive north through the community of Vantage to Ginkgo State Park.

Day 1

Cumulative		
Miles	(km)	Description
0.0	(0.0)	Stop 1-1.

Stop 1-1. Ginkgo State Park and Vicinity

This stop will include a general discussion of the basalt stratigraphy (Fig. 2) as exposed at the Vantage section, a look at some of the outstanding examples of petrified logs displayed at the park museum, and a visit to the type locality of the Vantage Interbed.

From this vantage point, one may observe the entire section of the Wanapum Basalt and the upper part of the Grande Ronde Basalt. The museum building rests on the eroded colonnade of the Museum flow, the uppermost flow of the Grande Ronde, an erosional landform known as the Museum platform. Interstate 90 is sited on this platform on the opposite bank and follows up the backslope of the Frenchman Hills anticline.

The route returns to Interstate 90 and crosses the Columbia River (Lake Wanapum).

Cumulative		
Miles	(km)	Description
3.5	(5.6)	On the east side of the river, turn right (south) onto SR 26, and a mile farther, keep right on SR 243, following the reservoir south.
7.5	(12.0)	Turn left, and follow the Bonneville Power substation road ~0.3 mi to the overlook near the top of slope.
7.8	(12.5)	Turn right to the Wanapum Dam overlook.
7.9	(12.6)	Stop 1-2.

Stop 1-2. Wanapum Dam

Wanapum Dam is the upstream unit of the two-dam Priest Rapids project constructed by the Public Utility District of Grant County during the late 1950s and early 1960s. This dam is built across a section of the valley that was deeply scoured by late Pleistocene cataclysmic glacial flooding. The scoured bedrock surface 50–150 ft (15–46 m) below the valley floor is developed

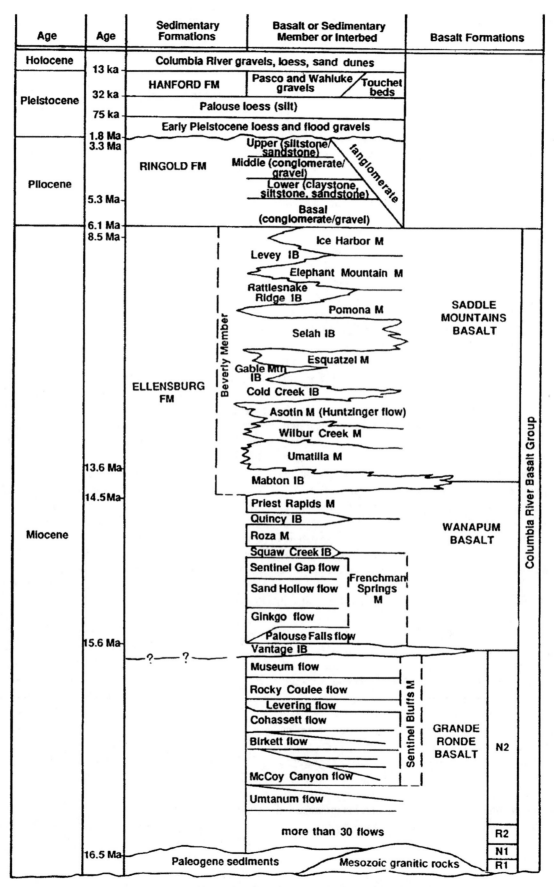

Figure 2. Stratigraphic column for the middle Columbia River region (from Galster and Coombs, 1989; modified from Mackin, 1961; Reidel and Fecht, 1981; and Swanson et al., 1979).

in a sequence of basalt flows and sediments that characterizes the Wanapum–Grande Ronde boundary in the central plateau region. At this location, a SSE-plunging anticline (Ryegrass structure) passes beneath the valley; it has a high-angle reverse fault on its east flank (Figs. 3 and 4). This combination of stratigraphy, structure, and erosional history made the site one of the more difficult to develop on the river. The complex geology resulted in the "dog leg" configuration of the dam structures.

Basalts exposed in the valley walls are the upper Frenchman Springs and Roza Members of the Wanapum Basalt. The lower flows of the Frenchman Springs Member (Sand Hollow and Ginkgo) form the bedrock surface beneath both sides of the valley floor. At mid-valley, the anticline and fault bring the upper two flows of the Grande Ronde Basalt (Museum and Rocky Coulee) higher, and the bedrock surface at mid-valley is formed on these two flows. Stratigraphically between the Ginkgo and Museum flows is the pillow-palagonite facies of the Ginkgo and the underlying Vantage interbed, both highly erodible and unsuitable for concrete dam foundations. Deep channels were eroded where these units subcrop beneath the valley and glacial flood alluvium and along the subcrop of the fault. Thus, the powerhouse/intake and spillway dams were sited on the shallowest bedrock at mid-valley; the former is oriented parallel to the former river channel. These structures are founded on the eroded tops of the Museum and Rocky Coulee flows. Long, flanking embankment sections founded

on highly pervious valley-fill gravels tie the concrete sections to rock abutments. Beneath the embankments, permanent slurry walls 12 ft (3.6 m) wide extend to the bedrock surface to control leakage through the gravels (Fig. 5). Where the depth to bedrock was greater than 80 ft (23 m) below the water table beneath the left embankment, a grouted section was installed between the bottom of the slurry wall and the bedrock surface. Grout and drainage curtains were installed beneath the heel of the intake and spillway structures (Galster and Imrie, 1989; Galster, 1989a).

Return to SR 26 intersection. Turn right on SR 26 and park on the shoulder ~500 ft (150 m) beyond the intersection. WATCH FOR TRAFFIC!

Cumulative Miles	(km)	Description
12.0	(19.3)	Stop 1-3.

Stop 1-3. Sand Hollow

The rock slope exposes the pillow-palagonite facies of the Ginkgo flow of the Frenchman Springs Basalt. In this part of the Columbia Basin, the Ginkgo is the basal flow of the Wanapum Basalt. It rests directly on the Vantage interbed and flowed into the lake in which the Vantage was deposited. It is locally invasive into the interbed and remnants of the interbed may be seen within the pillow-palagonite tuff-breccia.

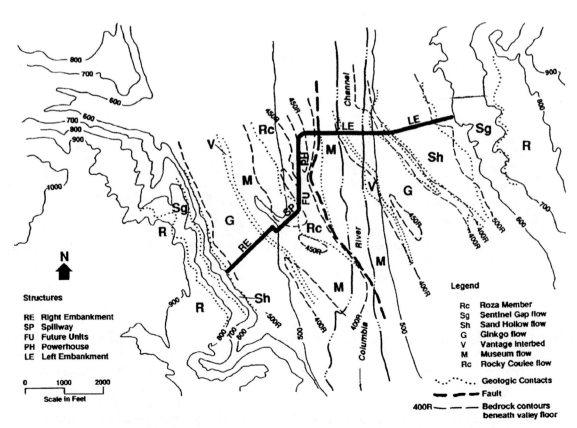

Figure 3. Generalized geologic plan at Wanapum Dam (from Galster, 1989a).

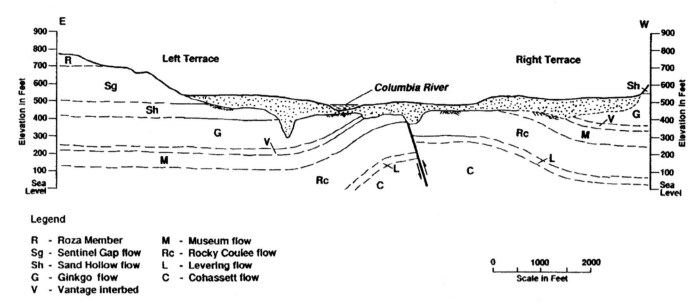

Figure 4. Generalized geologic section across Columbia River valley at Wanapum Dam (from Galster, 1989a).

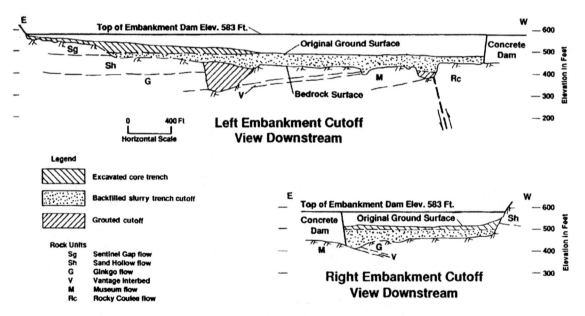

Figure 5. Wanapum Dam: Geologic sections along slurry cutoff trenches beneath left and right embankment sections (from Galster, 1989a).

We will turn our vehicles around here, and proceed west and north on SR 26.

Cumulative Miles	(km)	Description
13.0	(20.9)	Rejoin Interstate 90 and proceed north(east). Over the next several miles, the interstate

will follow the stripped structural surface of the Museum Platform, climbing up the south limb of the Frenchman Hills anticline, then dropping off the steeper, faulted north limb to the surface of the Quincy Basin.

19.0	(30.5)	Turn off on exit 143 and turn left (north) onto Silica Road; pass under Interstate 90.

19.7 (31.7)   Turn left (west) on Old Vantage Road, across from North Frontage Road. The Old Vantage Road is a section of former U.S. Highway 10. The light colored excavations to the north (right) are remnants of exploration and mining of the Quincy Diatomite, which stratigraphically lies between the base of the Priest Rapids Member and the underlying Roza Member. This segment of the old highway follows the south side of Frenchman Springs Coulee, one of the classic headward-eroding "dry falls" coulees that characterize the western edge of the Quincy Basin. These coulees were formed by one or more of the great outburst floods (jokulhlaups) from Glacial Lake Missoula. The road cuts expose the section through the Priest Rapids, Roza, and Frenchman Springs Members of the Wanapum Basalt. This is the type area of the Frenchman Springs Member. The floor of the coulee is essentially the eroded top of the Museum flow of the Grande Ronde Basalt overlain by flood gravels.

21.0 (33.7)   Stop 1-4.

Stop 1-4. Roza Columns

These columns are remnants of the basal colonnade of the Roza Basalt. The Roza is one of the most easily recognized flows of the Columbia River Basalt Group, owing to its porphyritic character and well-developed colonnade. At this location, the columns are a favorite practice area for rock climbers.

Cumulative Miles	(km)	Description
23.0	(37.0)	Return to Interstate 90 and head east.
28.6	(46.0)	Take exit 149 at George and turn left (north) onto SR 281 toward Quincy and Wenatchee. SR 281 crosses the West Canal of the Columbia Basin (irrigation) Project is located shortly after the turnoff.
39.5	(63.5)	At the center of Quincy, turn left (west) onto SR 28. We travel west from Quincy and drop off the plateau into the Columbia River valley.
48.1	(77.4)	Left of Crescent Bar, view one of the glacial outburst flood bars in the Columbia Valley. Mega ripples are well preserved on its surface.
50.6	(81.4)	Exposures of foreset bedding in palagonite facies may be seen to the right.
55.0	(88.5)	SR 28 crosses the mouth of Moses Coulee, a major outwash flood channel from the Okanogan lobe on the Waterville Plateau. Outwash was deposited in the valley both upstream and downstream of the coulee mouth, which periodically dammed the river.
57.9	(93.1)	Stop 1-5.

Stop 1-5. Rock Island Dam and State Route 28 Widening

Rock Island Dam was originally constructed between 1930 and 1933 near the downstream end of Rock Island Rapids. It was the first dam constructed on the main stem of the Columbia River. Here, the post-basalt Columbia River departed from its lava-marginal position and was diverted onto the lava plateau, probably at the beginning of the Wenatchee Mountain uplift. The river is against the east (left) wall of the valley, and a large, glacial outwash terrace occupies much of the valley bottom (Fig. 6). The terrace, rising 200 ft (60 m) above the right bank, is unique, its most recent Pleistocene outwash having been deposited in an up-valley direction. This outwash likely originated from Moses Coulee, which enters the Columbia Valley from the left bank a short distance downstream. The surface and riverward slopes of the terrace are littered with ice-rafted basalt blocks up to 20 ft (6 m) in diameter. Internally, the terrace is composed of sand and gravel, although lacustrine deposits are also present. The bedrock surface beneath the terrace is not defined. A highly irregular bedrock surface is exposed in the channel, developed on nearly flat-lying, unnamed flows within the R-2 section of the Grande Ronde Basalt. Because of the proximity of the site to the margin of the lava field, the basalt flows here exhibit invasive relationships with sedimentary units that were being concurrently

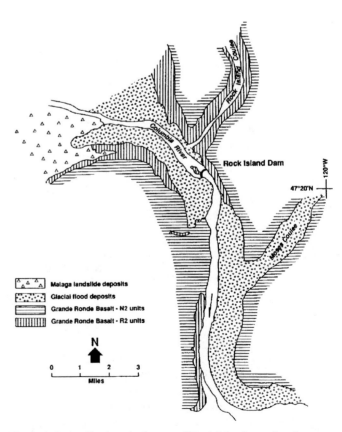

Figure 6. Generalized geologic map of Rock Island area showing configuration of glacial flood terrace deposits and bedrock canyon walls (from Galster, 1989b).

deposited along the lava margin. Thus, the basalts at this site are commonly found mixed with tuffaceous sandstone and siltstone as well as sediment-rich lava breccias and highly vesicular or scoriaceous zones common to flow contacts elsewhere.

As originally constructed, the dam consisted of a 37-bay spillway across the southern (right) part of the channel and a non-overflow intake dam and powerhouse in the north (left) bank channel. The powerhouse was founded on sound basalt, but the spillway section was founded on a highly complex and irregular section of sediments, breccias, and basalt (Fig. 7). A grout curtain was installed from the exterior heel of the structures, but no drainage system was installed below the foundation contact. The south (right) bank powerhouse was constructed during the late 1970s, and the original concrete structures were reinforced with 163 cable tendons anchored at various depths into bedrock. The deep powerhouse excavation exposed intercalated basalt and sediments. Although the powerhouse is founded mostly on dense or vesicular basalt, substantial portions are founded on siltstone and flow breccia. A strong artesian circuit was discovered beneath the powerhouse. This required installation of an extensive drainage system consisting of 48 shallow drains flowing through a

collector system to a dewatering sump, from whence drainage is pumped to tailwater (Galster and Imrie, 1989; Galster, 1989b).

Downstream of Rock Island Dam (elev. 600 ft), State Route 28 skirts the base of a large talus slope and a midslope band of basalt cliffs (Fig. 8) between mileposts 11 and 12. This intermediate cliff band forms a midslope bench around elevation 1000 ft. Another much larger talus slope laps onto this bench. The Waterville Plateau (elev. ~2500 ft) lies above the uppermost section of basalts. Near milepost 12, the lower talus slope ends, and the midslope basalt cliffs jut toward the river. For nearly 1000 ft (300 m), the highway is sited below a 300-ft (100-m)-high rock slope. Frequent rockfall below the talus slopes to the north and the large rock slope in the south pose a significant hazard to both the highway and the adjacent railroad.

Projected economic growth and increased commerce traffic in the region have prompted a multistaged widening project of the two-lane highway to four lanes, concurrently mitigating for the existing rockfall hazard. The adjacent railroad links Seattle and Spokane and currently handles ~24 trains per day, including a daily Amtrak passenger train. The proximity of this high-volume rail line and the river to the highway has restricted widening

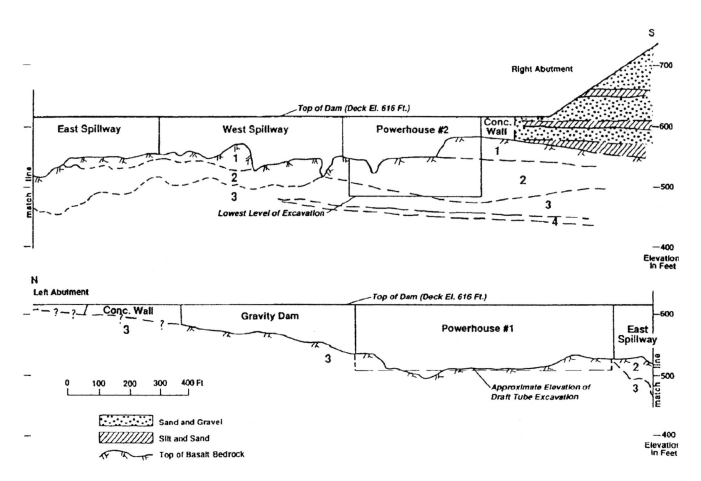

Figure 7. Geologic section at Rock Island Dam, view downstream. Numbered bedrock units are as follows: (1) columnar basalt; (2) invasive vesicular to highly scoriaceous basalt, tuffs, and breccia with highly irregular, invasive characteristics; (3) generally columnar to "massive" basalt; (4) tuff (from Galster, 1989b).

Figure 8. Looking north, the photograph shows the left bank just downstream of Rock Island Dam (left edge of photo). The highway and railroad are located at the base of the slope. Stages 1 and 2 of the widening project are located in the lower right of the photo near MP 12.

alternatives to the uphill side. The widening project as it is currently defined entails three stage: Stage 1, which is near completion, involves excavating the southern half of the rock slope; Stage 2, scheduled for 2005 construction, will excavate the northern half of the rock slope where cut heights exceed 300 ft; and Stage 3, currently unscheduled, will involve construction of an elevated two-lane roadway section in the adjacent talus slopes.

The basalt cliffs that form the midslope bench comprise the invasive flow of Hammond and belong to the reverse polarity (r2) flows of the Miocene-age Grande Ronde Basalt (Tabor et al., 1983). The southern portion of the outcrop near milepost 12 is cut near its base by a fault, juxtaposing highly fractured tuffaceous/sedimentary rocks and basalt. The project area lies within the northern portion of the Yakima fold belt (Reidel and Campbell, 1989); consequently, bedrock has been uplifted, gently folded and faulted. This deformation produced a laterally continuous fracture set that dips adversely out of the rock slope 30°–50°. High shear strengths on these fractures, however, have permitted the construction of near-vertical cut slopes for the Stage 1 project in excess of 150 ft high.

Cumulative Miles	(km)	Description
59.8	(96.2)	Left abutment of Rock Island Dam is on the left.
60.3	(97.0)	Rock Island Coulee, a lesser outwash channel from the Waterville Plateau, is to the right.
62.7	(100.8)	To the left is the Columbia River crossing of the Burlington Northern–Sante Fe Railroad

main transcontinental line. Both abutments of the bridge are founded in large basalt blocks of the Malaga landslide. The Malaga slide is probably early post-glacial, one of the several major slides that failed as the swollen, post-glacial Columbia River was reaching its current grade. The slide has impressive dimensions: it has a toe length of 7.5 mi along the river, a nominal width of 5.5 mi, and extends ~10 mi to a headscarp on Mission Ridge. The slide dominates the landscape on the west side of the river.

Day 2

Cumulative Miles	(km)	Description
70.0	(112.6)	Turn left and cross the Columbia River into Wenatchee.

East Wenatchee

The second day of the field trip will start with travel northward up the west bank of the Columbia River valley, with stops to discuss landslides and earthquake hazards. We will pass Rocky Reach Dam, Lake Chelan, and Wells Dam and stop at Chief Joseph Dam for a discussion of dam construction and the Bridgeport slide. From here, we will climb the northern side of the Waterville Plateau section of the Columbia Plateau, traverse the central and western parts of the plateau, and stop to discuss landslide and rockfall problems along U.S. Highway 2 as we drop down through Corbaley and Pine Canyons back into the Columbia Valley. We will return via Wenatchee, then head west on U.S. Highway 2 with a stop in Tumwater Canyon to discuss fire-related slope hazards. We will return to Seattle via Stevens Pass.

The day begins on the northern outskirts of Wenatchee. On Mission Street, cross the Wenatchee River and follow signs to Okanogan and Spokane. Turn right on U.S. Highway 2/97/28 and reset odometer.

Cumulative Miles	(km)	Description
0.8	(1.3)	Turn right on the off-ramp from U.S. 2/97/28 to U.S. 97A, then turn left (north) at the bottom of the off-ramp onto U.S. 97A toward Entiat. The highway crosses the Entiat fault in this location. The fault forms the northern

boundary of the Chiwaukum Graben, separating the Eocene Chumstick Formation and the Precambrian Swakane Biotite Gneiss.

3.5	(5.6)	Stop 2-1.

Stop 2-1. U.S. Highway 97A, Morrill Gravel Pit Slope Failure

On May 19, 1995, a catastrophic slope failure occurred in a gravel pit located ~2.5 mi north of Wenatchee on State Route 97A (Fig. 9). The pit was being mined at the time of the failure, which included two crusher/screening plants in the pit floor. The 1100-ft (350-m)-wide slope failure progressed very rapidly from the deepest portion of the excavation in the south to the north. While some workers were able to "ride out" or outrun the slide, two fatalities occurred. A forensic investigation was performed by the Mine Safety and Health Administration (MSHA) shortly after the accident (Ferriter and Ropchan, 1995).

The mine was exploiting alluvial and glacial outwash gravels. Deformation entailed ~200 ft (60 m) of vertical displacement

Two lost in rock slide

■ **Road buried:** It's unlikely a 5-year-old boy and a quarry worker caught in the slide could have survived, rescuers say.

By Aviva L. Brandt
The Associated Press

WENATCHEE — Rescue workers and bulldozers labored Friday to find two people — a 5-year-old boy and a quarry worker — buried in a rock slide that dumped tons of debris on a gravel pit and a nearby highway.

"Everyone else got out," said Chelan County Deputy Sheriff John Sanborn.

There was no warning, said 23-year-old worker Matt Bakken of Spokane: "I didn't hear anything. The crusher was running and we all had earplugs in."

The mid-morning slide buried 200 feet of the southbound lanes of Highway 97A 30 feet deep where it runs through hills along the Columbia River north of this Chelan County community.

No passing vehicles were believed caught in the slide. The focus was on the gravel pit, owned by Morrill Asphalt of Wenatchee, where subcontractor Lloyd Logging of Twisp had been crushing rock.

The 5-year-old, son of a woman employed at the site, was last seen in a modified bus being used as a lab and office, said Deputy Sheriff Mike Harum.

The boy's mother "was standing out by the road and when it started to slide she turned around and saw the child standing in the doorway of the bus," he said.

"She just couldn't get to him," Harum said, adding that the mother had been treated for shock.

Also missing was the quarry manager, last seen in the control room of a rock crusher at the opposite end of the work site, Harum said. The Wenatchee World identified the man as Tim Grace of Spokane.

Asked whether they could have survived, Harum said: "At this point, it's a very slight possibility."

The highway was closed from a point just north of Wenatchee to a point near Chelan, about 30 miles north.

AERIAL VIEW: Rescue workers gather at the site of a rock slide that buried a quarry worker and a 5-year-old boy under tons of debris. The slide closed Highway 97A near Wenatchee for about 30 miles.

Figure 9. From the front page of the May 20, 1995 edition of *The Olympian* newspaper (Olympia, Washington).

on the headscarp, 40 ft (10 m) of uplift, and 150 ft (45 m) of lateral movement in the terminus, with debris covering a portion of the highway. The headscarp formed on a steep bedrock (Swakane Gneiss) slope that had been covered in talus. The slide debris contained large blocks of rhythmically bedded silts and clays that were subjacent to the gravel deposits and previously unexposed in the pit. Decades earlier, investigations for the Rocky Reach Dam (located ~1.5 mi upstream) encountered silt/clay beds underlying 100 ft of gravel deposits (Stone and Webster Engineering, 1956), and similar deposits have been found downstream as much as 300 ft above current river level. Index tests by MSHA revealed these materials to be low plasticity silts and clays (ML and CL) with moisture contents in the clays near their liquid limits.

Based upon their geologic investigation and back-analyses of the slope failure, MSHA concluded the following:

1. Several hundred thousand cubic yards of material had been removed from the pit over a period of many years, and possibly as much as 160,000 cubic yards since 1992.

2. Mining at the base of the slope coupled with the driving stress of the talus "created the conditions for the slope failure."

3. The failure mechanism was a deep, rotational type failure within the weak, saturated silt/clay deposits, evidenced by an uplifted toe area and blocks of disturbed silt/clay within the slide debris. The upper, thin mantle of soil and vegetation from the talus slope "slid almost intact" in a "carpet-like movement" over the top of the displaced alluvium and talus.

4. There is no reference in the report to whether the crusher/screening plants were running at the time of the failure, but MSHA concluded that "man made vibration sources could not have had a significant effect on stability."

| Cumulative | | |
Miles	(km)	Description
5.4	(8.7)	Rocky Reach Dam.

Rocky Reach Dam

The Columbia River valley at Rocky Reach Dam is characterized by steep rock walls rising some 2000 ft (610 m) above the 4000-ft (1200-m)-wide valley floor. The river channel hugs the west side of the valley, and a composite glacial outwash terrace, rising 140 ft (43 m) above the river, dominates the eastern side of the valley bottom. The rock walls and the bedrock beneath the valley floor are Precambrian (?) Swakane Biotite Gneiss. The highest bedrock elevations are in and west of the river

channel, dictating the Z-shaped configuration of the concrete structures, all founded on the gneiss bedrock. During foundation preparation, numerous gently dipping foliation planes and faults required over-excavation and dental concrete.

Beneath the left bank terrace, the bedrock surface drops to elevations as much as 130 ft (40 m) lower than the rock surface in the river channel. Three geologic units were identified in the terrace (Fig. 10): a lower, highly pervious gravel, overlain by a thick varved clay with thin sand beds, and an upper sand and gravel unit. Leakage control was established by installing a random fill cutoff trench with a filtered impervious blanket on the upstream side up to the top of the clay. The lower gravel was grouted, and the length of the cutoff is 1950 ft (594 m). A collector drain in the lower gravels was installed downstream from the left abutment (Galster and Imrie, 1989; Coombs, 1989).

| Cumulative | | |
Miles	(km)	Description
19.5	(31.4)	Stop 2-2.

Stop 2-2. Ribbon Cliffs Landslide and Earthquake Point

On the night of December 14, 1872, a large earthquake (M ~6.5–7.0) occurred in the Pacific Northwest with an epicenter near the south end of Lake Chelan (Bakun et al., 2002). Sparse population and limited reliable eyewitness accounts have prevented precise location of the epicenter or determination of the intensity of shaking. Historical accounts reported damming of the Columbia River during the night of December 14, 1872, and that downstream near Wenatchee, the river was virtually dry on the following morning. It was around this time that Ribbon Cliff Landslide was believed to have occurred (Madole et al., 1995).

The Ribbon Cliff Landslide is situated on the west side of the Columbia River, ~3 mi (5 km) north of Entiat, Washington. The landslide occurred within the talus and colluvial apron formed during the middle and late Holocene (Madole et al., 1995). The landslide debris extends along the base of the slope for a distance of more than 3000 ft (900 m) and forms two small islands along the west bank of the river. Kienle et al. (1978) estimated the volume of the slide at 13 million cubic yards (10,000,000 m³).

Early studies based on eyewitness accounts concluded that the earthquake triggered the landslide, which in turn dammed the river. Kienle et al. (1978) looked at the rings of trees they believed were growing on the landslide deposits more than a century prior to the 1872 earthquake. Because tree rings did not show signs of

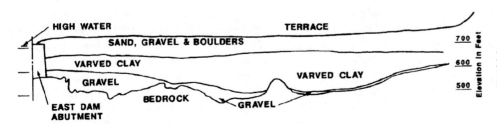

Figure 10. Geologic section along left bank cutoff of Rocky Reach Dam (upstream view) (from Coombs, 1989).

distress or disturbance to the trees, they felt the landslide could not have translocated the trees. Madole et al. (1995) employed tephrochronology, dendrochronology, radiocarbon dating, lichenometry, as well as studied the age and origin of surface deposits, to constrain the age of landsliding. They concluded that the landslide occurred within a 14 year period around the 1872 earthquake, and thus, the earthquake remains a potential trigger.

Recent work by Badger (2003) identified a mile-long lineament, oriented N75°E and located ~6 mi (10 km) upriver on the opposite bank, that is in direct alignment with the Ribbon Cliffs. On the same trend to the ENE, this lineament aligns with an ~500-ft (150-m)-wide zone of comminuted bedrock exposed along U.S. Highway 97 at milepost 227.1. A repeating set of very persistent, near-vertical fractures with similar strike is present in the highway cut slope to the west, immediately adjacent to this comminuted zone. Also notable is that the river is confined to a straight, 4-mi-long reach that is mostly bedrock and coincident with this lineament. Collectively, this lineament is more than 6 mi (10 km) in length. Bakun et al. (2002) summarized earthquake records between 1976 and 2001, and identified a zone of active seismicity in the immediate area. This lineament appears to be of tectonic origin and may be a source of this recent seismicity and possibly the 1872 earthquake as well.

Cumulative Miles	(km)	Description
25.8	(41.5)	Knapps Hill Tunnel with views of the Great (fill) Terrace (Russell, 1893) across the river on left bank. North of the tunnel, the highway follows Knapps Coulee, a late Pleistocene channel eroded by glacial meltwater from the Lake Chelan valley.
34.7	(55.8)	Lake Chelan: turn left to Chelan City Center (Wooden Ave.); follow signs to U.S. 97A and Okanogan.
44.8	(72.0)	At the intersection with SR 97, turn left (north) on SR 97 toward Pateros and Okanogan.
51.3	(82.5)	Stop 2-3.

Stop 2-3. Wells Dam Overlook

The Columbia River valley at Wells Dam is ~4000 ft (1220 m) wide, and the river channel is against the east valley wall (Fig. 11). The east valley wall rises as a series of narrow terraces backed by rocks of the Chelan–Colville granitic complex. More extensive glacial terraces are exposed on the right (west) side of the valley, partly burying bedrock knobs. The valley floor is underlain by an unevenly stratified sequence of glacial outwash and fluvial sediments consisting of gravel and sand with local cobble and boulder units and lenticular units of fine sand and silt representing glacial lake deposits. The site is close to the distal margin of the last advance of the Pleistocene Okanogan ice lobe. Till-like units pervasively overlie bedrock beneath the eastern part of the valley floor, and glacial lake deposits underlie the

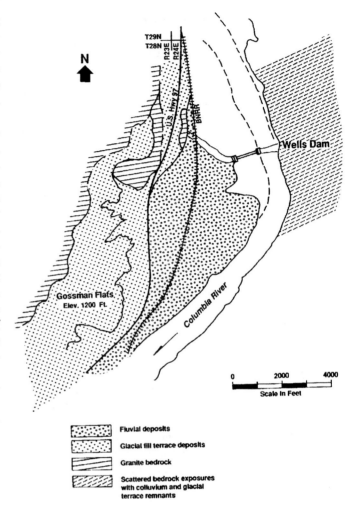

Figure 11. General geologic map of Wells Dam area (from Galster, 1989c).

Columbia River gravels. Total overburden thickness ranges from 50–100 ft (15–30 m) (Fig. 12). The outwash and fluvial units are highly pervious. West of the right abutment, a deep paleochannel in the bedrock surface extends to below elevation 354 ft (108 m), 200 ft (60 m) lower than the present channel of the river. This paleochannel is largely filled with glacial lake deposits.

The dam was designed as a hydrocombine with the spillway over the powerhouse in order to take advantage of shallow bedrock adjacent to the river channel. The foundation was sculpted by controlled blasting, and only two narrow zones required additional excavation and concrete backfill. A grout curtain was installed to a depth of 40 ft (12 m) below the foundation. The west embankment dam was constructed in the dry with an impervious core trench extending down to between elevations 720 and 740 ft (219–225.5 m). Most of the east embankment was constructed without dewatering, using underwater placing methods for the lower part of the embankment. As no compaction was thus possible, significant settlement of the overlying subaerial embankment

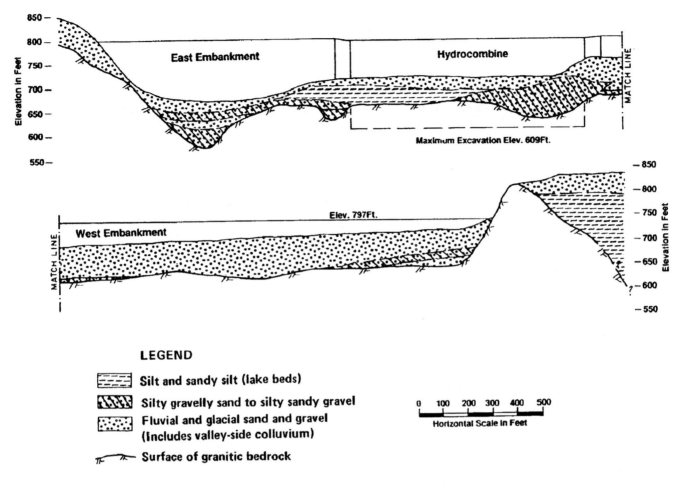

LEGEND

☰ Silt and sandy silt (lake beds)

▨ Silty gravelly sand to silty sandy gravel

⦂⦂ Fluvial and glacial sand and gravel
  (includes valley-side colluvium)

⌁ Surface of granitic bedrock

Figure 12. Generalized geologic section at Wells Dam (from Galster, 1989c).

was both predicted and experienced. The bedrock in both abutments was grouted (Galster and Imrie 1989; Galster 1989c).

Cumulative		
Miles	(km)	Description
58.0	(93.3)	Cross Methow River into Pateros.
65.0	(105.0)	Pass through Brewster.
69.0	(111.0)	Cross the Okanogan River.
70.0	(112.6)	Turn right (east) onto SR 17 toward Bridgeport.
77.0	(124.0)	Turn left (east) on the right bank road toward Chief Joseph Dam.
78.0	(125.5)	Turn left to North Viewpoint for Chief Joseph Dam.

Stop 2-4. North Viewpoint for Chief Joseph Dam

Chief Joseph Dam lies in a section of the Columbia River valley that is largely marginal to the north edge of the lava plateau (Fig. 13). It is here that the valley enters the southeast corner of the Okanogan Trench. Bedrock in the valley is granitic and

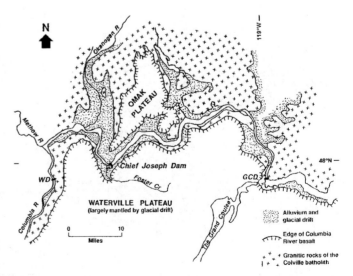

Figure 13. Chief Joseph Dam regional geologic setting (from Eckerlin and Galster, 1989).

locally gneissic rocks that characterize the Colville batholith and the Okanogan gneiss dome. Basalt of the Columbia River Basalt Group overlies the granitic complex higher on the slope and is locally interbedded with sediments of the Latah Formation. Because this east-west–trending segment of the valley is essentially transverse to the direction of flow of the Okanogan lobe of the Pleistocene continental ice sheet, a somewhat jumbled sequence of glacial deposits has been deposited at lower elevations in the valley. The complexities of interglacial erosion and post-glacial downcutting have left numerous fill and cut terraces at various elevations. Fields of basalt "lag" blocks plucked by the ice from the Omak Plateau to the north litter extensive areas of the valley at all elevations.

At the dam site, the post-glacial "inner" valley of the Columbia River is against the south side of the 12,000-ft (3650-m)-wide, 1000-ft (305-m)-deep, older, pre-glacial valley. A broad glacial terrace ranging in elevation from 1000 to 1200 ft (335–366 m) occupies the northern half of the pre-glacial valley

(Fig. 14). The terrace is composed, in descending order, of till, "dump moraine," and proglacial gravel overlying the bedrock surface, which rises gently northward. The right abutment of the dam is founded in this terrace (Fig. 15). The concrete structures of the dam are all founded on sound granitic rock characterized by varied joint patterns and minor faults. The most common joint set strikes N60°W and dips NE at a high angle. Joints are usually slickensided although healed with chlorite, and faults usually exhibit only a few inches (cm) of gouge. The rock surface at the left abutment of the dam rises to elevations above the reservoir. Just upstream, post-glacial downcutting against the south valley wall resulted in a large landslide involving glacial deposits, basalt, and associated sediments. This is the Bridgeport slide, covering an area of 1.25 mi^2 (3.2 km^2). The toe of the slide is in the reservoir, and the head of the slide is as much as 500 ft (150 m) above the reservoir. Parts of the slide are periodically active. The main failure zone is in weak, saturated sediments overlying the bedrock surface.

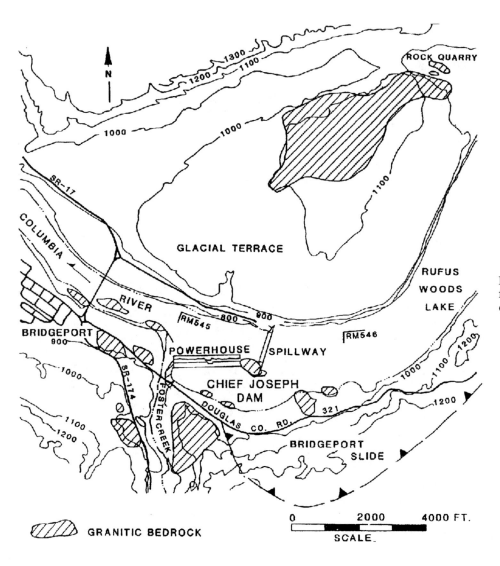

Figure 14. Generalized geologic map for Chief Joseph Dam area (from Eckerlin and Galster, 1989).

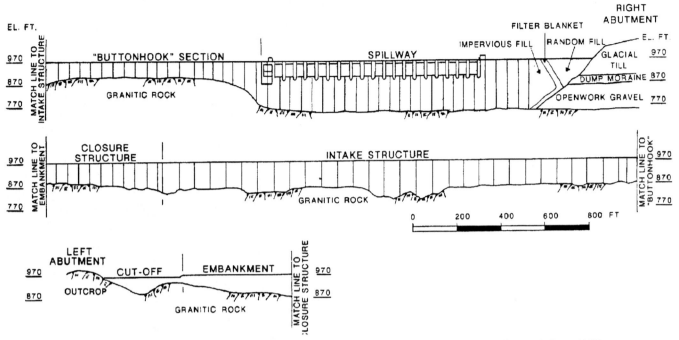

Figure 15. Geologic profile along axis of Chief Joseph Dam, view downstream (from Eckerlin and Galster, 1989).

The dam design for the right abutment entailed wrapping an embankment section around three concrete monoliths. The embankment is tied to an impervious blanket that extends 2000 ft (610 m) upstream. A 1000-ft (305-m)-long drainage relief adit and well system was driven into the highly pervious proglacial gravel, with its portal immediately downstream of the dam. At initial filling of the reservoir in 1955 to elev. 946 ft (288.3 m), the adit discharged 93 cfs (2.6 m³/s). Over the years, this has reduced to ~25 cfs (0.7 m³/s).

Original construction also included a grout curtain extending 80 ft (24 m) below the spillway foundation, 50–75 ft (15–23 m) below the intake dam foundation and 55 ft (17 m) below the foundations of the buttonhook and closure dams. A drain curtain extends 40 ft (12 m) below the foundation of the spillway dam and 28 ft (8.5 m) below the buttonhook and closure dams. Fan drains within the penstock openings and exterior drains from the downstream rock face provide drainage from beneath the intake dam.

During the late 1970s, the dam and reservoir were raised 10 ft (3 m), and the powerhouse was fully completed to its designed 27 units. Raising of the spillway dam required removal of large parts of the spillway gate piers and rebuilding of the spillway piers to higher elevations. This work was accomplished by controlled blasting behind floating cofferdams without any lowering of the reservoir (Eckerlin and Galster, 1989; Eckerlin, 1989).

Cumulative		
Miles	(km)	Description
79.0	(127.0)	Turn left (south) on SR 17 and cross the Columbia River.
81.6	(131.3)	At the intersection of SR 17 and Bridgeport Hill Road, turn right (south) onto Bridgeport Hill Road. Climb the north side of the Waterville Plateau along the West Fork Foster Creek.
92.9	(149.5)	At the T-intersection with SR 172, turn right (west) on SR 172. This road travels along the glacial drift plain of the Okanogan Lobe. Note the lag blocks littering the surface of the ground moraine. The road swings south in ~5 mi (8 km).
99.3	(159.8)	SR 172 drops down off the terminal moraine of the Okanogan Lobe to a glacial outwash plain blanketed with loess.
108.1	(173.9)	At Farmer (twin elevators), SR 172 intersects U.S. Highway 2. Turn right (west) toward Waterville.
125.0	(201.1)	Center of Waterville, county seat of Douglas County.
130.2	(209.5)	Turn left on the paved road and former alignment of U.S. 2; cross a creek and proceed ~0.4 mi (0.6 km) to a pullout/overlook on the right.
130.6	(210.1)	Stop 2-5.

Stop 2-5. Corbaley and Pine Canyons

U.S. Highway 2 drops steeply into Pine Canyon through loess-capped, Grand Ronde (N2) basalt underlain by weak sandstones of the Ellensburg Formation and the invasive flow of Hammond (Grand Ronde [R2]) basalt (Tabor et al., 1987). This sequence unconformably overlies pre-Tertiary marble, schist, and plutonic rocks of the Mad River terrane and Entiat Pluton. Near the base of Corbaley Canyon, Swakane Biotite Gneiss outcrops. Regional extension during the Eocene resulted in the emplacement of a roughly east-west–trending, near-vertically oriented, rhyolitic dike swarm in these pre-Tertiary rocks.

Rockfall is a significant highway hazard in Corbaley and the lower portion of Pine Canyons. In the last two decades, several rockslides have occurred that exceeded 1000 cubic yards in volume, causing multiday highway closures. The pre-Tertiary rocks commonly display large blocky fracture, and stability of cut slopes is dominantly controlled by adversely oriented, persistent fractures resulting in planar, wedge, and toppling failures. The rhyolitic dikes are often highly fractured and generate minor, raveling-type slope failures. Talus and colluvium are also a common source of raveling-type rockfall. Within much of this section, substandard ditch width and irregular slopes result in poor rockfall catchment, and limited sight distances exacerbate the hazard. Since 1997, numerous rockfall mitigation projects have been completed within this section of the highway. The stabilization work has included scaling loose debris, placement of wire mesh and/or net fabrics on highly fractured bedrock and across chutes and narrow ravines, and the installation of rock anchors and shotcrete to reinforce larger masses.

Landslides are pervasive on both sides of upper Pine Canyon (Fig. 16); most appear dormant and prehistoric. Slope failure initiates within the weak sedimentary rocks of the Ellensburg Formation, driven by loading from overlying basalt, creek incision, and possibly adverse groundwater conditions. The failure mode of the larger, now mostly dormant landslides appears to be translational. Landsliding has locally reactivated in lower portions of the slopes due to creek erosion and highway activities. The failure mechanism of these lower failures consists of both translational and rotational failure modes.

Figure 16. Vertical air photo of Pine Canyon and U.S. Highway 2. Hachured symbol denotes landslide limits based on air photo interpretation and limited field verification.

| Cumulative | | |
Miles	(km)	Description
131.0	(210.7)	Rejoin U.S. Highway 2; turn left down Pine Canyon and then Corbaley Canyon along an intermittent but high gradient drainage. Note the caliber and quantity of detritus exposed.
135.2	(217.5)	At the Y-intersection of U.S. 2 with U.S. 97, turn left (south), following the river (Lake Entiat reservoir) downstream back to Wenatchee.
142.8	(229.8)	Rocky Reach Dam, left bank.
147.5	(237.3)	Turn right (west) on U.S. 2/97 and cross the Columbia River; continue west toward Leavenworth and Stevens Pass. The highway crosses the Entiat fault into the Chiwaukum Graben through folded sandstone and shales of the Eocene Chumstick Formation. Note the terraces cut by the Wenatchee River during the gradual Late Tertiary uplift of the Cascade Range.
168.5	(271.1)	Center of Leavenworth: a Bavarian motif village.
169.3	(272.4)	Cross Leavenworth fault and enter Tumwater Canyon. U.S. 2 follows the original route of the Great Northern Railroad through the canyon.
173.6	(279.3)	Stop 2-6.

Stop 2-6. Tumwater Canyon and Hatchery Creek Fire

More than 40,000 acres of forest land around Leavenworth burned in the late summer of 1994. In Tumwater Canyon, the fires were mostly low intensity, and much of the forest did not burn. However, several hundred acres situated in the upper watersheds of five drainages on Tumwater Mountain experienced high-intensity burns (Fig. 17). A joint geotechnical evaluation of post-fire slope hazards was carried out by the U.S. Forest Service and the Washington State Department of Transportation (Lowell, 1994). Of principal interest, in order of decreasing priority, were debris flows, snow avalanches, and rockfall and their potential impacts to U.S. Highway 2 and a small community (The Alps), both located in the base of the canyon.

Four of the five watersheds that experienced high-intensity burns were judged to pose the greatest debris-flow hazard, based on activity and recentness of debris-flow deposition as well as

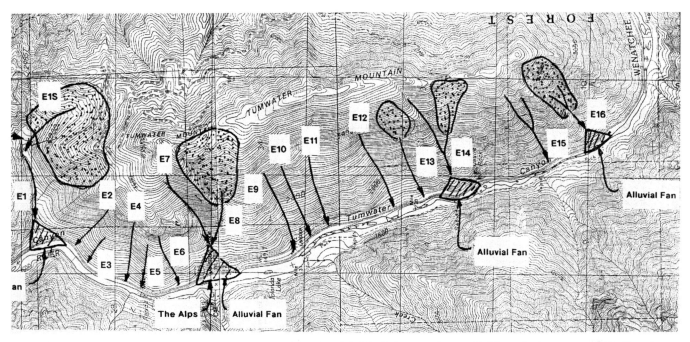

Figure 17. Map of Tumwater Mountain, showing drainages labeled as E#. Stippled pattern denotes high intensity burn areas, and hachured areas are alluvial fans. The Alps is a small community located along the highway in the base of the canyon.

the extent of surficial deposits within channels and upper basins. Remotely monitored piezometers installed within several slump features in drainage E-8 (Fig. 17) and a weather station on Tumwater Mountain have been used as predictive tools for increased debris-flow hazard.

Initiation zones for snow avalanches high on Tumwater Mountain are rare, small in size, and confined to narrow, steep drainages; much of this upper area was densely forested prior to the fires. Increased snow avalanche hazard is now expected to develop in drainages E-8, E-14, and E-16, due to the complete tree mortality and steep slopes in these high-intensity burn areas.

Rockfall within the canyon has been a historic problem. Rockfall frequency increased both during and shortly after the fires but dropped to background levels within several weeks following burning.

Cumulative Miles	(km)	Description
177.8	(286.0)	Cross the Wenatchee River. The highway follows the deep, glaciated valley of Nason Creek toward Stevens Pass.
197.4	(317.7)	Great Northern Railroad and Cascade Tunnel.
205.0	(330.0)	Stevens Pass, elevation 4061 ft (1238 m), 36 mi (58 km) west of Leavenworth. The route drops from Stevens Pass to the Tye and Skykomish River valleys. U.S. 2 passes through the communities of Skykomish, Index, Gold Bar, and Sultan, turns southwest on to SR 522 at Monroe, crosses the Snohom-

ish River, and returns to Seattle by way of Bothell and Interstate 5. From U.S. 2, turn on to SR 522 and follow to Interstate 5.

REFERENCES CITED

Badger, T.C., 2003, A topographic and structural lineament near Chelan, Washington: Geological Society of America Abstracts and Program, v. 35 (in press).

Bakun, W.H., Haugerud, R.A., Hopper, M.G., and Ludwin, R.S., 2002, The December 1987 Washington State Earthquake: Bulletin of the Seismological Society of America, v. 92, no. 8, p. 3239–3258.

Cheney, E.S., 2000, Tertiary geology of the eastern flank of the Central Cascade Range, Washington, *in* Woodsworth, G.L., Jackson, L.E., Jr., Nelson, J.L., and Ward, B., eds, Guidebook for Geological Field Trips in Southwestern British Columbia and Northern Washington, Annual Meeting of the Cordilleran Section of the Geological Society of America in Vancouver, April 2000: Geological Association of Canada, Cordilleran Section, Vancouver B.C., p. 205-227.

Coombs, H.A., 1989, Rocky Reach Dam, *in* Galster, R.W., Chairman, Engineering Geology in Washington: Washington Division of Geology and Earth Resources Bulletin 78, v. I, p. 391–396.

Eckerlin, R.D., 1989, The Bridgeport slide, *in* Galster, R.W., Chairman. Engineering Geology in Washington: Washington Division of Geology and Earth Resources Bulletin 78, v. II, p. 921–926.

Eckerlin, R.D., and Galster, R.W., 1989, Chief Joseph Dam, *in* Galster, R.W., Chairman, Engineering Geology in Washington: Washington Division of Geology and Earth Resources Bulletin 78, v. I, p. 405–416.

Ferriter, R.L., and Ropchan, D.M., 1995, Investigation of massive slope failure portable crusher no. 2 (ID no. 45-0326): Denver, Colorado, U.S. Department of Labor Mine Safety and Health Administration, 48 p.

Frecht, K.R., Reidel, S.P., and Tallman, A.M., 1987, Paleodrainage of the Columbia River system on the Columbia Plateau of Washington State—A summary, *in* Schuster, J. E., ed., Selected papers on the geology of Washington: Washington Division of Geology and Earth Resources Bulletin 77, p. 219–248.

Galster, R.W., 1989a, Wanapum Dam, *in Galster*, R.W., Chairman, Engineering Geology in Washington: Washington Department of Geology and Earth Resources Bulletin 78, v. I, p. 377–382.

Galster, R.W., 1989b, Rock Island Dam, *in* Galster, R.W., Chairman, Engineering Geology in Washington: Washington Division of Geology and Earth Resources Bulletin 78, v. I, p. 383–390.

Galster, R.W., 1989c, Wells Dam, *in* Galster, R.W., Chairman, Engineering Geology in Washington: Washington Division of Geology and Earth Resources Bulletin 78, v. I, p. 397–402.

Galster, R.W., and Coombs, H.A., 1989, Dams of the middle Columbia River: Introduction and geologic setting, *in* Galster, R.W., Chairman, Engineering Geology in Washington: Washington Division of Geology and Earth Resources Bulletin 78, v. I, p. 367–370.

Galster, R.W., and Imrie, A.S., 1989, Engineering geology of major dams on the Columbia River, International Geological Congress Field Trip Guidebook T382: Washington, D.C., American Geophysical Union, 69 p.

Gresens, R.L., 1987, Early Cenozoic geology of central Washington State: I. Summary of sedimentary, igneous and tectonic events, *in* Schuster, J. E., ed., Selected papers on the geology of Washington: Washington Division of Geology and Earth Resources Bulletin 77, p. 169–177.

Huntting, M.T., Bennett, W.A.G., Livingston, V.E., Jr., and Moen, W.S., 1961, Geologic map of Washington: Washington Division of Mines and Geology Geologic Map, scale 1:500,000, 1 sheet.

Kienle, C.F., Jr., Farooqui, S.M., Strazer, R.J., and Hamill, M.L., 1978, Investigation of Ribbon Cliff landslide, Entiat, Washington: Shannon & Wilson, Inc., unpublished Report in Washington Public Power Supply System PSAR for WPPSS Nuclear Project No. 1 and 4, Docket 50-460 and 50-513, Amendment 23, Nuclear Regulatory Commission.

Lowell, S.M., 1994, Hatchery Creek fire slope stability assessment—Project status, internal memorandum Washington State Department of Transportation.

Mackin, J.H., 1961, A stratigraphic section in the Yakima Basalt and the Ellensburg Formation in south-central Washington: Olympia, Washington, Washington Division of Mines and Geology Report of Investigations 19, 45 p.

Madole, R.F., Schuster, R.L., and Sarna-Wojcicki, A.M., 1995, Ribbon Cliff landslide, Washington, and the earthquake of 14 December 1872: Bulletin of the Seismological Society of America, v. 85, no. 4, p. 986–1002.

Reidel, S.P., and Campbell, N.P., 1989, Structure of the Yakima Fold Belt, central Washington, *in* Joseph, N.L., et al., eds., Geologic guidebook for Washington and adjacent areas: Olympia, Washington, Washington Division of Geology and Earth Resources Information Circular 86, p. 275–303.

Reidel, S.P., and Fecht, K.R., 1981, Wanapum and Saddle Mountains Basalts of the Cold Creek syncline area, *in* Subsurface geology of the Cold Creek Syncline: Richland, Washington, Rockwell Hanford Operations RHO-BWI-ST-14, p. 3-1–3-45.

Russell, I.C., 1893, A geological reconnaissance in central Washington: U.S. Geological Survey Bulletin 108, 108 p.

Stone and Webster Engineering Corporation, 1956, Preliminary studies of left abutment, Rocky Reach Hydroelectric Project, Public Utility District No. 1 of Chelan County, Washington.

Swanson, D.A., Wright, T.L., Hooper, P.R., and Bentley, R.D., 1979. Revisions in the stratigraphic nomenclature of the Columbia River Basalt Group: U.S. Geological Survey Bulletin 1457G, 59 p., 1 plate.

Tabor, R.W., V.A., Frizell, Jr., J.T., Whetten, R.B., Waitt, D.A., Swanson, G.R., Byerly, D.B., Booth, M.J., Herrington, and R.E., Zartman, 1987, Geologic map of the Chelan 30-minute by 60-minute quadrangle, Washington: U.S. Geological Survey Map I-1661, scale 1:100,000, 33 p., 1 sheet.

Tabor, R.W., Waitt, R.B., Frizell, Jr., V.A., Swanson, D.A., Byerly, G.R., and Bentley, R.D., 1983, Geologic map of the Wenatchee quadrangle, Washington: U.S. Geological Survey Map I-1311, scale 1:100,000, 26 p., 1 sheet.

Geological Society of America
Field Guide 4
2003

Regional Tertiary sequence stratigraphy and regional structure on the eastern flank of the central Cascade Range, Washington

Eric S. Cheney

Department of Earth and Space Sciences, Box 351310, University of Washington, Seattle, Washington 98195-1310, USA

ABSTRACT

The Tertiary sedimentary and volcanic rocks of the Cascade Range unconformably overlie a crystalline basement of previously accreted terranes. The Tertiary strata are parts of four synthems, or interregional unconformity-bounded sequences of tectonic origin. Thus, the formations in these synthems were not deposited in local basins.

The 55–38 Ma Challis synthem has five regional unconformity-bounded formations; the names with precedence are (from the base up) Swauk, Taneum, Teanaway, Roslyn, and Naches. Near Blewett Pass (nee Swauk Pass), the Challis fluvial and arkosic Swauk Formation is ~5 km thick and has several members in a generally upward-fining succession. The members of the Swauk do not interfinger, and some are separated by unconformities.

The Oligocene to mid-Miocene andesitic and rhyolitic Kittitas synthem is almost absent in the area. The most voluminous lithostratigraphic unit in the mid-Miocene to Pliocene Walpapi synthem is the Columbia River Basalt Group. Clasts of Columbia River Basalt Group and older rocks in the ca. 4 Ma Thorp Formation of the High Cascade synthem record initial uplift of the Cascade Range to the west.

North of Blewett Pass, the northwesterly segment of the Leavenworth fault is the Camas Creek reverse fault that places Swauk and Teanaway in the Blushastin anticline over a syncline in the Roslyn Formation. Northerly striking faults in the Leavenworth fault zone are parts of a younger system that cuts the Camas Creek thrust and northwest-striking folds in Challis rocks.

In style, scale, and age, the Camas Creek fault resembles the Easton Ridge thrust south of Cle Elum, the Eagle Creek fault in the Chiwaukum graben, and the Seattle fault in the Puget Lowland. These faults are on the steeper northeastern limbs of major anticlines in Challis rocks. Down plunge, these folds are more gentle in Walpapi rocks. These folds and faults are part of the regional Seattle-Wentachee-Kittitas fold-and-thrust belt.

The Straight Creek fault is a major, north-south, dextral fault in the northern Cascade Range. The fault offsets all five of the Challis unconformity-bounded formations. The southeasterly curving discontinuity along which it was mapped east of Easton is due to unconformities at the base of the Taneum and Teanaway, not a fault. The Straight Creek fault is 2.7 km west of Easton and passes southward beneath Kittitas rocks. Although the fault dextrally displaces pre-Tertiary units ≥90 km, Tertiary displacement is ≤55 km. This may indicate two (or more) periods of displacement. Perhaps the displaced portion of the fault underlies Puget Sound.

Two sets of post-Walpapi folds deform the Tertiary synthems. The Seattle-Wentachee-Kittitas fold-and-thrust belt is part of a set of northwest-striking folds. One of several north-trending regional anticlines causes the Cascade Range. The Cascade Range anticline, with an amplitude of ~3.5 km, has risen in approximately the past 3.5 m.y. This anticline causes the plunges of the Seattle-Wentachee-Kittitas

Cheney, E.S., 2003, Regional Tertiary sequence stratigraphy and regional structure on the eastern flank of the central Cascade Range, Washington, *in* Swanson, T.W., ed., Western Cordillera and adjacent areas: Boulder, Colorado, Geological Society of America Field Guide 4, p. 177–199. For permission to copy, contact editing@geosociety.org. © 2003 Geological Society of America.

fold-and-thrust belt folds. The two sets of folds cause a regional interference, or "egg-crate," pattern that dominates the present topography of the Pacific Northwest.

Keywords: unconformities, synthem, fold-and-thrust belt, egg-crate structure, Cascade Range.

INTRODUCTION

General

Synthems are unconformity-bounded sequences of interregional extent and of tectonic origin. They provide a regional stratigraphy that permits the identification of regional structures. Four Cenozoic synthems dominate the geology of south-central Washington (Fig. 1), including the area of this trip (Fig. 2).

The eastern flank of the Cascade Range in Washington has a wide variety of rock types, structural relations, and scenic views. The geology is better exposed on the eastern flank of the range; the western flank and the Puget Lowland have more extensive and thicker vegetation, glacial deposits, asphalt, and salt water.

This two-day trip focuses on the regional stratigraphy and structure of Tertiary sedimentary and volcanic rocks between Interstate 90 on the south and U.S. Highway 2 on the north. The trip starts in Seattle and crosses the Cascade Range on Interstate 90 to Cle Elum (exit 84). The first day examines Eocene units north of Cle Elum along U.S. Highway 97 near Swauk Pass and then visits Miocene units along State Route 10 between Cle Elum and Ellensburg. Overnight is in Ellensburg (Fig. 2). The second day includes scenic and structural overviews near Interstate 90 from Ellensburg to Easton. The trip terminates in Seattle in the evening of the second day.

Because of their 1:100,000 mapping of this region and beyond, Tabor et al. (1982, 1984, 1987, 2000) dominate the recent literature. After their mapping in the 1970s and early 1980s, 1:24,000 topographic maps became available, and access via logging roads improved substantially. However, since the mid-1990s, many logging roads have deteriorated. My mapping near Easton (Cheney, 1999) and Swauk Pass (still in progress) benefited from the new maps and roads. This more detailed work revises some of the map patterns and conclusions of Tabor et al. (1982, 2000). My information about the much larger intervening area between Easton and Swauk Pass is primarily from the literature.

Stratigraphic Models

Tabor et al. (1982, 1984), Taylor et al. (1988), and Evans and Johnson (1989) concluded that the Tertiary sedimentary and volcanic units in the Cascade Range were deposited in local basins. Specifically, Taylor et al. (1988) stated that the Swauk, Teanaway, and Roslyn formations were deposited in the Swauk basin bounded by the Straight Creek and Leavenworth faults. Most of this field trip is in the Swauk "basin."

An alternative stratigraphic model (Wheeler and Mallory, 1970; Cheney, 1994, 1997; Cheney and Sherrod, 1999) recognizes that Cenozoic rocks occur in four synthems (Fig. 2). Each of the synthems rests upon crystalline basement somewhere in Washington (Cheney, 1994), and each extends beyond the Pacific Northwest (Hanneman et al., 2002). Additionally, most of the formations within these synthems are bounded by regional unconformities.

In the alternative model, sedimentary and volcanic units are laterally discontinuous *not* due to deposition in separate (local) basins but due to differential *local preservation* below various unconformities. That is, due to uplift and *erosion*, regional unconformity-bounded formations are discontinuously (locally) *preserved* in separate *structural lows or structural basins*.

Compilation of numerous formations into synthems reveals the regional structures. Northwest-trending folds are cut by southwest-dipping reverse faults and by north- to northwest-trending strike-slip faults. A north-trending set of folds deforms rocks as young as Pliocene. The intersection of the two sets of folds creates an "egg-crate" pattern, which causes the present topography of the region (Cheney and Sherrod, 1999).

Age Constraints

The paucity of reliable age determinations is a problem. Because all of the Cenozoic rocks on the eastern flank of the Cascade Range are non-marine, detailed paleontological ages are lacking. The marine-nonmarine transition in most of the synthems is near the axis of the Puget Lowland (the "Interstate 5 Effect"). The combination of magnetostratigraphic work and the biostratigraphy of planktonic microfossils in the marine portion of the Tertiary rocks has revealed that previously recognized

SYNTHEM	Ma	EXAMPLES LITHOSTRA-TIGRAPHY	LITHOL-OGIES
High Cascade		Logan Hill Tieton	glacial & andesitic
	4		
Walpapi		Columbia River	basaltic
	20		
Kittitas		Fifes Peak Ohanapecosh	felsic & andesitic
	37		
Challis		L. Wenatchee Naches Roslyn Teanaway Taneum Swauk	arkosic, felsic & basaltic
	55		

Figure 1. Cenozoic synthems in the Pacific Northwest. Wavy lines are interregional unconformities.

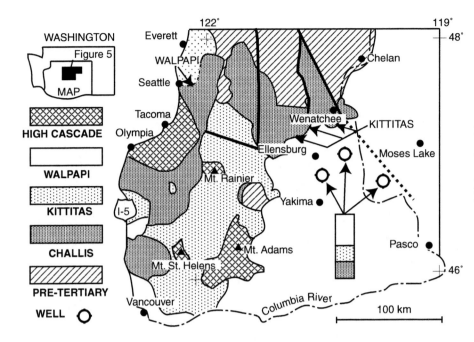

Figure 2. Distribution of Cenozoic synthems in south-central Washington (after Cheney, 1994, fig. 4). The black area in the index map is the area of the field trip. Note that the Challis and Kittitas occur in the subsurface in wells (dry holes) northeast of Yakima and that other bold arrows indicate areas of some synthems that are too small to show.

zones based on mollusks and benthic foraminifera have time-transgressive boundaries and are full of hiatuses and overlaps (Prothero, 2001). That is, previous paleontological correlations are suspect. It is hoped that new magnetostratigraphic and biostratigraphic studies in the marine rocks eventually will result in detailed subdivisions that can be projected eastward to aid correlation with and among the nonmarine rocks.

Radiometric dating in the Tertiary rocks of the central Cascade Range is sparse and of limited quality. Published fission-track ages from zircons in volcaniclastic units in the Swauk appear to be discordant and younger than those from the overlying Taneum felsic rocks (Cheney, 1994, fig. 8). The oldest age in the Taneum is 54.1 ± 2.1 Ma. Another example is the andesitic rocks of Mount Persis and the intrusions in them on the western flank of the Cascade Range. The volcanic rocks and the Index batholith have K/Ar ages of 38.1 ± 3.3 Ma and 34 Ma, respectively (Tabor et al., 2000); however, U-Pb ages from zircons in these units are 38.9 ± 0.3 and 37.0 ± 0.2 Ma, respectively (Smithson et al., 2003).

The discordant ages may due to increased geothermal gradients during burial to 5–10 km or to Challis, Kittitas, and Walpapi volcanism. Because the five unconformity-bounded formations of the Challis synthem probably only span 17 m.y. (55–38 Ma), K/Ar and fission-track ages may not be precise enough to distinguish some formations from others.

Sauk Pass

Alas, tradition requires an explanatory note about Swauk Pass northwest of Ellensburg. U.S. Highway 97 originally traversed Blewett Pass and was, therefore, the Blewett Pass Highway. In the 1960s, the highway was relocated eastward to Swauk Pass, but it was still called the Blewett Pass Highway. The Washington Department of Transportation solved this apparent problem by renaming Swauk Pass as Blewett Pass and by renaming Blewett Pass as Old Blewett Pass. However, Taylor et al. (1988) had already named a unit within the Eocene Swauk Formation as the sandstone facies of Swauk Pass, which is a name worth retaining. In 1989, the U.S. Geological Survey reissued the Swauk Pass 7.5 minute topographic quadrangle as Blewett Pass (and changed the name of the pass accordingly). Still unchanged, however, was the name of the original Blewett Pass on the Liberty 7.5 minute quadrangle to the west. The name Swauk Pass is lovingly retained herein.

Two Conclusions

Two conclusions can be made with some certainty about the Cascade Range and most of Washington. First, the Tertiary history is difficult to decipher precisely due to the paucity of radiometric and paleontologic dating, the lithologic similarity of some formations, the paucity of marker units, extensive cover by the High Cascade synthem, and limited detailed mapping. Second, and most important, faults and folds are *bigger and younger* than commonly supposed a decade ago.

REGIONAL GEOLOGIC UNITS

Pre-Tertiary Basement

The pre-Tertiary rocks of the Cascade Range are mostly terranes that accreted to North America during the Cretaceous (Dragovich et al., 2002). Accretion of these terranes extended North America westward, so that Tertiary rocks in the Cascade Range were deposited upon a cratonic-like basement.

The Ingalls Tectonic Complex of Jurassic and Cretaceous rocks is the basement north of Ellensburg and Swauk Pass (Tabor et al., 1982; Harper et al., this guidebook; Miller et al., this guidebook). The complex includes peridotite, serpentinized peridotite, diabase, gabbro, basaltic greenstone with minor felsic metavolcanic rocks (Hawkins Formation), and phyllitic argillite (Peshastin Formation). The 93–90 Ma Mount Stuart batholith intrudes the complex.

The Easton Metamorphic Suite is the pre-Tertiary basement west of Ellensburg. It consists of the Shuksan Greenschist, an albite-chlorite-epidote phyllite, and the Darrington phyllite, a pelitic unit with numerous dismembered quartz veinlets. The intervening belt of Challis rocks, which is a major focus of this field trip, obscures the relation of the Easton Metamorphic Suite to the Ingalls Tectonic Complex.

Challis Synthem

The 55–38 Ma Challis synthem consists of sedimentary rocks that are mostly arkosic and of volcanic rocks that are mostly basaltic or felsic. The synthem is widespread but discontinuous throughout Washington (Figs. 2 and 3). Five unconformity-bounded formations (Fig. 4, columns 50 and 56) occur in the area of the field trip, and at least eight occur in Washington. Because the Challis in the area has a duration of ~17 m.y. and is punctuated by six major unconformities, "punctuated equilibrium" describes its evolution.

Because previous workers did not recognize that various Challis successions are erosional remnants of unconformity-bounded sequences, synonymous formational names obscure the fairly simple regional stratigraphy and structure. With 1:24,000 mapping, Cheney (1999) confirmed the synonyms that Tabor et al. (1984) had inferred. The names with precedence are in bold font below (with synonyms in parentheses):

TOP

    Naches Formation: felsite, andesitic volcaniclastic rocks, and arkose

    Roslyn Formation (Chumstick): predominantly arkosic

    Teanaway Formation (basalt of Frost Mountain): basalt, with some felsite, felsic volcaniclastic rocks, and minor arkose

    Taneum Formation (Silver Pass, Peoh Point, Mount Catherine): felsite

    Swauk Formation (Manastash, Guye): predominantly arkosic

BOTTOM

The names with precedence are used hereafter in the area of the field trip (Fig. 5).

The cross sections of Tabor et al. (1982, 1987, 2000) and of Cheney (1999) show that the maximum thickness of each forma-

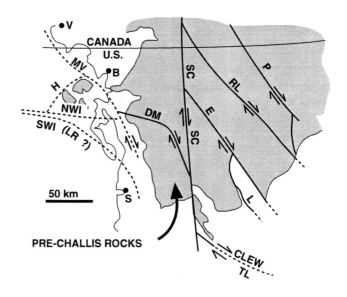

EC2-12/15/94

Figure 3. Major post-Challis faults in the northern Cascade Range. The areas of Cenozoic synthems are blank. The field trip is in the blank (Cenozoic) area between faults SC, L, and TL. The area between E and L is the Chiwaukum graben. Cities are: B—Bellingham, S—Seattle, and V—Vancouver, Canada. Abbreviations for faults are: CLEW—Cle Elum–Wallow; NWI—Northern Whidbey Island; DM—Devils Mountain; P—Pasayten; E—Entiat; RL—Ross Lake; H—Haro; SC—Straight Creek; L—Leavenworth; SWI—South Whidbey Island; LR—Leech River; TL—Taneum Lake; MV—Mount Vernon.

tion is >1 km. However, beneath the various Challis unconformities, or below the sub-Walpapi unconformity, each formation somewhere thins to zero.

The basaltic Teanaway Formation is the marker unit in the Challis in this area. The Swauk and Roslyn (and synonyms) are predominantly arkosic, and even the Naches and Teanaway contain mappable arkosic units (Cheney, 1999). The Taneum, Teanaway, and Naches (and synonyms) have felsic ash-flow tuffs that are difficult to distinguish from each other. Only the Teanaway Formation is predominantly basaltic. The Swauk and Taneum are below the Teanaway, whereas the Roslyn and Naches are above it. Additionally, a swarm of Teanaway dikes occurs in the area (Tabor et al., 1982; Dragovich et al., 2002); arkosic rocks intruded by basaltic dikes are Swauk, whereas arkosic rocks without dikes may be Swauk or Roslyn.

The Columbia River Basalt Group of the Walpapi synthem is a well-known flood basalt. The Teanaway may be part of Washington's other flood basalt. Northeast of Easton (Fig. 5), the Teanaway is at least 2500 m thick (Tabor et al., 2000). It has a discontinuous strike length of nearly 60 km. To the south, it passes beneath the Columbia River Basalt Group and the Kittitas (Fig. 5). Additionally, the swarm of Teanaway dikes extends at least 15 km beyond the present northern limit of the outcrop (Tabor et al., 1982; Dragovich et al., 2002). Perhaps the Teanaway correlates

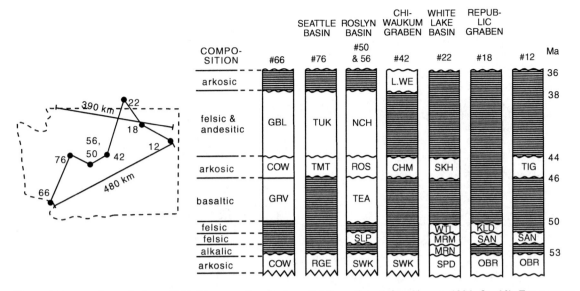

Figure 4. Proposed correlations of Challis unconformity-bounded formations (after Cheney, 1994, fig. 10). For sources of data and additional sections with additional formations, see Cheney (1994, fig. 6). Wavy lines between formations are unconformities; straight lines indicate that no local unconformity is yet known, and saw teeth indicate that the bottom of the formation is not exposed. Numbers on the columns correspond to locations on the map. Abbreviations for formations are: CHM—Chumstick; OBR—O'Brien Creek; SWK—Swauk; COW—Cowlitz; REN—Renton; TEA—Teanaway; GBL—Goble; RGE—Raging River; TIG—Tiger; GRV—Grays River; ROS—Roslyn; TMT—Tiger Mountain; KLD—Klondike Mountain; SAN—Sanpoil; TUK—Tukwila; MRM—Marama; SKH—Skaha; WTL—White Lake; MRN—Marron; SLP—Silver Pass; L.WEN—lower Wenatchee; NCH—Naches; SPB—Springbrook.

with other Challis basaltic rocks, such as the Grays River (column 66 of Fig. 4) or the Northcraft (between columns 66 and 76). If so, or if it correlates with some of the Challis-age Crescent basalt of westernmost Washington, it has a regional extent.

Tabor et al. (1982, 1984, 1987, 2000), Taylor et al. (1988), and Evans and Johnson (1989) thought that the various lithofacies in the Challis formations interfinger abruptly, both vertically and laterally. Specifically, no laterally persistent internal stratigraphy had been recognized in the Naches formations until Cheney (1999) recognized five laterally persistent lithostratigraphic units near Easton. Detailed mapping shows that the lithostratigraphic members of the Swauk Formation (Table 1) also are laterally persistent and do not interfinger at scales of 1:24,000–1:100,000. Because some of these members of the Swauk are bounded by unconformities, they cannot interfinger. Paleocurrent indicators show that the arkosic rocks of the Challis synthem generally had easterly to northerly sources.

The sources of the welded and devitrified tuffs in the Taneum, Teanaway, and Naches are unknown. Because some of the tuffs in the Naches have lithic clasts, their sources must have been nearby. Challis-aged plutons, which might be sources of the tuffs, are conspicuously absent in the area, but younger synthems could obscure such plutons.

An alternative interpretation is that the Naches is not Challis, but, rather, basal Kittitas, and, thus, is unrelated to Challis plutons. The major unconformity at the base of the Naches in the

Easton area and the andesitic nature of some of its volcaniclastic rocks (Cheney, 1999) imply that it could be Kittitas. To test this alternative, the 40–45 Ma fission-track ages of zircons from Naches rhyolites should be checked by modern methods, such as Ar-Ar or U-Pb on zircons, to see if they are ≤38 Ma.

Kittitas Synthem

The Kittitas synthem of intermediate to felsic volcaniclastic rocks is best exposed in the southern Cascade Range (Fig. 2) as the Ohanapecosh and Fifes Peak Formations, but it is absent from most of the area of this field trip. The Ohanapecosh has K/Ar dates as old as 36 Ma, the Fifes Peak is as young as 22 Ma, and a conspicuous unconformity occurs between them ca. 30 Ma (Tabor et al., 2000). To the west, Kittitas marine volcaniclastic rocks occur in the Puget Lowland and on the Olympic Peninsula (Cheney, 1994). In the marine rocks, the hiatus between the arkosic Cowlitz Formation (Challis) and the volcaniclastic Lincoln Creek Formation (Kittitas) is 37–36 Ma (Prothero, 2001). Cheney (1994) tentatively proposed the name Kittitas for the synthem.

The sources of the Kittitas synthem were quartz dioritic plutons in the Cascade Range, several of which intrude the faults shown in Figure 3 (Dragovich et al., 2002, fig. 5). Primarily on the basis K/Ar ages, the plutons were thought to range from 35 to 20 Ma (Dragovich et al., 2002). However, U/Pb ages of zircons of the Index batholith are ≥37 Ma (Smithson et al., 2003).

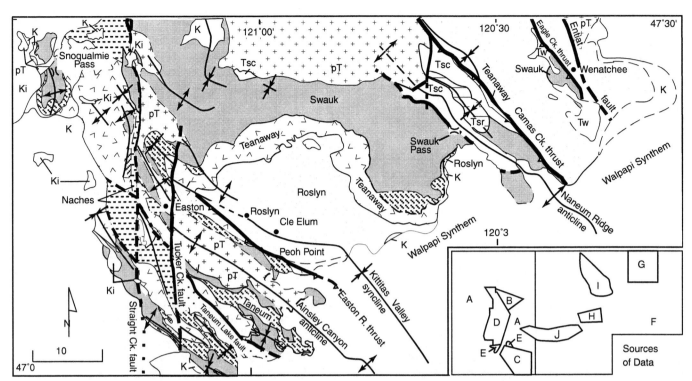

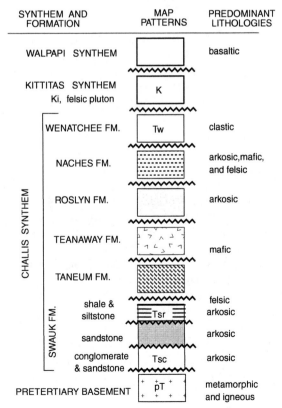

EXPLANATION

SYNTHEM AND FORMATION	MAP PATTERNS	PREDOMINANT LITHOLOGIES
WALPAPI SYNTHEM		basaltic
KITTITAS SYNTHEM Ki, felsic pluton	K	
WENATCHEE FM.	Tw	clastic
NACHES FM.		arkosic, mafic, and felsic
ROSLYN FM.		arkosic
TEANAWAY FM.		mafic
TANEUM FM.		
shale & siltstone	Tsr	felsic arkosic
sandstone		arkosic
conglomerate & sandstone	Tsc	arkosic
PRETERTIARY BASEMENT	pT	metamorphic and igneous

CHALLIS SYNTHEM

SWAUK FM.

Formational names used are those with precedence.
Wavy lines are unconformities.

Figure 5. Regional geology. Sources of data: A—Tabor et al. (2000); B—Foster (1960); C—Hammond (1980); D—Cheney (1999); E—Cheney (unpublished 1:24,000 mapping); F—Tabor et al. (1982); G—Gresens (1983); H—Margolis (1994); I—Figure 7; J—Walker (1980).

Walpapi

The synthem above the interregional mid-Miocene unconformity is the Walpapi (Wheeler and Mallory, 1970). In the Cascade Range and in eastern Washington, the most voluminous unit of the Walpapi is the Columbia River Basalt Group (Reidel et al., this guidebook). Figure 6 shows that the Columbia River Basalt Group has magnetostratigraphic units of reversed and normal polarity. N_0, R_1, and N_1 are absent in the area of Figure 5.

The basalts issued from a dike swarm near the common corner of Washington and Oregon with Idaho (Wells et al., 1989, fig. 1). They extend to coastal Washington and offshore Oregon (Wells et al., 1989; Yeats et al., 1998). In the marine section, the basal arkosic Astoria Formation is ≥19.2 Ma (Prothero, 2001). Walpapi basalts also occur in Idaho, California, and British Columbia (Wheeler and Mallory, 1970; Cheney, 1994).

On the eastern flank of the Cascade Range, several felsic volcaniclastic and several siliciclastic units occur within and above the Columbia River Basalt Group. Cross-beds in the volcaniclastic interbeds indicate a westerly source, whereas those in siliciclastic units imply an easterly source (Smith, 1988). The Ellensburg Formation in its type area is 280–350 m of volcaniclastic material

TABLE 1. STATIGRAPHIC UNITS OF THE SWAUK FORMATION

Informal name and symbol	Original name and reference	Maximum thickness	Conglomerate	Volcaniclastics	Other characteristics	Origin	Stratigraphy
Sandstone of Sand Creek Tsk	This paper	0.7 km x-sec E–F	None	None	Similar to Tsr except sandstones are cross-bedded and olive siltstone is minor	Fluvial and lacustrine	Unconformable on Tsa and Tsr
Sandstone of Red Hill Tsa	Sandstone facies of Red Hill of Taylor et al. (1988)	0.6 km x-sec E–F	≤ 2% clasts ≤ 10 cm, matrix-supported, beds ≤ 0.9 m	None	Similar to Tss, including thin conglomerates and thick sandstones	Fluvial	Conformable on Tsr
Siltstone of Tronsen Ridge Tsr	Shaly facies of Tronsen Ridge of Tabor et al. (1982), Taylor et al. (1988)	0.5 km on southwe st side of Tronsen Ridge	None	None	White sandstone, black siltstone, and olive siltstone in planar beds mostly 0.4 to 4 m thick	Lacustrine	Unconformable on Tss and Tsc
Sandstone of Swauk Pass Tss	Sandstone facies of Swauk Pass of Taylor et al. (I987)	0.9 km x-sec G–H	≤ 2% clasts ≤ cobbles, matrix-supported, beds ≤ 0.9 m	Minor	Sandstone (commonly cross-bedded and tan) and black siltstone, beds mostly < 8 m thick, but some > 8 and > 16 m thick	Fluvial	Unconformable on Tsg, Tsc, and regionally on pre-Tertiary
Green Siltstone Tsg	This paper	0.2 km x-sec G–H	None	Common	Green siltstone and sandstone, minor black siltstone	Volcaniclastic	Conformable on Tsc?
Conglomerate of Tronsen Creek Tsc	Conglomeratic facies of Tronsen Creek of Taylor et al. (1988)	0.5 km above and 1.3 km below Tscd, x-secs C–D, A–B	~20% clasts ≤ boulders, clast- and matrix-supported, polymict	None	Predominantly sandstone (poorly cross-bedded) and black siltstone, some beds > 20 m thick	Fluvial and mud flow?	Nonconformable on pre-Tertiary
Diamictite of Devils Gulch Tscd	Breccia of Devils Gulch of Taylor et al. (1988)	0.3 km x-sec C–D	~99% clasts < 3m, monomict	None	99% unbedded and unstratified, clasts rounded to angular, mostly tonalitic clasts but locally 0.1 to 100% ultramafic clasts	Mud-flow, or rock avalanche?	Only mappable unit in Tsc
Ironstone Tsci	Ironstone deposits of Lamey (1950), Tabor et al. (1982)	20 m?	Clasts ≤ boulders, matrix-supported	None	Ultramafic and other pre-Tertiary clasts in fine-grained, brown, and magnetic matrix, grades downward into laterite	Re-worked laterite	Discontinuous basal unit of Tsc

Notes. Stratigraphically upward is up in the table; x-sec—cross section in Figure 8.

above the youngest basalt. Unfortunately, this type section and all other volcaniclastic and siliciclastic interbeds in the Columbia River Basalt Group, regardless of their stratigraphic position, are now lumped as Ellensburg (Smith, 1988, fig. 2).

Significant unconformities exist in the Walpapi (Fig. 6). Locally, the type Ellensburg rests on the lowest unit of the Columbia River Basalt Group, the Grand Ronde Basalt Formation, and even on the Fifes Peak Formation of the Kittitas (Smith, 1988). The correlative Blakely Harbor Formation in the Seattle area also rests upon Kittitas strata (Blakely et al., 2002). These relations illustrate that intra-Walpapi erosion removed the Columbia River Basalt Group from some areas of the future Cascade Range and the Puget Lowland.

High Cascade Synthem

The informal name of the synthem above the Walpapi is High Cascade. It includes andesitic rocks of the Cascade volcanoes

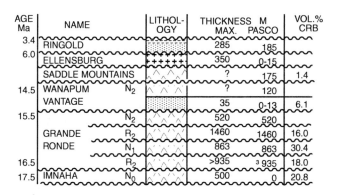

AGE Ma	NAME		LITHOL-OGY	THICKNESS M MAX. PASCO		VOL.% CRB
3.4	RINGOLD			285	185	
6.0	ELLENSBURG			350	0-15	
	SADDLE MOUNTAINS			?	175	1.4
14.5	WANAPUM	N₂		?	120	
	VANTAGE			35	0-13	6.1
15.5		N₂		520	520	
	GRANDE	R₂		1460	1460	16.0
	RONDE	N₁		863	863	30.4
16.5		R₂		>935	³935	18.0
17.5	IMNAHA	N₀		500	0	20.8

Figure 6. The Walpapi synthem in southeastern Washington. Sources of data are Smith (1988) and Cheney (1997, fig. 4). Wavy lines are unconformities. N_0, R_1, N_2, R_2, and N_2 are magnetostratigraphic units in the Columbia River Basalt (CRB) Group.

and widespread glaciogene sediments in the Cascade Range and elsewhere, including the Puget Lowland. The oldest known volcanism of the Cascade magmatic arc is the 3.72 ± 0.02 Ma Hannegan caldera (Hildreth et al., 2003).

The oldest unit of the High Cascade synthem in the area of the field trip is the Thorp Formation. It is unconformable on the Walpapi and dips easterly off the Cascade Range. Because the Thorp contains clasts of most of the resistant rocks in the Cascade Range to the west (including the Columbia River Basalt Group), it documents uplift and the resultant unroofing of the range. Tuffs in the Thorp have fission-track dates of 3.8 and 4.4 Ma and a K/Ar date of 4.5 Ma (Tabor et al., 1982). Thus, volcanism and the topographic rise of the Cascade Range began ca. 4 Ma.

STRUCTURE

Introduction

The major structural feature of the area is a belt of northwest-trending, southeast-plunging folds that affect the Challis rocks more than the Walpapi rocks. However, at least four ages of folding exist.

The synonymous formational names and the concept of small depositional basins hindered recognition of the major folds. For example, the reputed Eocene Swauk depositional basin of Taylor et al. (1988) is on-strike with the Kittitas Valley syncline in the Columbia River Basalt Group to the southeast. Thus, rather than extend the Kittitas Valley syncline northwestward through the syncline marked by the Roslyn Formation in the Swauk "basin," Tabor et al. (1982) terminated the Kittitas Valley syncline at the Columbia River Basalt Group/Challis contact.

Periods of Folding

Northwest-trending folds with kink-shaped crests and sinuous traces in the Swauk Formation north of Swauk Pass (Fig. 7)

record the earliest period of folding. The sinuous traces of the folds imply that they are refolded. Because the northeast-trending Teanaway dike swarm intrudes these folds, and because the Teanaway Formation is less intensely folded than the underlying Swauk and Taneum Formations (Tabor et al., 1982), these folds are pre-Teanaway in age. Thus, the sub-Teanaway unconformity probably is the major one in this area of the Challis synthem.

A second period of folding along northwest-trending axes created steep dips in rocks as young as the Teanaway and Roslyn (Chumstick) Formations (but not in the Naches and the Walpapi). An example north of Swauk Pass is the Blushastin anticline (Fig. 7). Other examples in Figure 5 are the Ainsley Canyon anticline and Kittitas Valley syncline near Cle Elum. This period of folding may have caused the sinuous traces of the pre-Teanaway folds.

The Yakima fold belt in the Columbia River Basalt Group south of Interstate 90 (Reidel et al., this guidebook) records a third period of folding. The Blushastin and Ainsley Canyon anticlines and the Kittitas Valley syncline in Challis rocks pass down-plunge to the southeast into more gentle folds or monoclines in the Columbia River Basalt Group. The Ainsley Canyon anticline is part of the Yakima fold belt.

North-trending regional anticlines and synclines record the fourth period of folding. The regional dip of the Columbia River Basalt Group on the eastern flank of the Cascade Range is eastward. Thus, the Cascade Range is a post–Columbia River Basalt Group, north-south anticline, and the Pasco basin to the east and the Puget Lowland to the west are post–Columbia River Basalt Group synclines (Cheney and Sherrod, 1999). Because dips are gentle and the pattern is regional, the synclinal nature of the formations of the Columbia River Basalt Group in the Pasco Basin (Cheney, 1997, fig. 3) is not obvious.

Near Cle Elum, the structural relief on the Columbia River Basalt Group (caused by the eastward regional dip and by the northwest-trending Naneum Ridge, Kittitas Valley, and Ainsley Canyon folds) is >1200 m (Tabor et al., 1982, fig. 3). Nonetheless, the Columbia River Basalt Group is commonly supposed to have ponded in the Pasco basin against the Cascade Range to the west and to have flowed down the ancestral gorge of the Columbia River (and other gorges) to the sea.

The paleontology of the Walpapi synthem provides additional proof of the post–Columbia River Basalt Group uplift of the Cascade Range (Cheney, 1997). The Vantage, Ellensburg, and Ringold units of the Walpapi (Fig. 6) contain flora and fauna that are typical of a warm-temperate, summer-wet climate, unlike the present steppe and local grassland caused by the rain shadow of the Cascade Range.

Faults

Tabor et al. (1982) and Taylor et al. (1988) believed that the Chumstick is a local formation deposited in the Chiwaukum graben when the bounding Leavenworth fault zone was active. Specifically, Tabor et al. (1982) believed that a bouldery deposit along U.S. Highway 97 is a Chumstick fanglomerate that flowed off the

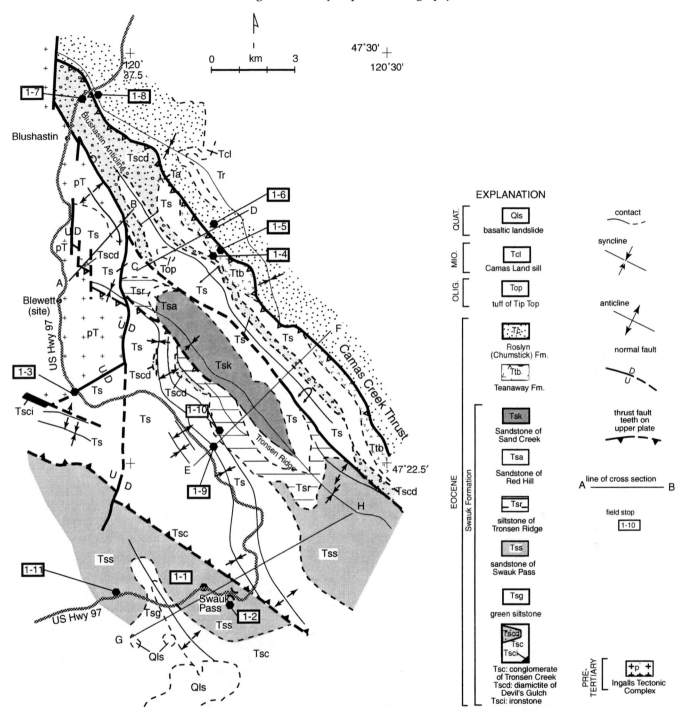

Figure 7. Tectonic map of the Swauk Pass area based on unpublished mapping at 1:24,000. Cross sections are in Figure 8.

fault into the Chiwaukum graben. However, the deposit (Tscd in Fig. 7) is neither Roslyn (Chumstick) nor a fanglomerate. Furthermore, the map of Tabor et al. (1987) shows that the southwestern part of the graben is another major northwest-trending syncline.

The 1:24,000 mapping, upon which Figure 7 is based, shows that the northwestern portion of the Leavenworth fault adjacent to Tscd dips almost vertically to ~70° southwestward and cuts across the entire southwestern limb of a syncline in the Roslyn. Thus, the fault places topographically higher and older Swauk and Teanaway in the Blushastin anticline over a syncline in the Roslyn. This segment of the Leavenworth fault is here called the Camas Creek thrust. Because the thrust eliminates the >1.5-km-wide limb of the syncline in the Roslyn (Fig. 7), it must have a displacement >3 km.

The Easton Ridge thrust south of Cle Elum dips 70°–80° to the southwest (Walker, 1980, map 1A). It thrusts Teanaway and Taneum in the Ainsley Canyon anticline over the Roslyn in the Kittitas Valley syncline to the northeast. Because the fault truncates the entire 3 km thickness of the Roslyn, it must have 3–5 km of displacement. In contrast, the structural relief on the Columbia River Basalt Group in the vicinity of the Taneum monocline on strike with the fault to the southeast is only ~0.5 km (Tabor et al., 1982, fig. 3). The two different amounts of displacement and the gentler dips in the Columbia River Basalt Group imply post-Walpapi (post-Thorp) reactivation of pre-Walpapi (pre-Naches) folds.

The northwest-striking Eagle Creek anticline in the Chiwaukum graben (Fig. 5) folds Roslyn and is similar to the Blushastin and Ainsley Canyon anticlines. Northwest of Wenatchee, the Eagle Creek anticline has pre-Tertiary metamorphic rocks in its core (Tabor et al., 1987). The fault on the northeastern limb is high-angle northwest of Wenatchee (Tabor et al., 1987) and is a thrust south of Wenatchee (Patton and Cheney, 1971). The northwest-trending Colokum Creek and Laurel Hill monoclines in the Columbia River Basalt Group on strike to the southeast (Tabor et al., 1982) imply that this fold also was reactivated in post–Columbia River Basalt Group time.

The major north-trending faults are the Straight Creek fault and northern segments of the Leavenworth fault mapped by Tabor et al. (1982, 1987). These northern faults cut the northwest-striking faults and folds that involve the Challis formations (Figs. 3 and 7), but they do not cut post-Naches rocks.

Some faulting in the region is demonstrably young. The post–Columbia River Basalt Group Ringold Formation is folded and faulted in the Yakima fold-and-thrust belt (Reidel et al., 1994, fig. 7). Scarps in the ca. 4 Ma Thorp Formation may be faults (Waitt, 1979). Some faults in the Yakima fold-and-thrust belt cut Pleistocene and Holocene sediments (Reidel et al., 1994, table 4).

TECTONICS

Basin Analysis

Taylor et al. (1988, p. 1032 and fig. 17) believed that the Swauk and Chumstick Formations were deposited syntectonically in the Swauk "basin" and Chiwaukum graben, respectively. They cited the following features as indicative of deposition in local strike-slip (pull-apart) basins:

1. The formations are between strike-slip faults.

2. Sediment accumulation rates (of nonmarine rocks) were high.

3. Lateral and vertical facies changes are abrupt.

4. Variations in paleocurrent data indicate disrupted and reorganized drainage systems.

5. Intermittent internal drainage systems existed (as shown by the presence of lacustrine units and abrupt changes of facies).

6. Interbedded volcanic and intrusive rocks imply extension.

7. The strike of many folds is acute to bounding folds.

8. Pulses of sedimentation alternated with compressional deformation.

However, in the areas of the Swauk Formation in Figure 9 and near Easton (Cheney, 1999, plate 1), the bounding faults (feature 1) are post-Swauk. Both the Swauk (Cheney, 1999) and the Chumstick (Tabor et al., 1982) occur beyond (outside) their reputed bounding faults. Feature 6 is minor, and feature 8 does not exist. Although the Swauk and Roslyn (Chumstick) appear to be thick, feature 2 has not been quantified. Convincing evidence for feature 3 (lateral and vertical interfingering of facies) and feature 5 (internal drainage, such as evaporitic deposits) has not yet been documented. Sigmoidal folds representative of feature 7 have not yet been mapped. The presence of diamictite and very coarse conglomerate in the Swauk Formation near Swauk Pass is compatible with (but does not demand the existence of) pull-apart basins.

Although the depositional settings of the Swauk and Roslyn (Chumstick) Formations may be somewhat uncertain, the existence of major folds (with dips higher than those generated during deposition) and of post-depositional faults indicate that the original extents of the Swauk and Roslyn were different than their present geographies. The unconformity-bounded nature of some members in the Swauk Formation and of other formations in the Challis synthem indicates that these members and formations are remnants of larger depositional systems. The generally arkosic and basaltic composition of the Challis synthem, its great extent (Fig. 4), and the seemingly great thickness of most of its formations suggest that the Challis behaved as though it were on a continental trailing edge or rifted margin.

OLYMPIC-WALLOWA LINEAMENT AND SEATTLE-WENATCHEE-KITTITAS FOLD-AND-THRUST BELT

The northwest-striking Olympic-Wallowa lineament (also known as OWL) of Raisz (1945) extends from northwestern Washington to northeastern Oregon. Southeast of Cle Elum, the lineament is better defined than to the northwest. This southeastern segment is almost entirely in the Columbia River Basalt Group and is known as the Cle Elum–Wallowa lineament. The Yakima fold-and-thrust belt is the portion of the Cle Elum–Wallowa lineament southeast of Cle Elum.

The segment of the Olympic-Wallowa lineament northwest of Cle Elum is coincident with the belt of northwest-trending folds that involve Challis rocks, extend down plunge into less steeply folded Walpapi rocks, and have reverse faults on their steeper northeastern limbs. This belt extends from Seattle (Newport Hills anticline and Seattle fault), to Wenatchee (the Blushastin anticline and Camas Creek fault and the Eagle Creek anticline and fault), to Kittitas County (Ainsley Canyon and Easton Ridge fault). This is the Seattle-Wenatchee-Kittitas fold-and-thrust belt (also known as SWIFT) sector of the Olympic-Wallowa lineament (Cheney, 2002). Southeasterly, the Seattle-Wenatchee-Kittitas fold-and-thrust belt passes into the Yakima fold-and-thrust belt in the Cle Elum–Wallowa lineament.

The seismic hazards of the Seattle basin and Seattle fault have been intensely studied in recent years (Blakely et al., 2002). These are not a local depositional basin and an isolated fault, as commonly supposed. Instead, they are the synthems and structures of the Seattle-Wentachee-Kittitas fold-and-thrust belt. Much could be learned by studying the analogs at Cle Elum and elsewhere, where vegetation, glacial deposits, asphalt, and salt water are less extensive than in the Puget Lowland.

Regional Egg-Crate Pattern

The map pattern of the Walpapi and other synthems in the Pacific Northwest outlines major north-striking folds in addition to the Cascade Range anticline (Cheney and Sherrod, 1999). The Coast Range (especially in Oregon but also including the Olympic Mountains) is anticlinal. The Puget-Willamette Lowland is the syncline between the Coast Range and the Cascade Range anticlines.

The north-striking, antiformal metamorphic core complexes of northeastern Washington and adjacent British Columbia and Idaho formed in the mid-Eocene. However, the salients of the Columbia River Basalt Group between the southern ends of the core complexes imply that some of the uplift of these complexes also is post-Walpapi

The intersection of the north-striking set of folds, which includes the Seattle-Wentachee-Kittitas fold-and-thrust belt, with the large, and more widely spaced, north-trending folds produces a regional interference, or "egg-crate," pattern (Cheney and Sherrod, 1999). Mount Stuart in the pre-Tertiary rocks northwest of Swauk Pass is the highest mountain in the central Cascade Range; as noted below, it is near the intersection of the Cascade Range anticline with the northwest-striking portion of the Naneum Ridge anticline. Uplift of the Cascade Range anticline caused the Naneum Ridge, Kittitas Valley, and Ainsley Canyon folds to plunge southeastward.

Plate Tectonic Models

Presumably, the two sets of folds of the egg-crate did not form simultaneously. The north-striking set probably is an "edge-effect" on the continent related to the off-shore Cascadia subduction zone. The post-Walpapi portion of the west- to northwest-striking set may be due to the drift of California to the northwest at millimeters/year, which causes the rotation of Oregon into Washington (Wells et al., 1998).

A similar model may explain Challis arkosic sedimentation and intermittent bimodal volcanism, as well as the pre-Walpapi portion of the northwest-striking folds. Perhaps the tailing edge-like sedimentation of the Swauk and the Roslyn (and correlative formations) occurred as the Farallon and Kula plates migrated northward in strike-slip mode relative to North America. A slab window, caused by intermittent subduction of the spreading center between the Farallon and Kula plates, may have caused the basaltic and felsic Challis volcanism (Breitsprecher et al., 2003).

ROAD LOG

Introduction

This road log uses mile posts (MP) on paved highways to identify most stops and points of interest. Although this trip begins and ends in Seattle, the road log begins just west of Swauk Pass (now Blewett Pass) at MP 163.4 on U.S. Highway 97; it ends near Easton on MP 67.4 of Interstate 90. The maps of Tabor et al. (1982, 2000) display the geology from North Bend on the western edge of the Cascade Range, to Swauk Pass, and to Easton.

The first part of Day One (Stops 1-1 to 1-10) uses 1:24,000 field sheets of mapping still in progress near Swauk Pass and Figures 7 and 8, which are the results of that mapping. Previous mappers of the area were Tabor et al. (1982), Taylor et al. (1988), and others they cite. Stop 1-4 includes a major discussion of the stratigraphy of the Swauk Formation, and Stop 1-10 has a major discussion of the structure. The last part of the day uses Tabor et al. (1982). Day Two uses Tabor et al. (1982, 2000) and Cheney (1999). These maps and this field guide have quite different stratigraphic and structural interpretations.

Some caveats are obvious. To complete the ambitious number of stops for both days, extended discussions must be held at night, rather than on the outcrops. Please be conscious of traffic at stops on paved roads. Adverse weather or darkness could cause the cancellation of some stops. Toilet facilities on Day One are limited to Stop 1-2 and after Stop 1-10: bushes beware.

Day One: Cle Elum to Ellensburg

From Cle Elum on Interstate 90, proceed northeastward on State Route 970 to the junction with U.S. Highway 97. Continue straight (north) on U.S. Highway 97 (MP 149.7) toward Wenatchee. Beyond the road to Liberty (MP 152.5) the rocks are mostly sandstone of the Swauk Pass Member of the Swauk Formation, cut by rusty-weathering dikes of the Teanaway swarm.

MP 163.4 (U.S. Highway 97)

Park in the pull-out on the right (south) side of the highway (0.5 mi before crossing the summit of Swauk Pass).

Stop 1-1. Sandstone of the Swauk Pass Member of the Swauk Formation

The sandstone and black siltstone (Tss of Table 1 and of Figs. 7 and 8) are typical of the sandstone of Swauk Pass (Evans and Johnson, 1989, fig. 4). Tss is the areally most extensive member of the Swauk (Tabor et al., 1982; Taylor, 1988). Taylor et al. (1988) estimated the thickness of the Tss to be 2300 m, but in the Swauk Pass area it is only 0.9 km thick (Table 1).

Here, the three beds of massive sandstone may be crevasse splays on a flood plain represented by the black siltstone. Elsewhere, this member has sandstone exceeding 25 m in thickness and rare pebble conglomerate up to 0.5 m thick. Fossil palm fronds have been vandalized by previous geologists and the

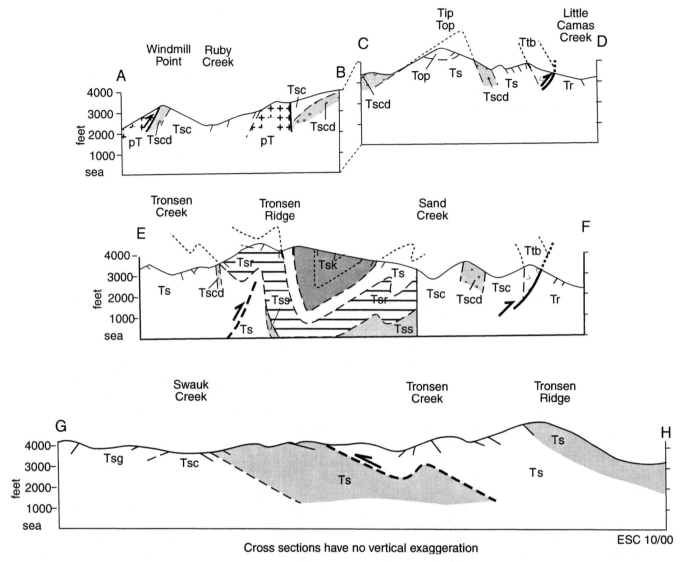

Figure 8. Cross sections of the Swauk Pass area. Lines of sections are on Figure 7. Scale of these cross sections is different than the scale of Figure 7. Cross sections have no vertical exaggeration. Explanation for Figure 7 also is the explanation for Figure 8. Because B of cross section A–B is on strike with C of C–D, the two cross sections can be joined. Due to the general absence of marker units, only the dips of bedding at the surface are shown (by ticks) in most places. Abrupt variations in these dips suggest that more faults or folds occur than are shown. See the discussion of Stop 1-10 for additional explanation.

Washington Department of Transportation, but fragments may remain. A sample of the sandstone should be collected to compare with other arkosic rocks at other stops.

Continue northbound (here eastward) on U.S. Highway 97.

MP 163.9 (U.S. Highway 97)

Swauk Pass (relabeled Blewett Pass). Turn right (south) and drive 0.4 mi to the U.S. Forest Service toilet and interpretive trail.

Stop 1-2. Scenic View of Tronsen Ridge

The real reason for this stop is to use the U.S. Forest Service toilet (which we will do again after Stop 1-10). However, 100 m

back down the road is a fine view of Tronsen Ridge and much of the Swauk Formation. To the southeast are well-forested, flat-topped ridges capped by nearly flat lying Columbia River Basalt Group; the sub-Walpapi unconformity can be visualized by extrapolating the Columbia River Basalt Group northward over the Swauk, which dips ~40° to the northeast. To the east, the thick-bedded portion on the northern part of Tronsen Ridge is the conglomerate of Tronsen Creek (Tsc in Table 1 and Fig. 7). The middle portion of the ridge is underlain by Tss, which also has thick beds in it (and which is indistinguishable from Tsc at this distance). On the barren slopes to the north are the thinner and planar-bedded siltstone and sandstone of the shale of Tronsen

Ridge (Tsr of Table 1 and Fig. 7). From here, it is not obvious that Tsr dips ≥50° northeastward and is unconformable on Tss and Tsc (Fig. 7). Teanaway dikes are conspicuous in Tsr.

Return to Swauk Pass at MP 163.9 on U.S. Highway 97 and turn right (northbound, but here eastward) and continue down the valley of Tronsen Creek.

MP 172.1 (U.S. Highway 97)

Stop at the large parking area on the southwestern (left) side of the highway.

Stop 1-3. Conglomerate of Tronsen Creek and the Magnet Creek Fault Zone

Walk ~200 m south (back up U.S. Highway 97). On the eastern side of the highway is a photogenic exposure: a small normal fault in bouldery Tsc is occupied by a Teanaway dikelet. The polymict conglomerate contains clasts of pre-Tertiary greenschist facies rocks and tonalite like those in the Ingalls Tectonic Complex. More discussion of the stratigraphy of the Swauk will occur in the relative quiet and safety of Stop 1-4.

Return to the parking area. At the northwestern end of the parking area is an altered Teanaway dike. Because it contains minor sulfides (<1%), is slightly magnetic, greenish, and has a felty texture, it resembles pre-Tertiary greenstone

The northeast-striking Magnet Creek fault passes just north of the parking area. It juxtaposes the pre-Tertiary Ingalls Tectonic Complex against the Swauk Formation to the southeast. Sheep Mountain above and northeast of the highway is mostly pre-Tertiary greenstone. On the opposite (east) side of the highway, black serpentinite is against Sauk strata, which are intruded by variably altered and fractured Teanaway dikes.

The Magnet Creek fault zone is part of the contact of the Swauk with pre-Tertiary rocks that extends for 30 km west of here (Fig. 5), and which is generally considered to be the edge of the Swauk basin (Taylor et al. 1988; Evans and Johnson, 1989). The southeasterly dip of the Swauk strata in and south of the Magnet Creek fault zone indicates that this fault does not mark the depositional edge of the basin. To the west, the contact is discontinuously marked by paleolaterite on ultramafic rocks of the Ingalls Tectonic Complex (Lamey, 1950; Tabor et al. 1982, 2000; Fig. 7). Because the dips of Swauk strata along this portion of the contact are steep to overturned (Tabor et al., 1982), this contact was the floor, not the edge, of the Swauk depository. Because the Ingalls Tectonic Complex is unconformably overlain by the Swauk, the pre-Tertiary clasts in Tsc could not have been derived from this portion of the Ingalls Complex and Mount Stuart batholith. Continue north on U.S. Highway 97 through the Ingalls Tectonic Complex and back into the Swauk Formation at Blushastin (MP 178).

MP 179.9 (U.S. Highway 97)

Turn right (east) on Camas Creek Road and drive past massive white outcrops of the Roslyn (Chumstick) Formation. At 2.8 mi on the right (south) is a quarry in the columnarly jointed Camas Land diabase,

which probably is an invasive sill from once overlying Columbia River Basalt Group. At the end of the pavement at 3.1 mi, bear right. Proceed another 3.5 mi to Stop 1-4 on an outward curve in the road.

Stop 1-4. Conglomerate of Tronsen Creek of the Swauk Formation

Stops 1-4, 1-5, and 1-6 are Swauk, Teanaway, and Roslyn (Chumstick), respectively. The strata at these stops dip to the northeast and strike to the northwest. The Swauk is cut by Teanaway dikes, but the Roslyn (Chumstick) is not.

Stops 1-3 and 1-4 are in the conglomerate of Tronsen Creek of the Swauk Formation (Tsc). However, even in Tsc, most outcrops are fluvial, feldspathic to lithofeldspathic sandstone and black siltstone, with minor or no polymictic conglomerate (Table 1). The most common lithology in most members of the Swauk may be black siltstone, but it weathers recessively.

At Stop 1-4, upward-fining intervals are bounded by minor unconformities at the bases of the conglomeratic beds. Because many of the conglomerates are matrix-supported and seem to have a bimodal distribution of sizes of clasts, some may have originated as mudflows.

The most conspicuous clasts are dioritic to granitic. Taylor et al. (1988) reported that the average composition from six localities is 42.9% plutonic, 26.2% volcanic, 20.5% quartz, 5.4% metamorphic, 2.8% chert, and 2.2% sedimentary. Very rare clasts of coal also occur at Stop 1-4.

The stratigraphic order of the members of the Swauk Formation (Table 1) is displayed in a previously unrecognized syncline that underlies Tronsen Ridge (Fig. 7). The syncline is so obscured by forest and by the lithologic similarities of the Swauk units that Tabor et al. (1982) were unable to map it from trails and roads. The southeastern nose of the syncline is only obvious from the air, from trail-less ridges to the south, and by mapping at 1:24,000.

Major criteria (Table 1) for distinguishing the members of the Swauk are the amount of conglomerate in the unit, the thickness of the conglomeratic beds, and the size of the clasts in the conglomerates. Other criteria are the style and thickness of bedding, the presence of map-scale unconformities, and, especially, stratigraphic position with respect to other members in the syncline.

The thickness of members of the Swauk (Table 1) is measured from cross sections (Fig. 8) and assumes no significant tectonic modification. Because at least some faults are thrusts and because most intraformational folding is impossible to map, the true thicknesses are probably less than those shown on the cross sections. The total apparent thickness of Tsc (including Tscd) is ~2.1 km; Taylor et al. (1988) estimated that it is >0.9 km thick.

Because the members are composed of various amounts of siltstone, sandstone, and conglomerate of variable thickness, they can be indistinguishable in some outcrops. Thus, Tabor et al. (1982) and Taylor et al. (1988) believed that the members abruptly interfinger laterally and vertically. Lithologies probably do interfinger on a scale of meters to a few tens of meters. However, interbedding and interfingering of lithologies is not the same as interfingering of lithostratigraphic units. Map-scale interfingering is undocumented. Because some of the members of the Swauk are bounded by unconformities (Table 1, Fig. 7), they cannot interfinger.

The stratigraphic position of units beyond the syncline on Tronsen Ridge is uncertain. A major assumption is that the conglomeratic unit in the core of the anticline southwest of Swauk Pass (Fig. 7) is the same unit as the more extensive conglomeratic unit in the syncline at Tronsen Ridge. Another assumption is that the sandstone above the conglomerate in the anticline (Tss of Stop 1-1) is the same unit as the sandstone above the conglomeratic unit on Tronsen Ridge. Because Tsc is in the core of the anticline southwest of Swauk Pass and is overlain by Tss in the syncline at Tronsen Ridge, Tsc is inferred to be the oldest of the member of the Swauk Formation. (Table 1). If only one unit of both Tsc and Tss exists, the Tsc/Tss contact north of Swauk Pass is a fault.

Detrital zircons do not resolve the relative ages of Tsc and Tss. J.A. Vance (2002, personal commun.) reported that the youngest peak of fission-track ages from detrital zircons from Tss at Stop 1-1 and from Tsc near Stop 1-9 is ca. 55 Ma.

If only one unit of Tsc and one of Tss exist, the some major generalities about the Swauk Formation follow:

1. It consists of 3 major unconformity-bounded sequences (UBS): Tsc, Tss, and the rocks above Tss at Tronsen Ridge (Tsr, Tsa, and Tsk).

2. These three UBS are a crudely upward-fining succession.

3. Due to its stratigraphic position in the syncline at Tronsen Ridge, Tsc is the oldest UBS. Tabor et al. (1982) and Taylor et al. (1988) believed that Tsc overlies Tss and underlies the strata of Tronsen Ridge.

4. Tsc, the oldest UBS, is limited to the area north of Swauk Pass (does not occur in the Cle-Elum–Easton area) because it is unconformably overlain by Tss (cut out by the sub-Tss unconformity). If Tsc were younger than Tss, remnants of such a thick unit might be expected to be preserved between Tss and unconformably overlying Taneum and Teanaway in the Cle Elum–Easton area.

5. Due to overlapping UBS, the thickness of the Swauk at a given locality may be substantially less than its composite stratigraphic thickness. Tabor et al. (1984) and Taylor et al. (1988) estimated the thickness of the Swauk to be <4800 m and >4800 m, respectively. Where Tsc and the UBS above Tsc at Tronsen Ridge are thickest, Tss is thinnest and locally is zero. The maximum thickness of the Swauk is in the vicinity of Tronsen Ridge, where Tsc is ~2 km thick and the UBS above Tss is <2 km thick (Table 1). Elsewhere, as in the Cle Elum–Easton area, where only Tss occurs, the preserved thickness of the Swauk must be less than the stratigraphic thickness of Tss, which Taylor et al. (1988) estimated to be 2300 m.

Before departing this stop, do not forget to collect a sample of arkosic sandstone.

Turn the vehicles around (toward U.S. Highway 97) and drive 0.8 mi back to the next major outward curve.

Stop 1-5. Basalt-bearing Conglomerate (Teanaway Formation)

This polymictic conglomerate (Ttb), unlike Stops 1-3 and l-4, contains clasts of basalt. Elsewhere, the Ttb contains <1% basaltic clasts, some basalt flows, and black siltstone. Tabor et al. (1982) recognized that the clasts are Teanaway but placed this conglomerate in the Roslyn (Chumstick) Formation (their unit Tcr), which had serious consequences. However, they did note (1982, p. 10) that this conglomerate could be Teanaway.

At the north end of the outcrop, small normal faults cut minor beds of sandstone and siltstone. Is this a model for the regional structure?

Tabor et al. (1982) believed that the contact of this conglomerate with Tsc of Stop 1-4 is a fault in the Leavenworth fault zone. Figure 7 shows that along strike to the southeast, Ttb cuts through Tsc into the underlying Tscd; that is, the contact probably is a disconformity.

Continue 0.6 mi toward U.S. Highway 97.

Stop 1-6. Roslyn (Chumstick) Sandstone

This is typical, white Chumstick sandstone (Tabor et al., 1982). It has a few pebble lags. In the absence of Teanaway dikes, this could be either Swauk or Chumstick. Nonetheless, Tabor et al. (1982) and previous investigators concluded that, compared to the generally tan Swauk, the Chumstick is white and is more massive (has fewer interbeds, including a lesser range of grain sizes) in a given outcrop.

To the northwest and southeast, this sandstone also truncates the basalt-bearing conglomerate and is in contact with Tsc (Fig. 7). Thus, Stops 1-4, 1-5, and 1-6 are a seemingly homoclinal succession of unconformable formations, Swauk, Teanaway, and Roslyn (Chumstick), respectively. Subsequent stops test this hypothesis.

Return to U.S. Highway 97. At the junction with U.S. Highway 97, prepare to turn left (southbound).

MP 179.9 (U.S. Highway 97)

Before the junction with U.S. Highway 97 and after the left turn (southbound) on to U.S. Highway 97, note the granitic boulders. These are erratics derived from the Mount Stuart batholith via the deeply glaciated valley of Ingalls Creek (which is tributary to Peshastin Creek ~3.7 km southwest of here). The erratics along Camas Creek Road are at an altitude of 460 m. On the ridge to the south, erratics attain an altitude of 780 m. South of Ingalls Creek, the erratics extend up Peshastin Creek ~1.7 km beyond Ruby Creek. Thus, the Ingalls Creek glacier was thick enough to flow >3 km *up* the valley of Peshastin Creek. As a result, only some portions of the area of Figure 7 *below* 780 m were glaciated. Presumably, the moraine at 780 m is pre-Wisconsin (Swanson and Porter, 1997).

Complete the left turn and proceed southbound on U.S. Highway 97.

MP 178.7 (U.S. Highway 97)

Turn left (southwest) on Old Blewett Highway and park.

Stop 1-7. Swauk Diamictite (Breccia of Devils Gulch)

This virtually monomictic diamictite (Tscd in Tsc of Table 1) has numerous slickensided joints. The diamictite is composed of rounded clasts of tonalite that could have been derived from the Mount Stuart batholith. The sand-sized to microscopic matrix is composed of mineral grains of the same tonalitic rock. At present, the Mount Stuart batholith crops out ≥4 km to the west, but similar plutons might be under the Chumstick to the east or beneath more distant Columbia River Basalt Group to the southeast. On strike to the southeast, parts of Tscd are 0.1%–100% rounded clasts of ultramafic rocks. Cashman (1974) described some of these as conglomerate. Ultramafic clasts are typically ellipsoidal and impart a crude foliation to Tscd. One outcrop of Tscd west of Peshastin Creek is composed almost entirely of clasts of pelitic phyllite.

Tabor et al. (1982) regarded Tscd as a fanglomerate, but here, and in almost all other outcrops, Tscd is unstratified, unsorted, and virtually monomict. Taylor et al. (1988) correlated Tscd with the breccia of Devils Gulch (however, Tscd is fragmental, not fragmented). Perhaps Tscd is a rock-avalanche deposit or a debris flow.

Tabor et al. (1982) assigned Tscd to the Chumstick Formation, but Figure 7 shows that to the southeast it is bounded above and below by the conglomeratic unit of the Swauk Formation (Tsc). Furthermore, elsewhere, Teanaway dikes occur in Tscd.

Tscd also is in fault contact (in the Leavenworth fault zone) with pre-Tertiary rocks to the southwest (Fig. 8; Tabor et al., 1982). Tabor et al. (1982, p. 8) stated that the "abundance [of diamictites] in the Leavenworth fault zone, especially those on the now downthrown side of the faults, suggests that they flowed off rapidly rising fault-block mountains." However, because this diamictite is in the Swauk, not the Chumstick, it does not indicate that the Chiwaukum graben was a syndepositional basin.

This outcrop has numerous slickensided joints. Figure 9 (by Nicholas W. Hayman) shows that many of the straiae trend NNE and have moderate plunges to either the NE or the SW. Step like surfaces on some joints imply that the hanging wall moved northeast. The preponderance of moderate plunges virtually precludes the normal fault proposed by Tabor et al. (1982). Stop 1-8 provides a possible explanation of what caused the straiae.

Continue north on Old Blewett Highway ~1 mi through outcrops of Tscd on the east and granitic erratics on the west.

Stop 1-8. Camas Creek Thrust

On the valley wall to the west, thick beds of Chumstick sandstone dip to the southwest toward Tscd. Unlike Stop 1-6, this relationship requires a fault between the Roslyn (Chumstick) and the Swauk. This fault barely "Vs" upstream, indicating that it dips steeply southwestward, but to the southeast (Fig. 7), "Vs" imply dips ≤70°. Locally, this fault truncates the axis of the syncline in the Roslyn (Fig. 7). Thus, the fault places a topographically high anticline of Swauk and Teanaway over a syncline in the Roslyn. This northwest-striking segment of the Leavenworth fault zone is here called the Camas Creek thrust. Because the thrust truncates the Roslyn (Chumstick), the Roslyn was not deposited syntectonically in the Chiwaukum graben.

A possible explanation for the trend and plunge of the straiae in Figure 9 is that they record northeastward transport on the Camas Creek thrust.

The Camas Creek thrust and the Eagle Creek thrust near Wenatchee dip to the southwest. The presence of these thrusts suggests that the great thickness of the Roslyn (Chumstick) Formation in the Chiwaukum graben may be tectonic.

Continue northward on Old Blewett Highway to the intersection of U.S. Highway 97 and turn left (southbound). Pass Stop 1-7 at MP 178.7 and continue southbound.

At about MP 179, the highway crosses another northwest-striking fault of the Leavenworth system and enters a gorge in pelitic rocks of the Ingalls Tectonic Complex. For the next 3 km, most of the erratics of the Ingalls Creek moraine have been removed by post-glacial erosion (note the V-shape of the inner gorge), construction of the Old Blewett Highway, and construction and widening of U.S. Highway 97.

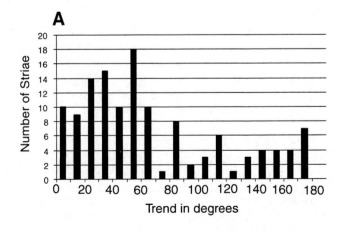 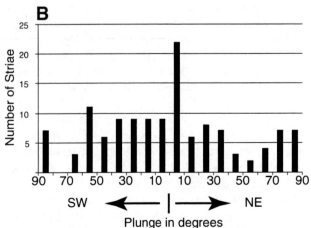

Figure 9. Trend (A) and plunge (B) of straiae on slickensided surfaces in Swauk diamictite (Tscd) at Blushastin.

MP 174.0 (U.S. Highway 97)

This is Blewett! The historical marker relives the glory days of in Culver Gulch, west of the highway. An arrastra is alongside the creek on the opposite (western) side of the highway (at the bottom of decaying wooden stairs). Placer mining began on Peshastin Creek in 1860; lode mining was mostly between 1877 and 1910 (Margolis, 1994; Woodhouse et al., 2002). Gold mineralization is in sulfide-bearing veins ≤5 m wide in serpentinized peridotite of the Ingalls Tectonic Complex (Margolis, 1994). Woodhouse et al. (2002) provided historic photographs and recounted the history of Blewett, Liberty, and other mining districts in the area. Recreational placer mining continues at Blewett and Liberty.

Continue southbound on U.S. Highway 97. Pass Stop 1-3 (Magnet Creek) at MP 172.1.

MP 168.8 (U.S. Highway 97)

Turn left (northeast) on Five Mile Road. Proceed 0.45 mi (past the second major curve).

Stop 1-9. Folded Swauk Lithologies

Walk eastward on the road for 0.2 mi to see typical (and typically poorly exposed) lithologies of Tsc, including float of clasts from conglomerate. The abrupt reversal in dip of the strata (see cross section E–F of Figure 8) indicates a syncline with a very sharp hinge. Such kink-like folds are typical of fold-and-thrust belts.

Continue up Five Mile Road almost 3.0 mi to the fourth major curve.

Stop 1-10. Tronsen Ridge Member of the Swauk Formation and Folds

This is Tsr (Table 1). The three components of Tsr (massive white sandstone, black siltstone, and olive siltstone) have planar bedding. The white sandstone may be due to failures of levees adjacent to flood plains represented by the black and olive siltstones.

In the syncline underlying Tronsen Ridge, Tsr is unconformable on Tsc and Tss, conformably overlain by Tsa, and unconformably overlain by Tsk (Fig. 7).

Because Tsr dips to the northeast, an anticlinal axis exists between here and Stop 1-9, <0.6 km to the southwest (cross section E–F of Fig. 8). The same anticline strikes through the ridge to the northwest, which is underlain by thickly bedded Tsc and a conspicuously rusty-weathering Teanaway dike. The conglomerates on the ridge and in the hogbacks across Tronsen Creek to the southwest are darker weathering than interbedded sandstone because lichen grow on the clasts in conglomerate but not on friable sandstone.

The folds of Stops 1-9 and 1-10 are part of a sinuous set (Fig. 7). Neither Tabor et al. (1982) nor Taylor et al. (1988) recognized the continuity and geometry of these folds. Dikes of the Teanaway swarm generally strike northeasterly (perpendicular to the folds). Sinuous folds also occur in Tss east of Tronsen Ridge (Fig. 7). The sinuous traces imply that these folds are refolded.

A major fault appears to occur near Stop 1-10. Northwest of here the southwestern limb of the syncline under Tronsen Ridge is mostly missing (Fig. 7). The contact with Tsc is interpreted (Figs. 7 and 8) to be a southwest-dipping thrust fault, but the dip

of the fault is unknown. Farther to the northwest at Windmill Point (at a deeper structural level), this fault cuts the pre-Tertiary basement (Fig. 7, cross section C–D of Fig. 8). However, near Stop 1-10 and to the southeast, the fault appears to die out in an anticline in Tsr (is a blind thrust).

Juxtaposition of cross section A–B with C–D reveals the asymmetry of the Blushastin anticline at Tip Top (Fig. 8).

The folds in Figures 7 and 8 may not have retained their original orientations. The overlying tuff at Tip Top (Fig. 8) has some crude sorting that dips ≤60°; these dips may indicate that the tuff is folded and that the folded rocks in the underlying Swauk Formation were refolded. The area of Figure 7 also is near the crest of the Naneum Ridge anticline (Fig. 5), which gently folds the Columbia River Basalt Group. Additionally, the Columbia River Basalt Group on the eastern flank of the Cascade Range has a regional eastward dip of a few degrees.

On a clear day, Stop 1-10 has scenic views. To the WNW is the Mount Stuart Range (≥18 km distant) underlain by the 90–93 Ma Mount Stuart batholith. To the SW (4 km) is flat-topped Diamond Head, which is nearly horizontal Columbia River Basalt Group unconformably overlying folded Swauk near Stop 1-2.

Return downhill to U.S. Highway 97 and turn left (southbound). If necessary, at MP 163.9 at Swauk Pass, turn left to use the U.S. Forest Service toilet at Stop 1-2. Pass Stop 1-1 at MP 163.4.

MP 161.2 (U.S. Highway 97)

Pull over on the very narrow shoulder and climb to the quarry on the right (north) side of the highway.

Stop 1-11. Teanaway Dike Cutting Volcaniclastic Unit in Sandstone of Swauk Pass

An intermediate volcaniclastic rock with felsic clasts occurs in Tss. Tabor et al. (1982) and Taylor et al. (1988) mapped the volcaniclastic rock as the Silver Pass member of the Swauk. However, the Silver Pass at its type locality northeast of Easton is unconformable upon the Swauk (Foster, 1960) and is equivalent to the Taneum Formation. Volcaniclastic rocks such as this are minor interbeds in Tss. This one has a strike length of 8 km (Taylor et al., 1988).

Remnants of a Teanaway dike occur on the wall of the quarry. The back-fill includes serpentinite of the Ingalls Tectonic Complex.

At the eastern end of the quarry, beds of Tss are nearly vertical and somewhat boudinaged. Clearly, the original geometry and extent of the "Swauk basin" is not preserved.

Continue southbound on U.S. Highway 97.

MP 152.5 (U.S. Highway 97)

On the left is the road to Liberty. Gold was discovered along Swauk Creek in 1873. Production from the Liberty district, which includes Williams and Swauk creeks, was from placer mining, "hard-rock" mining, and dredging (Jordan, 1967; Woodhouse et al., 2002). "Boom time" was 1891 to 1901. Some production continues today. The source of the gold is quartz veinlets in the Swauk and Taneum Formations and in Teanaway dikes. Swauk sandstone is hydrothermally altered and weakly mineralized for >9 km along a WNW-trending anticline (Margolis, 1994).

MP 151.9 (U.S. Highway 97)

Notice the "windrows" of gravel from dredging for gold in 1926 or 1940 (Jordan, 1967). "Volunteer" vegetation makes these windrows barely noticeable today.

MP 151.2 (U.S. Highway 97)

Park at the Liberty Café on the western side of the highway. The toilets are not available to non-customers.

Stop 1-12. Teanaway Formation

The major lithology of the Teanaway Formation is basalt (commonly rusty-weathering, black, and vesicular or amygdaloidal). Opposite the northern end of the parking lot, sparsely amygdaloidal basalt occurs below bedded, felsic volcaniclastic rock. Walk southward to see other lithologies and to note the southward dip toward the axis of the Kittitas Valley syncline. Try to distinguish pillows from concentric weathering and dikes from flows.

Continue southbound on U.S. Highway 97.

MP 149.7 (U.S. Highway 97)

At the Junction of U.S. Highway 97 with State Route 970, turn left on U.S. Highway 97 and proceed to the pass at MP 147.1

Stop 1-13. Gently Dipping Thorp Conglomerate

We interrupt our fairly orderly progression up-section to note here that the mostly basaltic conglomerate of the Thorp Formation overlies the Grande Ronde Basalt of the Columbia River Basalt Group (Tabor et al., 1982). By comparison, at Stop 1-17, the Thorp overlies a thinned interval of Ellensburg Formation. Thus, the sub-Thorp contact is a major unconformity. It is the basal unconformity of the High Cascade synthem.

Here, both the Columbia River Basalt Group and the Thorp dip southeastward and project northwestward over red weathering ridges of the Teanaway Formation.

Return downhill to the junction of U.S. Highway 97 and State Route 970, and turn left (west) on State Route 970 toward Cle Elum.

MP 9.0 (State Route 970)

The road ascends the pre-Wisconsin Swauk Prairie moraine (Swanson and Porter, 1997). The ice issued from the Teanaway Valley. The wheat fields and bison pasturage on the crest of the moraine may be the farthest ones west in Washington. Within 2 mi, the highway descends the arcuate interior portion of the moraine.

MP 6.9 (State Route 970)

Turn right (north) on Teanaway Road. Proceed ~0.4 mi and turn left (west) on Old Bridge Road. Cross the Teanaway River and stop just beyond the western end of the bridge.

Stop 1-14. Roslyn Formation

This white arkosic sandstone dips southwesterly toward the axis of the Kittitas Valley syncline. The sandstone is like the Chumstick at Stop 1-6.

Correlation of the Roslyn with the Chumstick is controversial but is made less so by considering their locations with respect to the Naneum Ridge anticline and the true position of the trace of the anticline. The trace of the Naneum Ridge anticline in Figure 5 is taken from Tabor et al. (1982). Tabor et al. (1982) drew the northwestern end of the axis of the anticline through the topographically highest portion of the Columbia River Basalt Group at Mission Ridge. However, their map shows that the basal contact of the Columbia River Basalt Group at Mission Ridge dips to the northeast; so, the axis of the anticline is farther southwest. The most obvious location for the axis is ~2 km to the southwest in the structurally higher (but topographically lower) southeast-trending salient of Swauk. This salient is between the Columbia River Basalt Group on Mission Ridge to the northeast and the Columbia River Basalt Group on Table Mountain and Diamond Head to the southwest. To the northwest, this axis would transit the main body of Tsc north of Swauk Pass and the crystalline rocks of the Mount Stuart area.

The Roslyn and the Chumstick occur above Teanaway basalt on opposite limbs of the Naneum Ridge anticline. Because Tabor et al. (1982) showed the basalt-bearing conglomerate of Stop 1-5 as Chumstick, not Teanaway, this relationship is obscure on their 1:100,000 map. Furthermore, because the Swauk is several unconformities below the Columbia River Basalt Group, the structure within the Swauk does not mimic that of the Columbia River Basalt Group. Nonetheless, this regional anticline is reasonably well illustrated (Fig. 5) by the unconformity-bounded formations of the Challis.

Continue south and take the first paved road to the left. In ≤0.1 mi, at State Route 970, turn right (southbound).

MP 2.6 (State Route 970)

At the intersection of State Route 970 with State Road 10, turn left (eastbound) on State Route 10 toward Ellensburg. Proceed 2.0 mi (3.2 km) eastbound.

MP 90.2 (State Route 10)

Just east of the junction with Taylor Road on the north are volcaniclastic strata, mapped as Ellensburg Formation by Tabor et al. (1982), who showed these as being below magnetostratigraphic unit R2 of the Grande Ronde. The most conspicuous clasts in this "Ellensburg" are hornblende-bearing dacites. As the road proceeds eastward (up-section in Columbia River Basalt Group) for the next 8 mi to Stop 1-15, note the number of stratigraphic positions at which Tabor at al. (1982) mapped Ellensburg and compare these with Figure 7.

MP 92.2 (State Route 10)

The quarry on the right is in R2 of the Grande Ronde (Tabor et al., 1982).

MP 92.6 (State Route 10)

Outcrops of "Ellensburg" between R2 and N2 occur for the next 0.5 mi, but parking is impossible.

MP 96.0 (State Route 10)

Park on the south (right) side of the highway, next to the Yakima River.

Stop 1-15. Anatomy of a Columbia River Basalt Flow

Note the three portions of this Grande Ronde flow: a basal pillow-palagonite unit, a middle portion of columns (collonade), and an upper portion of closely spaced fractures ("brickbat" texture), known as the entablature. Radial columns within the collonade may mark a former lava tube but probably were formed by non-planar cooling of the flow (without a lava tube). Entablatures are thought to form by catastrophic cooling beneath lakes that formed on top of still hot flows by the fluvial systems originally disrupted by the flows. Many Columbia River Basalt Group flows do not have a basal pillow-palagonite, but most have a collonade and an entablature. Note that here the Grande Ronde has a shallow dip to the southeast.

Continue eastward on State Route 10.

MP 98.3 (State Route 10)
_____

Pull off onto the wide shoulder on the south (right) side of the highway.

Stop 1-16. Ellensburg Formation

This is a photogenic exposure of the Ellensburg Formation. The massive (unbedded and poorly sorted) volcaniclastic material in the lower part of the cliff is underlain and overlain by bedded units. Note the huge size of the clast on the upper contact of the massive unit. Cross-bedding in the bedded units implies a western source (the site of the present Cascade Range). Many clasts are hornblende dacite. The Ellensburg is noted for its abundance of pumice. Elsewhere, the upper part of the Ellensburg contains minor clasts of the Columbia River Basalt Group (Smith, 1988).

Note the gentle dip of the Ellensburg to the southeast.

The orange-weathering Thorp Formation in the cliffs above the Ellensburg is at the level of the road at Stop 1-17.

The stratigraphic definition of the Ellensburg is a problem. Here it is on the Grande Ronde, but in the last 13 km, it is below and in the Grande Ronde. A similar interbed up to 62 m thick, the Selah unit, occurs in the Saddle Mountains Basalt (Smith, 1988). In the type area 30 km to the south, Ellensburg is above the stratigraphically highest Saddle Mountains Basalt and is 280–350 m thick (Smith 1988). Here at Stop 1-16, Ellensburg is >62 m thick on the Grande Ronde. All of these volcaniclastic units are considered to be "Ellensburg" (Waitt, 1979; Tabor et al., 1982; Smith, 1988, fig. 2). Consider the unconformities in Figure 6 and solve the "Ellensburg problem(s)."

Continue east on State Route 10.

MP 100.8 (State Route 10)
_____

Pull off onto the wide shoulder on the south side of the highway.

Stop 1-17. Thorp Formation

Most of the Thorp weathers orange. Whiter units in the cliff above are tuffs that have fission-track dates of 3.8 and 4.4 Ma and a K-Ar date of 4.5 Ma (Tabor et al., 1982). These dates should be rechecked. The Thorp is younger than, or correlates with, the upper part of the Ringold of Figure 7 (Waitt, 1979).

Four major types of clasts occur in the Thorp. In the talus at the side of the highway, these are: (1) hornblende dacite and other felsic volcanic rocks; (2) granite (presumably Oligocene to Miocene); (3) pre-Tertiary metamorphic rocks (mostly greenschist facies); and (4) basalt. All of these lithologies currently crop out in the Cascade Range to the west. Clasts of sedimentary rocks (such as Challis arkose) are exceedingly rare. The basalt is both Teanaway and the Columbia River Basalt Group (Waitt, 1979). The presence of clasts of Columbia River Basalt Group provides a maximum age for initial uplift of the Cascade Range.

Continue east on State Route 10.

MP 105.3 (State Route 10)
_____

Turn right (southeast) at the 4-way stop and follow signs for U.S. Highway 97 for 1.2 mi to the Interstate 90 intersection west of Ellensburg. End of Day One.

Day Two: Ellensburg to Easton

From Ellensburg, proceed eastbound on Interstate 90.

MP 110.1 (Interstate 90)
_____

Exit Interstate 90 onto Interstate 82 southbound toward Yakima.

MP 7.4 (Interstate 82)
_____

Exit to viewpoint on Manastash Ridge (and anticline) near Vanderbilt Gap. Walk to the outcrop next to the second (upper) parking lot.

Stop 2-1. Walpapi Stratigraphy and Structure

The Manastash Ridge anticline is the northeasternmost of a set of closely spaced folds in the Columbia River Basalt Group known as the Yakima fold belt. In addition, Stop 2-1 is a photogenic example of the Squaw Creek sandstone and bounding Columbia River Basalt Group basalts. The Squaw Creek ranges in various places from a few centimeters to 16 m thick (Smith, 1988, fig. 4). Beneath the Squaw Creek is the Frenchman Springs flow. Above it is the Roza flow with a well-developed pillow-palagonite zone at its base. Both basalts are units of the Wanapum Basalt Formation of Figure 6.

The composition of the Squaw Creek is remarkably similar to the Challis arkoses. The cross-bedding implies a source to the east, the opposite of that of the Ellensburg at Stop 1-16. The nearest possible source for the Squaw Creek must be beyond the present (and former) geographic extent of the Columbia River Basalt Group.

Similar sandstones range from the Vantage sandstone (at the base of the Wanapum and unconformable on the Grande Ronde) to the Rattlesnake in the Saddle Mountains Basalt, to the Snipes Mountain conglomerate above the Saddle Mountains (Smith, 1988). Some of these units also contain dacitic clasts.

The Vantage has silicified ginkos and other trees, especially at Ginko State Park ~37 km east of here. The Vantage is ≤35 m thick in eastern Washington (Smith, 1988), but in southwestern Washington, the marine arkose in the same stratigraphic position is at least a kilometer thick. This marine arkose is called the Astoria Formation, but so too is equally thick marine arkose beneath the Grande Ronde in southwestern Washington and adjacent Oregon. Solve the Astoria(s) problem. Contemplate the original extent and setting of the Walpapi.

All of the predominantly siliciclastic interbeds, like the Squaw Creek, and all of the predominantly volcaniclastic units, are lumped as the Ellensburg Formation (Smith, 1988, fig 2; Reidel, 1994). If you thought you had solved the Ellensburg problem at Stop 1-16, try again (with the aid of the unconformities in Fig. 6).

The view, clouds permitting, is most instructive. Lookout Mountain is the low, wooded plateau in the Kittitas Valley 30 km to the northwest; there along State Route 10 and farther east at Stops 1-15 and 1-16, Walpapi rocks dip gently southeastward. The distant slope of the Kittitas Valley to the northeast is underlain by Grande Ronde, which dips several degrees to the southwest toward the units here at Stop 2-1, which dip more steeply and dip to the northeast. Thus, the Kittitas Valley syncline is asymmetric and plunges southeast.

Mount Stuart, which is the highest peak in the central part of the Cascade Range, may be visible to the northwest; it is underlain by pre-Tertiary rocks. The Columbia River Basalt Group on the skyline northeast of Kittitas Valley has a regional southeastward dip, which, if extended northwestward, projects above Mount Stuart. That is, regionally, the Columbia River Basalt Group dips to the east off the Cascade Range as it does at Stop 1-13. Thus, the ages of uplift of the Cascade Range and of the southeast plunge of the Kittitas Valley syncline are post–Columbia River Basalt Group. The eastward dip of the Thorp at Stop 1-13 shows that much of the uplift is <4 Ma.

The north-trending Cascade Range anticline and the Kittitas Valley syncline represent two sets of intersecting folds that are a post-Walpapi structural "egg-crate" (Cheney and Sherrod, 1999). The plunge of the Kittitas Valley syncline shows that the two sets of folds did not form simultaneously. Interestingly, the Cascade Range anticline plunges southward at the Washington-Oregon border where it is crossed by the west-trending The Dalles-Umatilla syncline (Wheeler and Mallory, 1979; Cheney and Sherrod, 1999); thus, some west-trending folds are younger than the Cascade Range anticline. The Columbia River crosses the Cascade Range by following The Dalles–Umatilla syncline.

Presumably, the north-striking set of folds is related to the Cascadia subduction zone and is an edge-effect on the continent. The west-trending set may be geologic evidence for the suggestion of Wells et al. (1998) that California is drifting northward at millimeters/year and rotating Oregon "into the soft underbelly of Washington."

Return to Interstate 82 and continue southeastward toward Yakima. Note the variable dips of the Squaw Creek and of basalt in the first few road-cuts in Vanderbilt Gap. The tower on the hill 8 km to the south is on Umtanum Ridge, the anticline at Stop 2-2.

MP 11.4 (Interstate 82)

Take exit 11; pass under Interstate 82, and take the westbound lanes of Interstate 82 back toward Ellensburg.

MP 3.0 (Interstate 82)

Take the exit for Thrall and proceed westward to State Route 821. Turn left (southward) on State Route 821 and proceed down scenic Yakima Canyon.

Between MP 19 and 21 are several debris flows of basaltic blocks that buried the highway and built fans into the Yakima River on 3 July 1998. A cloudburst dumped 7.6 cm of rain on ~12 km^2 in less than an hour. The largest resultant debris flow was 37,000 m^3 and 4.5 m deep on State Route 821 (Kaatz, 2001).

MP 12.4 (State Route 821)

Stop at the entrance of Lmuma Creek (formerly the Squaw Creek Recreation Site) to enjoy the view. Then drive downhill to the parking lot and toilets.

Stop 2-2. Yakima Fold-and-Thrust Belt

Look southeastward down the canyon to Umtanum Ridge. The topographic relief is ~500 m. Enjoy (possibly even sketch or photograph) the northeastward-verging, asymmetrical Umtanum Ridge anticline in the Columbia River Basalt Group in the Yakima fold belt. Infer the thrust at the base of the nearly vertical northeastern limb. Price and Watkinson (1989, fig. 16) drew such a cross section of this fold 45 km to the southeast; there, the fold involves Grande Ronde and Wanapum. Reidel et al. (1994, fig. 7) showed that the post–Columbia River Basalt Group Ringold Formation is involved in the Yakima folds.

110 km to the northwest, the active Seattle fault is increasingly recognized as a fold (Newport Hills anticline) and thrust (Blakely et al., 2002; Sherrod et al., this guidebook). These are on strike with, and the same age as, the Yakima fold-and-thrust belt and the Seattle-Wentachee-Kittitas fold-and-thrust belt. In the southern Cascade Range between Easton and the Puget Lowland, folds in Kittitas rocks (Tabor et al., 2000) may be other examples of the Seattle-Wentachee-Kittitas fold-and-thrust belt. The Yakima folds and the Seattle-Wentachee-Kittitas fold-and-thrust belt are representatives of the northwest-trending folds of the egg-crate.

Across the valley, volcanic ashes are exposed in the terraces. Clearly, multiple periods of erosion and deposition have occurred in the valley in response to tectonism and/or climate change.

Return to Ellensburg via State Route 821 and proceed westbound on Interstate 90 toward Seattle.

MP 89.5 (Interstate 90)

Exit into the aptly named The Indian John Hill Rest Area.

Stop 2-3. Scenic View from Indian John Hill

The rest area has a splendid view of Mount Stuart to the northwest. To the west is the forested dip slope of the Roslyn Formation on the northern limb of the Kittitas Valley syncline. Westward, Interstate 90 follows the axial trace of the Kittitas

Valley syncline (Tabor et al., 1982, Figure 3) and descends the western erosional edge of the Columbia River Basalt Group to Challis rocks in the Yakima River valley.

Continue westbound on Interstate 90. Exit 84 leads to Stops 2-4 to 2-7, but if weather conditions force cancellation of Stops 2-4 to 2-7, proceed to Stop 2-8 via exit 78.

MP 84.4 (Interstate 90)

Take exit 84 to Cle Elum. Cle Elum and the neighboring towns of Roslyn (a.k.a. Cicely, Alaska, in the TV series "Northern Exposure") and Ronald mined coal (mostly underground) from the Roslyn Formation from 1882 to 1963. Production was 64 million tons (Walker, 1980). Mining ceased when railroad locomotives evolved to diesel.

Exit 84 also ultimately leads to spectacular Peoh Point, which should be visited on a clear day when neither snow nor mud is on the dirt roads. Peoh Point is the high point (1225 m) on the southern side of the valley. Follow the signs to South Cle Elum. To get there, go into Cle Elum, turn west, cross under Interstate 90 and over the Yakima River. The main road then jogs through South Cle Elum.

About a mile southwest of South Cle Elum and south of the road is an overgrown dump of an abandoned coal mine in the uppermost part of the Roslyn Formation. Strata in the mine reportedly dip ~70° southward (Walker, 1980; Tabor et al., 1982). These rocks are in the footwall of the Easton Ridge thrust.

Approximately 1.7 mi southwest of South Cle Elum, the road crosses the covered trace of the Easton Ridge thrust (Walker, 1980, map 1A) and curves southward into a canyon. East of the road, the hanging wall of the thrust is marked by inconspicuous outcrops of felsic volcanic rocks (Taneum Formation) and by northward dipping arkose of the Swauk Formation, mapped as Manastash Formation (Tabor et al., 1982). Juxtaposition of the Taneum in the hanging wall against uppermost Roslyn indicates that displacement on the Easton Ridge thrust is greater than the thickness of the Roslyn Formation, which is ~3 km.

About 0.8 mi beyond the curve, turn left (south) on U.S. Forest Service Road 3350 (dirt). Follow U.S. Forest Service Road 3350. Minor outcrops are black Darrington Phyllite. At ~5 mi, notice the extensive hummocky topography. At ~6 mi, turn right and proceed <0.2 mi on U.S. Forest Service Road 211.

Stop 2-4. Grande Ronde Formation

Here, the Grande Ronde is at an altitude of 1250 m. Notice fragments of tuffaceous material intermixed with the basalt. Nearby, the tuffaceous material was mapped as Ellensburg Formation (Tabor et al., 1982), but it could be Kittitas. The hummocky topography is part of a large landslide (Tabor et al., 1982); however, we shall consider the basalt to be virtually in place.

Return to U.S. Forest Service Road 3550 and continue eastward 0.5 mi. Turn left onto U.S. Forest Service Road 115 and stop.

Stop 2-5. Darrington Phyllite

The pre-Tertiary Darrington Phyllite is in the core in the Ainsley Canyon anticline. The phyllite characteristically has numerous, predominantly concordant quartz veinlets. Stops 2-4 and 2-5 demonstrate that at least locally the Walpapi unconformably overlies crystalline basement.

Continue 0.5 mi northeastward on U.S. Forest Service Road 15.

Stop 2-6. Swauk (Manastash) Formation

Here, an underwhelming outcrop of northward-dipping Swauk (Manastash) demonstrates that the Challis also nonconformably overlies Darrington Phyllilite. An interesting exercise is to draw these two unconformities on a single cross section.

Two synthemal basal unconformities may be wonderful, but the best is yet to come. Continue about a mile on U.S. Forest Service Road 115 (but NOT if the road is covered with mud or snow) to the locked gate and parking lot for Peoh Point. Peoh Point has a stunning view; take Figure 5, Tabor et al. (1982), and a camera. The <0.5 mi road and Peoh Point are in Taneum-equivalent, hornblende-bearing felsite.

Stop 2-7. Scenic View from Peoh Point

The view extends to Alaska: Cicely and its former coal-wash piles from the Roslyn Formation are only 8 km to the northwest.

The structural relief of the Columbia River Basalt Group in the Kittitas Valley syncline and the Naneum Ridge anticline on the distant skyline to the northeast is 1200 m.

The long, descending ridge to the west that trends toward Peoh Point is Easton Ridge. It is underlain by northeast-dipping Teanaway basalt on the southwestern limb of the Kittitas Valley syncline in the hanging wall of the Easton Ridge thrust. The rusty-weathering rib ~2 km east of the Peoh Point is the same Teanaway basalt that passes beneath the Columbia River Basalt Group. At the base of the cliff below Peoh Point, the Easton Ridge thrust virtually eliminates the Teanaway. The Easton Ridge thrust is an analog of the Seattle fault and of the Camas Creek thrust of Stop 1-8; all are in the Seattle-Wentachee-Kittitas fold-and-thrust belt.

Rusty-weathering hills ~15 km to the northeast are underlain by Teanaway basalt on the northeastern limb of the Kittitas Valley syncline.

Mount Stuart dominates the skyline 23 km to the north.

Moraines dam Cle Elum Lake 15 km to the northwest.

The range of hills that crosses the Yakima Valley ~25 km to the northwest marks the trace of the Straight Creek fault at Stop 2-12.

Retrace the route to Interstate 90 and continue westbound toward Seattle.

If the trek to Peoh Point was cancelled, use Stop 2-8 to inspect the Darrington Phyllite by taking exit 78. Go left (south) on Golf Course Road for 0.8 mi and turn left. Proceed 0.9 mi to the junction of West Side Road and Fawler Creek Road.

Stop 2-8. Darrington Phyllite

This pelitic phyllite has folded and disrupted quartz veinlets. Please do not hammer them. The protolith probably was Jurassic and the metamorphism Cretaceous (Tabor et al., 2000; Tabor, 1994).

Return to Interstate 90 and continue westward past Easton. Take exit 70, turn left (south) over Interstate 90, and follow the frontage road (southeastward) toward Easton <0.5 mi to the entrance of Easton State Park.

Stop 2-9. Sub-Teanaway Unconformity

Northeastward across Interstate 90 and the Yakima Valley is a ridge with >600 m of relief. The brown-weathering, bedded unit to the right is Teanaway basalt, which dips moderately northeastward. At the base of the ridge and up valley (to the left) is a lighter colored and more steeply dipping unit, the Taneum Formation (here, a.k.a. Silver Pass volcanic rocks). Rarely are intra-Challis unconformities so obvious.

Because Easton is the beginning of the rain shadow on the eastern side of the Cascade Range, this park is popular with many residents of soggy Pugetopolis. The lake is the intake for the irrigation canal(s) of the Kittitas Valley.

Enter the park and turn right at the first stop sign. Proceed along the northern shore of Lake Easton for l mi and turn into the picnic area (with toilets). Walk to the outcrop 100 m west of the swimming beach.

Stop 2-10. Shuksan Greenschist and Straight Creek Fault

The Shuksan is a handsome green phyllite. On the hill 100 m to the northwest, quartz veinlets are more numerous than here. The foliation strikes to the northwest and is nearly vertical. The Shukan and Darrington are in the core of the Ainsley Canyon anticline. Teanaway basalt on the northeastern side of the valley dips northeastward (Stop 2-9). The railroad grade on the southern shore of Lake Easton is in arkosic Swauk sandstone on the southwestern limb of the anticline.

Another reason for this stop is to contemplate the location of the Straight Creek fault (SC in Fig. 3). The Darrington and the Shuksan are distinctive and widespread units in the northern Cascade Range west of the Straight Creek fault. Dextral displacement of 90–190 km on Straight Creek fault offsets these units to the Easton–Peoh Point area (Tabor, 1994; Tabor et al., 2000).

Tabor (1994, fig. 6) and Tabor et al. (2000) mapped the trace of Straight Creek fault in the Easton area as a southeast-striking discontinuity that separates Naches Formation on the southwest from the Swauk, Taneum, Teanaway, and Roslyn on the northeast. More detailed mapping (Cheney, 1999) demonstrated that much of the Naches mapped by previous investigators was a "wastebasket" composed of Naches plus unrecognized portions of all other Challis formations, which occur on both sides of the southeastern discontinuity (Fig. 5). Thus, the major criterion for locating the Straight Creek fault is invalid.

Furthermore, Tabor et al. (2000) placed the Shuksan Greenschist at Easton west of the Straight Creek fault and extended the fault southeastward on the northeastern side of the Yakima Valley along a discontinuity that eliminates the Swauk Formation. This southeastward curving discontinuity is caused by the sub-Taneum and the sub-Teanaway unconformities on the northwestern limb of the Ainsley Canyon anticline (Fig. 5; Cheney, 1999).

The greatest structural discontinuity is a south-striking one 2.7 km west of Easton. Challis formations do occur on both sides of it. This discontinuity is inferred to be the Straight Creek fault. To the north, the Straight Creek fault truncates the Ainsley Canyon anticline (Fig. 5). To the south, the Straight

Creek fault is unconformably overlain by the Kittitas (Fig. 5; Cheney, 1999).

Cheney (1999) showed that the Straight Creek fault cuts the Taneum Lake fault (TLF in Fig. 10B). Tabor (1994) inferred that Taneum Lake fault and the Darrington–Devils Mountain fault zone probably are the same dextral fault, with 112 km of displacement. If so, Taneum Lake fault and the Darrington–Devils Mountain fault zone, both of which cut Naches or equivalent rocks, are offset ~55 km by the Straight Creek fault (Fig. 10B). In contrast, pre-Tertiary units are offset 90–190 km (Tabor, 1994). This difference implies that Straight Creek fault had pre-Tertiary and post-Naches dextral movements (Fig. 10 B–D). The portion of Straight Creek fault dextrally offset by the Darrington–Devils Mountain fault zone might be one of the north-trending faults in the Puget Lowland (Figs. 3 and 10C).

Retrace the route back to the westbound entrance ramp of Interstate 90. Turn left on the frontage road just past the ramp. At 0.7 mi past this turn, note the old concrete bridge on the shore of the Lake Easton south of Interstate 90. Green phyllite at the eastern end of this bridge is the westernmost known outcrop of Shuksan Greenschist; so the Straight Creek fault is somewhere to the west of here. In another 0.4 mi, turn left, pass under both lanes of Interstate 90, proceed to the turn-around at the end of the pavement, and return to the northern underpass of the westbound lanes of Interstate 90.

Stop 2-11. Arkosic Naches or Swauk?

This outcrop of arkosic sandstone and black siltstone was mapped by Tabor et al. (2000) as Naches. Compare the arkose with samples of the Swauk and Roslyn. Look for leaf fossils. Cheney (1999) considered this as Swauk on the southwestern limb of the Ainsley Canyon anticline. If this is Swauk, it is west of the trace of the Straight Creek fault mapped by Tabor et al. (2000).

Return to the ramp at exit 70 of Interstate 90 and proceed westbound on Interstate 90. Once past MP 69, navigate into the left lane while going uphill and prepare to pull off into a turn-around between the westbound and eastbound lanes at MP 67.4.

MP 67.4 (Interstate 90)

Park in the turnaround. Very carefully dash across the eastbound lanes of Interstate 90.

Stop 2-12. Straight Creek Fault

These rocks are limonitic, aphanitic, white-weathering felsite with disseminated pyrite. The eastern end of the exposure has centimeter-scale felsic clasts in a felsic matrix (which is characteristic of Naches felsite), dark volcaniclastic rocks, and black siltstone. Similar felsite occurs on the hillsides north and south of the freeway.

Carefully dash back to the north side of the eastbound lanes and walk downhill ~0.4 mi past the covered interval to the long road-cut. The rocks strike to the northwest and dip steeply to the southwest. The upper unit, at the western end, is black felsite, which Cheney (1999) erroneously mapped as basalt. The lower

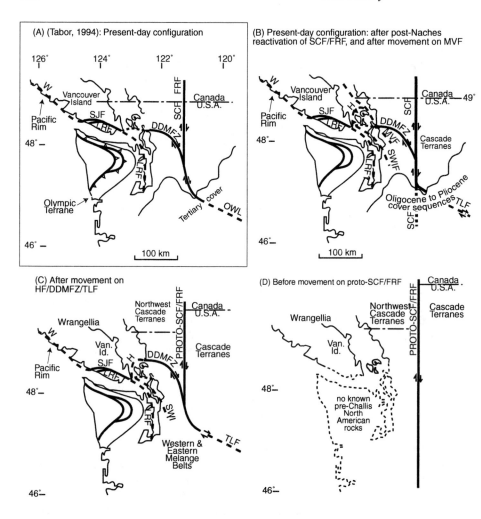

Figure 10. Alternative interpretations and histories of the Straight Creek fault. See the discussion of Stop 2-10 for additional explanation. In Canada, the northern continuation of Straight Creek fault is FRF. Abbreviations for faults: DDMFZ—Darrington–Devils Mountain fault zone; OWL—Olympic-Wallowa lineament; SJF—San Juan fault; FRF—Fraser River fault; SWIF—Southern Whidbey Island fault; HF—Haro fault; LRF—Leech River fault; TLF—Taneum Lake fault; MVF—Mount Vernon fault; WCF—West Coast fault.

unit is a hornblende-bearing, felsic, fragmental unit in which welding and quartz grains are most visible at the eastern end of the road-cut. These units were mapped as Naches (Tabor et al., 2000) but are Taneum. Note that they project southwest of the strata of Stop 2-11. The two units (Taneum felsite and Swauk arkosic rocks) strike into the Naches felsite ≤0.6 km to the northwest. The intervening covered interval is part of the largest structural discontinuity in the area and is inferred to be trace of Straight Creek fault (Cheney, 1999).

Return to the vehicles and very carefully re-enter the westbound lanes of Interstate 90.

After doing so, Seattle is ~67 minutes to the west.

End of Day Two and of the field trip.

ACKNOWLEDGMENTS

I thank B.L. Sherrod and N.W. Hayman for numerous insightful discussions. T. Bush, B. Byers, I.R. Cheney, N.I. Chutas, R.B. Frost, M.W. Hawkes, S. Moon, S. Petrisor, C. Rhone, B.L. Sherrod, and D. Trippett kindly assisted in the field work near Swauk Pass. N.W. Hayman rescued the figures from e-oblivion and revised many of them. T.W. Swanson kindly reviewed the manuscript.

REFERENCES CITED

Blakely, R.J., Wells, R.E., Weaver, C.S., and Johnson, S.Y., 2002, Location, structure, and seismicity of the Seattle fault zone, Washington: Evidence from aeromagnetic anomalies, geologic mapping, and seismic-reflection data: Geological Society of America Bulletin, v. 114, p. 169–177.

Breitsprecher, K., Thorkelson, D.J., Groome, W.G., and Dostal, J., 2003, Geochemical confirmation of the Kula-Farallon slab window beneath the Pacific Northwest in Eocene time: Geology, v. 31, p. 351–354.

Cashman, S.M., 1974, Geology of the Peshastin Creek area, Washington [M.S. thesis]: University of Washington, 29 p., 1 plate.

Cheney, E.S., 1994, Cenozoic unconformity-bounded sequences of central and eastern Washington: Washington Division of Geology and Earth Resources Bulletin 80, p. 115–139.

Cheney, E.S., 1997, What is the age and extent of the Cascade magmatic arc?: Washington Geology, v. 25, no. 2, p. 28–32.

Cheney, E.S., 1999, Geologic map of the Easton area, Kittitas County, Washington: Washington Division of Geology and Earth Resources Open File Report 99-4, scale 1:31,680, with 11 p. text.

Cheney, E.S., 2002, The SWIFT sector of OWL: Geological Society of America Abstracts with Programs, v. 34, no. 6, p. 27.

Cheney, E.S., and Sherrod, B.L., 1999, The egg-crate of the Pacific Northwest: Geological Society of America Abstracts with Programs, v. 31, no. 5, p. A44.

Dragovich, J.D., Logan, R.L., Schasse, H.W., Walsh, T.J., Lingley, W.S., Jr., Norman, D.K., Gerstel, W.J., Lapen, T.J., Schuster, J.E., and Meyers, K.D., 2002, Geologic map of Washington—Northwest Quadrant: Washington Division of Geology and Earth Resources Geologic Map GM-50, scale 1:250,000, 3 sheets, 72 p. text.

Evans, J.E., and Johnson, S.Y., 1989, Paleogene strike-slip basins of central Washington: Swauk Formation and Chumstick Formation: Washington Division of Geology and Earth Resources Information Circular 86, p. 215–237.

Foster, R.J., 1960, Tertiary geology of a portion of the central Cascade Mountains, Washington: Geological Society of America Bulletin, v. 71, p. 99–126.

Gresens, R.L., 1983, Geology of the Wenatchee and Monitor Quadrangles, Chelan and Douglas Counties, Washington: Washington Division of Geology and Earth Resources Bulletin 75, 75 p.

Hammond, P.E., 1980, Reconnaissance geologic map of southern Washington Cascade Range: Portland State University Department of Earth Sciences, 31 p., 2 plates, scale 1:125,000.

Hanneman, D.L, Cheney, E.S., and Wideman, C.J., 2002, Cenozoic sequence stratigraphy of northwestern USA, *in* Raynolds, R.G., and Flores, R.M., eds., Cenozoic systems of the Rocky Mountain Region: Denver, Colorado, Society of Economic Paleontologists and Mineralogists, Rocky Mountain Division, p. 1–21.

Hildreth, W., Fierstein, J., and Lanphere, M., 2003, Eruptive history and geochronology of the Mount Baker volcanic field, Washington: Geological Society of America Bulletin, v. 115, p. 729–764.

Jordan, J., 1967, You're at Liberty here, mines and miners of the Swauk: Yakima, Washington, Franklin Press, 103 p.

Kaatz, M.R., 2001, 1998 debris flows near the Yakima River, Kittitas County, Washington—Some geomorphic implications: Washington Geology, v. 29, no. 3/4, p. 3–10.

Lamey, C.A., 1950, The Blewett iron-nickel deposit, Chelan County, Washington: U.S. Geological Survey Bulletin 969D, p. 87–101.

Margolis, J., 1994, Epithermal gold mineralization, Wenatchee and Liberty Districts, Central Washington: Society of Economic Geologists Guidebook Series, v. 20, p. 31–34.

Patton, T.C., and Cheney, E.S., 1971, L-D mine, Wenatchee, Washington: New structural interpretation and its utilization in exploration: Transactions of the Society of Mining Engineers, v. 250, p. 6–11.

Price, E.H., and Watkinson, A.J., 1989, Structural geology and strain distribution within eastern Umtanum fold ridge, south-central Washington, *in* Reidel, S.P., and Hooper, P.R., eds., Volcanism and tectonism in the Columbia River Flood-Basalt Province: Boulder, Colorado, Geological Society of America Special Paper 239, p. 265–281.

Prothero, D.R., 2001, Chronostratigraphic calibrations of the Pacific Coast Cenozoic: A summary, *in* Prothero, D.R., ed., Pacific Coast Cenozoic: Los Angeles, California, Society of Economic Paleontologists and Mineralogists, Pacific Section, Book 91, p. 377–394.

Raisz, E., 1945, The Olympic-Wallowa lineament: American Journal of Science, v. 243-A, p. 479–485.

Reidel, S.P., Campbell, N.P., Fecht, K.R., and Lindsey, K.A., 1994, Late Cenozoic structure and stratigraphy of south-central Washington: Washington Division of Geology and Earth Resources Bulletin 80, p. 159–180.

Smith, G.A., 1988, Neogene synvolcanic and syntectonic sedimentation in central Washington: Geological Society of America Bulletin, v. 100, p. 1479–1492.

Smithson, D.M., Rowins, S.M., Mortensen, J.K., and Newport, G.R., 2003, Late Eocene felsic magmatism and reduced porphyry copper-gold mineralization at the North Fork deposit, Central Cascade Mountain Range, Washington: Geological Association of Canada–Mineralogical Associa-tion of Canada–Society of Economic Geologists Joint Annual Meeting Abstracts, v. 28, PDF abstract no. 169.

Swanson, T.W., and Porter, S.C., 1997, Cosmogenic isotope ages of moraines in the southeastern North Cascade Range: Guidebook for Pacific Northwest Friends of the Pleistocene Field Excursion, September 19–21, 1997, 18 p.

Tabor, R.W., 1994, Late Mesozoic and possible early Tertiary accretion in western Washington State: The Helena-Haystack mélange and Darrington–Devils Mountain fault zone: Geological Society of America Bulletin, v. 106, p. 217–232.

Tabor, R.W., Waitt, R.B., Jr., Frizzell, V.A., Jr., Swanson, D.A., Byerly, G.R., and Bentley, R.D., 1982, Geologic map of the Wenatchee 1:100,000 Quadrangle, Central Washington: U.S. Geological Survey Miscellaneous Investigations Series Map I-1311, 26 p, 1 plate, scale 1:100,000.

Tabor, R.W., Frizzell, V.A., Jr., Vance, J.A., and Naeser, C.W., 1984, Ages and stratigraphy of lower and middle Tertiary sedimentary and volcanic rocks of the central Cascades, Washington—Application to the tectonic history of the Straight Creek fault: Geological Society of America Bulletin, v. 95, p. 26–44.

Tabor, R.W., Frizzell, V.A., Jr., Whetten, J.T., Waitt, R.B., Swanson, D.A., Byerly, G.R., Booth, D.B., Hetherington, M.J., and Zartman, R.E., 1987, Geologic map of the Chelan 30-Minute by 60-Minute Quadrangle, Washington: U.S. Geological Survey Miscellaneous Investigations Series Map I-1661, 1 sheet, scale 1:100,000, with 29 p. text.

Tabor, R.W., Frizzell, V.A., Jr., Booth, D.B., and Waitt, R.B., 2000, Geologic map of the Snoqualmie Pass 30′ × 60′ Quadrangle, Washington: U.S. Geological Survey Miscellaneous Investigations Series Map I-2538, scale 1:100,000, with 48 p. text.

Taylor, S.B., Johnson, S.Y., Fraser, G.T., and Roberts, J.W., 1988, Sedimentation and tectonics of the lower and middle Eocene Swauk Formation in eastern Swauk Basin, central Cascades, central Washington: Canadian Journal of Earth Sciences, v. 25, p. 1020–1036.

Waitt, R.B., Jr., 1979, Late Cenozoic deposits, landforms, stratigraphy, and tectonism in Kittitas Valley, Washington: U.S. Geological Survey Professional Paper 1127, 18 p.

Walker, C.W., 1980, Geology and energy resources of the Roslyn–Cle Elum area, Kittitas County, Washington: Washington Division of Geology and Earth Resources Open File Report 80-1, scale 1:24,000, with 59 p. text.

Wells, R.E., Simpson, R.W., Bentley, R.D., Beeson, M.H., Mangan, M.T., and Wright, T.L., 1989, Correlation of Miocene flows of the Columbia River Basalt Group from the central Columbia River Plateau to the coast of Oregon and Washington, *in* Reidel, S.P., and Hooper, P.R., eds., Volcanism and tectonism in the Columbia River Flood-Basalt Province: Boulder, Colorado, Geological Society of America Special Paper 239, p. 113–129.

Wells, R.E., Weaver, C.S., and Blakely, R.J., 1998, Fore-arc migration in Cascadia and its neotectonic significance: Geology, v. 26, p. 759–762.

Wheeler, H.E., and Mallory, V.S., 1970, Oregon Cascades in relation to Cenozoic stratigraphy, *in* Gilmore, E.H., and Stradling, D.F., eds., Proceedings of the Second Columbia River Basalt Symposium: Eastern Washington State College Press, p. 97–124.

Woodhouse, P., Jacobson, D., Cady, G., and Pisoni, V., 2002, Discovering Washington's Historic Mines, Volume 2: The East Central Cascade Mountains and the Wenatchee Mountains: Arlington, Washington, Oso Publishing Co., 336 p.

Yeats, R.S., Kulm, L.D, Goldfinger, C., and McNeill, L.C., 1998, Stonewall anticline: An active fold on the Oregon continental shelf: Geological Society of America Bulletin, v. 110, p. 572–587.

Geological Society of America
Field Guide 4
2003

Biogeochemical processes at ancient methane seeps: The Bear River site in southwestern Washington

James L. Goedert

Burke Museum of Natural History and Culture, University of Washington, Seattle, Washington 98195, USA

Steven R. Benham

Department of Geosciences, Pacific Lutheran University, Tacoma, Washington 98447, USA

Keywords: hydrocarbon seeps, limestone, Tertiary, paleontology, Washington.

INTRODUCTION

The main objective of this one-day trip will be to visit the late Eocene Bear River cold-methane-seep deposit in Pacific County, southwestern Washington, providing an opportunity to collect samples from the richly fossiliferous deep-water limestone. To and from the Bear River deposit we will traverse marine Tertiary basalts and volcaniclastic and siliciclastic strata. If weather and road conditions permit, we intend to make brief stops at other seep deposits, rock outcrops, and sites of local historic importance. The trip is intended to not only provide a geographic context for the complex geology of southwestern Washington, but also some sense of the difficulties, posed by both climate and vegetation, that are encountered by geologists and paleontologists when working in this region.

Two of the stops on this field trip (Fig. 1), the Naselle River bridge and the Bear River site, were also visited as part of a previous Geological Society of America field trip (Nesbitt et al., 1994). The Menlo site has not received any study subsequent to brief analyses in the early 1990s (Goedert and Squires, 1990; Squires and Goedert, 1991; Campbell and Bottjer, 1993). In Pacific County, megapaleontological studies (other than those dealing with cold-seep assemblages) have concentrated mainly on a limited number of localities in only a few rock units (e.g., Moore, 1984); for some rock units, published megapaleontological data is nonexistent. One late Eocene and newly recognized cold-seep limestone near Knappton is introduced for the first time herein.

The abbreviation UWBM, used for specimen and locality numbers, is from the Burke Museum of Natural History and Culture, University of Washington, Seattle.

Tectonic Setting and Cold Seeps

Eocene to early Miocene strata in southwestern Washington were deposited in a forearc basin. Subduction of the Juan de Fuca plate beneath the North American plate is an ongoing process, with its beginnings in late Eocene time (Vance et al., 1987). Since then, abyssal plain, outer slope, and outer fan-fringe sediments have been, and are being, scraped off of the subducting plate, forming the Cascadia accretionary prism (Ritger et al., 1987; Kulm and Suess, 1990). In some areas of southwestern Washington, up to 7 km of deep-water marine sediments are now exposed (Wells, 1989). Tectonic compression of these strata have resulted in a decrease in sediment porosity with a concomitant overpressure in pore fluids (Ritger et al., 1987; Campbell, 1992), which then move to the surface via fault zones, fractures, and dipping permeable strata.

Hydrocarbon seeps are relatively common in the subduction zone offshore of Oregon and Washington (Kulm et al., 1986; Ritger et al., 1987). Some conspicuous features of these submarine hydrocarbon seeps are anomalous carbonate chimneys, crusts, and "doughnut" structures (Schroeder et al., 1987; Kulm and Suess, 1990). Deep-water authigenic limestone deposits indicate that hydrocarbons have been venting to the surface since the onset of subduction during the Eocene in western Washington, and even longer elsewhere along the Pacific slope of North America (Campbell and Bottjer, 1993).

Frequently, chemosymbiotic invertebrate communities develop around seeps. Chemosymbiotic biota, commonly bivalves and tube-dwelling worms (Arp and Childress, 1981; Felbeck et al., 1981; Rau, 1981; Southward et al., 1981; Childress and Mickel, 1982; Cavanaugh, 1983), and host endosymbiotic bacteria. These chemolithotrophic or methanotrophic bacteria are able to oxidize the reduced compounds, such as methane and hydrogen sulfide in seepage fluids, and supply their hosts with energy and nutrients (Paull et al., 1984; Kennicutt et al., 1985; Suess et al., 1985).

Ancient vent- and/or seep-associated palaeocommunities, ranging in age from Silurian to Pliocene, have been reported from many parts of the world (Campbell and Bottjer, 1995). Campbell and Bottjer (1993) proposed a "seep-search strategy" that

Goedert, J.L., and Benham, S.R., 2003, Biogeochemical processes at ancient methane seeps: The Bear River site in southwestern Washington, *in* Swanson, T.W., ed., Western Cordillera and adjacent areas: Boulder, Colorado, Geological Society of America Field Guide 4, p. 201–208. For permission to copy, contact editing@geosociety.org. © 2003 Geological Society of America.

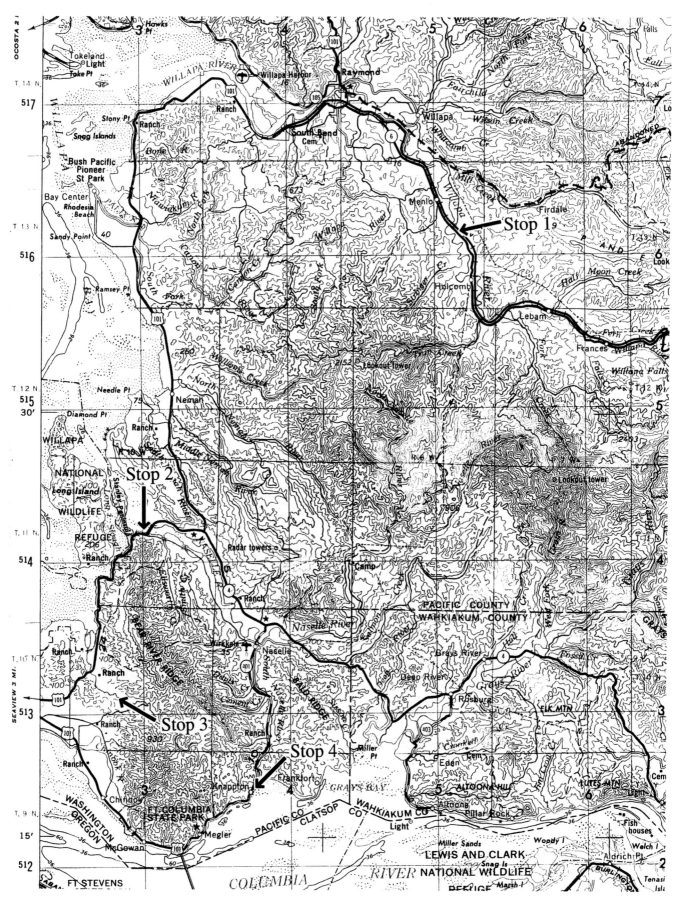

Figure 1. Index map of southwestern Washington State. Base map U.S. Geological Survey, Hoquiam, Washington; Oregon, 1:250,000, 1958 edition (revised 1974).

employs palaeontologic, isotopic, and petrographic analyses to recognize ancient hydrocarbon seep deposits. Other features that are typical of ancient seep deposits were identified by Kelly et al. (1995). Biomarker analyses are yet another tool that can be used to identify ancient seep deposits (Peckmann et al., 1999, 2002, 2003; Thiel et al., 1999) by revealing ^{13}C-depleted molecular fossils. These "chemofossils" are distinct isoprenoid hydrocarbons, diagnostic for the past presence of Archaea and associated sulphate-reducing bacteria. They indicate that the anaerobic oxidation of methane was an important biogeochemical process in the past, just as in methane-rich environments today (Elvert et al., 1999, 2000; Boetius et al., 2000; Hinrichs et al., 2000; Pancost et al., 2000; Valentine and Reeburgh, 2000; Thiel et al., 2001a, b; Teske et al., 2002; Michaelis et al., 2002). Some seep sites in late Eocene and Oligocene strata just north of the study area were recently shown to contain chemofossils in addition to fossils thought to be microbial filaments (Peckmann et al., 2002).

Cold-seep carbonates have now been reported from Eocene, Oligocene, and Miocene-Pliocene strata in Washington and northwestern Oregon (Goedert and Squires, 1990; Campbell, 1992; Campbell and Bottjer, 1993; Peckmann et al., 2002). In most cases, they are enclosed within thick-bedded mudstones and siltstones that were deposited in deep water. Most of these seep assemblages are bivalve-dominated. Many of the seep assemblages also include worm tubes (Goedert and Squires, 1990; Goedert et al., 2000), hexactinellid sponges (Rigby and Goedert, 1996), gastropods (Squires, 1995; Goedert and Kaler, 1996; Goedert and Benham, 1999), polyplacophorans (Goedert and Campbell, 1995; Squires and Goedert, 1995), and in rare cases, corals (Peckmann et al., 2002). Invertebrate fecal pellets, foraminiferans, sponge spicules, echinoid fragments, plant debris, and microbial filaments are visible in thin sections (Goedert and Campbell, 1995; Peckmann et al., 2002). Precipitation of carbonate at some sites was rapid (Peckmann et al., 2002) because rarely preserved fossils such as articulated polyplacophorans and delicate sponges can be found.

Some small carbonate deposits in Washington appear to be allochthonous (Goedert and Campbell, 1995; Goedert et al., 2000, 2003), similar to the "Type 3" deposits of Conti and Fontana (1998) that are found in the Apennine Mountains. These types of seep deposits appear to have slid or slumped downslope from where the actual seeps were located. Some seep carbonates found in exposures of the Lincoln Creek Formation near Knappton (Goedert and Squires, 1993) are now known to be Type 3 deposits.

THE BEAR RIVER METHANE-SEEP DEPOSIT

Prior to the discovery of recent seep-associated chemosymbiotic communities in the deep sea (Lonsdale, 1977), the paleoecology of fossil bivalve-dominated invertebrate assemblages that were preserved in carbonate deposits in deep-marine strata could not be correctly interpreted (Campbell et al., 2002; Goedert et al., 2003). An excellent example of this is the Bear River site, a localized, relatively small limestone deposit within nearly barren late Eocene, deep-water strata, which preserves an anomalous and relatively diverse invertebrate assemblage. The only detailed study of this site previous to 1990 was an analysis of the limestone by Danner (1966, p. 433), and he simply referred to it as a "reef." Fossils of sponges from the Bear River site were described by Rigby and Jenkins (1983), but the authors did not discuss the associated fauna or speculate as to the paleoecology of this site. The Bear River assemblage was only recently interpreted as an ancient cold-seep community (Goedert and Squires, 1990; Squires and Goedert, 1991), and this was largely on the basis of the bivalve taxa that are present.

Petrographic, isotopic, and biomarker analyses have not yet been published for the Bear River site (K.A. Campbell, 2002, personal commun.), nor have they been performed on the Menlo and Knappton seep carbonates. Because of this, the focus here will necessarily be on the preserved megafauna and how the assemblage compares with other cold-seep assemblages, modern and fossil.

According to Danner (1966) the limestone was 15 m thick, 68 m long, and 38 m wide in 1954, when the quarry was last active. It is mapped as being within the "Siltstone of Cliff Point" (Wells, 1989), and an exposure in the limestone quarry yielded calcareous nannofossils and benthic foraminifers that indicate a late Eocene age (Goedert and Squires, 1990). Invertebrate megafossils are rare in the siltstone, and are mostly unidentifiable, poorly preserved mollusks, and some tube-like fossils that may be *Bathysiphon*. A locality 1 km south of the Bear River deposit yielded a single large concretion that contained several small, articulated fish skeletons (Natural History Museum of Los Angeles vertebrate paleontology collections).

Bivalvia are the dominant molluscan component of the Bear River seep assemblage in terms of the number of taxa. In this assemblage, solemyids are relatively rare. Most bivalves are articulated and seem to be both randomly oriented and distributed. Some bivalves were very large individuals, up to 13 cm long (Goedert and Squires, 1990). They appear to be *Acharax dalli* (Fig. 2F), a species found at many outcrops of Eocene and Oligocene strata in the Pacific Northwest. A few large bivalves, up to 9 cm long and identified as pitarids by Goedert and Squires (1990, their Fig. 2t), are herein referred to the genus *Cryptolucina* Saul et al., 1996. Specimens (Fig. 2A) of *Vesicomya* (*Calyptogena*) *chinookensis* Squires and Goedert, 1991, are fairly common, up to 9 cm long, and in many cases found in clusters. Specimens of the mytilid *Modiolus* (*Modiolus*) *willapaensis* Squires and Goedert, 1991, abound (Fig. 2B); most are closed valved and appear to be in random orientations. Only one small specimen (Fig. 2C) of the thyasirid *Conchocele bisecta* (Fig. 2D) has been found so far.

Gastropods are common, but most are small and inconspicuous. Specimens of *Provanna antiqua* Squires, 1995, are common. These were some of the first recognized fossils for the genus *Provanna*, found today at cold seeps, hydrothermal vents, and on sunken wood. The trochid *Margarites* (*Pupillaria*) *columbiana* Squires and Goedert, 1991, is relatively rare. Other

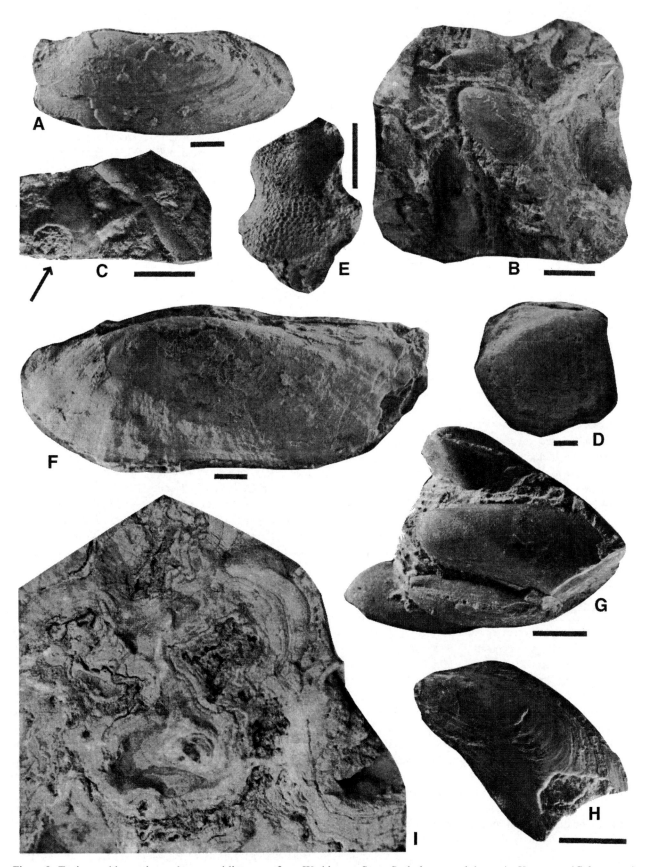

Figure 2. Tertiary cold-seep invertebrates and limestone from Washington State. Scale bars equal 1 cm. A: *Vesicomya (Calyptogena) chinookensis*, UWBM 97513, Bear River seep site. B: Cluster of *Modiolus (Modiolus) willapaensis*, UWBM 97514, Bear River seep site. C: Bear River limestone with worm tube and fragmentary bivalve (arrow), *Conchocele bisecta*, Conrad, UWBM 97515. D: *Conchocele bisecta*, UWBM 97308, Whiskey Creek deposit, Clallam County, Washington. E: Hexactinellid sponge, *Aphrocallistes polytretos*, UWBM 97516, Bear River seep site. F: *Acharax dalli*, UWBM 97307, Whiskey Creek deposit, Clallam County, Washington. G and H: Unidentified modiolids from allochthonous seep carbonate in the Lincoln Creek Formation northeast of Knappton. G: Cluster of several individuals with same orientation, UWBM 97517. H: Single large specimen, UWBM 97518. I: Field photo of seep limestone from Silt-stone of Shoalwater Bay at Knappton, Washington; note the wavy, laminated crusts that are typical of many seep carbonates.

gastropods are present, and some represent unnamed taxa. These are the subject of ongoing studies.

Fossils of the hexactinellid sponge *Aphrocallistes polytretos* (Rigby and Jenkins, 1983) are abundant and uncrushed (Fig. 2E). A few worm tubes (Fig. 2C) have been found (Goedert and Squires, 1990) and are probably those of pogonophorans.

Comparisons of the Menlo, Bear River, and Knappton Assemblages with Those from Other Ancient and Modern Seeps

The Bear River bivalve assemblage included solemyids, modiolids, lucinids, vesicomyids, and rare thyasirids, and that from the Menlo deposit included only modiolids, vesicomyids, and thyasirids (Goedert and Squires, 1990). Seep carbonates in the Lincoln Creek Formation at Knappton yielded solemyids, modiolids (Fig. 2G and H), thyasirids, and vesicomyids (Goedert and Squires, 1993). The seep carbonate (Fig. 2I) from the Siltstone of Shoalwater Bay at Knappton appears to contain only modiolids. These taxa are constituents of many living (e.g., Sibuet and Olu, 1998) and ancient seep assemblages. In the eastern Pacific Ocean, species of the cosmopolitan solemyid genus *Acharax* are most common below 400 m (Coan et al., 2000), and at cold seeps and vents they have been observed in areas of low fluid flow, commonly at depths >2000 m (Sibuet and Olu, 1998). The genus *Acharax* has been reported from Eocene, Oligocene, and Miocene-Pliocene seep assemblages in Washington (Goedert and Squires, 1990; Campbell, 1992; Peckmann et al., 2002).

The term "*Thyasira-Lucinoma-Solemya* Communities" was coined by Hickman (1984) to describe a recurring association of solemyids, lucinids, and thyasirids (= *Conchocele*) found in deep-water Eocene and Oligocene strata of the Pacific Northwest. Some of these are now known to be cold-seep assemblages (e.g., Campbell and Bottjer, 1993; Nesbitt et al., 1994). Although *Lucinoma* appears to have been absent from the Bear River assemblage, lucinids of the genus *Cryptolucina* were present. The rarity of the thyasirid genus *Conchocele* in the Bear River deposit is puzzling, because it has been reported from Eocene and Oligocene cold-seep sites in Washington (Goedert and Squires, 1990, 1993; Goedert et al., 2003; Peckmann et al., 2002), Oregon (Campbell and Bottjer, 1993; Nesbitt et al., 1994), and California (Squires and Gring, 1996) and is the dominant bivalve taxon at some sites. *Lucinoma* has not been reported from sites where *Cryptolucina* is found (Goedert et al., 2003). Species of *Lucinoma* have not been reported from water deeper than 665 m (Coan et al., 2000) in the northeastern Pacific Ocean, hence it is possible that the depositional setting for the Bear River site was deeper than this. Lucinids and thyasirids have been found in a number of ancient seep assemblages in Washington and Oregon (Goedert and Campbell, 1995; Peckmann et al., 2002; Goedert et al., 2003), and in living seep assemblages elsewhere (Sibuet and Olu, 1998; Kamenev et al., 2001), yet they have not been found living at seep or vent sites in the northeastern Pacific Ocean (Sibuet and Olu, 1998).

Species of the large lucinid *Cryptolucina* were originally based on specimens from Eocene deep-water mudstones and cold-seep deposits on the south side of the Olympic Mountains (Saul et al., 1996). Recently, Kelly et al. (2000) reported another species of *Cryptolucina* from Cretaceous seep deposits in Greenland, and Goedert et al. (2003) reported *Cryptolucina* from late Eocene seep deposits on the north side of the Olympic Peninsula. Bivalves that may be species of *Cryptolucina* have also been reported from Late Jurassic and Early Cretaceous seep deposits in California (Stanton, 1895; Saul et al., 1996), Tertiary seep deposits in Peru (Saul et al., 1996; cf. Olsson, 1931), and Cretaceous and Miocene seeps in Japan (Kanie and Sakai, 1997).

Most known seep sites are inhabited by at least one vesicomyid species (Sibuet and Olu, 1998). Vesicomyids are geographically widespread, apparently due to their ability to survive fluctuating fluid-flow and variable sulphide concentrations (Sibuet and Olu, 1998). In the eastern North Pacific, vesicomyids live at water depths >370 m (Coan et al., 2000). Vesicomyid shells commonly occur in aggregations at hydrocarbon seeps and hydrothermal vents (Sibuet and Olu, 1998). Vesicomyids have been reported from Eocene and Oligocene cold-seep sites in Washington (Goedert and Squires, 1993; Goedert and Campbell, 1995; Peckmann et al., 2002; Goedert et al., 2003); Oligocene, Miocene, and Pliocene seep sites in Japan (listed in Campbell and Bottjer, 1995); Eocene or Oligocene seeps in Peru (Olsson, 1931); an Eocene seep site in California (Squires and Gring, 1996); Miocene or Pliocene rocks, possibly seeps, in Panama (Olsson, 1942); and Oligocene seep deposits in Alaska (Moore, 1969; Kanno, 1971; Miller, 1975).

According to Sibuet and Olu (1998), living seep mytilids are more restricted in their geographic and bathymetric distribution than are vesicomyids. They do not inhabit seeps or vents in the eastern Pacific Ocean today (Sibuet and Olu, 1998). Frequently found at cold seeps in areas of high fluid flow (Aharon, 1994; MacDonald et al., 1989), they can host both methanotrophic and sulfur-oxidizing endosymbiotic bacteria (Fisher et al., 1993). All of the living vent and/or seep mytilids appear to be species of the genus *Bathymodiolus*. Fossil mytilids from the Menlo, Bear River, and Knappton seep sites appear instead to be species of the genus *Modiolus* (Goedert and Squires, 1990; Squires and Goedert, 1991). In the Bear River deposit, they appear to make up most of the preserved biomass. For unknown reasons, in Washington, mytilids have not been found in seep sites younger than Oligocene.

Provannid gastropods have been associated with hydrocarbon seeps since at least Late Jurassic time (K. Campbell, unpublished data, cited in Warén and Bouchet, 2001). Provannids are also a common element of some living vent and/or seep assemblages (Sibuet and Olu, 1998).

Sponges have been reported as being abundant near seeps off Barbados (Le Pichon et al., 1990; Sibuet and Olu, 1998) and have been reported attached to carbonates from seeps off Oregon (Kulm and Suess, 1990). In some cases, sponges have been found attached to worm tubes at seeps off Oregon (Suess et al., 1985) and in the Gulf of Mexico (MacDonald et al., 1989; Harrison et al., 1994). Sponges have been reported from very few ancient hydrocarbon seep deposits. Other than the Bear River site, at which only *Aphrocallistes polytretos* has been found, there is only one additional Oligocene seep site in Washington State that yielded numerous, well-preserved fossils of hexactinellid sponges (Rigby and Goedert, 1996).

Taxa common to other seep sites in Washington, such as small infaunal nuculanid bivalves, have not yet been found in the Menlo, Bear River, or Knappton limestones. The Bear River seep assemblage included more than 11 symbiont-containing species (solemyids, vesicomyids, mytilids, lucinids, and thyasirids), grazers (margaritid), scavengers, and predators (naticid and turrid gastropods). Seep communities with this kind of diversity are termed "mature" (sensu Juniper and Sibuet, 1987; Olu et al., 1996; Sibuet and Olu, 1998).

ROAD LOG

Begin this trip by leaving the Seattle Convention Center and proceed south on Interstate 5 to Chehalis, Washington. Use exit 77 to head west, toward Raymond, on State Highway 6 (this is mile 0 of this field guide). You will follow the valley of the Chehalis River, going upstream, through farm and timber country until reaching the town of Pe Ell, where the highway turns more to the west to cross the Coast Range.

Cumulative Miles	(km)	Description
44.4	(71.0)	Stop 1.

Stop 1. The Menlo Cold-Seep Limestone at the Niemcziek Ranch

This site is on private property and permission from the Niemcziek family is required. *The slope is steep, and when wet the footing may be poor.* This deposit is poorly exposed, but is visible from the highway, situated on the slope below and left of the house. Although it is somewhat concealed presently, Danner (1966) reported that this deposit was 2 m thick, 7.5 m long, and 3 m wide. The first paleontologic studies of this limestone (Goedert and Squires, 1990; Squires and Goedert, 1991) reported bivalves and one gastropod. It is in deep-water strata of the Lincoln Creek Formation, and is late Eocene in age (Rau, 1951). It takes its name from the nearby town of Menlo, Washington.

Continue west on Highway 6 to Raymond, and at the junction with Highway 101 (mile 52.2; km 83.5) turn left and proceed southward toward Illwaco and Astoria. At the junction of Highway 101 and State Highway 4 (mile 85.7; km 137.1), turn west at the stop sign, following Highway 101 south to Stop 2 at the south end of the bridge over the Naselle River.

Cumulative Miles	(km)	Description
87.0	(139.2)	Stop 2.

Stop 2

At the south end of the Naselle River bridge is a roadcut exposure of the lowermost part of the Lincoln Creek Formation (the Knappton Sandstone unit) and the contact with the Siltstone of Shoalwater Bay (Wells, 1989). Concretions from the Lincoln Creek Formation at this locality commonly contain fossils of the crabs *Pulalius vulgaris* (Rathbun, 1926) and rarely the portunid *Minohellenus triangulum* (Rathbun, 1926). The sandstone also contains shallow-water mollusks (Nesbitt et al., 1994). The Siltstone of Shoalwater Bay is nearly barren of megafossils; however, at a few localities farther south, toward Bear River, some concretions contain fish bones. Both the Lincoln Creek Formation (at Knappton and elsewhere) and the Siltstone of Shoalwater Bay contain cold-seep deposits. It is important to understand that the Naselle River bridge outcrop is fairly typical for this part of the Pacific Northwest. In most cases, outcrops like this have a short lifespan because of landslides, rapid weathering, and the growth of vegetation.

Continue south on Highway 101 toward Astoria.

Cumulative Miles	(km)	Description
98.1	(156.9)	Turn left on Highway 101, again toward Astoria.
98.7	(157.9)	Turn left again, following Highway 101 toward Astoria.
99.0	(158.4)	Turn left on Chinook Valley Road.
101.4	(162.2)	Turn left on Walberg Road. This is an unimproved, narrow road with several secondary roads branching off for logging operations. Stay on the well-traveled main course of this road until passing the pumping station for Ilwaco's water supply.
103.1	(164.9)	Left on narrow road, graveled partly with limestone from the Bear River site.
103.4	(165.4)	Stop 3.

Stop 3. Bear River Quarry

Unfortunately, this quarry is used for target practice by gun enthusiasts. *Watch for broken glass and metal debris.* Return to Chinook Valley Road.

Cumulative Miles	(km)	Description
105.4	(168.6)	Turn left on Chinook Valley Road.
108.3	(173.3)	Veer left on Chinook Valley Road.
109.2	(174.7)	Chinook. Turn left on Highway 101.
119.6	(191.4)	Astoria Bridge. Outcrops on the north side of the highway are the "basalt boulder conglomerate at Pt. Ellice" (Wells, 1989). A small number of the clasts in this conglomerate are concretions, some of which contain fossil crabs, and more rarely, mollusks. Continue east on Highway 401.
124.1	(198.6)	Stop 4.

Stop 4. Townsite of Knappton (Cementville)

This townsite was founded by Jabez B. Knapp. In 1868, Knapp built a kiln and barrel factory at this site for the production of cement (Appelo, 1975) but the resource was too limited. In 1869, Knapp built a sawmill, and this supported the town well into the 1920s.

Approximately 300 m north of the Knappton monument, a small amount of limestone can be found on both sides of the highway, and it is derived from outcrops of the late Eocene Siltstone of Shoalwater Bay. A few blocks of limestone have fallen from the forested hillside on the west side of Highway 401, and a few larger boulders are on the beach on the east side of the highway. The limestone does not appear to contain megafossils, but does have well-developed calcitic laminations. Another block of limestone, very similar to this limestone but containing abundant fossils of small modiolids, is lying on the beach approximately another 300 m north of this location.

Outcrops of the Oligocene part of the Lincoln Creek Formation along the north side of the bay just to the northwest of the Knappton townsite have yielded a diverse invertebrate assemblage (Moore, 1984). Cold-seep limestones have been found here as well (Goedert and Squires, 1993), and these appear to be allochthonous, or Type 3 deposits (Conti and Fontana, 1998).

End of trip. Return to Seattle via Highway 401 to Highway 4 (Longview/Kelso) to Interstate 5.

REFERENCES CITED

Aharon, P., 1994, Geology and biology of modern and ancient submarine hydrocarbon seeps and vents: An introduction: Geo-Marine Letters, v. 14, p. 69–73.

Appelo, C.E., 1975, Knappton, the first 50 years, Pacific County, Washington: Deep River, Washington, Privately published by Carlton E. Appelo, 88 p.

Arp, A.J., and Childress, J.J., 1981, Blood function in the hydrothermal vent vestimentiferan tube worm: Science, v. 213, p. 342–344.

Boetius, A., Ravenschlag, K., Schubert, C.J., Rickert, D., Widdel, F., Gieseke, A., Amann, R., Jørgensen, B.B., Witte, U., and Pfannkuche, O., 2000, A marine consortium apparently mediating anaerobic oxidation of methane: Nature, v. 407, p. 623–626.

Campbell, K.A., 1992, Recognition of a Mio-Pliocene cold seep setting from the northeast Pacific convergent margin, Washington, U.S.A.: Palaios, v. 7, p. 422–433.

Campbell, K.A., and Bottjer, D.J., 1993, Fossil cold seeps: National Geographic Research and Exploration, v. 9, p. 326–343.

Campbell, K.A., and Bottjer, D.J., 1995, Brachiopods and chemosymbiotic bivalves in Phanerozoic hydrothermal vent and cold seep environments: Geology, v. 23, p. 321–324.

Campbell, K.A., Farmer, J.D., and Des Marais, D., 2002, Ancient hydrocarbon seeps from the Mesozoic convergent margin of California: Carbonate geochemistry, fluids and palaeoenvironments: Geofluids, v. 2, p. 63–94.

Cavanaugh, C.M., 1983, Symbiotic chemoautotrophic bacteria in marine invertebrates from sulfide-rich habitats: Nature, v. 302, p. 58–61.

Childress, J.J., and Mickel, T.J., 1982, Oxygen and sulfide consumption rates in the vent clam *Calyptogena pacifica*: Marine Biology Letters, v. 3, p. 73–79.

Coan, E.V., Scott, P.V., and Bernard, F.R., 2000, Bivalve seashells of western North America: Santa Barbara Museum of Natural History Monographs, v. 2, p. 1–764.

Conti, S., and Fontana, D., 1998, Recognition of primary and secondary Miocene lucinid deposits in the Apennine Chain: Memorie di Scienze Geologiche, v. 50, p. 101–131.

Danner, W.R., 1966, Limestone resources of western Washington: State of Washington Division of Mines and Geology Bulletin, v. 52, 474 p.

Elvert, M., Suess, E., and Whiticar, M.J., 1999, Anaerobic methane oxidation associated with marine gas hydrates: Superlight C-isotopes from saturated and unsaturated C^{20} and C^{25} irregular isoprenoids: Naturwissenschaften, v. 86, p. 295–300.

Elvert, M., Suess, E., Greinert, J., and Whiticar, M.J., 2000, Archaea mediating anaerobic methane oxidation in deep-sea sediments at cold seeps of the eastern Aleutian subduction zone: Organic Geochemistry, v. 31, p. 1175–1187.

Felbeck, H., Childress, J.J., and Somero, G.N., 1981, Calvin-Benson cycle and sulphide oxidation enzymes in animals from sulphide rich habitats: Nature, v. 293, p. 291–293.

Fisher, C.R., Brooks, J.M., Vodenichar, J.S., Zande, J.M., Childress, J.J., and Burke, R.A., Jr., 1993, The co-occurrence of methanotrophic and chemoautotrophic sulfur-oxidizing bacterial symbionts in a deep-sea mussel: Marine Ecology, v. 14, p. 277–289.

Goedert, J.L., and Benham, S.R., 1999, A new species of *Depressigyra*? (Gastropoda: Peltospiridae) from cold-seep carbonates in Eocene and Oligocene rocks of western Washington: The Veliger, v. 42, p. 112–116.

Goedert, J.L., and Campbell, K.A., 1995, An early Oligocene chemosynthetic community from the Makah Formation, northwestern Olympic Peninsula, Washington: The Veliger, v. 38, p. 22–29.

Goedert, J.L., and Kaler, K.L., 1996, A new species of *Abyssochrysos* (Gastropoda: Loxonematoidea) from a Middle Eocene cold-seep carbonate in the Humptulips Formation, western Washington: The Veliger, v. 39, p. 65–70.

Goedert, J.L., and Squires, R.L., 1990, Eocene deep-sea communities in localized limestones formed by subduction-related methane seeps, southwestern Washington: Geology, v. 18, p. 1182–1185.

Goedert, J.L., and Squires, R.L., 1993, First Oligocene records of *Calyptogena* (Bivalvia: Vesicomyidae): The Veliger, v. 36, p. 72–77.

Goedert, J.L., Peckmann, J., and Reitner, J., 2000, Worm tubes in allochthonous cold seep carbonates from Lower Oligocene rocks of western Washington: Journal of Paleontology, v. 74, p. 992–999.

Goedert, J.L., Thiel, V., Schmale, O., Rau, W.W., Michaelis, W., and Peckmann, J., 2003, The late Eocene "Whiskey Creek" methane-seep deposit (western Washington State) Part I: Geology, Paleontology, and Molecular Geobiology: Facies, v. 48, p. 223–240.

Harrison, F.W., Gardiner, S.L., Rützler, K., and Fisher, C.R., 1994, On the occurrence of endosymbiotic bacteria in a new species of sponge from a hydrocarbon seep community in the Gulf of Mexico: Transactions of the American Microscopy Society, v. 113, p. 419–420.

Hickman, C.S., 1984, Composition, structure, ecology, and evolution of six Cenozoic deep-water mollusk communities: Journal of Paleontology, v. 58, p. 1215–1234.

Hinrichs, K.U., Summons, R.E., Orphan, V., Sylva, S.P., and Hayes, J.M., 2000, Molecular and isotopic analysis of anaerobic methane-oxidizing communities in marine sediments: Organic Geochemistry, v. 31, p. 1685–1701.

Juniper, S.K., and Sibuet, M., 1987, Cold seep benthic communities in Japan subduction zones: Spatial organization, trophic strategies and evidence of temporal evolution: Marine Ecology Progress Series, v. 40, p. 115–126.

Kamenev, G.M., Nadtochy, V.A., and Kuznetsov, A.P., 2001, *Conchocele bisecta* (Conrad, 1949) (Bivalvia: Thyasiridae) from deep-water methane-rich areas of the Sea of Okhotsk: The Veliger, v. 44, p. 84–94.

Kanie, Y., and Sakai, T., 1997, Chemosynthetic thraciid bivalve *Nipponothracia*, gen. nov. from the Lower Cretaceous and Middle Miocene mudstones in Japan: Venus, Japanese Journal of Malacology, v. 56, p. 205–220.

Kanno, S., 1971, Tertiary molluscan fauna from the Yakataga District and adjacent areas of southern Alaska: Paleontological Society of Japan Special Papers, v. 16, p. 1–154, plates 1–18.

Kelly, S.R.A., Ditchfield, P.W., Doubleday, P.A., and Marshall, J.D., 1995, An Upper Jurassic methane-seep limestone from the fossil Bluff Group forearc basin of Alexander Island, Antarctica: Journal of Sedimentary Research, v. A65, p. 274–282.

Kelly, S.R.A., Blanc, E., Price, S.P., and Whitham, A.G., 2000, Early Cretaceous giant bivalves from seep-related limestone mounds, Wollaston Forland, northeast Greenland, *in* Harper, E.M., Taylor, J.D., and Crame, J.A., eds., The evolutionary biology of the Bivalvia: Geological Society Special Publication 177, p. 227–246.

Kennicutt, M.C., II, Brooks, J.M., Bidigare, R.R., Fay, R.R., Wade, T.L., and MacDonald, T.J., 1985, Vent-type taxa in a hydrocarbon seep region on the Louisiana slope: Nature, v. 317, p. 351–353.

Kulm, L.D., and Suess, E., 1990, Relationship between carbonate deposits and fluid venting: Oregon accretionary prism: Journal of Geophysical Research, v. 95, p. 8899–8915.

Kulm, L.D., Suess, E., Moore, J.C., Carson, B., Lewis, B.T., Ritger, S.D., Kadko, D.C., Thornburg, T.M., Embley, R.W., Rugh, W.D., Massoth, G.J.,

Langseth, M.G., Cochrane, G.R., and Scamman, R.L., 1986, Oregon subduction zone: Venting fauna and carbonates: Science, v. 231, p. 561–566.

Le Pichon, X., Foucher, J.-P., Boulègue, J., Henry, P., Lallemant, S., Benedetti, M., Avednik, F., and Marriotti, A., 1990, Mud volcano field seaward of the Barbados accretionary complex: A submersible survey: Journal of Geophysical Research, v. 95, p. 8931–8943.

Lonsdale, P., 1977, Clustering of suspension-feeding macrobenthos near abyssal hydrothermal vents at oceanic spreading centers: Deep-Sea Research, v. 24, p. 857–863.

MacDonald, I.R., Boland, G.S., Baker, J.S., Brooks, J.M., Kennicutt, M.C., II, and Bidigare, R.R., 1989, Gulf of Mexico hydrocarbon seep communities II. Spatial ditribution of seep organisms and hydrocarbons at Bush Hill: Marine Biology, v. 101, p. 235–247.

Michaelis, W., Seifert, R., Nauhaus, K., Treude, T., Thiel, V., Blumenberg, M., Knittel, K., Gieseke, A., Peterknecht, K., Pape, T., Boetius, A., Amann, R., Jørgensen, B.B., Widdel, F., Peckmann, J., Pimenov, N.V., and Gulin, M.B., 2002, Microbial reefs in the Black Sea fuelled by anaerobic oxidation of methane: Science, v. 297, p. 1013–1015.

Miller, D.J., 1975, Geologic map and sections of the central part of the Katalla District, Alaska: U.S. Geological Survey, Miscellaneous Field Studies Map MF-722, sheets 1–2.

Moore, E.J., 1984, Molluscan paleontology and biostratigraphy of the lower Miocene upper part of the Lincoln Creek Formation in southwestern Washington: Contributions in Science, v. 351, p. 1–42.

Moore, G.W., 1969, New formations on Kodiak and adjacent islands: U.S. Geological Survey Bulletin, v. 1274-A, p. A27–A35.

Nesbitt, E.A., Campbell, K.A., and Goedert, J.L., 1994, Paleogene cold seeps and macroinvertebrate faunas in a forearc sequence of Oregon and Washington, *in* Swanson, D.A., and Haugerud, R.A., eds., Geologic field trips in the Pacific Northwest, volume 1: Geological Society of America Guidebook, p. 1D1–1D11.

Olsson, A.A., 1931, Contributions to the Tertiary paleontology of northern Peru: Part 4, the Peruvian Oligocene: Bulletins of American Paleontology, v. 17, p. 99–264.

Olsson, A.A., 1942, Tertiary and Quaternary fossils from the Burica Peninsula of Panama and Costa Rica: Bulletins of American Paleontology, v. 27, p. 1–106.

Olu, K., Duperret, A., Sibuet, M., Foucher, J.-P., and Fiala-Médioni, A., 1996, Structure and distribution of cold seep communities along the Peruvian active margin: Relationship to geological and fluid patterns: Marine Ecology Progress Series, v. 132, p. 109–125.

Pancost, R.D., Sinninghe Damste, J.S., de Lint, S., van der Maarel, M.J.E.C., Gottschal, J.C., and the Medinaut Shipboard Party, 2000, Biomarker evidence for widespread anaerobic oxidation in Mediterranean sediments by a consortium of methanogenic archaea and bacteria: Applied and Environmental Microbiology, v. 66, p. 1126–1132.

Paull, C.K., Hecker, B., Commeau, R., Freeman-Lynde, R.P., Neumann, A.C., Corso, W.P., Golubic, S., Hook, J., Sikes, E., and Curray, J., 1984, Biological communities at Florida Escarpment resemble hydrothermal vent communities: Science, v. 226, p. 965–967.

Peckmann, J., Thiel, V., Michaelis, W., Clari, P., Gaillard, C., Martire, L., and Reitner, J., 1999, Cold seep deposits of Beauvoisin (Oxfordian; southeastern France) and Marmorito (Miocene; northern Italy): Microbially induced authigenic carbonates: International Journal of Earth Sciences, v. 88, p. 60–75.

Peckmann, J., Goedert, J.L., Thiel, V., Michaelis, W., and Reitner, J., 2002, A comprehensive approach to the study of methane-seep deposits from the Lincoln Creek Formation, western Washington State, USA: Sedimentology, v. 49, p. 855–873.

Peckmann, J., Goedert, J.L., Heinrichs, T., Hoefs, J., and Reitner, J., 2003, The late Eocene 'Whiskey Creek' methane seep deposit (western Washington State), Part II: Petrology, stable isotopes, and biogeochemistry: Facies, v. 48, p. 241–254.

Rathbun, M.J., 1926, The fossil stalk-eyed Crustacea of the Pacific Slope of North America: United States National Museum Bulletin, v. 138, p. 1–155.

Rau, G., 1981, Hydrothermal vent clam and tube worm $^{13}C/^{12}C$: Further evidence of non-photosynthetic food sources: Science, v. 213, p. 338–340.

Rau, W.W., 1951, Tertiary foraminifera from the Willapa River valley of southwestern Washington: Journal of Paleontology, v. 25, p. 417–453.

Rigby, J.K., and Goedert, J.L., 1996, Fossil sponges from a localized cold-seep limestone in Oligocene rocks of the Olympic Peninsula, Washington: Journal of Paleontology, v. 70, p. 900–908.

Rigby, J.K., and Jenkins, D.E., 1983, The Tertiary sponges *Aphrocallistes* and *Eurete* from western Washington and Oregon: Contributions in Science, v. 344, p. 1–13.

Ritger, S., Carson, B., and Suess, E., 1987, Methane-derived authigenic carbonates formed by subduction-induced pore-water expulsion along the Oregon/Washington margin: Geological Society of America Bulletin, v. 98, p. 147–156.

Saul, L.R., Squires, R.L., and Goedert, J.L., 1996, A new genus of cryptic lucinid? bivalve from Eocene cold seeps and turbidite-influenced mudstone, western Washington: Journal of Paleontology, v. 70, p. 788–794.

Schroeder, N.A.M., Kulm, L.D., and Muehlberg, G.E., 1987, Carbonate chimneys on the outer continental shelf: Evidence for fluid venting on the Oregon margin: Oregon Geology, v. 49, p. 91–96.

Sibuet, M., and Olu, K., 1998, Biogeography, biodiversity and fluid dependence of deep-sea cold-seep communities at active and passive margins: Deep-Sea Research II, v. 45, p. 517–567.

Southward, A.J., Southward, E.C., Dando, P.R., Rau, G.H., Felbeck, H., and Flügel, H., 1981, Bacterial symbionts and low $^{13}C/^{12}C$ ratios in tissues of Pogonophora suggest unusual nutrition and metabolism: Nature, v. 293, p. 616–620.

Squires, R.L., 1995, First fossil species of the chemosynthetic-community gastropod *Provanna*: Localized cold-seep limestones in upper Eocene and Oligocene rocks, Washington: The Veliger, v. 38, p. 30–36.

Squires, R.L., and Goedert, J.L., 1991, New late Eocene mollusks from localized limestone deposits formed by subduction-related methane seeps, southwestern Washington: Journal of Paleontology, v. 65, p. 412–416.

Squires, R.L., and Goedert, J.L., 1995, An extant species of *Leptochiton* (Mollusca: Polyplacophora) in Eocene and Oligocene cold-seep limestones, Olympic Peninsula, Washington: The Veliger, v. 38, p. 47–53.

Squires, R.L., and Gring, M.P., 1996, Late Eocene chemosynthetic? bivalves from suspect cold seeps, Wagonwheel Mountain, Central California: Journal of Paleontology, v. 70, p. 63–73.

Stanton, T.A., 1895, Contributions to the Cretaceous paleontology of the Pacific Coast: The fauna of the Knoxville beds: U.S. Geological Survey Bulletin, v. 133, p. 1–132.

Suess, E., Carson, B., Ritger, S.D., Moore, J.C., Jones, M.L., Kulm, L.D., and Cochrane, G.R., 1985, Biological communities at vent sites along the subduction zone of Oregon, *in* Jones, M.L., ed., Hydrothermal vents of the eastern Pacific: An overview: Bulletin of the Biological Society of Washington, v. 6, p. 475–484.

Teske, A., Hinrichs, K.-U., Edgcomb, V., de Vera Gomez, A., Kysela, D., Sylva, S.P., Sogin, M.L., and Jannasch, H.W., 2002, Microbial diversity of hydrothermal sediments in the Guaymas Basin: Evidence for anaerobic methanotrophic communities: Applied and Environmental Microbiology, v. 68, p. 1994–2007.

Thiel, V., Peckmann, J., Reitner, J., Seifert, R., Wehrung, P., and Michaelis, W., 1999, Highly isotopically depleted isoprenoids-molecular markers for ancient methane venting: Geochimica et Cosmochimica Acta, v. 63, p. 3959–3966.

Thiel, V., Peckmann, J., Richnow, H.H., Luth, U., Reitner, J., and Michaelis, W., 2001a, Molecular signals for anaerobic methane oxidation in Black Sea seep carbonates and a microbial mat: Marine Chemistry, v. 73, p. 97–112.

Thiel, V., Peckmann, J., Schmale, O., Reitner, J., and Michaelis, W., 2001b, A new straight-chain hydrocarbon biomarker associated with anaerobic methane cycling: Organic Geochemistry, v. 32, p. 1019–1023.

Valentine, D.L., and Reeburgh, W.S., 2000, New perspectives on anaerobic methane oxidation: Environmental Microbiology, v. 2, p. 477–484.

Vance, J.A., Clayton, G.A., Mattison, J.M., and Naeser, C.W., 1987, Early and middle Cenozoic stratigraphy of the Mount Rainier–Tieton River area, southern Washington Cascades: Washington State Division of Geology and Earth Resources Bulletin, v. 77, p. 269–290.

Warén, A., and Bouchet, P., 2001, Gastropoda and Monoplacophora from hydrothermal vents and seeps: New taxa and records: The Veliger, v. 44, p. 116–231.

Wells, R.E., 1989, Geologic map of the Cape Disappointment-Naselle River area, Pacific and Wahkiakum counties, Washington: U.S. Geological Survey, Miscellaneous Investigations Series, Map I-1832.

Geological Society of America
Field Guide 4
2003

Holocene lahars and their by-products along the historical path of the White River between Mount Rainier and Seattle

Paul H. Zehfuss

Department of Earth and Space Sciences, University of Washington, Box 351310, Seattle, Washington 98195-1310, USA

Brian F. Atwater

U.S. Geological Survey at Department of Earth and Space Sciences, University of Washington, Seattle, Washington 98195-1310, USA

James W. Vallance

U.S. Geological Survey, Cascades Volcano Observatory, 1300 SE Cardinal Court, B10, S100, Vancouver, Washington 98683, USA

Henry Brenniman

AMEC Earth and Environmental, 11335 NE 122nd Way 100, Kirkland, Washington 98034-6919, USA

Thomas A. Brown

Center for Accelerator Mass Spectrometry, Lawrence Livermore National Laboratory, Livermore, California 94550, USA

ABSTRACT

Clay-poor lahars of late Holocene age from Mount Rainier change down the White River drainage into lahar-derived fluvial and deltaic deposits that filled an arm of Puget Sound between the sites of Auburn and Seattle, 110–150 km downvalley from the volcano's summit. Lahars in the debris-flow phase left cobbly and bouldery deposits on the walls of valleys within 70 km of the summit. At distances of 80–110 km, transitional (hyperconcentrated) flows deposited pebbles and sand that coat terraces in a gorge incised into glacial drift and the mid-Holocene Osceola Mudflow. On the broad, level floor of the Kent Valley at 110–130 km, lahars in the runout or streamflow phase deposited mostly sand-sized particles that locally include the trunks of trees probably entrained by the flows. Beyond 130 km, in the Duwamish Valley of Tukwila and Seattle, laminated andestic sand derived from Mount Rainier built a delta northward across the Seattle fault. This distal facies, warped during an earthquake in A.D. 900–930, rests on estuarine mud at depths as great as 20 m.

The deltaic filling occurred in episodes that appear to overlap in time with the lahars. As judged from radiocarbon ages of twigs and logs, at least three episodes of distal deposition postdate the Osceola Mudflow. One of these episodes occurred ca. 2200–2800 cal. yr B.P., and two others occurred ca. 1700–1000 cal. yr B.P. The most recent episode ended by about the time of the earthquake of A.D. 900–930. The delta's northward march to Seattle averaged between 6 and 14 m/yr in the late Holocene.

Keywords: lahar, delta, Holocene, Mount Rainier, Seattle.

INTRODUCTION

This field trip guide describes natural and man-made exposures of lahar deposits along the White River. Its focus is on clay-poor lahar deposits that postdate the Osceola Mudflow of middle Holocene age and which change downstream into fluvial and deltaic facies. The itinerary highlights preliminary results of work intended to flesh out the history of post-Osceola lahars and to explore volcanic hazards of the valley floor between Auburn and Seattle.

GEOLOGIC SETTING

At 4392 m, Mount Rainier is the tallest volcano in the Cascade Range. On sunny days the mountain looms over Seattle, 90 km to the northwest (Fig. 1). Its glacial ice amounts to 4.4 km³ (Driedger and Kennard, 1986). Valleys that head on the volcano's flanks convey meltwater and debris flows toward the Puget Sound lowland and the Columbia River, via the Puyallup, Carbon, Nisqually, White, and Cowlitz Rivers. All but the Cowlitz flow into Puget Sound (Fig. 1).

The trip follows the historical path of the White River from the foothills near the volcano to the Duwamish River delta. The White River drains the northern and northeastern side of the mountain, then turns westward toward Auburn, where it exits a gorge and flows onto a broad, flat valley bottom (Fig. 1). As recently as 1907, the White River continued northward to Seattle,

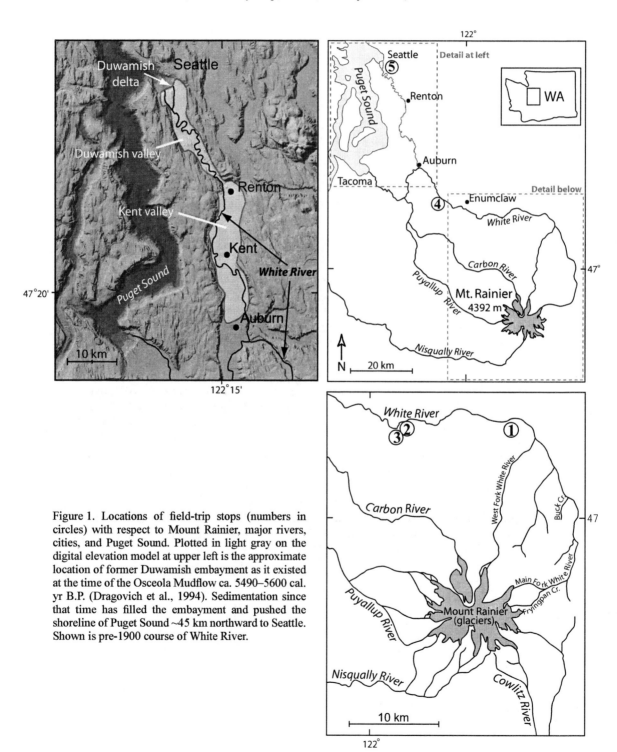

Figure 1. Locations of field-trip stops (numbers in circles) with respect to Mount Rainier, major rivers, cities, and Puget Sound. Plotted in light gray on the digital elevation model at upper left is the approximate location of former Duwamish embayment as it existed at the time of the Osceola Mudflow ca. 5490–5600 cal. yr B.P. (Dragovich et al., 1994). Sedimentation since that time has filled the embayment and pushed the shoreline of Puget Sound ~45 km northward to Seattle. Shown is pre-1900 course of White River.

joining the Green and Black Rivers along the way. The combined waterway, named the Duwamish River, is now fed only by the Green River. Human modifications diverted the other tributary rivers, including the White River, which now flows south and then west to Tacoma (Fig. 1).

The broad shoulders of Mount Rainier rise above folded volcanic and volcaniclastic rocks of Eocene and Miocene age (Fiske et al., 1963). After dike intrusion and volcanic eruptions as early as 26 Ma, these rocks were intruded by the Tatoosh pluton 14–18 Ma (Swanson et al., 1989). Eroded remnants of the Lily Creek Formation (1.2 and 1.3 Ma; Sisson and Lanphere, 2000) probably represent early products of the modern volcano (Crandell, 1963; Mattinson, 1977). The present volcanic cone, chiefly andesitic in composition, began forming ca. 0.5 Ma (Sisson and Lanphere, 2000).

At least 11 tephra-producing eruptions occurred at Mount Rainier during the Holocene (Mullineaux, 1974). The most recent eruption, in the 1840s, produced only scattered deposits of pumice (Mullineaux, 1974; Sisson, 1995). Greater volumes of tephra erupted 2200–2600 ^{14}C yr B.P. (ca. 2200–2800 cal. yr B.P.) and ca. 1000 ^{14}C yr B.P. (ca. 1000 cal. yr B.P.) (Vallance and Donoghue, 2000). Because these voluminous tephra layers contain charcoalized twigs, pyroclastic flows probably accompanied their eruption.

DEFINITIONS

A "lahar" is a gravity-driven mixture of sediment and water that originates from a volcano (Vallance, 2000). Lahars are common occurrences at many volcanoes, particularly during eruptions. Because lahars can travel many tens of kilometers from their sources, some have resulted in great loss of life and property. For example, a 1985 eruption at the Colombian volcano Nevado del Ruiz generated a lahar that devastated the town of Armero, killing more than 23,000 people.

As observed at Mount St. Helens (Scott, 1988), lahars can be divided into those that are rich in clay (clay-rich lahars) and those with little clay (clay-poor lahars). Clay-rich lahars typically initiate as flank failures and leave diamictic deposits. Clay-poor lahars (also called "non-cohesive lahars"), generally consisting of gravel- and sand-sized clasts, originate as water floods that entrain material. Flood waters leading to the formation of clay-poor lahars have been produced during lake breakout (Pierson, 1999; Pringle and Cameron, 1999), intense rainfall (Rodolfo and Arguden, 1991; Hodgson and Manville, 1999; Lavigne et al., 2000; Lavigne and Thouret, 2003), and from melting of snow and ice by hot pyroclastic flows (Eppler, 1987; Scott, 1988; Major and Newhall, 1989).

A lahar can vary in character with time and distance downstream. It may comprise one or more flow phases, which include debris-flow phase, transitional- or hyperconcentrated-flow phase, and stream-flow phase (Vallance, 2000). In a debris-flow phase the solid and liquid fractions of the lahar have about equal volume and the two fractions move approximately in unison in a vertical section. In a stream-flow phase, water transports the lahar's fine-grained sediment in suspension (suspended load) and moves its coarse-grained sediment along the bed at discrete intervals (bed load). A transitional flow phase, commonly known as "hyperconcentrated flow," is intermediate between debris flow and stream-flow. In a transitional phase, a lahar carries higher sediment loads than does stream-flow, but it vertically sorts solids by size and density more than does a debris flow. Although the literature distinguishes the hyperconcentrated-flow phase from more dilute and more concentrated phases in terms of solids fraction, transitions are gradational and dependent on other factors like sediment-size distribution and energy of the flow. Thus, flow-phase transitions cannot be precisely defined.

The height above the channel bottom of the flowing lahar is the stage. A lahar commonly has an initial rising or waxing stage, a peak-inundation stage, and a relatively prolonged falling or waning stage.

In proximal or medial reaches, clay-poor lahars commonly leave poorly sorted, massive or crudely stratified deposits. Such deposits can be inversely graded. In distal reaches, the flows leave voluminous, sandy deposits that may extend tens of kilometers beyond the main body of the lahar (Scott, 1988). These distal facies are commonly referred to as "lahar-runout" deposits. They commonly exhibit bedding and better sorting than that of the proximal or medial deposits.

Radiocarbon ages in this guide are reported in radiocarbon years before A.D. 1950 (^{14}C yr B.P.), in calibrated years before A.D. 1950 (cal. yr B.P.), or both. For an age in cal. yr B.P., we report the range at two standard deviations, computed with the INTCAL98 calibration data of Stuiver et al. (1998) and an error multiplier of 1.0. Most of the ranges are treated as limiting maximum ages because they were measured on detrital wood or charcoal that predates the time of deposition. In cases where the outer preserved rings of a detrital tree were dated, the resulting age may approximate the time when the tree was knocked down, entrained, and deposited during a laharic episode.

PREVIOUS WORK

Lahars of Mount Rainier

Mount Rainier readily generates lahars. It has an enormous volume of snow and ice available for melting during an eruption. It also stores water beneath its glaciers, which sometimes release outburst floods. Its huge mass of hydrothermally altered rock weakens the edifice and contributed to the enormity of the Osceola Mudflow, which had a volume of 3.8 km[3] (Crandell, 1971; Vallance and Scott, 1997).

Lahars are thought to represent the greatest hazard from Mount Rainier (Driedger and Scott, 2002). Several hundred thousand people live in lowland areas underlain by Holocene lahars or laharic deposits derived from the volcano. In the 1990s, the U.S. Geological Survey installed a monitoring system in the Carbon and Puyallup Valleys that is intended to detect and warn residents of the impending arrival of a lahar.

Concern about future lahars on Mount Rainier spurred the mapping and dating of the volcano's Holocene lahar deposits (Crandell, 1971; Scott et al., 1995). This work shows that Mount Rainier produced both clay-rich and clay-poor flows during the Holocene. The largest of these was the clay-rich Osceola Mudflow, which flowed down the White River drainage ca. 5490–5600 cal. yr B.P. (Vallance and Scott, 1997). Lesser, mainly post-Osceola lahars were catalogued by Crandell (1971) and Scott et al. (1995), who used radiocarbon ages and constraining ash layers to place them in time. They identified at least five clay-poor lahar deposits in the White River drainage, deposited since the time of the Osceola Mudflow. The largest of these flows, found as much as 60 m above present river level, form part of a unit named the Deadman Flat lahar assemblage (Scott et. al, 1995). Wood from within a deposit of the assemblage near the confluence of Fryingpan Creek and the White River (Fig. 1) gave an age of 800–1260 cal. yr B.P. (Scott et al., 1995).

Deposits younger than the Osceola Mudflow define at least three episodes of clay-poor lahars in the White River drainage. This inferred history is based on previous work by Crandell (1971), Scott et al. (1995), and Pringle (2000), and on new results from our study (Table 1, Fig. 2). The oldest of the three episodes coincided with volcanism of Summerland age (ca. 2200–2600 ^{14}C yr B.P. or 2200–2800 cal. yr B.P.), when as many as five eruptions may have occurred (Vallance and Donoghue, 2000). Next came two episodes that correspond to the Deadman Flat assemblage of Scott et al. (1995). The oldest of these, provisionally called the Twin Creek episode, dates to ca. 1350–1700 cal. yr B.P. (range of three ages on detrital wood and charcoal from deposits along White River). The youngest, provisionally called the Fryingpan Creek episode, occurred ca. 800–1260 cal. yr B.P. (wood in lahar deposit; Scott et al., 1995).

Lahar Deposits Downstream from Auburn

Most of the foregoing history has been inferred from deposits exposed on valley walls on or near the mountain. Farther downstream, exposures of lahar deposits are rare because stream gradients are low and river incision shallow. However, runout of large clay-poor lahars from Mount Rainier reached the Puget Sound lowland (Scott et al., 1995; Pringle, 2000; Pringle and Scott, 2001). Examples include valley-floor deposits of andesitic sand (informally called "black sand") between Auburn and Seattle that were probably derived from lahars in the White River drainage (Cisternas, 2000; Pringle et al., 1997; Pringle, 2000; Pringle and Scott, 2001).

Formerly an arm of Puget Sound, the valley floor near Auburn and Kent was successively filled by the Osceola Mudflow, deltaic and fluvial deposits, and andesitic sand (Dragovich et al., 1994). The Osceola Mudflow flowed onto the floor of Puget Sound from Auburn to Renton (Fig. 1) (Luzier, 1969; Mullineaux, 1970; Dragovich et al., 1994). The thickness of post-Osceola deposits in this area affords estimates of average rates of sedimentation and delta-front migration for the past 5490–5600 cal. yr. Today, the

Duwamish River delta lies within the city limits 2 km southwest of downtown Seattle, ~35 km north of Auburn (Fig. 2). Estimated using that 35 km distance, post-Osceola sedimentation resulted in progradation at an average long-term rate of ~7 m/yr during the late Holocene (Dragovich et al., 1994). Higher rates can now be estimated for laharic episodes, as discussed below in the section on Stop 5.

NEW WORK

To further explore the history of lahars and lahar-derived deposits from the White River drainage basin, we are examining deposits along a profile that extends from the flanks of Mount Rainier to Puget Sound. Close to the volcano, we are studying the sedimentology and age of debris-flow and transitional facies of lahars. Far from the volcano, we are making parallel studies of fluvial and deltaic facies of andesitic deposits, as well as materials stratigraphically beneath them, as sampled in engineering borings and exposed in excavations. To clarify lahar hazards, we hope to identify which deposits represent lahars or their runout and which resulted from later fluvial recycling of lahar-derived sediment.

Our new ages show that andesitic sand accumulated in the Kent and Duwamish Valleys around the times of the Summerland, Twin Creek, and Fryingpan Creek episodes of clay-poor lahars in the White River drainage. Twigs, as well as detrital and buried trees, sampled from within sand units have yielded ages similar to those for upstream lahar deposits (Table 1, Fig. 2).

Our findings also clarify downstream changes in the geomorphic setting and sedimentary facies of lahar and lahar-derived deposits (Fig. 2). Close to the volcano, lahar deposits plaster steep valley walls incised into the flanks of the volcano and into older bedrock. At the margin of the Puget Sound lowland, lahar deposits coat fluvial terraces situated in a gorge carved into late Pleistocene glacial drift and Holocene mudflow deposits. In distal areas, lahar-derived deposits comprise the floors of broad valleys once filled with marine waters.

Like the field-trip route, our discussion of lahar-related deposits now moves downstream from Mount Rainier to Puget Sound—from the White River valley, through White River gorge and Kent Valley, to the Duwamish Valley.

White River Valley

Post-Osceola lahar deposits in the White River valley between Mount Rainier and Enumclaw (Fig. 2) occupy terraces and valley walls as much as 60 m above the current river. Deposits exposed near river level are clast supported and extremely poorly sorted, consisting of boulders (up to 2 m diameter) and cobbles in a matrix of pebbles and sand. Deposits high on valley walls and near the back edges of terraces are finer grained and better sorted, mostly composed of pebble- and sand-sized particles. Deposits upstream from Enumclaw were probably left by lahars in the debris-flow phase.

Tephra layers provide ways to estimate relative and numerical ages of deposits at Mount Rainier (Mullineaux, 1974) and,

TABLE 1. RADIOCARBON AGES FROM THIS STUDY AND APPLICABLE AGES FROM PREVIOUS STUDY

Lab ID	Age (^{14}C yr B.P.)	Age* (cal. yr B.P.)	Location	Deposit	Sample name	Sample material
Non-laharic deposits						
CAMS 88808	750 ± 40	570 – 710	White River	overbank	RC-S85-1	charcoal
Beta 146056	920 ± 40	740 – 930	Seattle	marsh	Q23A	herbaceous stem
Beta 146057	1120 ± 40	950 – 1140	Seattle	marsh	Q23B	herbaceous stem
Beta 146058	890 ± 50	690 – 930	Seattle	marsh	Q23C	herbaceous stem
Beta 146059	990 ± 40	790 – 960	Seattle	marsh	Q23D	herbaceous stem
Beta 145728	1890 ± 60	1700 – 1960	Seattle	bay mud	Q15F	charcoal
Beta 145729	1730 ± 60	1520 – 1810	Seattle	bay mud	Q17B	wood
Beta 145730	1990 ± 70	1810 – 2120	Seattle	bay mud	Q15G	wood
Fryingpan Creek episode						
CAMS 88809	1090 ± 40	930 – 1170	Kent/Auburn	runout sand	RC-MC3-S14	wood
Beta 106603	1280 ± 60	1070 – 1300	Seattle	runout sand	N.A.§	twig
N.A.	1120 ± 80	800 – 1260	White River	lahar	N.A.	wood
N.A.	1255 ± 130	930 – 1410	White River	lahar	N.A.	stump on lahar deposit
Twin Creek episode						
WW3353	1560 ± 40	1350 – 1530	White River	lahar	N.A.	charcoal
Beta 169386	1560 ± 60	1320 – 1560	Kent/Auburn	runout sand	RC 277-4	buried tree
WW3782	1615 ± 35	1410 – 1610	White River	lahar	N.A.	charcoal
CAMS 88805	1655 ± 30	1420 – 1690	Seattle	runout sand	DGS-6A	twig
Beta 171509	1670 ± 50	1430 – 1700	White River	lahar	RC-S132-A	charred wood
CAMS 88806	1790 ± 40	1570 – 1820	Seattle	runout sand	DGS-6B	twig
CAMS 88801	1850 ± 45	1630 – 1880	Kent/Auburn	runout sand	RC-MC6-S10	wood
WW3796	1761 ± 39	1308[†] – 1565[†]	Kent/Auburn	runout sand	755-A	inner rings detrital tree
WW3797	1783 ± 40	1371[†] – 1624[†]	Kent/Auburn	runout sand	755-B	inner rings detrital tree
WW3798	1773 ± 41	1420[†] – 1671[†]	Kent/Auburn	runout sand	755-C	inner rings detrital tree
WW3799	1561 ± 44	1250[†] – 1438[†]	Kent/Auburn	runout sand	755-D	inner rings detrital tree
Summerland episode						
Beta 169383	2110 ± 60	1970 – 2310	Boeing Field	runout sand	MF 28B	outer rings detrital tree
CAMS 88807	2115 ± 40	1950 – 2300	South Park	runout sand	DGS-9M-1	twig
CAMS 91123	2170 ± 35	2060 – 2310	South Park	runout sand	DGS-9N	twig
Beta 169384	2180 ± 60	2000 – 2340	Boeing Field	runout sand	MF 29A1	outer rings detrital tree
Beta 169385	2180 ± 60	2000 – 2340	Boeing Field	runout sand	MF 29B4	outer rings detrital tree
CAMS 88800	2260 ± 50	2150 – 2350	Boeing Field	runout sand	B410	twig
CAMS 88799	2265 ± 40	2150 – 2350	Seattle	runout sand	RC-H14	twig
CAMS 88804	2500 ± 40	2360 – 2740	Boeing Field	runout sand	B407	twig
CAMS 88803	2510 ± 40	2610 – 2740	Kent/Auburn	runout sand	RC-B5-S10	wood
Beta 172998	2540 ± 60	2370 – 2770	White River	lahar	RC-S115-1	charcoal

*Dates calibrated using CALIB program of Stuiver et al. (1998) and are calibrated years before 1950 (cal. yr B.P.).
[†]Dates were adjusted after calibration by subtracting number of rings between sample and outer rings of tree. Error expanded to account for uncertainty in ring counts (± 3 years).
§N.A. indicates no data available.

to a lesser extent, downstream along the White River drainage. Among widespread tephra layers from post-Osceola time is Mount St. Helens tephra set Y, which erupted in the centuries between 2470–3700 and 3640–4240 cal. yr B.P. (each of these ranges represents a bounding radiocarbon age; calibrating these ages with data available in the early 1970s, Mullineaux (1974) assigned set Y an approximate age of 3000–3900 cal. yr B.P.). Next, from Mount Rainier itself, is set C, which dates to the interval between 1530–2690 and 1900–2780 cal. yr B.P. (Mullineaux (1974): ca. 2.2 ka). Pumice from set C appears not only at Mount

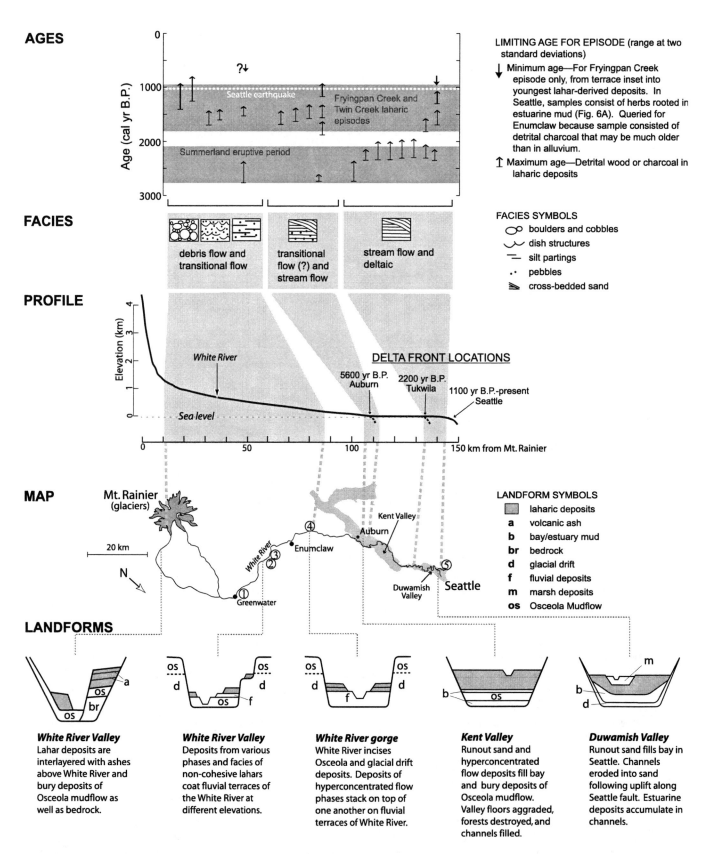

Figure 2. Geometry and generalized interpretations of lahar and lahar-runout deposits in the White River system of Mount Rainier and the Duwamish Valley. Ages for laharic and non-laharic sediments (from Table 1) in the White River and Duwamish Valleys. Age of sample 277-D omitted because it is discordant with three other ages on logs from same deposit (277-A, B, C; Table 1).

Rainier but also, recycled, in andesitic deposits as far downstream as Seattle. Tephra set W, which erupted from Mount St. Helens and blanketed much of Mount Rainier, is unrecognized around Puget Sound. The largest of the set W eruptions occurred in A.D. 1479 (Fiacco et al., 1993).

Radiocarbon ages from woody debris within lahar deposits suggest that a previously unrecognized lahar, or lahar episode (Twin Creek), occurred ca. 1500 cal. yr B.P. A chunk of charred tree found in a deposit ~30 m above the White River near the confluence with Buck Creek yielded an age of 1320–1560 cal. yr B.P. Charcoal from a lahar deposit near the town of Greenwater gave ages of 1350–1520 and 1410–1610 cal. yr B.P. (Table 1).

White River Gorge

Just east of the present location of Enumclaw, the Osceola Mudflow left the confines of bedrock valleys and spread out over the rolling surface of the Puget Sound lowland. The White River has since cut a gorge 40 m deep by incising through deposits of the Osceola Mudflow and into the underlying glacial drift. Terraces formed during the incision provided platforms for subsequent lahar deposits. Lahar deposits found here occur as much as 30 m above current river level.

At least three lahars successively covered terraces in the gorge. Charred wood in the oldest of the three yielded an age of 2370–2770 cal. yr B.P. (Table 1). The deposits are granular and poorly sorted, with mean grain sizes in the sand-size range. While sand-sized particles are predominantly andesitic, pebble-sized clasts vary in lithologic composition and may be granitic, metamorphic, or andesitic. Deposits are mostly massive; however, in some places they are normally graded at their tops and inversely graded at their bottoms. Each of the deposits resembles those left by transitional flows, a resemblance that suggests dilution of the lahars as they descended the gorge.

As with deposits seen in the White River valley, lahar deposits near river level in the gorge are substantially coarser than those flanking terraces. The deposits attain thicknesses up to 3 m and contain abundant clasts up to 20 cm in diameter, with occasional boulders up to 1 m in diameter. The matrix, which is comprised of pebble- and sand-sized clasts, resembles material observed in deposits on the terraces.

Kent Valley

At the city of Auburn, the White River exits the post-Osceola gorge and spills onto the valley bottom of the White and Green Rivers (Fig. 1). For simplicity, we refer to this portion of the valley between Auburn and Renton as the Kent Valley.

The Osceola Mudflow in the vicinity of Auburn, as identified in geotechnical borings, underlies as much as 80 m of deltaic and laharic deposits (Dragovich et al., 1994). The White River's post-Osceola incision of the drift plain east of the Kent Valley by the White River delivered sediment to the delta at Auburn while forming the gorge. Lahars flowing through the White River gorge buried the deltaic sediments with andesitic sand and gravel. From this point, laharic debris has traveled both northward toward Seattle and southward toward Tacoma.

Shallow deposits near Auburn are composed chiefly of laharic sediment, as judged from boreholes logs, borehole samples, and excavations from engineering projects. Construction projects and cross sections drawn from lines of geotechnical borings provide views of these deposits and means for sampling dateable material and sediment. Uninterrupted layers of andesitic, moderately to poorly sorted sand and gravel are as much as 10 m thick and are overlain by overbank silt deposits up to 6 m thick. Buried logs, common near the tops of sand deposits, yield ages similar to that of the lahar dated ca. 1300–1600 cal. yr B.P. in the White River valley. Wood fragments found low in two, separate sand deposits gave ages of 930–1170 and 2610–2740 cal. yr B.P. (Table 1), similar to the ages of the Fryingpan Creek episode and Summerland eruptive period, respectively.

Duwamish Valley

West of Renton, below the former confluence of the White River with the Black River, the Duwamish River flows through a narrow bedrock gap and then continues northward into a widening valley that leads to Puget Sound (Fig. 1).

To study andesitic deposits in the Duwamish Valley, we are using geotechnical borings, construction site excavations, and peels made from vertical slices (geoslices). The slices, 0.5 m wide and up to 8 m long, were peeled with hydrophilic grout that reveals sedimentary structures and liquefaction features (Atwater et al., 2001).

Ice-sheet cover and retreat, marine inundation, and the arrival of sediment from lahars produced a diverse array of subsurface sedimentologic units observed in borings from this area. The Duwamish Valley itself was carved by subglacial meltwater streams during the Last Glacial Maximum (Crandell, 1963; Booth 1994). During recession, glacial ice dammed marine waters at the Strait of Juan de Fuca, flooding much of the area that is now Puget Sound with freshwater lakes, where silt and clay accumulated (Bretz, 1913; Thorson, 1989). Continued ice retreat eventually opened the strait, allowing marine waters to invade the region. In the Duwamish Valley, the resulting embayment produced tens of meters of mud that contains marine shells. As laharic debris built northward through the valley from Auburn, the White River delta overrode this mud.

The delta reached Tukwila by Summerland time (Fig. 2). By Fryingpan Creek time (ca. 1100 cal. yr B.P.), it had crossed the Seattle fault and probably came to within a few kilometers of the site of downtown Seattle. Then, or soon afterward, the floodplain behind the delta front was warped by an earthquake on the Seattle fault, dated elsewhere to A.D. 900–930 (Bucknam et al., 1992; Atwater, 1999). As discussed at Stop 5, the Fryingpan Creek episode probably ended around the time of this earthquake, and no subsequent lahars in the White River drainage have managed to leave much if any stratigraphic record in Seattle since 780–930 cal. yr B.P.

SUMMARY

Ages for lahar deposits in the White River system suggest at least three episodes of post-Osceola clay-poor lahars. One episode was concurrent with volcanism of Summerland age (2200 to 2600 yr B.P. or 2200 to 2800 cal. yr B.P.; Vallance and Donoghue, 2000). The following episodes occurred ca. 1500 (Twin Creek) and 1100 cal. yr B.P. (Fryingpan Creek).

The timing of lahar and lahar-derived sand deposition in the Kent and Duwamish Valleys as far as the city of Seattle appears to coincide with the timing of clay-poor lahars in the White River system. Sand deposits correlative with the Summerland episode occur near the surface at Tukwila, suggesting that delta progradation had extended at least to that point by ca. 2200 cal. yr B.P. Deposits correlative with Twin Creek and Fryingpan Creek episodes occur throughout the length of the field area between Mount Rainier and Seattle and contributed to further delta progradation as well as floodplain aggradation.

ROAD LOG

The sequence of stops moves downstream along the White and Duwamish River valley areas beginning at a site ~50 km downstream of Mount Rainier (Fig. 1). Distances are given in miles and kilometers (in parentheses).

Stop 1. Weyerhaeuser Gravel Quarry

Note: The first stop is located on Weyerhaeuser land and advanced permission is required in order to gain access.

Cumulative Miles	(km)	Description
0.0	(0.0)	On Interstate 5 south out of Seattle set odometer to 0.0 at milepost 165.
23.0	(37.0)	Exit right to Highway 18 east (exit 142A).
27.5	(44.3)	Exit right onto State Route 164 east (Auburn Way/Enumclaw).
42.4	(68.2)	Turn left at the stoplight onto SR 410 east.
54.8	(88.2)	Turn right down small gravel road and drive through Weyerhaeuser gate #54). Just beyond gate, turn left onto gravel road.
55.4	(89.1)	Turn right onto small dirt road.
55.5	(89.4)	Arrive at quarry.

Highlights

Sedimentary structures of a clay-poor lahar exposed in quarry walls.

Description

As of July 2003, the walls of this active quarry formed a large (30 m diameter) horseshoe-shaped exposure made up entirely of a single lahar deposit, locally at least 3 m thick. The White River flows 300 m to the north and 13 m below, meandering around the deposit. The stop is 7 km downstream from the town of Greenwater and 50 km from the summit of Mount Rainier, as measured down the Main Fork of the White River.

The exposure provides excellent views of clay-poor lahar deposits. In grain size and sedimentary structures, the deposits closely resemble those of the lahar floodplain facies described by Scott (1988) for modern lahars in the Toutle-Cowlitz river system of Mount St. Helens. Overall, the portion of the deposit in view is normally graded and composed of sand- and pebble-sized material. The upper meter of the exposure contains more pumice lapilli (layer C) than does the rest of the section.

Especially striking are dish and pillar structures. These form chains of concave-upward silt partings commonly broken from one another at their ends (Fig. 3). The partings themselves are millimeters in thickness and are spaced tens of centimeters apart vertically. Although typically associated with subaqueous sediment gravity flows, these structures also occur in hyperconcentrated-flow deposits (Scott et al., 1995). They apparently begin to form soon after deposition as water is expelled upward out of the deposit (Lowe and LoPiccolo, 1974). Continued translocation of clay and silt enhances the structures (Scott et al., 1995).

Charcoal collected from within the deposit gave ages of 1350–1520 and 1410–1610 cal. yr B.P. The dates are similar to that of a lahar deposit near the confluence of the White River with Buck Creek (1320–1560 cal. yr B.P.; charred wood). The ages are also similar to those of logs in fluvial deposits near Kent (1320–1560 cal. yr B.P.; Pringle, 2000; Table 1). Thus far, no eruptive products that correlate with this lahar episode (Twin Creek) have been identified on Mount Rainier volcano itself.

Stops 2a and 2b. Mud Mountain Dam

Cumulative Miles	(km)	Description
55.5	(89.4)	Return to small dirt road that led to the quarry.
55.7	(89.6)	Turn left onto gravel road.
56.3	(90.5)	Turn right and proceed through Weyerhaeuser gate #54.
63.6	(102.4)	Turn left onto Mud Mountain Dam Road.
64.3	(103.5)	Turn left into gravel parking area and park. On foot, follow trail south out of parking area. After ~100 m the trail will meet a gravel road. Follow the road past two metal gates and down the hill. Stop 2a is a small outcrop of lahar deposits on the left side of the road and is 0.5 mi (0.8 km) from the parking lot. Continue down the road for an additional 0.4 mi (0.6 km) to Stop 2b, a large outcrop of lahar deposits.

Highlights

Two more examples of clay-poor lahar deposits. Morphology of the White River valley.

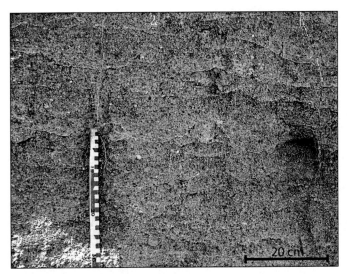

Figure 3. Dish structures exposed in the quarry wall at Stop 1.

Figure 4. Rip-up clasts in a lahar deposit exposed in walls of the river-cut exposure at Stop 2b. Pencil shows scale.

Description

On the right hand side of Mud Mountain Dam Road, as you leave the highway, is a patch of clear-cut land exposing several mounds, or hummocks. The surface here is underlain by deposits of the Osceola Mudflow, which flowed through this part of the White River valley at a width of ~3 km and maximum depth of 130 m (Scott and Vallance, 1995; Vallance and Scott, 1997). Most hummocks of the Osceola Mudflow consist of car- to house-sized pieces of the volcanic edifice that survived transport and deposition and became stranded along the margins of the flow as the flow waned (Scott et al., 1995; Vallance and Scott, 1997).

A clay-poor lahar deposit forms an exposure a few meters high at Stop 2a. Grain size ranges from fine sand to gravel. Although the deposit is undated, it contains pumice from layer C and therefore postdates 2200 yr B.P. (age from Mullineaux, 1974). Horizontal bands of brownish silt (millimeters in thickness) are also present in the outcrop. The bands appear genetically related to dish structures but lack their characteristic concavity. Using a hand auger, we bored 5 m beneath ground surface without reaching the base of the lahar deposit.

As the road descends out of the forest and toward Stop 2b, the White River valley comes into view. During rainy winter months, this part of the valley is often inundated by water impounded by Mud Mountain Dam, situated 1 km to the west. Downstream, the river flows through a narrow notch incised into bedrock.

Where the road begins to flatten on the modern floodplain of the White River, a clay-poor lahar deposit crops out in an exposure 30 m long and 3 m tall (Stop 2b). The deposit is made of andesitic sand and occasional pebble laminae. A particularly impressive feature of the exposure is the presence of large "rip-up" clasts (up to 0.5 m) that float in the sand matrix (Fig. 4). The composition of clasts resembles that of laminated ice-sheet drift exposed in nearby roadcuts. Suspended in the deposit are cobbles and even boulders up to 40 cm in diameter. The age of

this deposit is uncertain, however, it is probably younger than the Osceola Mudflow, which it appears to overlie. Deposits of the Osceola Mudflow crop out near river level at this locality.

Stop 3. Mud Mountain Dam Recreational Facility

Cumulative Miles	(km)	Description
64.3	(103.5)	Turn left out of the parking area onto Mud Mountain Dam Road.
65.8	(105.9)	Arrive at Mud Mountain Dam Recreation Area.

Highlights

Lunch at the Chinook shelter. Public restrooms.

Description

Mud Mountain Dam was completed in 1949 with the purpose of reducing flood hazards on the valley floor between Auburn and Tacoma. During wet, winter months in the late 1800s and early 1900s, farmland in the lower White River valley in the vicinity of

Auburn and Kent often flooded. During that period, local farmers feuded over the course of the flooding channel. Several clandestine attempts were made to redirect floodwaters by blowing up portions of the channel with dynamite. One such attempt backfired when explosions generated a landslide that redirected water in the direction of the demolitionists' land, and completely dried up the original channel of the White River (*Valley Daily News*, June 9, 1991, p. 29–30). Today the White River, which used to flow northward to Seattle, instead flows southward to Tacoma.

While intended for flood control, Mud Mountain Dam might provide some defense against lahars traveling down the White River valley. The dam is composed of local, unconsolidated mudflow and glacial till deposits, sand and gravel, and an inner concrete wall. The total storage capacity of the dam, 130 million m³, equals about half the volume of lahars that Scott et al. (1996) assigned an average recurrence interval of 500–1000 yr. The estimated volume (3.8 km³) of the Osceola Mudflow, a singular postglacial event, is roughly 30 times that of the dam's storage capacity.

Stop 4. Golden Valley Terrace Sequence

Cumulative Miles	(km)	Description
65.8	(105.9)	Leave Mud Mountain Dam Recreation Area and follow Mud Mountain Dam Road back to SR 410.
68.0	(109.4)	Turn left onto SR 410.
76.5	(123.1)	At stoplight, turn right onto Park Avenue.
76.9	(123.8)	Road turns left and becomes Naches St.
77.2	(124.2)	Turn right at stop sign onto West Mason St.
78.6	(126.5)	Turn right at stop sign onto Sumner-Buckley Highway east.
81.0	(130.4)	Turn right onto Buckley-Tapps Highway.
81.9	(131.8)	Turn right at sign for Golden Valley and follow the drive down a steep hill.
82.3	(132.4)	Arrive at Stop 4.

Highlights
Backhoe pit into as many as three successive lahar deposits.

Description
Now 17 km downstream from Mud Mountain Dam, we stand inside the margin of the Puget Sound lowland. At the sign to Golden Valley, we dropped off the plain of till-mantled outwash that dominates the lowland's skyline (Booth, 1994). Below the sign, the White River has cut a gorge into the Osceola Mudflow and underlying Pleistocene glacial drift. Prominent terraces capped by younger lahar deposits are preserved in pockets on either side of the river (Fig. 5A).

Backhoe trenches and hand auguring reveal a sequence of at least three lahar deposits that may be traced between different terraces on either side of the White River (Fig. 5A). The lower-

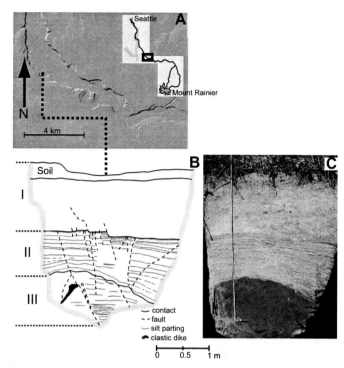

Figure 5. Location of backhoe trench at Stop 3. Box on inset map shows area of digital elevation model. Note prominent terraces along White River, which are covered with lahar deposits. Trench log (B) and photo (C) showing three individual lahar deposits (I, II, and III). Sedimentology of the deposits suggests deposition in a hyperconcentrated-flow facies. The age is from a sample collected in a trench 150 m to the southeast that showed identical stratigraphy.

most deposit (at least 1 m thick) is dark gray in color and composed primarily of poorly sorted andesitic sand. Charcoal from this unit (Fig. 5B and 6) yielded an age of 2370–2770 cal. yr B.P., similar to that of the Summerland eruptive episode. The middle unit (~1 m thick) is the most poorly sorted in the section and contains mostly dark gray andesitic sand with abundant pebbles of varying lithologies. The coarsest material is concentrated in the middle 0.5 m of the unit. Also present are horizontal silt laminae similar to those observed at Stop 2a. The uppermost deposit (~1 m thick) is massive and also primarily composed of andesitic sand. A higher fraction of silt than underlying deposits gives it a distinct yellowish hue. Fragments of pumice (up to 1 cm) from layer C are also common in this unit. The two upper units must postdate 2370–2770 cal. yr B.P., and they predate 570–710 cal. yr B.P. if that is the approximate age of the next lowest terrace.

That terrace was built by ordinary fluvial deposits of the White River. Well-exposed in the modern river bank, these deposits consist of cobble gravel capped with 1–2 m of overbank silt. Charcoal from the base of the overbank deposits gave an age of 570–710 cal. yr B.P. If that charcoal is similar in age to its time of deposition, no large lahars have descended this part of the White River in the past six centuries.

Stop 5. Terminal 107, Seattle

Cumulative		
Miles	(km)	*Description*
82.3	(132.4)	Turn around and return to the drive, which leads out of Golden Valley.
82.7	(133.0)	Turn left onto Buckley-Tapps Highway.
84.0	(135.2)	Turn right onto Sumner-Buckley Highway.
86.1	(138.6)	Take a left at stoplight onto 214th Avenue east.
86.9	(139.9)	Take a right at the stoplight onto Highway 410 west.
93.4	(150.3)	Take the exit for Highway167 north (Seattle).
112.8	(181.5)	Take the exit for I-405 south.
115.1	(185.2)	Take the exit for I-5 north (Seattle).
123.3	(198.4)	Take the exit for West Seattle Bridge, Columbian Way (exit 163). Follow signs for West Seattle Bridge.
125.6	(202.1)	Take the exit for Port of Seattle terminals 5–115 (Delridge Way SW, South Seattle Community College, SW Spokane St.). At the bottom of the ramp, follow signs for West Marginal Way.
126.8	(204.1)	Turn left at Edmunds St. into the Duwamish Public Access Area and park.

Highlights

Peels of sediment slice into deltaic deposits probably derived from lahars of Mount Rainier.

Description

This site was recently made into a park as part of an effort to restore natural environments along the Duwamish waterway. The bottom of the waterway is a Superfund site. Lahar-derived sand, which underlies most of the rest of the valley floor, contains polluted groundwater.

The waterway originated in 1914 through hydraulic dredging across former meanders of the Duwamish River (Fig. 6). Some of the dredge spoils were pumped eastward to fill tide flats and marshes. The park itself occupies a natural terrace that the Port of Seattle had planned to make into a terminal. Discovery of an archaeological site led the Port to abandon that plan.

Cross section A–A'

Deposits above a bedrock surface record the postglacial history of this part of Duwamish Valley (cross section A–A' in Fig. 6).

That history begins with glacial ice or subglacial streams that cut the bedrock surface and covered it with compact drift. The drift was then covered by clay that probably accumulated in the proglacial lake during retreat of the Puget lobe of the Cordilleran Ice Sheet.

Shell-bearing deposits in the middle of the section were deposited after marine waters inundated the area when the ice sheet retreated beyond the Straight of Juan de Fuca. Those waters arrived before 11,280–11,940 cal. yr B.P., the age of a log in peat between two units of bay mud. Above this peat, shell-rich units of

sand and gravel represent deep-water currents or a beach. These coarse-grained deposits were then covered by estuarine mud.

The uppermost part of section A–A' records the arrival of the White River delta. Stratigraphic units containing intercalated mud and andesitic sand probably represent bottomset beds of the lahar-fed delta. Andesitic sand of the delta is 20 m thick across much of the section. The sand is capped by mud and peat as much as 5 m thick—probably the deposits of floodplains and marshes of the past two millennia. Radiocarbon ages of logs in the uppermost part of the sand near the eastern end of cross section A–A' suggest that the delta built past this area in Summerland time (1970–2310 and 2000–2340 cal. yr B.P.; Table 1, Fig. 6). The samples came from sand less than 2 m below the mud cap.

Cross Section B–B'

Subsurface deposits in Seattle's part of the Duwamish Valley show evidence for further progradation of the delta fed by lahars from Mount Rainier and for subsequent vertical displacement during an earthquake. A valley cross section drawn from geotechnical borings shows that andesitic sand deposits bury shell-bearing estuarine deposits to depths of 20 m or more (B–B', Fig. 6). At 4th Avenue south, 0.9 km east of Stop 5, a twig in andesitic sand ~13 m below the top of the sand gave an age of 2150–2350 cal. yr B.P. (Table 1), in the range of the Summerland eruptive episode (Fig. 6). Twigs and a stick higher in the andesitic sand gave ages of 1470–1820, 1420–1690, and 1070–1300 cal. yr B.P. (Table 1). If these younger ages date deposition of the delta's topset beds, an arm of Puget Sound persisted off the site of Terminal 107 through the Summerland eruptive period; the White River delta did not build past the site of Terminal 107 until the Twin Creek or Fryingpan Creek lahar episodes.

The andesitic sand along cross section B–B' probably predates 1020–1050 cal. yr B.P. because it contains burrows at elevations at or above modern high tides. Uplift along the Seattle fault ca. A.D. 900–930 (Bucknam et al., 1992; Atwater, 1999) raised this burrowed sand ~5 m, thereby forming a single valley-floor terrace that stood above the level of historical floods.

Non-laharic deposits in contact with the sand provide additional ages that limit the time when the White River delta built past the site of Terminal 107. At Terminal 107, the sand overlies shell-bearing bay deposits from which sticks have given ages of 1520–1810, 1700–1960, and 1810–2120 cal. yr B.P. (Table 1). Marsh deposits inset into the andesitic sand began forming by 780–930 cal. yr B.P. (Table 1). These deposits lack sand layers other than sand-blow lenses connected to feeder dikes. The marsh deposits thus imply that no sandy lahar runout has approached the site of Seattle in the past eight to ten centuries.

In fall of 2000, a team of Japanese, American, and Chilean scientists collected giant vertical slices (geoslices) of deltaic, andesitic sand in the lower Duwamish Valley. Peels made from these slices show cross-bedding and parallel lamination. Most of the sand in the peels is moderately well to moderately sorted, medium to coarse sand. Rip-up mud clasts and planar-laminated sand (Fig. 7) suggest energetic flow. Sand dikes and convoluted laminae perhaps resulted from the earthquake of A.D. 900–930.

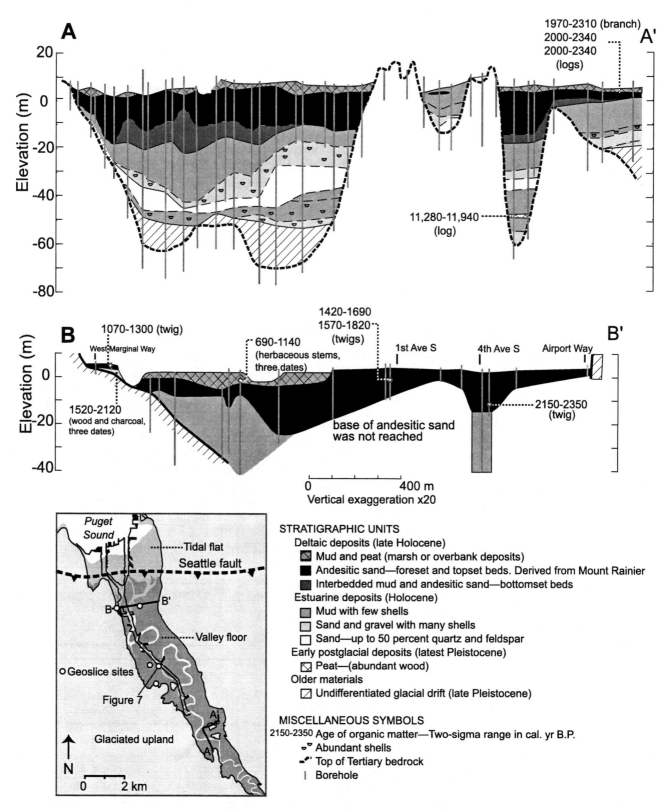

Figure 6. Map showing lower Duwamish Valley (Fig. 1), general physiographic features, and the location of cross sections A–A′ and B–B′. Also shown are the former course of the Duwamish River (in white) and the present, straightened course (outlined in black). Cross sections A–A′ and B–B′ are drawn from geotechnical borings. All ages are cal. yr B.P. Italicized ages are from non-laharic deposits, shell-bearing mud from intertidal or subtidal bay-bottom deposits, and mud and peat from a tidal marsh.

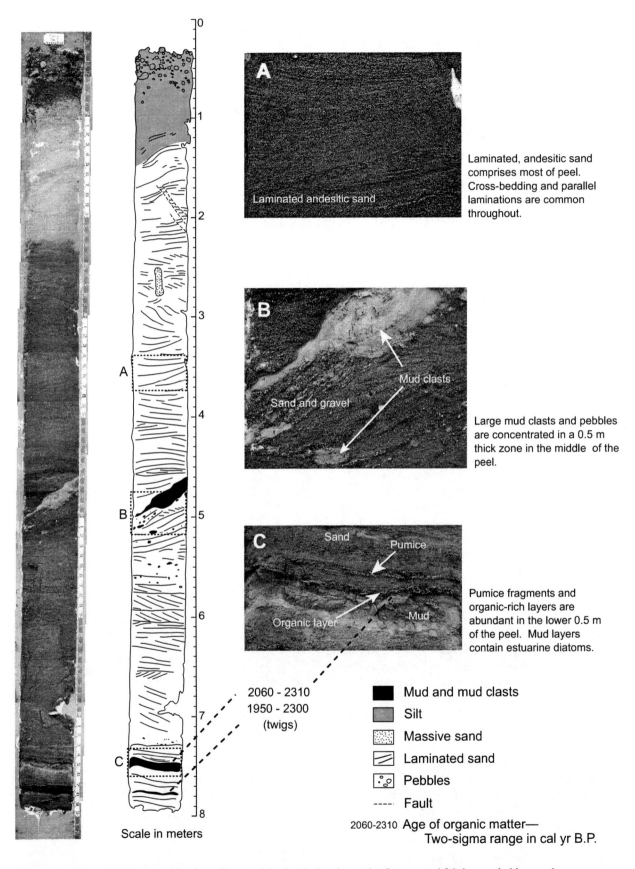

A

Laminated andesitic sand

Laminated, andesitic sand comprises most of peel. Cross-bedding and parallel laminations are common throughout.

B

Mud clasts

Sand and gravel

Large mud clasts and pebbles are concentrated in a 0.5 m thick zone in the middle of the peel.

C

Sand

Pumice

Organic layer

Mud

Pumice fragments and organic-rich layers are abundant in the lower 0.5 m of the peel. Mud layers contain estuarine diatoms.

0

1

2

3

A

4

B

5

6

2060 - 2310
1950 - 2300
(twigs)

7

C

8

Scale in meters

◼ Mud and mud clasts

▧ Silt

▨ Massive sand

▨ Laminated sand

▨ Pebbles

---- Fault

2060-2310 Age of organic matter—
Two-sigma range in cal yr B.P.

Figure 7. Photo-mosaic of geoslicer peel 9; sketch showing major features and fabric revealed in sample.

Delta Progradation Rates

The White River delta built northward from Auburn at an average rate of 6.9 m/yr in the late Holocene according to estimates by Dragovich et al. (1994). Dragovich and his coworkers assumed that the delta prograded 35 km since Osceola time, which they assigned to 5700 cal. yr B.P. Average progradation rates can now be computed for subdivisions of post-Osceola time, by means of new evidence reported in this field guide.

Between the time of the Osceola Mudflow and the approximate end of the Summerland episode (ca. 2200 cal. yr B.P.), rates of delta progradation between Auburn and Tukwila were similar to or slightly higher than the longer-term average calculated by Dragovich et al. (1994). At Tukwila, floodplain mud began to cover deltaic sand ca. 2000–2340 cal. yr B.P., as judged from the age of logs and a branch in the eastern part of cross section A–A′ (described above and plotted in Figs. 2 and 6). The delta thus prograded at least 26 km between the time of the Osceola Mudflow and the end of the Summerland episode. Calculated using the age range of the Osceola Mudflow (5490–5600 cal. yr B.P; Vallance and Scott, 1997) and the age of the logs, the minimum long-term average delta progradation rate between Auburn and Tukwila in the intervening period was 7.2 to 8.3 m/yr.

The delta probably prograded at least this fast during the Twin Creek and Fryingpan Creek episodes, as the delta advanced to its present site in Seattle. As a starting point for this interval we use the age and location of the logs and branch on the eastern part of section A–A′ in Tukwila (2000–2340 cal. yr B.P.; Fig. 6). As a conservative end point, we use the age and location of a twig in andesitic sand on the terrace at Stop 5, on cross section B–B′ (1070–1300 cal. yr B.P.; Fig. 6). The distance between these sites is 7 km, for an average delta progradation rate of 5.5–10.0 m/yr. The rate was higher if during that time the delta built beyond Stop 5 to its present position, 3 km beyond Stop 5. In that case, the rate averaged 7.9–14.3 m/yr during the laharic episodes that began in Summerland time.

ACKNOWLEDGMENTS

Zehfuss received funding from the Geological Society of America, the U.S. Geological Survey, and the Department of Earth and Space Sciences of the University of Washington. Paula Zermano at Lawrence Livermore National Laboratory (LLNL) helped with processing of radiocarbon dates. AMEC and Sound Transit generously provided access to boring logs and samples used to construct the cross section A–A′ in Figure 6. Shannon and Wilson also generously provided access to boring samples and construction sites in the Kent Valley and along section A–A′, as did Dames and Moore (now URS) along section B–B′. The government of Japan funded the geoslicing under the U.S.-Japan Earthquake Disaster Mitigation Partnership. That work was arranged by Kenji Satake, directed by Tsuyoshi Haraguchi and Keita Takada, and carried out by staff of Bergerson Construction Co. Marco Cisternas collected the wood samples from the eastern part of section A–A′. We thank Joanne Bourgeois, Alan Gillespie, Patrick Pringle, and Ron Sletten for discussions; Nathan Chutas and Elizabeth Mahrt for field help; and Kevin Scott, Carolyn Garrison-Laney, and Terry Swanson for manuscript reviews. This work was performed under the auspices of the U.S. Department of Energy by the University of California, Lawrence Livermore National Laboratory under contract no. W-7405-Eng-48.

REFERENCES CITED

Atwater, B.F., 1999, Radiocarbon dating of a Seattle earthquake to A.D. 900–930: Seismological Research Letters, v. 70, no. 2, p. 232.

Atwater, B.F., Burrell, K.S., Cisternas, M., Higman, B., Barnhardt, W.A., Kayen, R.E., Minasain, D., Satake, K., Shimokawa, K., Haraguchi, T., Takada, K., Baker, D., and Nakata, T., 2001, Grouted sediment slices show signs of earthquake shaking: Eos (Transactions, American Geophysical Union), v. 82, no. 49, p. 603, 608.

Booth, D.B., 1994, Glaciofluvial infilling and scour of the Puget lowland, Washington, during ice-sheet glaciation: Geology, v. 22, p. 695–698.

Bretz, J.H., 1913, Glaciation of the Puget Sound region: Washington Geological Survey Bulletin, v. 8, 244 p.

Bucknam, R.C., Hemphill-Haley, E., and Leopold, E.B., 1992, Abrupt uplift within the past 1700 years at southern Puget Sound, Washington, *in* Proceedings of Conference LXII; eighth joint meeting of the U.S.-Japan conference on natural resources (UJNR), panel on earthquake prediction technology: U.S. Geological Survey Open File Report 93-0542, p. 75–78.

Cisternas, M.V., 2000, Preliminary findings about the "black sand" in the lower Duwamish River valley, Seattle, Washington, *in* Palmer, S.P., ed., Final report—Program announcement no. 98-WR-PA-1023, geotechnical/ geologic field and laboratory project: Washington Division of Geology and Earth Resources contract report, 1 v.

Crandell, D.R., 1963, Surficial geology and geomorphology of the Lake Tapps Quadrangle, Washington: U.S. Geological Survey Professional Paper 388-A, 84 p.

Crandell, D.R., 1971, Postglacial lahars from Mount Rainier volcano, Washington: U.S. Geological Survey Professional Paper 677, 75 p.

Dragovich, J.D., Pringle, P.T., and Walsh, T.J., 1994, Extent and geometry of the mid-Holocene Osceola Mudflow in the Puget lowland—Implications for Holocene sedimentation and paleogeography: Washington Geology, v. 22, no. 3, p. 3–26.

Driedger, C.L., and Kennard, P.M., 1986, Ice volumes on Cascade volcanoes—Mount Rainier, Mount Hood, Three Sisters and Mount Shasta: U.S. Geological Survey Professional Paper 1365, 28 p.

Driedger, C.L., and Scott, K.M., 2002, Mount Rainier—Learning to live with volcanic risk: U.S. Geological Survey Fact Sheet 034-02.

Eppler, D.B., 1987, The May 1915 eruptions of Lassen Peak, II: May 22 volcanic blast effects, sedimentology and stratigraphy of deposits, and characteristics of the blast cloud: Journal of Volcanology and Geothermal Research, v. 31, p. 65–85.

Fiacco, R.J., Palais, J.M., Germani, M.S., Zielinski, G.A., and Mayewski, P.A., 1993, Characteristics and possible source of a 1479 A.D. volcanic ash layer in a Greenland ice core: Quaternary Research, v. 39, p. 267–273.

Fiske, R.S., Hopson, C.A., and Waters, A.C., 1963, Geology of Mount Rainier National Park, Washington: U.S. Geological Survey Professional Paper 444, 93 p.

Hodgson, K.A., and Manville, V.R., 1999, Sedimentology and flow behavior of a rain triggered lahar, Mangatoetoenui Stream, Ruapehu volcano, New Zealand: Geological Society of America Bulletin, v. 111, p. 743–754.

Lavigne, F., and Thouret, J.C., 2003, Sediment transportation and deposition by rain-triggered lahars at Merapi volcano, Central Java, Indonesia: Geomorphology, v. 49, no. 1–2, p. 45–69.

Lavigne, F., Thouret, J.C., Voight, B., Suwa, H., and Sumaryono, A., 2000, Lahars at Merapi volcano, Central Java: An overview: Journal of Volcanology and Geothermal Research, v. 100, p. 423–456.

Lowe, D.R., and LoPiccolo, R.D., 1974, The characteristics and origin of dish and pillar structures: Journal of Sedimentary Petrology, v. 44, p. 484–501.

Luzier, J.E., 1969, Geology and ground-water resources of south-western King County, Washington: Washington Department of Water Resources Water Supply Bulletin, v. 28, 260 p., 3 plates.

Major, J.J., and Newhall, C.G., 1989, Snow and ice perturbation during historical volcanic eruptions and the formation of lahars and floods: Bulletin of Volcanology, v. 52, p. 1–27.

Mattinson, J.M., 1977, Emplacement history of the Tatoosh volcanic-plutonic complex, Washington; ages of zircons: Geological Society of America Bulletin, v. 88, p. 1509–1514.

Mullineaux, D.R., 1970, Geology of the Renton, Auburn, and Black Diamond quadrangles, King County, Washington: U.S. Geological Survey Professional Paper 672, 92 p.

Mullineaux, D.R., 1974, Pumice and other pyroclastic deposits in Mount Rainier National Park, Washington: U.S. Geological Survey Bulletin, v. 1326, 83 p.

Pierson, T.C., 1999, Transformation of a water flood to debris flood to debris flow following the eruption-triggered transient-lake breakout of the crater on March 19, 1982, *in* Pierson, T.C., ed, Hydrologic consequences of hot-rock/snowpack interactions at Mount St. Helens volcano, 1982–84: U.S. Geological Survey Professional Paper 1586, p. 19–36.

Pringle, P., 2000, Buried forests of Mount Rainier volcano—Evidence for extensive Holocene inundation by lahars in the White, Puyallup, Nisqually, and Cowlitz River Valleys [abs.]: Washington Geology, v. 28, p. 28.

Pringle, P.T., and Cameron, K.A., 1999, Eruption-triggered lahar on May 14, 1984, *in* Pierson, T.C., ed, Hydrologic consequences of hot-rock/snowpack interactions at Mount St. Helens volcano, 1982–84: U.S. Geological Survey Professional Paper 1586, p. 81–103.

Pringle, P.T., and Scott, K.M., 2001, Postglacial influence of volcanism on the landscape and environmental history of the Puget lowland, Washington—A review of geologic literature and recent discoveries, with emphasis on the landscape disturbances associated with lahars, lahar runouts, and associated flooding, *in* Puget Sound Research 2001, Proceedings: Olympia, WA, Washington State Puget Sound Water Quality Action Team, 23 p.

Pringle, P.T., Boughner, J.A., Vallance, J.W., and Palmer, S.P., 1997, Buried forests and sand deposits containing Mount Rainier andesite and pumice show evidence for extensive laharic flooding from Mount Rainier in the lower Duwamish Valley, Washington [abs.], *in* Washington Department of Ecology; Washington Hydrological Society, Abstracts from the 2nd symposium on the hydrogeology of Washington State: Olympia, WA, Washington Department of Ecology, p. 5.

Rodolfo, K.S., and Arguden, A.T., 1991, Rain-lahar generation and sediment-delivery systems at Mayon volcano, Philippines, *in* Fisher, R.V., and Smith, G.A., eds, Sedimentation in volcanic settings: Society for Sedimentary Geology Special Publication 45, p. 71–87.

Scott, K.M., 1988, Origins, behavior, and sedimentology of lahars and lahar-runout flows in the Toutle-Cowlitz river system: U.S. Geological Survey Professional Paper 1447-A, 75 p.

Scott, K.M., and Vallance, J.W., 1995, Debris flow, debris avalanche, and flood hazards at and downstream from Mount Rainier, Washington: U.S. Geological Survey Hydrologic Investigations Atlas HA-729, 9 p.

Scott, K.M., Vallance, J.W., and Pringle, P.P., 1995, Sedimentology, behavior, and hazards of debris flows at Mount Rainier, Washington: U.S. Geological Survey Professional Paper 1547, 56 p.

Sisson, T.W., 1995, History and hazards of Mount Rainier, Washington: U.S. Geological Survey Open-File Report 95-0642.

Sisson, T.W., and Lanphere, M.A., 2000, The geologic history of Mount Rainier volcano, Washington [abs.]: Washington Geology, v. 28, no. 1/2, p. 28.

Stuiver, M., Reimer, P.J., Bard, E., Beck, J.W., Burr, G.S., Hughen, K.A., Kromer, B., McCormac, F.G., van der Plicht, J., and Spurk, M., 1998, INTCAL98 Radiocarbon age calibration 24,000–0 cal BP: Radiocarbon, v. 40, p. 1041–1083.

Swanson, D.A., Cameron, K.A., Evarts, R.C., Pringle, P.T., and Vance, J.A., 1989, Cenozoic volcanism in the Cascade Range and Columbia Plateau, southern Washington and northernmost Oregon: American Geophysical Union Field Trip Guidebook T106, p. 21–24.

Thorson, R.M., 1989, Glacio-isostatic response of the Puget Sound area, Washington: Geological Society of America Bulletin, v. 101, p. 1163–1174.

Vallance, J.W., 2000, Lahars, *in* Sigurdsson, H., Houghton, B.F., McNutt, S.R., Rymer, H., Stix, J., eds., Encyclopedia of volcanoes: San Diego, Academic Press, p. 601–616.

Vallance, J.W., and Donoghue, S., 2000, Holocene eruptive history of Mount Rainier [abs.]: Washington Geology, v. 28, p. 29.

Vallance, J.W., and Scott, K.M., 1997, The Osceola Mudflow from Mount Rainier: Sedimentology and hazard implications of a huge clay-rich debris flow: Geological Society of America Bulletin, v. 109, p. 143–163.

Geological Society of America
Field Guide 4
2003

Pleistocene tephrostratigraphy and paleogeography of southern Puget Sound near Olympia, Washington

Timothy J. Walsh
Michael Polenz
Robert L. (Josh) Logan

*Washington Department of Natural Resources, Division of Geology and Earth Resources,
POB 47007, Olympia, Washington 98504-7007, USA*

Marvin A. Lanphere
Thomas W. Sisson

U.S. Geological Survey, 345 Middlefield Rd., Menlo Park, California 94025, USA

ABSTRACT

Our detailed mapping in the south Puget Sound basin has identified two tephras that are tentatively correlated to tephras from Mount St. Helens and Mount Rainier dated ca. 100–200 ka and 200 ka, respectively. This, plus the observation that fluvial and lacustrine sediments immediately underlying the Vashon Drift of latest Wisconsin age are nearly everywhere radiocarbon infinite, suggests that glacial and nonglacial sediments of more than the past five oxygen-isotope stages are exposed above sea level. Distal lacustrine advance outwash equivalent to the Lawton Clay in the Seattle area is conspicuously absent. Instead, a thick (>120 ft) glaciolacustrine silt below the Vashon sediments contains dropstones and is radiocarbon infinite. Elsewhere, coarse-grained advance Vashon outwash rests unconformably on radiocarbon-infinite nonglacial sediments. These relationships may imply that late Pleistocene tectonic activity has modified the paleotopography and stratigraphy of the south Puget Sound area.

Keywords: Quaternary stratigraphy, tephrochronolgy, Puget Lowland, glacial geology.

INTRODUCTION

Late Wisconsinan-age Vashon Drift covers most of the southern Puget Lowland. Pre-Vashon units are generally exposed only along coastal bluffs, where mass wasting is common. Landslides and colluvium disrupt and obscure the continuity of exposures so that pre-Vashon geologic history is not easily deciphered. In the south Puget Lowland (south of Tacoma), all finite radiocarbon ages reported before 1966 are suspect due to laboratory contamination (Dorn et al., 1962; Fairhall et al., 1966). Stratigraphic assignments based on these invalid radiocarbon ages are now questionable and need to be re-evaluated. In addition, radiocarbon dating of the Salmon Springs Glaciation, which is subjacent to the Vashon Drift in the Puyallup Valley

(southeast of Tacoma; Fig. 1; Table 1) ca. 70 ka has been shown to be incorrect. The Salmon Springs has since been shown to be ca. 0.8–1 Ma on the basis of K-Ar dating of included tephra and reverse magnetization, suggesting deposition during the Matuyama Reversed Chron (Blunt et al., 1987; Westgate et al., 1987; Easterbrook, 1994). However, older geologic mapping has shown the drift below the Vashon to be Salmon Springs on the basis of stratigraphic position alone. Recent mapping and paleomagnetic analysis (Hagstrum et al., 2002) has demonstrated that this unconformity is probably due to the presence of an anticline that has been growing through at least the latter half of the Quaternary Period. Elsewhere in the Puget Lowland, particularly in the north, Easterbrook (1986, 1994) has demonstrated that glacial and nonglacial deposits of oxygen-isotope stage (OIS) 4-6,

Walsh, T.J., Polenz, M., Logan, R.L., Lanphere, M.A., and Sisson, T.W., 2003, Pleistocene tephrostratigraphy and paleogeography of southern Puget Sound near Olympia, Washington, *in* Swanson, T.W., ed., Western Cordillera and adjacent areas: Boulder, Colorado, Geological Society of America Field Guide 4, p. 225–236. For permission to copy, contact editing@geosociety.org. © 2003 Geological Society of America.

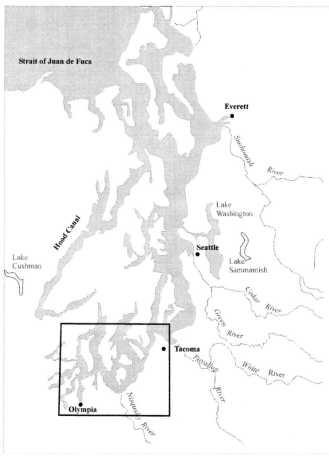

Figure 1. Location map for south Puget Sound field trip. Box shows area of trip, shown in detail in Figure 3.

a conceptual model for the more recent pre-Vashon geologic history that is consistent with our observations but by no means compelling.

The OIS 4 glaciation, called the Possession Glaciation, in the northern Puget Lowland was mild relative to stages 2 and 6 (Mix, 1987; Fig. 2), which are represented in the Puget Lowland by the Vashon and Double Bluff Drifts, respectively. The Possession Ice Sheet probably did not extend far south of Seattle (Lea, 1984; Troost, 1999). Because the ice sheet blocked drainage out of Puget Sound to the Strait of Juan de Fuca, a proglacial lake was impounded in most of the southern Puget Lowland. Streams flowing into this lake, such as the Nisqually, Puyallup, and Skokomish Rivers, formed an alluvial plain and deltas grading to lake level. These nonglacial sediments, deposited during stage 4, are all radiocarbon infinite and overlie and interfinger with Possession outwash deposits. Once Possession ice no longer impounded the lake (but sea level was still significantly below modern sea level), existing drainages, such as the Nisqually and Puyallup Rivers, deeply and rapidly incised into their former alluvial plains and became entrenched. At least initially, stage 3, called the Olympia Interglaciation (Armstrong et al., 1965), was characterized by downcutting and erosion. As sea level began to rise, most deposition was confined to the entrenched channels. Because stage 3 sea level was probably ~100 ft lower than modern sea level (Ludwig et al., 1996, and references therein), stage 3 deposits were areally restricted. As Vashon ice advanced and sea level fell again at the beginning of stage 2, these rivers preferentially downcut in the same channels, eroding the late Olympia deposits so that finite-aged Olympia deposits would be rare above sea level.

As Vashon ice moved southward and grounded across the Strait of Juan de Fuca during stage 2, it dammed the northern outlet of the Puget Sound basin. Proglacial streams carried fluvial sediments southward into the Puget Lowland, filling proglacial lakes and eventually the Puget Sound basin, first with silts, then with sands and gravels. These sediments form the "great lowland fill" of Booth (1994). Ice overrode these sediments, covering most of them with till, or scoured them away to deposit till directly onto pre-Vashon sediments. Subglacial channels were subsequently eroded into the fill. Proglacial lakes became impounded in these channels at different elevations above today's sea level as ice impinged on divides. The former lakebeds are presently the southernmost inlets of Puget Sound. (For a more thorough discussion of the subglacial channel network, see Booth [1994] and Booth and Goldstein [1994].) As these proglacial lakes spilled into lower-elevation basins and channels near the end of the Pleistocene, they deposited coarse, steeply dipping deltaic gravels along the margins of the channels and basins. Some of these deposits can be found near Steilacoom and Fort Lewis.

Much of the drainage originating from the ice sheet flowed southward and southwestward toward the Chehalis River. Some of the drainage probably occurred as glacial-lake outburst floods as valley-blocking ice dams breached during ice retreat. Deep

called Possession Drift, Whidbey Formation, and Double Bluff Drift respectively, are exposed above sea level. For these reasons, previous mapping of Salmon Springs in the south Puget Sound is almost certainly in error. We have also shown (Logan et al., 2003; Walsh et al., 2003) that deposits previously mapped as Salmon Springs (Noble and Wallace, 1966; Walters and Kimmel, 1968) in the Nisqually River valley were deposited by alpine glaciers from the south Cascade Mountains.

We have systematically sampled all datable material from nonglacial sediments subjacent to the Vashon Drift and found them to be older than previously reported. With a few exceptions, these sediments have been beyond the range of radiocarbon dating.

The antiquity of the pre-Vashon units causes radiocarbon dating to be of little help for making correlations, and abrupt facies changes within glacial and nonglacial units also render correlations tenuous. We have augmented radiocarbon dating with paleomagnetic analyses (J.T. Hagstrum, 2000, personal commun.) and chemical analyses of tephra (Table 2) to reconstruct the stratigraphy of south Puget Sound, although much remains to be done. Despite these difficulties, we have developed

TABLE 1. PUGET LOWLAND STRATIGRAPHIC NOMENCLATURE

Epoch	Age* (ka)	Puget Lowland Willis (1898), Bretz (1913)	Southeast Puget Lowland Crandell et al. (1958)	Central and northern Puget Lowland Armstrong et al. (1965), Mullineaux et al. (1965)	¹⁴C age (ka)	Tacoma area Walters and Kimmel (1968)	Puget Lowland Easterbrook (1986), Blunt et al. (1987), Troost (1994), Easterbrook (1999)	Southern Puget Lowland This report	Age (ka)
RECENT	10								
PLEISTOCENE — LATE		Vashon Glaciation	Vashon Drift	Fraser Glaciation — Sumas Stade	9–11				
				Everson Interstade (glaciomarine)	11–13.5				
				Vashon Stade	13.5–25	Vashon Drift	Vashon Drift	Tumwater Sand	
						Steilacoom Gravel	Steilacoom Gravel	Steilacoom Gravel	
				Vashon till		Vashon till	Vashon till	Vashon till	
				Esperance Sand		Colvos Sand	Esperance Sand	Colvos Sand	
				Lawton Clay			Lawton Clay	Lawton Clay?	
				Evans Creek Stade (alpine)	15–25				
		Puyallup Interglaciation	Unnamed erosional/nonglacial interval	Olympia Interglaciation	15–35	Kitsap Formation	Olympia beds	alpine drift? Olympia beds	15–60
							Possession Drift	reversely magnetized sediments	60–80; ca. 100
		Admiralty Glaciation					Whidbey Formation		100–250
PLEISTOCENE — MID	300						Double Bluff Drift	Unnamed beds on Ketron Island; alpine drifts from Mount Rainier?	60–500?
PLEISTOCENE — EARLY	800		Salmon Springs Drift	Salmon Springs Glaciation		Salmon Springs Drift	Salmon Springs Glaciation		800
			Puyallup Formation	Puyallup Interglaciation		Puyallup Formation	Puyallup Interglaciation		>1000
			Stuck Drift	Stuck Glaciation		Stuck Drift	Stuck Glaciation		ca. 1600
			Alderton Formation	Alderton Interglaciation		Alderton Formation	Alderton Interglaciation		>1600
	1800		Orting Drift	Orting Glaciation		Orting Drift	Orting Glaciation		>1600

*Not to scale. Modified from Borden and Troost (2001).

VARIATIONS $^{18}O/^{16}O$

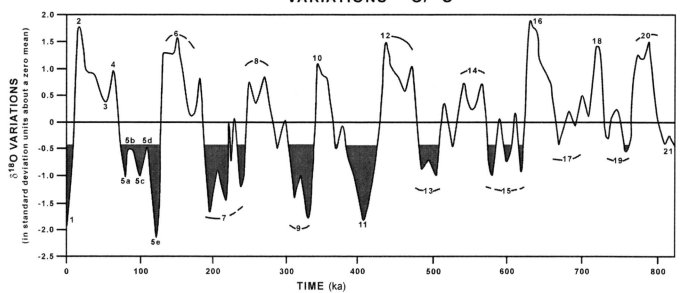

Figure 2. Marine oxygen-isotope stages (from Morrison, 1991). The numbers within the graph are stage numbers; the even-numbered peaks (at top) are glacial maxima, and the odd-numbered troughs (at bottom) are interglacial minima. The blue areas indicate interglacial episodes, based on a cutoff at –0.5 $\delta^{18}O$ oxygen-isotope values (equivalent to Holocene interglacial values).

troughs were carved out of the fill by subglacial fluvial erosion, and extensive and complex terraces and braided channels were formed. As the ice receded, northward-flowing streams near Olympia filled the deep troughs with sandy sediments characterized by northward-directed paleocurrent indicators. These sediments provide evidence that drainage reorganized to flow northward through the recently formed outwash plain. The thickness of these sediments varies substantially throughout the area, reaching more than 400 ft at the Port of Olympia.

In the waning stages of the Fraser Glaciation, Glacial Lake Russell covered a large area of the southern Puget Lowland and deposited a relatively thin layer (1–10 ft) of fine-grained varved sediments to an elevation of ~140 ft. These lacustrine silts (and rare clays and peats) usually overlie Vashon till, but rarely may overlie Vashon outwash. The latest Vashon sand is important because it is widespread throughout the populous South Sound area and appears to behave differently from the rest of the Vashon Drift during earthquakes (Palmer et al., 1999a, 1999b; Bodle, 1992; King et al., 1990).

The OIS 6 glaciation, called the Double Bluff Glaciation, in northern Puget Sound was probably as extensive as, or locally more extensive than, the stage 2 or Vashon Stade of the Fraser Glaciation (Mix, 1987; Lea, 1984). The end moraines of this glaciation lie a short distance beyond the inferred limit of the Vashon ice in the vicinity of Tenino (Lea, 1984). Subglacial erosion was probably similar to the erosion that Booth (1994) documented beneath Vashon ice and would have left more accommodation space for deposition during the interglacial time of OIS 5. For pre-Vashon nonglacial deposits that are radiocarbon infinite,

therefore, it is difficult to distinguish deposits of OIS 3 from deposits of OIS 5 or OIS 7.

In some outcrops, however, tephras are present that provide a tool for geochemical correlation to known eruptions of nearby Cascade stratovolcanoes. Tephra analyses from numerous sites (F.F. Foit, Jr.; A.M. Sarna-Wojcicki; and T.W. Sisson, 2000–2002, personal communs.) have begun to form a framework that will help to enable more precise correlations of similar lithofacies in south Puget Sound.

BOAT LOG

Leave Steilacoom floating dock (1 on Fig. 3) and travel southwest ~1.25 mi (2 km) to the north end of Ketron Island, and travel southward along the east shore. At the north end of the island, lodgment till of the Vashon Drift is exposed at sea level, making a wave-cut platform, and forms the surface of the island. As the surface elevation of the island increases to the south, the lower contact of the till can be seen to rise in the bluff approximately parallel to the surface. Exposures are spotty and change with sloughing of the bluff and vegetation, but Vashon till can be seen truncating alternating deposits of oxidized pebble-to-cobble gravel and fluvial sand with interbedded silt. Just south of the ferry terminal, a sand channel trending east-west can be seen near the top of the bluff. There are angular blocks of silt up to ~2 ft across that also conform to the margins of the sand channel. Detrital carbon from this channel is radiocarbon infinite. We interpret this as a subglacial channel below Vashon ice reworking older deposits.

Continue south to southern tip of Ketron Island.

TABLE 2. MICROPROBE ANALYSES OF TEPHRAS DISCUSSED IN TEXT

Sample #	Location	Analyst	n	SiO$_2$	TiO$_2$	Al$_2$O$_3$	Fe$_2$O$_3$	MnO	MgO	CaO	Na$_2$O	K$_2$O	P$_2$O$_5$	Cl	SO$_3$	Orig. total
93RW96	SA	S	16	74.81	0.30	13.58	1.51*	0.02	0.29	1.35	3.96	4.02	0.05	0.15	0.02	94.29
19-1E-11-73E-1	KI	S	26	74.13	0.32	13.76	1.79*	0.05	0.32	1.50	4.14	3.82	0.05	0.15	0.01	96.18
19-1E-11-73E-2	KI	S	21	74.82	0.30	13.54	1.50*	0.06	0.27	1.38	4.02	3.95	0.04	0.14	0.01	94.73
19-1E-11-73E-3	KI	S	15	74.59	0.39	13.56	1.55*	0.01	0.33	1.43	4.03	3.94	0.03	0.15	0.01	94.27
19-1E-11.73C	KI	F	16	75.04	0.29	13.42	1.68		0.30	1.41	3.71	3.91		0.17		
19-1E-11.73D	KI	F	17	75.05	0.29	13.56	1.71		0.30	1.43	3.63	3.84		0.16		
19-1E-11.73E	KI	F	16	74.85	0.31	13.65	1.74		0.31	1.42	3.69	3.83		0.17		
PLW8-11-01P4 T483-7	KI	F	16	75.27	0.28	13.54	1.66	0.04	0.27	1.28	3.79	3.87				90.34
19N-1W-23.85A	BC	F	16	76.59	0.11	14.27	1.07		0.33	1.80	3.45	2.26		0.09		
19N-1W-23.85A	BC	S	25	76.10	0.14	14.13	1.07*	0.05	0.35	1.77	3.93	2.29	0.05	0.08	0.02	94.32
PLW8/11/01P1 T483-8	BC mode 1	SW	9	76.14	0.09	14.37	1.07	0.02	0.35	1.82	4.03	2.10				86.54
PLW8/11/01P1 T483-8	BC mode 2	SW	2	73.47	0.19	15.00	1.75	0.05	0.36	1.48	4.63	3.07				94.77
19-1E-17.41A	AIS	F	15	76.43	0.12	14.17	1.11		0.35	1.84	3.65	2.21		0.09		
PLW8/11/01_P3 T487-4	AIS	SW	10	76.55	0.09	14.19	1.02	0.04	0.34	1.79	3.90	2.07				91.37
19/1E-18.94	AISW bulk	F	18	71.87	0.60	14.26	2.97		0.64	2.13	3.94	3.42		0.14		
19/1E-18.94	AISW glass 1	F	1	63.24	1.43	15.41	7.08		1.87	4.60	4.05	2.22		0.08		
19/1E-18.94	AISW glass 2	F	9	70.27	0.74	14.59	3.54		0.77	2.42	4.26	3.23		0.15		
19/1E-18.94	AISW glass 3	F	7	75.13	0.30	13.68	1.64		0.31	1.40	3.53	3.84		0.14		
19-1W-2.68	DH bulk	F	20	75.40	0.24	13.69	1.55		0.31	1.53	3.62	3.48		0.15		
19-1W-2.68	DH glass 1	F	5	76.50	0.11	13.98	1.10		0.33	1.80	3.74	2.33		0.08		
19-1W-2.68	DH glass 2	F	15	75.04	0.29	13.59	1.70		0.31	1.43	3.58	3.86		0.17		
PLW8/11/01P2ASH T487-3	DH mode 1	SW	8	75.23	0.30	13.81	1.63	0.02	0.29	1.38	3.82	3.54				92.63
PLW8/11/01P2ASH T487-3	DH mode 2	SW	4	74.43	0.32	14.01	1.86	0.03	0.46	1.59	3.97	3.35				95.42
19-2W-4.28B	TI bulk	F	20	72.16	0.59	14.21	2.90		0.62	2.09	3.79	3.45		0.16		
19-2W-4.28B	TI glass 1	F	2	64.83	1.24	15.44	6.12		1.54	4.21	4.07	2.41		0.11		
19-2W-4.28B	T I glass 2	F	9	70.55	0.74	14.61	3.49		0.75	2.37	4.06	3.24		0.16		
19-2W-4.28B	TI glass 3	F	9	75.42	0.29	13.53	1.60		0.29	1.34	3.45	3.88		0.17		

Note: Location abbreviations are SA—Sunset Amphitheater on Mount Rainier; KI—Ketron Island; BC—Butterball Cove; AIS—Anderson Island South at Thompson Cove; AISW—Anderson Island Southwest; DH—Devil's Head; TI—Totten Inlet. Analysts are S—Thomas W. Sisson, U.S. Geological Survey; F—Franklin F. Foit, Jr., Washington State University; SW—Andrei M. Sarna-Wojcicki, U.S. Geological Survey. All analyses recalculated to 100%.

*(reported as FeO).

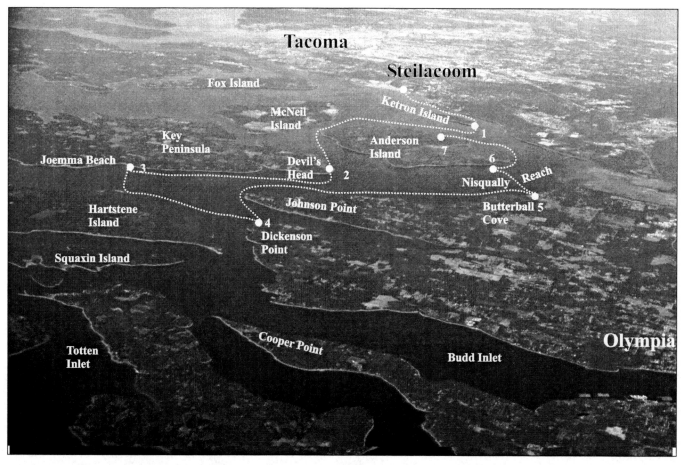

Figure 3. Field trip itinerary, looking east-northeast. At bottom are the peninsulas and inlets of the Olympia area. The Nisqually Delta is at the right middle. Scale variable.

Stop 1. Southern Tip of Ketron Island: Outcrop of Pumiceous Sands and Older Glacial Gravels of Probable OIS Stages 7 and 8

Vashon till and advance outwash cap the slope and can be accessed from above. Much of the section below is inaccessible here because of the unstable slope, but appears to be largely silt and fine sand. The lowest 75 ft of the section here is accessible from the beach. The unit at beach level is made up of sand, silt, and peat. The peat is radiocarbon infinite and the silt is normally magnetized (J.T. Hagstrum, 2000, personal commun.). Above this lower unit is gravel of mixed provenance, which we infer to be deposits of the Cordilleran Ice Sheet. Above the gravel (Figs. 4, 5) is a richly pumiceous sand with large amplitude cross-beds that imply transport toward the west and northwest. The pumice has been analyzed at three different labs and correlates strongly with the pumice at Sunset Amphitheater (Figs. 6, 7, 8; Table 2) on Mount Rainier, which is visible from here on a clear day. Lanphere and Sisson (unpublished data) have dated this pumice at Sunset Amphitheater (sample 93RW96, $^{40}Ar/^{39}Ar$ on plagioclase, plateau age 194 ± 12 ka [2-sigma], isochron age

184 ± 54 ka) to probable OIS 7 (Fig. 3). Above this is the inaccessible section. The presence of fluvial deposits from Mount Rainier shows that Ketron Island was not an island during OIS 7, and that, while the present-day configuration of islands, peninsulas, and channels may be a useful analogue for the nonglacial paleotopography, the present channel configuration was probably formed in Late Wisconsinan time and does not necessarily mimic any prior paleobathymetry and paleotopography.

From Ketron Island, travel north-northwest. Anderson Island is to the west. Exposures along the bluff are mostly blue-gray glaciolacustrine silt and fluvial fining upward sequences of sand, silt, and peat. At about the ferry terminal, the Vashon Drift becomes visible in the bluff. As on Ketron Island, the Vashon surface slopes to the north and Vashon Drift cuts out the older section. As we pass westward, north of Anderson Island, through Balch Passage, we travel between the small Eagle Island on the south and McNeil Island on the north. On Eagle Island, Vashon till makes up the wave-cut platform and forms the surface of the island. On McNeil Island, Vashon Drift forms the upper surface and older glacial and nonglacial sediments are exposed in the bluff below. Near the prison, to the west, Vashon till can be seen

Figure 4. Cross-bedded sand bearing the tephra correlated with the tephra of Sunset Amphitheater from Mount Rainier. The crossbeds here imply a paleocurrent direction to the west-northwest, implying that Cormorant Passage to the east of here was filled at the time of deposition of this tephra.

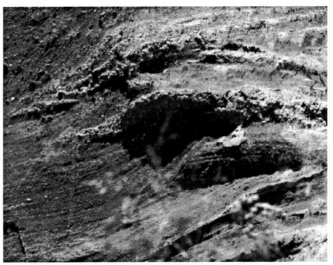

Figure 5. Close-up of outcrop adjacent to Figure 4, showing the high concentration of pumice clasts in this deposit as well as the high concentration of tephra making up the sand matrix.

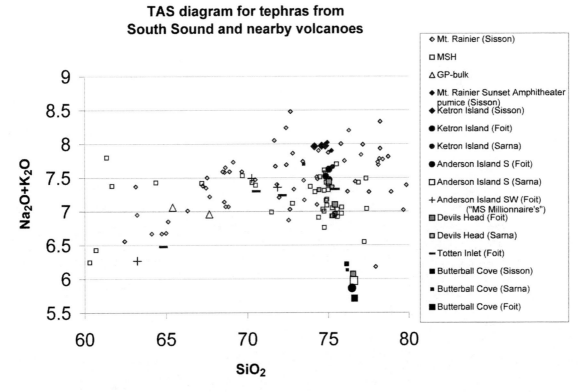

Figure 6. Plot of total alkalis vs. silica (TAS) for tephras from Mount Rainier, including Sunset Amphitheater, Mount St. Helens (MSH) and Glacier Peak (GP-bulk), the principal producers of late Pleistocene tephra in Washington. The other analyses plotted, with the analysts name in parentheses, are from this study.

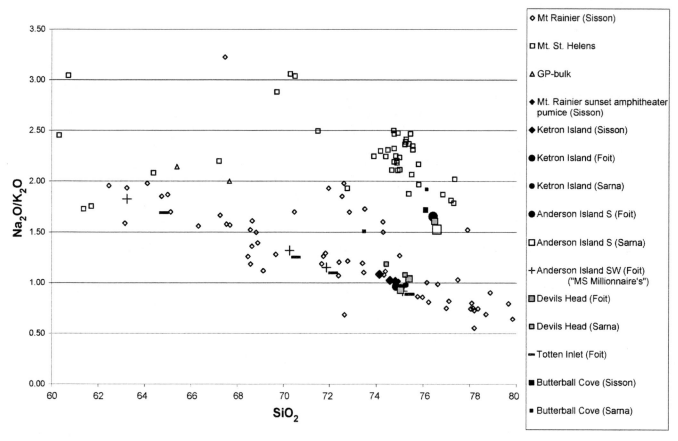

Figure 7. Plot of the ratio of sodium to potassium vs. silica for tephras from Mount Rainier, including Sunset Amphitheater, Mount St. Helens (MSH), and Glacier Peak (GP-bulk), the principal producers of late Pleistocene tephra in Washington. The other analyses plotted, with the analysts name in parentheses, are from this study.

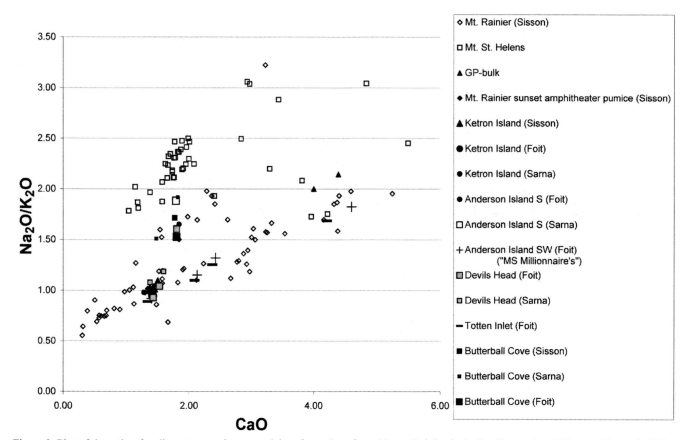

Figure 8. Plot of the ratio of sodium to potassium vs. calcium for tephras from Mount Rainier, including Sunset Amphitheater, Mount St. Helens (MSH), and Glacier Peak (GP-bulk), the principal producers of late Pleistocene tephra in Washington. The other analyses plotted, with the analyst's name in parentheses, are from this study.

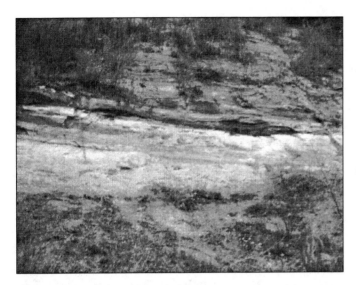

Figure 9. Nonglacial beds at Devil's Head, showing the lowermost several fining-upward sequences overlying older glacial gravels. A radiocarbon date of 50,500 ka and a normally magnetized silt analysis are from the center-right of this photo.

cutting through the older section down to sea level at the prison dock. As we round the north end of Anderson Island and travel southwest, the Vashon Drift rises up the bluff again and exposes fluvial and lacustrine sediments with a prominent diatomite exposed at about midbluff. As we approach Devil's Head, the southernmost tip of the Key Peninsula, a large deep-seated landslide is visible in the southeast-facing bluff.

Continue around the point and land at Devil's Head.

Stop 2. Devil's Head: Low-Energy Fluvial Sequence Underlying Vashon Drift with Two Reworked Tephras

The base of the section is a pebble-to-cobble gravel of mixed provenance, inferred to be of Cordilleran glacier origin. Above the gravel, there are seven sequences of fine-to-medium sand fining upward to silt and peat or peaty soils (Fig. 9, 10). Near the bottom of this unit is a peat bed that was radiocarbon dated ca. 50,500 yr B.P. by Minze Stuiver, University of Washington (Walsh, 1987). Sand below this peat contains a concentrated layer of reworked tephra that contains two distinct modes (Figs. 6, 7, 8). One of the modes is the Sunset Amphitheater tephra of Mount Rainier. The other cannot be confidently matched to a well-characterized tephra, but bears a strong resemblance to Mount St. Helens set Cy of Mullineaux (1996) (F.F. Foit, 2000, personal commun.; A.M. Sarna-Wojcicki, 2003, personal commun.). Silt from approximately this horizon is normally magnetized (J.T. Hagstrum, 2000, personal commun.). The uppermost of the fining-upward sequences is capped by a peaty soil (Fig. 10), from which we obtained an accelerator mass spectrometry radiocarbon age of 30,120 ± 250 ka. This is one of the few sections that we can confidently

Figure 10. Uppermost peaty soil in the sequence of Olympia beds at Devil's Head, from which an accelerator mass spectrometry radiocarbon age of 30,120 ± 250 ka was obtained.

assign to OIS 3 and consider to be Olympia beds (Table 1). Overlying this sequence and capping the bluff is the Vashon till overlying outwash sand and gravel, in turn overlying a diamicton that may be a flow till or that may imply a minor retreat and readvance of Vashon ice.

Travel north along west shore of Key Peninsula. The bluff all along here is capped by Vashon till. Underlying the till is advance outwash sand and gravel, which in turn overlies a thick sequence of Colvos Sand (Table 1). The Colvos is a distal advance outwash sand that correlates with the Esperance Sand. In the Seattle area, though, Esperance Sand overlies Lawton Clay, a distal glaciolacustrine silt deposited in the lake formed in Puget Sound when Vashon ice dammed the Strait of Juan de Fuca (Fig. 1). There is no equivalent of the Lawton Clay, however, below the Colvos. Rather, as along the coast of the Key peninsula, Colvos Sand either overlies a fluvial sequence similar to that exposed at Devil's Head or an older glaciolacustrine silt. We will see examples of these at Stops 3 and 7, respectively.

Continue north to Joemma Beach State Park.

Stop 3. Joemma Beach State Park: Colvos Sand Overlying Olympia Beds or Older Equivalent

Vashon Drift caps the bluff here. Advance outwash grades downward from sand and gravel to Colvos Sand, a fine-medium sand with low-amplitude sedimentary structures. These beds overlie fluvial sands, silts, and peat resembling those at Devil's Head, but which are radiocarbon infinite (Fig. 11).

Travel west toward Hartstene Island. Sand, silt, and peat exposed along the base of the bluff are similar to those at Joemma Beach, and are also radiocarbon infinite. Silt in an exposure due west of Joemma Beach is reversely magnetized (J.T. Hagstrum, 2000, personal commun.). Reverse magnetization is generally thought to imply deposition during the Matuyama Reversed Chron, and for nonglacial sediments, this would suggest correlation with the Puyallup or Alderton Formations (Table 1). However, because the evidence from included tephras suggests deposition during OIS 5 and 7, we suspect that this magnetization represents a shorter reversed interval, such as the Blake Reversed Subchron, which occurs within OIS 5. These sediments are more or less continuously exposed along the southeast shore of Hartstene Island.

Continue southeast to Dickenson Point.

Stop 4. Dickenson Point: Optional Stop

Glacial deposits older than the Vashon in this area generally lack till. Sand and gravel deposits containing clasts exotic to the area, such as high-grade metamorphic rocks or granitics, particularly those bearing pink feldspar, suggest northern provenance. Thick (>50 ft) lacustrine deposits, commonly bearing dropstones,

Figure 11. Colvos Sand exposure at Joemma Beach, grading upward to Vashon advance outwash sand and gravel. The sand overlies a sand and silt unit that may be the equivalent of the Olympia beds but are radiocarbon infinite.

also suggest deposition during continental glaciations. Deposits of OIS 4, called Possession Drift, in the northern Puget Lowland (Table 1) are generally less extensive than the OIS 6 deposits (Double Bluff Drift in the northern Puget Lowland) and Possession ice may never have occupied this area and may be represented here only by outwash deposits. At this locality, though, there is an older till exposed within a landslide deposit. This outcrop also indicates another difficulty in previous mapping, that landslides dropping Vashon till below its general elevation may be misinterpreted as older till.

Stop 5. Butterball Cove: Outcrop of Tephra Similar to Mount St. Helens Cy

At the top of this exposure are silts deposited in Glacial Lake Russell, which existed in Puget Sound during glacial retreat until an outlet to the Strait of Juan de Fuca was uncovered. These sediments are commonly found in south Puget Sound at elevations below ~140 ft, and are usually, as here, deposited directly on Vashon till. Underlying the till is 10 ft of Vashon advance outwash sand and gravel. The Vashon is underlain by ~50 ft of fluvial sediments. Peat in about the middle of these sediments, and wood from near the top are both radiocarbon infinite (Logan et al., 2003). The lower 5–10 ft of these fluvial sediments contain highly concentrated tephra that can be traced at low tide for ~200 ft along shore. The full extent of the tephra is unknown because the outcrop is covered by a landslide and retaining wall to the southeast and by bulkheads to the northwest. Chemical analysis (Figs. 6, 7, 8) does not precisely match this tephra with any known tephras, but it is almost certainly not from Mount Rainier, and it closely resembles Mount St. Helens set Cy (Mullineaux, 1996). According to A.M. Sarna-Wojcicki (2003, personal commun.), this tephra closely resembles other older Mount St. Helens tephras that he calls "proto-C tephra," such as at Carp Lake (Whitlock et al., 2000), which are probably 100–200 ka.

Travel north-northeast across Nisqually Reach to Thompson Cove (Figs. 12, 13).

Stop 6. Thompson Cove, Anderson Island: Optional Stop, Outcrop of Butterball Cove Tephra

The tephra here is equally concentrated as it is at Butterball Cove, but the sand here is interbedded with gravel. This outcrop was exposed by a slump caused by the M6.8 Nisqually earthquake. This sand and gravel can be traced more-or-less continuously to an outcrop labeled "MS millionaire's" on Figures 6, 7, and 8, although we have not found the tephra west of this outcrop. Tephra sampled there is ~25 ft higher in the section and correlates with tephra found to the southwest of here in Totten Inlet (lower left corner, Fig. 3). These tephras appear to be from Mount Rainier, but have not been correlated to any specific tephra there. Peat below the upper tephra there is radiocarbon infinite.

Travel east and then north around Lyle Point. Vashon till caps the bluff and cuts down to the west through the older sequence.

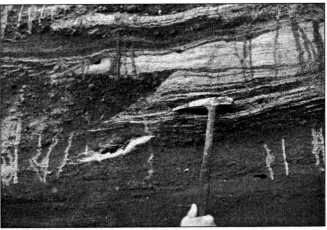

Figure 13. Close-up of Figure 12, showing abrupt truncation of lower part of tephra beds by gravelly sands. This bedding style is common here, showing abundant channel switching and high sediment loads. Paleocurrent indicators here are chaotic but generally have a northward component. This tephra is found at Thompson Cove on Anderson Island to the north-northwest of here in equally high concentration.

Figure 12. Tephritic sand overlying glacial gravel at Butterball Cove. At low tide, the tephra layer can be traced ~200' along shore, where it is cut off to the southeast by a landslide (and now a retaining wall), and to the northwest by bulkheads.

A thin silt near the top of the pre-Vashon section can be seen to thicken to the north and is exposed sporadically along the bluff to De Oro Bay, where Vashon till cuts down through the section to below sea level. At the north end of the bay, the till cuts upsection, again exposing thick silt along the bluff at Cole Point. Continue north to Country Club Beach.

Stop 7. Country Club Beach, Anderson Island: Thick Sequence of Pre-Vashon Glaciolacustrine Silt Overlying Older Till

This is the only in-place exposure of pre-Vashon till we have found in this area. It is overlain by gravel of mixed provenance, which is in turn overlain by a thick (>120 ft) sequence of blue-gray silt with dropstones. In the lower 20 ft of this silt there are abundant ptygmatic folds that are probably due to slumping caused by high sedimentation rates in a glacial lake. There is a thin sand capped by a paleosol about halfway up this section, overlain by another 60 ft of glaciolacustrine silt, suggesting that these may be two different glaciations. The upper unit is capped by small-scale channel deposits, sloping to the east and lined by layers of charcoal, which are radiocarbon infinite. We speculate that these may represent recessional glaciolacustrine deposits of Double Bluff (OIS 6) Drift overlain by advance glaciolacustrine deposits of Possession (OIS 4) Drift.

Return to Steilacoom. End of trip.

REFERENCES CITED

Armstrong, J.E., Crandell, D.R., Easterbrook, D.J., and Noble, J.B., 1965, Late Pleistocene stratigraphy and chronology in southwestern British Columbia and northwestern Washington: Geological Society of America Bulletin, v. 76, p. 321–330.

Blunt, D.J., Easterbrook, D.J., and Rutter, N.W., 1987, Chronology of Pleistocene sediments in the Puget Lowland, Washington, *in* Schuster, J.E., ed., Selected papers on the geology of Washington: Washington Division of Geology and Earth Resources Bulletin 77, p. 321–353.

Bodle, T.R., 1992, Microzoning the likelihood of strong spectral amplification of earthquake motions using MMI surveys and surface geology: Earthquake Spectra, v. 8, p. 501–527.

Booth, D.B., 1994, Glaciofluvial infilling and scour of the Puget Lowland, Washington, during ice-sheet glaciation: Geology, v. 22, p. 695–698.

Booth, D.B., and Goldstein, B.S., 1994, Patterns and processes of landscape development by the Puget lobe ice sheet, *in* Lasmanis, R., and Cheney, E.S., convenors, Regional geology of Washington State: Washington Division of Geology and Earth Resources Bulletin 80, p. 207–218.

Borden, R.K., and Troost, K.G., 2001, Late Pleistocene stratigraphy in the south-central Puget Lowland, Pierce County, Washington: Washington Division of Geology and Earth Resources Report of Investigations, v. 33, 33 p.

Bretz, JH., 1913, Glaciation of the Puget Sound region: Washington Geological Survey Bulletin, v. 8, 244 p., 3 plates.

Crandell, D.R., Mullineaux, D.R., and Waldron, H.H., 1958, Pleistocene sequence in southeastern part of the Puget Sound lowland, Washington: American Journal of Science, v. 256, p. 384–397.

Dorn, T.F., Fairhall, A.W., Schell, W.R., and Takashima, Y., 1962, Radiocarbon dating at the University of Washington I: Radiocarbon, v. 4, p. 1–12.

Easterbrook, D.J., 1986, Stratigraphy and chronology of Quaternary deposits of the Puget Lowland and Olympic Mountains of Washington and the Cascade mountains of Washington and Oregon: Quaternary Science Reviews, v. 5, p. 145–159.

Easterbrook, D.J., 1994, Chronology of pre-late Wisconsin Pleistocene sediments in the Puget Lowland, Washington, *in* Lasmanis, R., and Cheney, E.S., convenors, Regional geology of Washington State: Washington Division of Geology and Earth Resources Bulletin, v. 80, p. 191–206.

Fairhall, A.W., Schell, W.R., and Young, J.A., 1966, Radiocarbon dating at the University of Washington III: Radiocarbon, v. 8, p. 498–506.

Hagstrum, J.T., Booth, D.B., Troost, K.G., and Blakely, R.J., 2002, Magnetostratigraphy, paleomagnetic correlation, and deformation of Pleistocene deposits in the south central Puget Lowland, Washington: Journal of Geophysical Research, v. 107, no. B4, doi 10.1029/2001JB000557, paper EPM 6, 14 p.

King, K.W., Tarr, A.C., Carver, D.L., Williams, R.A., and Worley, D.M., 1990, Seismic ground-response studies in Olympia, Washington, and vicinity: Seismological Society of America Bulletin, v. 80, p. 1057–1078.

Lea, P.D., 1984, Pleistocene glaciation at the southern margin of the Puget lobe, western Washington: University of Washington Master of Science thesis, 96 p., 3 plates.

Logan, R.L., Walsh, T.J., Schasse, H.W., and Polenz, Michael, 2003, Geologic map of the Lacey 7.5-minute quadrangle, Thurston County, Washington: Washington Division of Geology and Earth Resources Open File Report 87-3, 1 plate, scale 1:24,000.

Ludwig, K.R., Muhs, D.R., Simmons, K.R., Halley, R.B., and Shinn, E.A., 1996, Sea-level records at ~80 ka from tectonically stable platforms-Florida and Bermuda: Geology, v. 24, p. 211–214.

Mix, A.C., 1987, The oxygen-isotope record of glaciation, *in* Ruddiman, W.F., and Wright, H.E., Jr., eds., North America and adjacent oceans during the last deglaciation: Boulder, Colorado, Geological Society of America, Geology of North America, v. K-3, p. 111–125.

Morrison, R.B., 1991, Introduction, *in* Morrison, R.B., ed., Quaternary nonglacial geology-Conterminous U.S.: Boulder, Colorado, Geological Society of America, Geology of North America, v. K-2, p. 1–12.

Mullineaux, D.R., 1996, Pre-1980 tephra-fall deposits erupted from Mount St. Helens, Washington: U.S. Geological Survey Professional Paper 1563, 99 p.

Mullineaux, D.R., Waldron, H.H., and Rubin, M., 1965, Stratigraphy and chronology of late interglacial and early Vashon glacial time in the Seattle area, Washington: U.S. Geological Survey Bulletin 1194-O, 10 p.

Noble, J.B., and Wallace, E.F., 1966, Geology and ground-water resources of Thurston County, Washington; Volume 2: Washington Division of Water Resources Water-Supply Bulletin 10, v. 2, 141 p., 5 plates.

Palmer, S.P., Walsh, T.J., and Gerstel, W.J., 1999a, Geologic folio of the Olympia-Lacey-Tumwater urban area, Washington-Liquefaction susceptibility map: Washington Division of Geology and Earth Resources Geologic Map GM-47, 1 sheet, scale 1:48,000, 16 p.

Palmer, S.P., Walsh, T.J., and Gerstel, W.J., 1999b, Investigation of earthquake ground motion amplification in the Olympia, Washington, urban area [abs.]: Seismological Research Letters, v. 70, no. 2, p. 250.

Troost, K.G., 1999, The Olympia nonglacial interval in the south-central Puget Lowland, Washington: University of Washington Master of Science thesis, 123 p.

Walsh, T.J., compiler, 1987, Geologic map of the south half of the Tacoma quadrangle, Washington: Washington Division of Geology and Earth Resources Open File Report 87-3, 10 p., 1 plate, scale 1:100,000.

Walsh, T.J., Logan, R. L, Polenz, M., and Schasse, H.W, 2003, Geologic map of the Nisqually 7.5-minute quadrangle, Thurston and Pierce Counties, Washington: Washington Division of Geology and Earth Resources Open File Report 2003-10, 1 plate, scale 1:24,000

Walters, K.L., and Kimmel, G.E., 1968, Ground-water occurrence and stratigraphy of unconsolidated deposits, central Pierce County, Washington: Washington Department of Water Resources Water-Supply Bulletin, v. 22, 428 p., 3 plates.

Westgate, J.A., Easterbrook, D.J., Naeser, N.D., and Carson, R.J., 1987, Lake Tapps tephra—An early Pleistocene stratigraphic marker in the Puget Lowland, Washington: Quaternary Research, v. 28, no. 3, p. 340–355.

Whitlock, C., Sarna-Wojcicki, A.M., Bartlein, P.J., and Nickmann, R.J., 2000, Environmental history and tephrostratigraphy at Carp Lake, southwestern Columbia Basin, Washington, USA: Palaeogeography, Palaeoclimatology, Palaeoecology, v. 155, no. 1–2, p. 7–29.

Willis, B., 1898, Drift phenomena of Puget Sound: Geological Society of America Bulletin, v. 9, p. 111–162.

Geological Society of America
Field Guide 4
2003

Recent geoarchaeological discoveries in central Washington

Gary Huckleberry*

Department of Anthropology, Washington State University, Pullman, Washington 99164-4910, USA

Brett Lenz*

Columbia Geotechnical Associates, 904 East Second Avenue, Ellensburg, Washington 98926, USA

Jerry Galm*

*Department of Geography and Anthropology, Rm. 105, Isle Hall, Eastern Washington University,
Cheney, Washington 99004-2420, USA*

Stan Gough*

*Archaeological and Historical Services, Rm. 105, Isle Hall, Eastern Washington University,
Cheney, Washington 99004-2420, USA*

ABSTRACT

Geoarchaeological research in the mid-Columbia region of central Washington over the past 10 yr has produced new information regarding Paleo-Indian archaeology and environmental change in the inland Northwest. Stratigraphic, sedimentological, and geomorphic studies provide important contextual information for locating and interpreting Washington's earliest archaeological sites and human remains. Recent discoveries increasingly point toward human occupation of the region during a time of post-glacial warming and reduced effective moisture 11,200–9000 ^{14}C yr B.P. This field guide presents recent research focusing on geoarchaeological studies at the Kennewick Man discovery site, at latest Pleistocene relict Channeled Scabland marsh sites, and at the recently excavated Sentinel Gap Paleo-Indian site.

Keywords: geoarchaeology, Columbia Basin, forensic geology, Channeled Scablands, Paleo-Indian, Sentinel Gap.

INTRODUCTION

Archaeological geology, or geoarchaeology, has long played an important part in understanding Washington State's natural and cultural prehistory. Geoarchaeology has provided insight not only into Washington's early residents, but also into the environmental dynamics that influenced human occupation and settlement. In the past 15,000 yr, central Washington has witnessed tremendous environmental changes associated with climatic variability as well as tectonism and vulcanism. Humans have been present in central Washington at least as early as 11.2 ka[1]

(Mehringer and Foit, 1990) and thus experienced a landscape recovering from the end of the last ice age and the numerous cataclysmic floods that scoured much of the Columbia Basin (Waitt, 1980; Baker and Bunker, 1985; Baker et al., 1991). This was a time of large-scale extinction of terrestrial megafauna, a time when plant communities were reshuffling in response to climatic change, and a time when floodplains and hillslopes were adjusting to post-glacial conditions. Washington's earliest archaeological sites thus contain a record of human adaptation to radically changing environments. To understand how humans learned to live in this dynamic landscape, one has to consider the big picture, including both material culture and environment.

---

*E-mail: Huckleberry—ghuck@wsu.edu; Lenz—blenz@geoscientists.org;
Galm—jgalm@ewu.edu; Gough—sgough@ewu.edu.

---

[1]All ages are given in uncorrected ^{14}C yr unless otherwise noted.

Huckleberry, G., Lenz, B., Galm, J., Gough, S., 2003, Recent geoarchaeological discoveries in central Washington, *in* Swanson, T.W., ed., Western Cordillera and adjacent areas: Boulder, Colorado, Geological Society of America Field Guide 4, p. 237–249. For permission to copy, contact editing@geosociety.org. © 2003 Geological Society of America.

Geoarchaeological studies emphasizing stratigraphy, sedimentology, and geomorphology help archaeologists locate these early sites, but more importantly, geoarchaeology helps to define the dynamic physical context of early human settlement as well as site-formation processes that have modified the cultural record.

This one-day field trip focuses on recent and ongoing geoarchaeological studies at three localities along the middle reach of the Columbia River between the towns of Vantage and Kennewick (Fig. 1). All three studies are associated with the early human inhabitants of the region during the Pleistocene-Holocene transition ca. 9–11 ka. The first stop is Columbia Park, the discovery site of the controversial 9000-yr-old Kennewick Man skeleton, the oldest, well-preserved human remains from the Pacific Northwest. While the fate of the skeleton remains to be determined by the courts, archaeologists and geologists have studied sediments at the discovery site and those adhering to the bones in order to better define the age and origin of the skeleton.

The second and third stops are at sites near the Columbia River at Sentinel Gap, where classic scabland topography and deposits formed by the numerous outburst floods of glacially dammed Lake Missoula provide a geological context for understanding latest Pleistocene–early Holocene human settlements and behavior. The BPA Springs and Sentinel Gap sites contain a suite of cultural and natural deposits, including in situ tephras and polygenic soils with buried A horizons, redoximorphic mottling, and carbonates. These and other early sites in central Washington are commonly associated with springs or watercourses, suggesting water was an important limiting resource during the post–Younger Dryas shift to warmer and drier conditions. An understanding of hydrological and geomorphic changes associated with the Pleistocene-Holocene transition in the mid-Columbia Basin is essential to locating Washington's earliest archaeological sites and understanding how prehistoric hunters and gatherers adapted to the new post-glacial world.

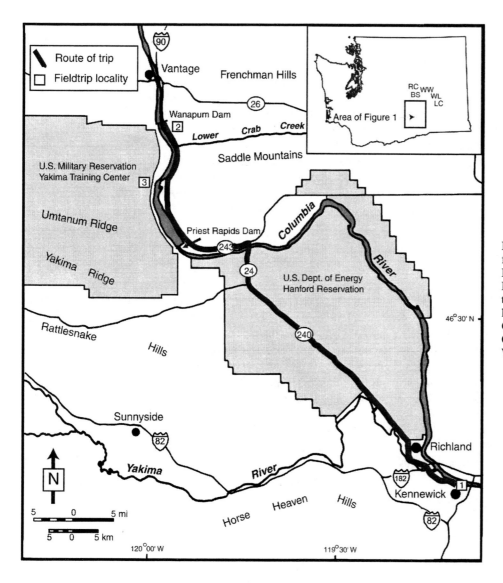

Figure 1. Map of mid-Columbia region, field trip routes, and features of interest. Field trip localities are 1—Columbia Park; 2—BPA Springs; and 3—the Sentinel Gap Paleo-Indian site. Index map localities discussed in text: RC—Richey Clovis; BS—Bishop Spring; LC—Lind Coulee; WW—Winchester Wasteway; WL—Willow Lake.

THE GEOLOGICAL CONTEXT OF KENNEWICK MAN

In July 1996, an ancient, disarticulated, but well-preserved skeleton was discovered along the shore of Lake Wallula (Columbia River) at Columbia Park in the town of Kennewick (Fig. 1). A single ^{14}C date on bone collagen yielded a $\delta^{13}C$ corrected age of 8410 ± 60 ^{14}C yr B.P. (Taylor et al., 1998), making it the oldest well-preserved skeleton in the Pacific Northwest and only one of approximately a dozen in all of the New World. Like several other Paleo-Indian skeletons, this specimen contains craniometric features quite different from most modern Native Americans. This skeleton, however, soon became a source of contention between scientists, government agencies, and Native American tribes. Most Native Americans want the skeleton returned for reburial, whereas many scientists would like to see the skeleton preserved for future study. Since the skeleton's discovery, several books (Downey, 2000; Thomas, 2000; Chatters, 2001), arguments in professional journals (Swedlund and Anderson, 1999; Owsley and Jantz, 2001), and countless letters to professional newsletters and newspaper editors have been written regarding the "Kennewick Man" conflict. The conflict is multifaceted, involving components of science, religion, law, and politics. The legal conflict concerns how the 1990 Native American Graves Protection and Repatriation Act (NAGPRA) has been implemented in this case and whether or not these skeletal remains can be culturally affiliated with a coalition of local tribes[2]. In 1996, a lawsuit was filed against the federal government by a group of scientists (Bonnichsen et al. vs. U.S. Government), and in 2002, a federal magistrate ruled in favor of the scientists. However, that decision has been appealed and is currently being reviewed by the U.S. Ninth Circuit Court of Appeals in San Francisco.

The focus here is on geological studies performed in association with the skeleton. Because the disarticulated skeleton was found scattered along a ~30 m section of shoreline, important contextual information was lacking. Three scientific studies have been performed in an effort to better define the physical and temporal context of Kennewick Man (Huckleberry et al., 1998; Wakeley et al., 1998; Chatters, 2000; Huckleberry et al., 2003). Two of these studies are reviewed below.

1997 Study of the Discovery Site

Columbia Park is located on land administered by the U.S. Army Corps of Engineers (ACE). A preliminary, non-invasive study of the shoreline was performed in December 1997 under the direction of the U.S. Army Corps Waterways Experiment Station (WES) in Vicksburg, Mississippi, that provided important information regarding the general stratigraphic context of the skeleton as well as the general geomorphic setting (Huckleberry et al., 1998; Wakeley et al., 1998).

Columbia Park is situated on a low terrace of the Columbia River, between the mouths of the Yakima and Snake Rivers. This low stream terrace is inset beneath a higher Pleistocene stream terrace associated with the numerous Glacial Lake Missoula outburst floods (Reidel and Fecht, 1994a). The skeletal remains were recovered at river mile 331.6, at an elevation of 104 m above sea level. This section of the terrace currently supports a wetland, receiving runoff from upslope canal irrigation. However, prior to historic agriculture, the site appears to have been relatively well drained. Historic aerial photography from the 1930s indicates that the locality was once cultivated (Wakeley et al., 1998), and historic trash deposits were found eroding out of the streambank. The farm was abandoned after construction of McNary Dam and the creation of Lake Wallula in the 1950s.

During the 1997 study, a 345 m section of streambank at Columbia Park was investigated around the area where the bones were recovered. Stratigraphic profiles, soil descriptions, and sediment samples were collected at specified intervals determined by ACE personnel. In addition, vibracore sampling and ground penetrating radar analysis was performed along the shore. Sediment analysis including granulometry, petrography, and pH was performed at the WES Geotechnical Laboratory in Vicksburg and the Geoarchaeology Laboratory at Washington State University in Pullman.

The terrace edge at Columbia Park provides an exposure of <2 m depth and reveals two lithostratigraphic units (Fig. 2). The surface unit, Lithostratigraphic Unit I, is composed of a loose to friable fine sand to very fine sand and ranges 25–80 cm in thickness. Many of the larger quartz sand grains are frosted, whereas many smaller grains have fresh, angular surfaces (Wakeley et al., 1998). Mazama tephra, dated at 6.7 ka (~7600 calendar yr B.P.), is discontinuously preserved at or near the base of Unit I and is best preserved at the western end of the study area where Unit I interfingers with alluvial fan gravels derived from the higher Pleistocene terrace. Only the upper ~1 m of Lithostratigraphic Unit II is exposed in the terrace edge, although vibracores provide additional subsurface information. Unit II is a finer-grained deposit characterized by weakly preserved, horizontally bedded, very fine sand and silt. In places, very fine, centimeter-scale graded bedding is evident in the upper deposit (Huckleberry et al., 1998); however, most of the primary bedding is not preserved in the upper 1 m, probably due to bioturbation. Bedding becomes more distinct below a 2 m depth and consists of several more distinct coarse-to-fine graded and cross-bedded sequences.

Both Lithostratigraphic Units I and II have been subjected to pedogenesis. Unit I contains a plowed, surface organic layer and is weakly calcareous at depth (Ap/Bk horizonation). In contrast, the upper part of Unit II contains more distinct carbonate development in the form of concretions (5–10 mm in diameter) with amorphous equant aggregate, tubule, and dendrite shapes. Many of these concretions form a lag deposit along the shoreline of Lake Wallula. These concretions are composed mostly of calcite,

---

[2]For contrasting opinions on the Kennewick Man controversy, see http://www.cr.nps.gov/aad/kennewick/, http://www.saa.org/repatriation/index.html, http://www.umatilla.nsn.us/ancient.html, and http://www.friendsofpast.org/. For a chronology of events related to the case, see http://www.kennewick-man.com/.

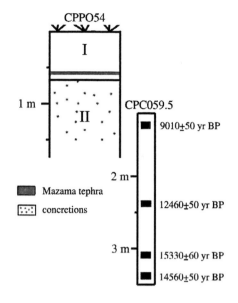

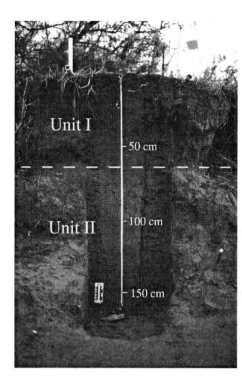

Figure 2. Stratigraphic column of terrace edge (CPP054) and vibracore (CPC059.5) at the Kennewick Man discovery site, Columbia Park. Photograph is from CPP044.

although silica content varies and increases to the west. In fact, secondary silica forms a distinct duripan in the uppermost 10 cm of Unit II, 200 m upstream (west) from the skeleton locality. The duripan is likely related to the weathering of overlying Mazama tephra and both the textural change and greater surface area of the Unit II sediments that would facilitate silica precipitation (Chadwick et al., 1987). In terms of matching the skeleton to the stratigraphy, the calcitic concretions are important signatures because similar concretions are found on the exterior surfaces of Kennewick Man (see below).

The presence of Mazama tephra indicates that Lithostratigraphic Unit II was deposited prior to 6.7 ka, and that Lithostratigraphic Unit I dates from sometime prior to 6.7 ka to the present. Further age control comes from ^{14}C-dated shell middens located in the middle to upper part of Unit I that date ca. 6.0 ka (Table 1). Radiocarbon dates on sediment organic extracts collected from Vibracore CPC059.5 provide numerical ages for Unit II. A sample from ~120 cm below the terrace tread in the upper part of Unit II yielded a date of 9010 ± 50 ^{14}C yr B.P.; sediment samples 230–350 cm below the terrace surface yielded ^{14}C dates ranging 12.4–15.3 ka. The lower two dates are reversed and coeval with the latest Missoula outburst floods (Baker et al., 1991; Clague et al., 2003). Wakeley et al. (1998) interpret these lower deposits that display repeated coarse to fine bedding as correlating to the common rhythmite beds associated with the glacial outburst floods (see Waitt, 1980). However, given the reversal of the lower two radiocarbon dates, it is possible that these lowermost deposits are post-glacial, overbank deposits formed by Columbia River vertical accretion, and that older carbon has been reworked in the system.

Regardless of the true age of the lowermost deposits, Mazama tephra, soil development, and ^{14}C dates on shells indicate that the surface and near-surface deposits comprise a fluvial terrace formed following the last of the Missoula outburst floods. Lithostratigraphic Unit II near the skeleton recovery site is indicative of a low-energy, flood basin environment characterized by repeated increments of overbank sediment subject to substantial root mixing. Wakely et al. (1998) interpret Lithostratigraphic Unit I as a continuation of low-energy, overbank alluvial deposition, whereas Huckleberry et al. (1998) consider the deposit to be mostly eolian in origin. Huckleberry et al. argue that the Columbia River downcut prior to the Mazama eruption at 6.7 ka, forming the terrace, a geomorphic event identified elsewhere in the region (Marshall, 1971; Hammatt, 1977; Cochran, 1988; Chatters and Hoover, 1992). Deposition resumed on the terrace primarily by eolian reworking of alluvium and exogenous sediment, forming Unit I. The surface stabilized sometime during the late Holocene (4–5 ka?), allowing for humification and calcification to develop a weak soil typical of regional late Holocene landforms (Busacca, 1989).

The skeleton was exposed by bank retreat during high water in the spring of 1996 and was eroded from the upper part of Lithostratigraphic Unit II, 60–150 cm below the surface. This inference is based largely on the presence of calcitic concretions in the soil and on the skeleton. The skeleton was therefore located stratigraphically beneath the Mazama tephra, suggesting an age congruent with the original 8410 ± 60 ^{14}C yr B.P. date on the skeleton. However, because this study was limited to the streambank, the three-dimensional architecture of the deposits could not

TABLE 1. RADIOCARBON DATES FROM THE KENNEWICK MAN SKELETON AND COLUMBIA PARK, WASHINGTON

Lab no.	Skeletal element	Uncorrected AMS ^{14}C measurement	δ^{13}C (PDB)
Beta-133993	Portion of right first metatarsal	8410 ± 40	−12.6
UCR-3807/CAMS-60684	Portion of right first metatarsal	8130 ± 40	−10.8
UCR-3476/CAMS-29578	Fifth left metacarpal	8370 ± 60	−14.9
UCR-3806/CAMS-60683	Portion of left tibial crest	6940 ± 30	−10.3
AA-34818	Portion of left tibial crest	5750 ± 100	−21.9

Lab no.	Columbia Park Location	Stratigraphic unit	Depth below surface (cm)	Material dated	Uncorrected AMS ^{14}C measurement	δ^{13}C (PDB)
Beta-113838	CPP005	I	60–80	Total shell carbonate	6230 ± 60	−8.2
Beta-113977	CPP005	I	60–65	Total shell carbonate	5820 ± 80	−8.3
WW 1626	CPC059.5	II	10–20	Bulk sediment organic carbon	9010 ± 50	−25
WW 1737	CPC059.5	II	130–138	Bulk sediment organic carbon	12460 ± 50	−25
WW 1627	CPC059.5	II	190–200	Bulk sediment organic carbon	15330 ± 60	−25
WW 1738	CPC059.5	II	220–229	Bulk sediment organic carbon	14560 ± 50	−25

Notes: Data from Taylor et al., 1998; Wakeley et al., 1998; http://www.cr.nps.gov/aad/kennewick.
^{13}C numbers with no decimal place are estimated values.

be defined, and it is not certain to what degree the Mazama tephra has been fluvially reworked in the immediate vicinity of the skeleton. Sampling of sediments was limited to preselected locations along the terrace edge, and any sampling within the terrace was prohibited by ACE. Both Huckleberry et al. (1998) and Wakeley et al. (1998) argued that subsequent subsurface work should be done at the site to better define deposit architecture and pinpoint the depth from which the skeleton was eroded. However, before subsequent work could be performed, the ACE buried the discovery site with tons of earth and rock debris in April 1998. The discovery site is now effectively sealed from future study.

Department of Interior Study of Skeleton Sediment

In 1998, the U.S. Department of Interior (DOI) began assisting the ACE in the Kennewick Man case and elected to perform a series of nondestructive tests on the skeleton in order to better determine the applicability of NAGPRA. This included analysis of sediments adhering to the skeleton in an effort to correlate the bones to a particular depth at the discovery site and affirm the early Holocene age of Kennewick Man. During this phase of the case, DOI was particularly interested in confirming a pre-Columbian age for the remains in order to confidently declare that the skeletal remains were Native American. The requested sedimentological study falls within the realm of forensic geology, whereby sediments are used to link a suspect to a crime scene (Murray and Tedrow, 1992). In this case, the physical and chemical properties of sediments on the skeleton were compared to sediments from Columbia Park in an effort to better define the original skeleton provenience in the streambank. In February 1999, sediments were extracted from the skeleton under the direct supervision of curators at the Burke Museum in Seattle, Washington, for a series of laboratory tests including granulometry, thin-section and micromorphology analysis, X-ray diffraction, thermogravimetry, and trace element analysis (Huckleberry et al., 2003). Columbia Park sediments collected by WES

personnel during the December 1997 study were analyzed for comparison. These sediments were from profiles and a vibracore located close to the skeleton recovery site. Granulometry data compiled by Chatters (2000) during his original study of the site was also used in the analysis.

The main hypothesis to test was whether or not sediments located on the surface of post-cranial bones and within the cranium matched those from the upper part of Lithostratigraphic Unit II (60–150 cm depth) containing the concretions. Thermogravimetric analysis demonstrated that concretions from both the skeleton and Unit II are chemically similar (~50% calcite), further supporting a correlation. Another goal was to determine if vertical changes in the physical and chemical properties of Unit II were sufficiently distinct to discern the depth of burial within a range of 20–30 cm. Chatters (2000) suggested that the granulometric data indicate that sediments on the skeleton best match terrace sediments at depths of 80–85 cm and 135–140 cm. However, differences in grain size and sorting within Unit II are statistically insignificant, a likely result of bioturbation (Huckleberry et al., 2003). Likewise, post-depositional processes and consequent homogenization of fluvial beds preclude any distinct matches in organic matter, calcium carbonate, mineralogy, or trace-element chemistry. A further test was performed to try to match skeleton and streambank sediments: micromorphological analysis of a ped removed from the cranium was compared in a blind test to that of thin sections of soil peds collected from the streambank by Huckleberry et al. (1998). Cranial sediments are dominated by silt-sized grains of quartz within a calcareous matrix, indicating that the sediments are derived from the original terrace (rather than shoreline deposits). However, in terms of correlation, there is inadequate differentiation of the fabric and mineralogy to pinpoint from exactly which stratum the cranium was derived.

In sum, analysis performed on sediments attached to Kennewick Man was nondestructive to bone and further supported the hypothesis that the skeleton was derived from the upper part of Lithostratigraphic Unit II at 60–150 cm depth. It was not possible to further distinguish the burial depth of the skeleton due to insufficient vertical anistropy of sediments. Also, correlation was hampered by the inability to perform more detailed sampling of the control section due to the 1998 burial of the site. Nonetheless, sedimentological analyses such as these may prove useful in future NAGPRA-related cases where nondestructive analyses are needed to help determine original provenience of human remains.

UPPER PLEISTOCENE–EARLY HOLOCENE GEOARCHAEOLOGY OF SCABLAND MARSH SYSTEMS

Large-scale efforts to locate archaeological sites on the Columbia Plateau have generally failed to identify early Paleo-Indian sites; as such, they remain underrepresented in the regional archaeological record. Understanding the nature of such surveys makes it clear why this is the case. Efforts by the federal government to dam the Columbia and Snake Rivers for hydroelectric power generation in the 1950s–1960s were the impetus for many large-scale archaeological projects, yet sediments immediately adjacent to reservoir margins that were surveyed are primarily Holocene in age. Similarly, large-scale surveys away from the river have failed to target early deposits through shovel probing or other subsurface methods. As an example, Axton et al. (1999) surveyed >13,000 acres in the central Columbia Plateau and did not record a single archaeological site. This is a perfect example of a landscape where the application of even rudimentary geoarchaeological analysis coupled with subsurface methodology is integral to site discovery. Speaking of the problems faced during the survey, Axton et al. (1999, 2.4) state,

Fluctuations in the reservoir level remove and deposit sediment, and the area of Upper Crab Creek is a shallow, meandering drainage. Wind and water continually erode and move unconsolidated sediment, causing constantly shifting landforms in some areas. Much of the study area is covered by volcanic ash and dense vegetation, which probably obscures some cultural remains.

Early Paleo-Indian archaeology on the plateau may only exist in locations that contain sediments of appropriate age and are conducive to burial and preservation. Therefore, searching for Paleo-Indian archaeology in Holocene sediments adjacent to reservoir margins is a fruitless task. Fortunately, the Missoula floods created a landscape away from the Columbia and Snake River systems that meets both requirements. When considered in light of the hydrogeologic setting that exists in scabland tracts of the plateau, the recurring geomorphic and stratigraphic sequences there can aid in the location of Paleo-Indian archaeological sites.

The geomorphic and hydrogeologic features that are successfully correlated with evidence of early humans on the plateau include (1) scoured and denuded bedrock adjacent to fluvial deposits; (2) extinct paleolakes and remnant landforms associated with high stands of these bodies of water; and (3)

alluvial terraces within scabland flood channels which formed as the result of post-flood dewatering and subsequent upper Pleistocene–earliest Holocene alluviation. With regard to hydrogeology, the Columbia Basin Irrigation Project has recharged Pleistocene waterways, seeps, and springs adjacent to the above-mentioned geomorphic features, approximating the former, late Pleistocene marsh systems, which existed after the final scabland flooding. Paleo-Indian sites are found in such marsh settings.

A critical factor in understanding the geoarchaeological context of early Columbia Plateau archaeological sites lies in the timing of the cross-scabland floods, which have effectively created "archaeological bedrock" throughout much of the heart of the Columbia Plateau. While glacial outburst flooding across the plateau is generally understood to have ceased ca. 13 ka, smaller floods restricted to the main stream Columbia River channel, no less catastrophic with regard to human occupation and archaeological site preservation, are known to have continued through ca. 11 ka (Gough, 1995). Within the mainstream Columbia River corridor these late floods likely scoured and re-deposited Paleo-Indian evidence. Compounding this factor is the onset of the early Holocene aggradation ca. 9 ka (Mierendorf, 1983; Chatters and Hoover, 1986, 1992; Huckleberry et al., 1998; Lenz et al., 2001), which likely eroded or disrupted early sites that were not already impacted by the late glacial flooding. In the uplands of the scabland, away from the river corridor, loess and other eolian deposition created an environment conducive to site preservation.

At present, the Richey Clovis site (Fig. 1) is the sole early Paleo-Indian site recognized in the literature of the plateau (Mehringer and Foit, 1990; Gramly, 1991). Latest Pleistocene stemmed point sites are presently unrecognized in the region, and early Holocene sites have been primarily recorded along the Snake River. Isolated early Paleo-Indian finds are summarized by Michael Avey (2000, personal commun.); none of them is reported from the river corridors. Isolated early artifacts found in the uplands and scabland tracts include the Badger Creek Spring Clovis Point, surface-collected northeast of Wenatchee, and the Beverly Clovis Point, surface-collected on the Lower Crab Creek drainage (Fig. 1) along the margin of an extinct paleolake (M. Avey, 2000, personal commun.). Recently discovered latest Pleistocene–earliest Holocene sites from scabland marsh and wetland settings are described below.

Bishop Spring

The Bishop Spring site is located on the western margin of the Columbia Plateau, in Grant County, Washington (Fig. 1). Limited geoarchaeological fieldwork by Lenz has confirmed the presence of archaeology ranging from latest Pleistocene to early Holocene in age. Artifacts recovered from controlled excavation and through collection by local residents include stemmed points, macroblades derived from prepared cores, blade core fragments, and isolated spur tools. Randall Schalk (2002, personal commun.) reports that faunal remains from the site include seven identified taxa: *Marmota* (marmot), *Canis* (dog), *Cervus*

TABLE 2. UPPERMOST PLEISTOCENE TEPHROCHRONOLOGY OF THE COLUMBIA PLATEAU

Tephra	Uncorrected ^{14}C measurement	Method of Determination	Citation
Mt. Mazama	7627 ± 150	Radiocarbon and multiparameter annual ice core signatures	Zdanowicz et al., 1999
Mt. St. Helens Layer J	12,000–11,500	Radiocarbon	Mullineaux, 1996
Glacier Peak	11,200 ± 100 to 12,290 ± 50	Radiocarbon Radiocarbon	Mehringer, 1977; Gough, 1995
Mt. St. Helens Set S	~13,000 12,800 ± 60	Radiocarbon Radiocarbon	Mullineaux, 1996; Lenz et al., 2002
Mt. St. Helens Set C	>35,000	Radiocarbon	Mullineaux , 1996

Note: Mazama age is calendric.

(elk), *Odocoileus* (deer), *Ovis* (sheep), *Bison* (bison), and Bovidae (sheep/bison family).

The site is located on a bedrock bench 300 m above the floodplain of the Columbia River. Preliminary stratigraphic investigations indicate the presence of a basal gleyed unit of flood sediment overlain by stacked A horizons that are separated by airfall tephras. Redox features are present in the overlying post flood sediment units, but only in those units that underlie Mazama tephra. Four successive tephras are present, beginning with St. Helens Set S (ca. 12.9 ka), Glacier Peak (11.2 ka), Mazama (6.7 ka), and Mount St. Helens 1980[3] (Table 2). The tephras bracket eolian, fluvial, and colluvially redeposited silt beds, into which paleosols formed and were buried. Paleosols are present as buried A horizons with abundant plant macrofossils, and photos taken by local collectors during the initial site discovery show apparently in situ artifacts more than 1 m below the Glacier Peak horizon, although this has yet to be confirmed through controlled excavation.

Lind Coulee

The Lind Coulee site is located along a scabland tract in central Grant County (Fig. 1). More than 2 m of cultural deposits and several dates centered ca. 9 ka are reported (Irwin and Moody, 1978). During excavations in the 1970s, Mount St. Helens Set J tephra was identified in stratigraphic sections across the site, and buried archaeology underlying the tephra horizon was reported. At the time of excavation, the age of set J was thought to lie between 8 and 9 ka (Irwin and Moody, 1978; Mullineaux et al., 1978). Radiocarbon dates including two by Daugherty (1956) at 9400 ± 940 ^{14}C yr B.P. and 8518 ± 400 ^{14}C yr B.P. and one by Fryxell at 8600 ± 65 ^{14}C yr B.P. (Sheppard and Chatters, 1976 *in* Irwin and Moody, 1978), conformed well to the tephrochronology. Since these early excavations, however, Mullineaux (1996) has adjusted St. Helens J dates, suggesting that the original dates

<10 ka are invalid. He suggests that the original samples were likely contaminated by younger carbon due to extended surface exposure. The current interpretation is that Set J probably was erupted between 12,000 and 11,500 ^{14}C yr ago (Table 2).

A bison scapula closely underlying the St. Helens J ash was dated by Moody (1978) at 12,830 ± 1050 ^{14}C yr B.P. but this was considered invalid due to inconsistency with the younger dates and correlation of the erroneously dated St. Helens J tephra. Three new accelerator mass spectrometry (AMS) ^{14}C dates from bone collagen samples associated with the lower levels at Lind Coulee are 10,060 ± 45 ^{14}C yr B.P. (CAMS-94856), 10,250 ± 40 ^{14}C yr B.P. (CAMS-94857), and 9810 ± 40 ^{14}C yr B.P. (CAMS-95524) (Craven, 2003). Hence, the earliest cultural materials from Lind Coulee are at least 1 k.y. older than originally interpreted, and may even be older in light of our present understanding of the age of St. Helens J tephra. Artifacts at the site include stemmed projectiles, eyed bone needles, and crescents, although the earliest horizons do not contain projectile points.

Winchester Wasteway

The Winchester Wasteway site (45GR156) is located in Grant County between George and Ephrata Washington (Fig. 1). The site consists of a Clovis fluted point, a stemmed point, and a large (~24 ha) but exceedingly sparse lithic scatter (<200 flakes) exposed on the surface of a cultivated field. The site lies within an irrigation overflow channel that drains agricultural lands of the Quincy Basin southward into the Drumheller Channels.

The Wasteway is a former scabland channel and is one of few scabland locales that display classic alluvial features, including a broad meander and multiple inset upper Pleistocene terraces, mapped as T4–T0. The geomorphic history of the site is quite similar to that of the Lind Coulee site 35 km to the east, serving initially during scabland floods as a drainage path (T4), during the waning stages of Glacial Lake Columbia as a waterway that partially drained Grand Coulee[4] flows (T3), and finally as an upper

---

[3]St. Helens Set S is identified based on stratigraphic position within outburst flood rhythmite beds; Glacier Peak was chemically sourced by electron microprobe at Washington State University GeoAnalytical Laboratory; Mazama and Mount St. Helens 1980 tephras are identified based on stratigraphic position.

---

[4]Moses Lake is the primary waterway through which glacial Lake Columbia drained; see Moody (1978) for discussion.

Pleistocene alluvial drainage (T2–T0). The Columbia Basin irrigation project has recharged the waterway, and the site is now dominated by wetland plant species. During the drier parts of the Holocene, the site was likely dry (Moody, 1978).

Willow Lake Site

The Willow Lake site lies in the Crab Creek drainage in central Grant County (Fig. 1). The lake occupies a relatively deep trench created by scabland flooding, and is flanked by mega-flood bars. While no formal subsurface investigations have taken place at the site, artifacts recovered by the original family to homestead the site area include stemmed and lanceolate projectiles, blade-based tools, and the left-side mandible of a *Smilodon* that displays parallel, incised marks that are consistent with butchering.

BPA Springs

Located on a bedrock bench near the Wanapum Dam, the BPA Springs site (Fig. 1) is a latest Pleistocene–early Holocene site discovered in late 2002 during a comprehensive archaeological survey that targeted early sites via systematic shovel probing (Lenz, 2003). The site had no surface expression, but a stemmed (Windust) point and flakes were recovered in shovel probes, with projectile point typology leading to the interpretation of this site as Paleo-Indian in age.

Sediments at the site consist of rhythmically bedded Pleistocene flood sediments with tephra within the final five flood units (i.e., rhythmites). Such a stratigraphic position implies that the tephra is St. Helens set S. The fine-grained flood sediments have prominent redoximorphic features extending through the base of the overlying unit, similar to deposits at the Sentinel Gap site (see below). Redox mottling suggests a fluctuating water table at the close of the Pleistocene, perhaps in response to unstable climatic conditions during the Pleistocene-Holocene transition (e.g., Davis et al., 2002). The overlying sediment is nearly homogenous eolian sand capped by a moderately deflated surface. Artifacts were located in the shovel probes overlying the rhythmites within the homogenous sands. Controlled excavations have yet to be performed.

Discussion

Early Paleo-Indian sites are underrepresented in the archaeological record of the Columbia Plateau. Recent investigations into interior plateau scabland environments show that (1) early Paleo-Indian sites are located in scabland tracts well away from the Columbia River, within the former path of Missoula flooding; (2) several of the sites display evidence for a high, fluctuating water table during the close of the Pleistocene but prior to deposition of Mazama tephra and in many places lie adjacent to or within marsh settings; (3) all of the sites are located in shallow groundwater environments that have been recharged by the

Columbia Basin Irrigation Project; and (4) each of the sites contains the similar recurrent scabland stratigraphic sequence that includes rhythmite beds with St. Helens Set S (ca. 12.9 ka), Set J (11.5–12.0 ka) and Glacier Peak (ca. 11.2 ka) tephras. Middle to late Holocene deposition is dominated by loess and pedogenesis under relatively arid conditions.

GEOARCHAEOLOGY OF THE SENTINEL GAP SITE

In 1997, routine archaeological inspection of an access road on the east side of the Department of the Army's Yakima Training Center (YTC) resulted in the discovery of the Sentinel Gap site (45KT1362). The site is at an elevation of 187 m on an unnamed tributary of Hanson Creek immediately west of the Columbia River valley (Fig. 1). Site deposits are situated within a gently dipping sandy landform. The Saddle Mountains anticline rises immediately north of the site, while the southern perspective opens onto the alluvial valley of Hanson Creek. A north-south oriented basalt escarpment bounds the site to the east and a parallel basalt ridge ~120 m to the west defines the limits of this tributary valley. Two ephemeral stream channels that join at the north end of the site area drain the smaller of two basins in this tributary valley. These two ephemeral channels drain an area of ~0.1 km² upslope of the site, greatly limiting fluvial deposition on the site landform. Minor fluvial deposits interbedded with thick massive eolian sediments in the Holocene section of the sandy landform are evidence of episodic flashy flow across the site area. This depositional regime has persisted since the terminal Pleistocene. Most recently, the ephemeral channel has shifted from a shallowly incised position on the sandy landform's eastern margin to its western edge.

Sentinel Gap investigations have focused on understanding the stratigraphic context of the cultural deposits, site structure and function, occupation surface content, and the site's geomorphic setting. Initiated with site testing in 1998 (Gough, 1999), these research priorities were continued during two field seasons of large-scale excavation (1999 and 2000).

The Archaeological Record

Test excavations at Sentinel Gap provided baseline data necessary for implementing large-scale excavations (Gough, 1999). The multiple accomplishments of the testing program include site boundary definition, correlation of the human occupation surface across the landform, and documentation of multiple cultural features and a high density and diversity of artifacts.

Large-scale excavations in the summers of 1999 and 2000 focused on the 8-cm-thick site-wide Paleo-Indian occupation surface and revealed a patterned distribution of artifacts and cultural features (Galm and Gough, 2000, 2002; Galm et al., 2002). The site occupation area is bounded on three sides (northeast-south) by an arcuate distribution of lithic debris features representing human translocation from primary work areas to a secondary waste disposal zone. The occupation surface contains 13 major lithic waste dump features, two centrally located, charcoal-rich burned areas, and a wide range and high density of tool types and functional object categories. Tool categories present within this diverse artifact assemblage include bifaces, representing a range of stages in at least two distinct manufacturing trajectories, projectile points, knives, scrapers, striated palette stones with an apparent surface covering of red ochre, spokeshaves, gravers, and a variety of bone tools and formed objects, including three styles of what appear to be bone "beads" (Galm et al., 2002; Gough and Galm, 2003). The sample of recovered faunal remains includes bison, elk, mountain sheep, badger, and salmon. The total site artifact assemblage numbers over 283,000 items. Charcoal samples derived from the two burns and other cultural contexts on the occupation surface yielded five dates, all in excess of 10 ka (Table 3).

The combined site evidence strongly suggests an affiliation with cultural expressions of late Paleo-Indian adaptations in the western United States in opposition to definition as a component of the coeval Windust phase reported from the Pacific Northwest and Northern Great Basin (Rice, 1972; Ames, 1988). This distinction is perhaps best reflected in (1) emphasis on the large size of lithic implements; (2) consistent use of a core-biface reduction system and a resulting de-emphasis on an economizing of toolstone materials; and (3) the poor representation of an expedient tool industry at Sentinel Gap. The sample of projectile points and knives and late stage bifaces from Sentinel Gap exhibit stylistic similarities to Haskett complex projectiles first reported from southeastern Idaho (Butler, 1965, 1967). Haskett projectile point forms in turn exhibit strong typological affinities to styles in the Agate Basin complex from the Great Plains (Frison, 1991).

Geomorphic Context of the Sentinel Gap Site

Site and landform deposits were investigated through deep testing within the site area, balk profiles surrounding the main excavation area, and a continuous stratigraphic profile cut along the west and south margins of the landform (Fig. 3). Stratigraphic descriptions and samples were obtained from all of these locations.

The sandy landform stratigraphy begins with deposition of fluvial Pleistocene gravels and ~5 m of rhythmically bedded medium sands to very fine sands comprising Lithostratigraphic Unit 1. One or more of the last glacial outburst floods of probable Lake Missoula origin deposited these sands. The inferred age of this suite of flood sediments is at or earlier than ca. 12.7–12.4 ka (Atwater, 1986). Floods moving down the Columbia River were primarily erosive in this reach, and accumulated evidence indicates repeated late Pleistocene flooding of this locality (Baker and Bunker, 1985; Waitt, 1985; Smith, 1993; Reidel and Fecht, 1994b).

The uppermost ~2 m of Lithostratigraphic Unit 1 exhibits post-depositional alteration in the form of cemented iron and manganese reduction-oxidation mottles. Present throughout this zone, these features indicate a locally fluctuating water table in the terminal Pleistocene. The vertical orientation of these redox features (≥ 2 m in length) suggests the growth of phreatophytic plants most likely within a subsequent period of stability represented by (Lithostratigraphic Unit 3) soil formation.

Eolian reworking of the outburst flood sands began immediately following the recession of the floodwaters. The source of the eolian sands is unknown but flood and loess deposits are the most plausible candidates (McDonald and Busacca, 1988). The first of the eolian sand deposits (Lithostratigraphic Unit 2; Fig. 3) incorporates two volcanic tephras, Mount St. Helens layer J (Mullineaux, 1996) overlain by Glacier Peak (layer G/B; Porter, 1978; Foit et al., 1993). Mullineaux (1996) places deposition of Mount St. Helens layer J at 12–11.5 ka. Late Pleistocene Glacier Peak eruptions occurred within an ~200–300 yr interval ca. 11.2 ka (Foit et al., 1993). Volcanic ash bed preservation in this eolian setting implies deposition in an aggradational environment, perhaps a lush riparian habitat. Combined with the redox features and buried soil A horizons, the evidence indicates generally wetter conditions than those that have existed at this location throughout most of the Holocene.

Landscape stability punctuated by periodic eolian deposition is indicated by soil development and repeated burial represented in Lithostratigraphic Unit 3 (Fig. 3). Three distinct buried soil A horizons are evidence of Younger Dryas interval soil formation (Galm et al., 2000; Holliday, 1997, 2000). Soil formation in this time interval, at the Sentinel Gap site and regionally, is indicative of wetter conditions at the Pleistocene-Holocene boundary. Two ^{14}C assays

TABLE 3. ^{14}C DATES FROM THE SENTINEL GAP SITE (45KT1362)

Lab no./method	Provenience	Material	^{14}C Age Uncorrected
Beta 133664/AMS	OS*, feature 99.6, flake concentration	Charcoal	10,010 ± 60 yr B.P.
Beta 133663/AMS	OS*, feature 99.1, burn	Charcoal	10,160 ± 60 yr B.P.
Beta 133650/radiometric	OS*, feature 99.1, burn	Charcoal	10,680 ± 190 yr B.P.
Beta 133665/AMS	OS*, feature 99.3, burn	Charcoal	10,130 ± 60 yr B.P.
Beta 124167/AMS	OS*, oxidized stain and artifacts	Charcoal	10,180 ± 40 yr B.P.
Beta 125771/AMS	OS*, feature 1	Bison bone collagen	6130 ± 40 yr B.P.
Beta 124168/radiometric	Upper paleosol†	Organic sediment	8880 ± 70 yr B.P.
SR 5500, CAMS 65979/AMS	Upper paleosol†	Humic acids	9630 ± 40 yr B.P.

Notes: *Paleoindian occupation surface
†Upper of three A horizons of Younger Dryas paleosols

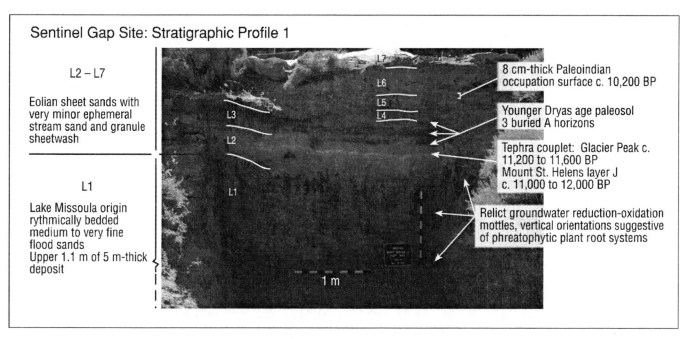

Figure 3. Lithostratigraphic units L1 through L7 at the Sentinel Gap site. Note L2 volcanic tephra beds, L3 buried soil A horizons, and Paleo-Indian occupation surface. Glacial outburst flood and tephra beds, as well as the three A horizons, dip to the southeast.

(Table 3) on the uppermost A horizon in this suite returned ages of 9630 ± 40 ^{14}C yr B.P. (Galm et al., 2000) and 8880 ± 70 ^{14}C yr B.P. (Gough, 1999). Both ages are considered too young considering the well-dated overlying cultural horizon (ca. 10.2 ka).

Approximately 60–80 cm of sand accumulated between the deposition of the tephra couplet and the formation of the occupation surface located at the boundary of Lithostratigraphic Units 5 and 6. Ca. 10.2 ka, site occupants conducted a variety of base camp activities while in residence, creating a discrete in situ archaeological record of unusual richness. The degree of bone preservation and orientations of artifacts indicates that the cultural materials were quickly buried by drifting sands following site abandonment. Occupation surface site deposits are capped by ~80 cm of sands up to the modern surface (Lithostratigraphic Unit 7). Of the ~3.3 m of terminal Pleistocene and Holocene eolian and minor fluvial deposits at Sentinel Gap, ≤ 80 cm accumulated in the Holocene after ca. 10.2 ka (Fig. 3). Varying degrees of calcium carbonate ($CaCO_3$) accumulation is present in all lithostratigraphic units. These carbonates encapsulate artifacts and weakly cement some occupation surface sediments. Human use appears to have enhanced carbonate precipitation on the occupation surface.

As noted earlier, site chronology is provided by a suite of radiocarbon assays on charcoal derived from the occupation surface (Table 3). An additional date of 6130 ± 40 ^{14}C yr B.P. was obtained on bone (bison) recovered from the occupation surface (Gough, 1999). This date is rejected as too recent considering the five other assays from the occupation surface and the inherent problems with accurately dating ancient bone (Stafford et al., 1991).

The rapid capping of the occupation surface evidences an abrupt change in climatic conditions in the early Holocene. Sand aggradation after ca. 10.2 ka not only preserved the rich archaeological record at Sentinel Gap but marks the onset of warmer and drier conditions in this region as well. All available evidence indicates the persistence of warm-dry conditions at this locale up to the modern day. The stratigraphic sequence at Sentinel Gap, including flood deposits, a tephra couplet, soil formation, Paleo-Indian site occupation surface, and the accumulating, yet thin, eolian sand cap, provides a valuable record of paleoclimatic and human history in the Pacific Northwest.

ROAD LOG

This 70 mi trip begins in Kennewick and ends at Wanapum Dam, south of Vantage. Much of the road log is borrowed from an earlier field guide by Campbell and Reidel (1991). Emphasis is on late Quaternary landforms and deposits as a context for understanding the archaeological record.

km	mi	Notes
0.0	0.0	Begin road log at Columbia Park (Stop 1), intersection of Columbia Drive and Edison Avenue.
0.2	0.1	Turn right onto freeway entrance to State Route 240 west. Highway follows along base of late Pleistocene fluvial terrace formed by outburst floods. Most of the town of Kennewick rests upon this terrace.

6.4	4.0	Cross Yakima River near its confluence with the Columbia River. The low surfaces comprise the Yakima River delta and represent an important winter camp area during the late prehistoric period. Area was visited by Captain William Clark in 1805.
6.8	4.2	Enter onramp for Interstate 182 westbound. Stay in right lane.
8.9	5.5	Take exit 4 onto State Route 240. Drive along west side of Richland, population 39,000, and home of the Pacific Northwest National Laboratories that oversees the Hanford Nuclear Reservation. In 1942, Richland was selected as part of the Manhattan Project as the place for production of weapons grade radioactive plutonium. Richland's population jumped from 247 to 11,000, nearly all of whom were employed by the federal government (see www.ci.richland.wa.us).
11.3	7.0	View to west (left) shows Pleistocene outburst flood terrace correlated with previous terrace in Kennewick (see Reidel and Fecht, 1994a).
15.9	9.9	Stoplight; turn left to continue on SR 240 west. Proceed northwest. For the next 12 mi, the highway crosses eolian deposits derived from outburst flood sediments laid down in waters pooled upslope from Wallula Gap. Sandy soils dominate elevations below 240 m, reflecting the patterns of slackwater deposition, and have been the source of sediment for an extensive sand sheet as well as a series of transverse and parabolic dune fields (Reidel and Fecht, 1994a). Given the abundant supply of loose sediment, eolian activity is controlled by wind speed and vegetative cover. Winds during the winter are predominantly out of the southwest and driven by synoptic scale pressure systems, whereas those of summer are northwesterly and dominated more by local daytime heating (Stetler and Gaylord, 1996). Range fires during the summer of 2000 burned across large parts of the Hanford Reservation, fueled in part by the invasion of exotic grasses. The loss of vegetation resulted in the mobilization of large amounts of wind-blown sediment. Eolian activity has varied in the past in response to late Quaternary climatic variability, with maximum dune activity associated with dry conditions during the early Holocene. Increased moisture during the late Holocene increased vegetative cover and helped to stabilize many of the dunes. The parabolic

dunes were utilized by prehistoric hunters in the procurement of bison as evidenced by the Tsulim site located on the Hanford Reservation (Chatters et al., 1995).

23.5	14.6	The ridge to the west is part of the Rattlesnake Hills, one of a series of anticlinal forms in the mid-Columbia region that comprise the Yakima fold belt. These structures generally trend east-west and are composed mostly of the Columbia River Basalt Group. Folded Pliocene gravel units overlying the basalts indicate that these uplifts are geologically recent.
28.8	17.9	To the SW is Horn Rapids Dam on the Yakima River. The Yakima River takes a detour to the north here due largely to the uplift of the Horse Heaven Hills anticline to the south. For the next 27 mi, State Route 240 crosses the Department of Energy Hanford Nuclear Reservation. The establishment of this reservation during the 1940s has resulted in preserving the most extensive and continuous shrub-steppe habitat remaining in the state of Washington.
38.8	24.1	Boulders derived as ice-rafted debris during the outburst floods are visible to the right (east) side of the highway.
61.2	38.0	Highway climbs onto the Cold Creek gravel bar. This landform was produced by the outburst floods as they passed through Sentinel Gap, a narrow gorge where the Columbia River cuts through the Saddle Mountains (another synclinal ridge), located 16 mi to the northwest.
62.3	38.7	Intersection with State Route 24. Proceed north on State Route 24.
65.2	40.5	As road curves to the left, the turnoff to the right provides a good view to the north of a giant gravel bar produced by the outburst floods. The white cliffs located to the east of the gravel bar are composed of the Pliocene Ringold Formation and are exposed by the Columbia River.
66.6	41.4	The abandoned buildings are several inactive reactors of the Hanford Project. This was the site of the first plutonium production that was used in the production of atomic bombs dropped on Japan in 1945.
70.1	43.6	Cross over the Columbia River. The Columbia River watershed is 670,000 km² and extends into parts of seven states and British Columbia. This is the last free-flowing segment of the Columbia River outside its headwaters. In 2000, President Clinton established

Hanford Reach National Monument in an effort to protect this relatively pristine stretch of the river. The monument is currently administered by the U.S. Fish and Wildlife Service.

70.9	44.1	Turn left onto State Route 243.
74.4	46.3	Road crosses under power lines. The hill to the north is another giant flood bar deposited as floodwater diverged downstream from the Sentinel Gap constriction.
77.0	47.9	To the S and SW is the Umtanum Ridge anticline. Like many of the Yakima folds, this anticline is broken by thrust faults on the north flank. Above the thrust fault, basalt flows are steeply dipping to nearly vertical.
82.2	51.1	To the NW is a good view of Priest Rapids Dam. To the NE is the Wahluke Slope forming the southern flank of the anticlinal Saddle Mountains.
90.7	56.4	Straight ahead is a view of Sentinel Gap. In contrast to the Horn of the Lower Yakima, this segment of the Columbia River is antecedent as it maintained its course as the Saddle Mountains rose thus cutting the gorge at Sentinel Gap.
91.5	56.9	Large basalt boulders on both sides of the highway are lag deposits from the outburst floods.
93.4	58.0	Turn right onto Mattawa Rd. and proceed east. Road climbs onto the Wahluke Slope.
94.0	58.4	Turn left onto Manson Lane.
94.3	58.6	Park in gravel parking lot at the end of Manson Lane by the large flagpole (Stop 2). View to the west across the Columbia River includes Hanson Creek on the U.S. Army Yakima Training Center where the Sentinel Gap Paleo-Indian site is located. Because of access restrictions, discussion of the site will be from this vantage point.
94.7	58.8	Return to Mattawa Rd. and turn right (east).
95.3	59.2	Turn right back onto SR 243 north.
102.3	63.6	Road cut on west side of highway reveals 6.7 ka (7600 calibrated yr B.P.) Mazama tephra that originated from what is today Crater Lake, Oregon. This is the most widespread and visible tephra on the Columbia Plateau and provides a useful stratigraphic marker for Holocene deposits.
103.4	64.3	Sand dunes to the east of the highway are composed of wind-reworked outburst flood sediment formed by strong winds blowing through Sentinel Gap.
105.5	65.6	Entering Schwana. View of Lower Crab Creek drainage to the east. This served as an

overflow channel for outburst floodwaters crossing the Columbia Basin to the northeast. The area contains classic scabland topography with scoured valley bottoms and discontinuous flood bars.

111.7	69.4	Road to Wanapum Dam and Tourist Center.
112.4	69.8	Turn right onto unmarked paved road.
112.7	70.0	Turn right to overlook located near the BPA Springs site (Stop 3 and end of Road Log).

REFERENCES CITED

Ames, K.M., 1988, Early Holocene Forager mobility strategies on the Southern Columbia Plateau, *in* Willig, J.A., Aikens, C.M., and Fagan, J.L., eds., Early human occupation in far Western North America: The Clovis-Archaic Interface: Nevada State Museum, Anthropological Papers No. 21, p. 325–360.

Atwater, B.F., 1986, Pleistocene glacial-lake deposits of the Sanpoil River Valley northeastern Washington: U.S. Geological Survey Bulletin 1661, 39 p.

Axton, S., Boreson, K., and Regan, D., 1999, A cultural resources survey of Potholes Reservoir, Grant County, Washington; Eastern Washington University Reports in Archaeology and History 100-113, Archaeological and Historical Services: Report prepared for U.S. Bureau of Reclamation, Pacific Northwest Region, Seattle.

Baker, V.R., and Bunker, R.C., 1985, Cataclysmic Late Pleistocene flooding from glacial Lake Missoula: A review: Quaternary Science Reviews, v. 4, p. 1–41.

Baker, V.R., Bjornstad, B.N., Busacca, A.J., Fecht, K.R., Kiver, E.P., Moody, U.L., Rigby, J.G., Stradling, D.F., and Tallman, A.M., 1991, Quaternary geology of the Columbia Plateau, *in* Morrison, R.B., ed., Quaternary nonglacial geology; coterminous U.S.: Boulder, Colorado, Geological Society of America, Geology of North America, v. K-2, p. 215–246.

Busacca, A.J., 1989, Long Quaternary record in eastern Washington, USA, interpreted from multiple buried paleosols in loess: Geoderma, v. 45, p. 105–122.

Butler, R.B., 1965, A report on investigation of an early man site near Lake Channel, southern Idaho: Tebiwa: v. 7, no. 2, p. 39–40.

Butler, R.B., 1967, More Haskett point finds from the type locality: Tebiwa, v. 10, no. 1, p. 25.

Campbell, N.P., and Reidel, S.P., 1991, Geologic guide for State Routes 240 and 243 in south-central Washington: Washington Geology, v. 19, p. 3–17.

Chadwick, O.A., Hendricks, D.M., and Nettleton, W.D., 1987, Silica in duric soils: a depositional model: Soil Science Society of America Journal, v. 51, p. 975–982.

Chatters, J.C., 2000, The recovery and first analysis of an early Holocene human skeleton from Kennewick, Washington: American Antiquity v. 65, p. 291–316.

Chatters, J.C., 2001, Ancient encounters: Kennewick Man and the first Americans: New York, Simon and Schuster, 303 p.

Chatters, J.C., and Hoover, K.A., 1986, Changing late Holocene flooding frequencies on the Columbia River, Washington: Quaternary Research, v. 26, p. 309–320.

Chatters, J.C., and Hoover, K.A., 1992, Response of the Columbia River fluvial system to Holocene climatic change: Quaternary Research, v. 37, p. 42–59.

Chatters, J.C., Campbell, S.K., Smith, G.D., and Minthorn, P.E., Jr., 1995, Bison procurement in the far west: a 2100-year-old kill site on the Columbia Plateau: American Antiquity, v. 60, p. 751–763.

Clague, J.J., Barendregt, R., Enkin, R.J., Foit, Jr., F.F., 2003, Paleomagnetic and tephra evidence for tens of Missoula floods in southern Washington: Geology, v. 31, p. 247–250.

Cochran, B., 1988, Significance of Holocene alluvial cycles in the Pacific Northwest interior [Ph.D. thesis]: Moscow, University of Idaho, 255 p.

Craven, S.L., 2003, Lithic variation in hafted bifaces at the Lind Coulee Site (45GR97), Washington [M.S. thesis]: Pullman, Washington State University, 103 p.

Davis, L.G., Muehlenbachs, K., Schweger, C.E., and Rutter, N.W., 2002, Differential response of vegetation to postglacial climate in the Lower Salmon River Canyon, Idaho: Palaeogeography, Palaeoclimatology, Palaeoecology, v. 185, p. 339–354.

Daugherty, R., 1956, Archaeology of the Lind Coulee Site, Washington: Proceedings of the American Philosophical Society, v. 100, p. 223–278.

Downey, R., 2000, Riddle of the bones: politics, science, race, and the story of Kennewick Man: New York, Springer-Verlag, 202 p.

Foit, F.F., Mehringer, Jr., P.J., and Sheppard, J.C., 1993, Age, distribution, and stratigraphy of Glacier Peak tephra in eastern Washington and western Montana, United States: Canadian Journal of Earth Science, v. 30, p. 535–552.

Frison, G.C., 1991, Prehistoric hunters of the High Plains: New York, Academic Press, 532 p.

Galm, J.R., and Gough, S., 2000, Site 45KT1362, ca. 10,000 yr B.P., occupation in central Washington: Current Research in the Pleistocene, v. 17, p. 29–31.

Galm, J.R., and Gough, S., 2002, Thick and thin biface production systems: Analysis and interpretation of the Sentinel Gap assemblage: 28th Great Basin Anthropological Conference: Elko, Nevada, Abstracts, p. 44.

Galm, J.R., Gough, S., and Nials, F.L., 2000, Project Fogoil alluvial chronology, Yakima Training Center, Kittitas and Yakima counties, Washington: Eastern Washington University Reports in Archaeology and History 100–114, 180 p.

Galm, J.R., Gough, S., and Nials, F.L., 2002, Archaeology and Paleoecology of the Sentinel Gap site: Society for American Archaeology, 67th Annual Meeting, Denver, Colorado, Abstracts, p. 112.

Gough, S., 1995, Description and interpretation of late Quaternary sediments in the Rocky Reach of the Columbia River Valley Douglas County, Washington [Masters thesis]: Cheney, Eastern Washington University, 112 p.

Gough, S., 1999, Archaeological test excavations at sites 45KT1362 and 45KT726, Yakima Training Center, Kittitas County, Washington: Eastern Washington University Reports in Archaeology and History 100-109, 68 p.

Gough, S., and Galm, J.R., 2003, Bone technology at the Sentinel Gap site: Current Research in the Pleistocene, v. 19, p. 27–29.

Gramly, R.M., 1991, The Richey Clovis Cache: Buffalo, Persimmon Press, 70 p.

Hammatt, H.H., 1977, Late Quaternary stratigraphy and archaeological chronology in the Lower Granite Reservoir area, lower Snake River, WA [Ph.D. thesis]: Pullman, Washington State University, 272 p.

Holliday, V.T., 1997, Paleoindian geoarchaeology of the Southern High Plains: Austin, TX, University of Texas Press, 297 p.

Holliday, V.T., 2000, Folsom drought and episodic drying on the Southern High Plains from 10,900–10,200 ^{14}C yr B.P.: Quaternary Research, v. 53, p. 1–12.

Huckleberry, G., Stein, J.K., and Goldberg, P., 2003, Determining the provenience of Kennewick Man skeletal remains through sedimentological analyses: Journal of Archaeological Science, v. 30, p. 651–665.

Huckleberry, G., Stafford, Jr., T.W., and Chatters, J.C., 1998, Preliminary geoarchaeological studies at Columbia Park, Kennewick, Washington, USA: Walla Walla, WA, U.S. Army Corps of Engineers, 23 p.

Irwin, A.M., and Moody, U., 1978, The Lind Coulee Site (45GR97): Report submitted to U.S. Bureau of Reclamation under terms of contract no. 14-06-100-8705, Seattle, 367 p.

Lenz, B.R., Gentry, H., and Clingman, D., 2001, Timing and characteristics of early Holocene aggradation, Columbia River, Washington State: Geological Society of America Abstracts with Program, v. 33, no. 6, p. A-312.

Lenz, B.R., 2003, Methodology for archaeological inventory, Priest Rapids Hydroelectric Project, FERC project 2114: Ephrata, Washington, Grant County Public Utility District.

Marshall, A., 1971, An alluvial chronology of the lower Palouse River Canyon and its relation to local archaeological sites [M.S. thesis]: Pullman, Washington State University, 73 p.

McDonald, E.V., and Busacca, A.J., 1988, Record of pre-late Wisconsin giant floods in the Channeled Scabland interpreted from loess deposits: Geology, v. 16, p. 728–731.

Mehringer, P.J., and Foit, Jr., F.F., 1990, Volcanic ash dating of the Clovis cache at East Wenatchee, Washington: National Geographic Research, v. 6, p. 495–603.

Mierendorf, R., 1983, Fluvial processes and prehistoric settlement patterns along the Rocky Reach of the Columbia River, *in* Shalk, R.F., and Mierendorf, R., eds., Cultural Resources of the Rocky Reach of the Columbia River, Center for Northwest Anthropology Project Report 1: Pullman, Washington State University, v. 2, p. 647–663.

Moody, U., 1978, Microstratigraphy, paleoecology, and tephrochronology of the Lind Coulee Site, Central Washington [Ph.D. dissertation]: Pullman, Washington State University, 273 p.

Mullineaux, D.R., Wilcox, R.E., Ebaugh, W.F., Fryxell, R., and Rubin, M., 1978, Age of the last major scabland flood of the Columbia Plateau in eastern Washington: Quaternary Research, v. 10, no. 2, p. 171–180.

Mullineaux, D.R., 1996, Pre-1980 tephra-fall deposits erupted from Mount St. Helens, Washington: U.S. Geological Survey Professional Paper 1563, 99 p.

Murray, R.C., and Tedrow, J.C.F., 1992, Forensic geology: Englewood Cliffs, N.J., Prentice Hall, 203 p.

Owsley, D.W., and Jantz, R.L., 2001, Archaeological politics and public interest in Paleoamerican studies: lessons from Gordon Creek Woman and Kennewick Man: American Antiquity, v. 66, p. 565–575.

Porter, S.C., 1978, Glacier Peak tephra in the North Cascade Range, Washington—stratigraphy, distribution, and relationship to late-glacial events: Quaternary Research, v. 10, p. 30–41.

Reidel, S.P., and Fecht, K.R., 1994a, Geologic map of the Richland 1:100,000 Quadrangle, Washington, Open File Report 94-8: Olympia, Washington, Division of Geology and Earth Resources, 1 sheet.

Reidel, S.P., and K.R., Fecht, 1994b, Geologic Map of the Priest Rapids 1:100,000 Quadrangle, Washington, Open File Report 94-13: Olympia, Washington Division of Geology and Earth Resources, 22 p., 1 plate.

Rice, D.G., 1972, The Windust phase in Lower Snake River region prehistory: Laboratory of Anthropology, Washington State University, Report of Investigations, no. 50, 225 p.

Sheppard, J., and Chatters, R., 1976, Washington State University natural radiocarbon measurements II: Radiocarbon, v. 18, p.143.

Smith, G.A., 1993, Missoula flood dynamics and magnitudes inferred from sedimentology of slack-water deposits on the Columbia Plateau, Washington: Geological Society of America Bulletin, v. 105, p. 77–100.

Stafford, T.W., Jr., Hare, P.E., Currie, L., Jull, A.T.J., and Donahue, D.J., 1991, Accelerator radiocarbon dating at the molecular level: Journal of Archaeological Science, v. 18, p. 35–72.

Swedlund, A., and Anderson, D., 1999, Gordon Creek Woman meets Kennewick Man: new interpretations and protocols regarding the peopling of the Americas: American Antiquity, v. 64, p. 569–576.

Stetler, L.D., and Gaylord, D.R., 1996, Evaluating eolian-climatic interactions using a regional climate model from Hanford, Washington (USA): Geomorphology, v. 17, p. 99–113.

Taylor, R.E., Kirner, D.L., Southon, J.R., and Chatters, J.C., 1998, Radiocarbon dates of Kennewick Man: Science, v. 280, p. 1171–1172.

Thomas, D.H., 2000, Skull wars: Kennewick Man, archaeology, and the battle for Native American identity: New York, Basic Books, 326 p.

Waitt, R.B., 1980, About forty last-glacial Lake Missoula Jokulhlaups through southern Washington: Journal of Geology, v. 88, p. 653–679.

Waitt, R.B., 1985, Case for periodic, colossal jokulhlaups from Pleistocene glacial Lake Missoula: Geological Society of America Bulletin, v. 96, p. 1271–1286.

Wakeley, L.D., Murphy, W.L., Dunbar, J.B., Warne, A.G., and Briuer, F.L., 1998, Geologic, Geoarchaeologic, and Historical Investigations of the Discovery Site of Ancient Remains in Columbia Park, Kennewick, Washington: U.S. Army Corps of Engineers Technical Report GL-98-13, 69 p.

Zdanowicz, C.M., Zielinski, G.A., Germani, M.S., 1999, Mount Mazama eruption: Calenderical age verified and atmospheric impact assessed: Geology, v. 27, p. 621–624.

Geological Society of America
Field Guide 4
2003

Evolution of a polygenetic ophiolite:
The Jurassic Ingalls Ophiolite, Washington Cascades

Gregory D. Harper*

Department of Earth and Atmospheric Sciences, State University of New York, Albany, New York 12222, USA

Robert B. Miller

Department of Geology, San Jose State University, San Jose, California 95192-0102, USA

James H. MacDonald, Jr.

Department of Earth and Atmospheric Sciences, State University of New York, Albany, New York 12222, USA

Jonathan S. Miller
Ante N. Mlinarevic

Department of Geology, San Jose State University, San Jose, California 95192-0102, USA

ABSTRACT

The Ingalls Ophiolite Complex is a suprasubduction-zone ophiolite formed largely in a fracture-zone setting. Mantle tectonites are cut by a large, high-T shear zone overprinted by sheared serpentinite. Mafic complexes of ca. 161 Ma gabbro, sheeted dikes, and pillow lava occur as large blocks in the sheared serpentinite. An overlying Late Jurassic argillite unit contains minor chert, graywacke, and pebble conglomerate, along with lenses of ophiolite breccias. Detrital serpentinite forms some of these breccias, and mafic blocks in other breccias range up to hundreds of meters in diameter. Older basement is locally present in the ophiolite complex, including (1) phyllite, metachert, and pillow basalt of the undated De Roux unit overlain by (2) Early Jurassic pillow lava, basalt breccia, and minor chert and oolitic limestone of the Iron Mountain unit (fossil seamount). This older basement indicates that at least part of the ophiolite is polygenetic. The presence of a high-T mantle shear zone and ophiolitic breccias containing clasts derived from the lower crust and upper mantle suggest formation of the ophiolite in a fracture zone setting. Late calc-alkaline dikes cut the various units, some of which are related to the middle Cretaceous Mount Stuart batholith, and others of which are probably Late Jurassic.

Keywords: ophiolite, Ingalls, Jurassic, Washington, sedimentary serpentinite, ophiolite breccia.

GEOLOGIC SETTING

The North Cascades and San Juan Islands of Washington and southwest British Columbia consist of numerous small tectonostratigraphic terranes that were involved in major Cretaceous orogenesis (e.g., Davis et al., 1978; Tabor et al., 1987, 1989; Brandon et al., 1988). These terranes are disrupted by the Paleogene Straight Creek–Fraser River fault (Fig. 1), which records dextral strike slip with estimated offsets from 90–190 km (e.g., Misch, 1977; Tabor et al., 1984). The Mesozoic geology of the North Cascades of Washington and southwest British Columbia is commonly described in terms of four tec-

*E-mail: gdh@albany.edu.

Harper, G.D., Miller, R.B., MacDonald, J.H., Jr., Miller, J.S., and Mlinarevic, A.N., 2003, Evolution of a polygenetic ophiolite: The Jurassic Ingalls Ophiolite, Washington Cascades, *in* Swanson, T.W., ed., Western Cordillera and adjacent areas: Boulder, Colorado, Geological Society of America Field Guide 4, p. 251–265. For permission to copy, contact editing@geosociety.org. © 2003 Geological Society of America.

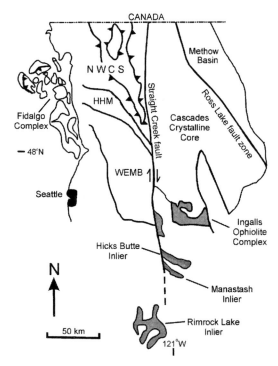

Figure 1. Simplified geologic map showing the tectonic elements of the North Cascades. NWCS—Northwest Cascade System; WEMB—western and eastern mélange belts. Shaded areas are Mesozoic rocks south of the Cascade crystalline core, which are also part of the Northwest Cascade System. Modified from Tabor et al. (1989) and Miller et al. (1993a).

tonic elements (Fig. 1; e.g., Tabor et al., 1989): the Northwest Cascades system, the western and eastern mélange belts, and the crystalline core of the North Cascades (Cascades core).

The Northwest Cascades system, which includes the Jurassic Ingalls Ophiolite Complex, and the structurally overlying western and eastern mélange belts comprise a stack of thrust sheets consisting mostly of low-grade, late Paleozoic and Mesozoic arc and oceanic assemblages (e.g., Misch, 1966; Brown, 1987; Brandon et al., 1988; Tabor, 1994). Most of the Northwest Cascades system and the eastern and western mélange belts lie west of the Straight Creek fault (Fig. 1), which separates these rocks from the Cascades core. This core consists of several tectonostratigraphic terranes amalgamated before Late Cretaceous and Paleogene plutonism and amphibolite-facies metamorphism (Tabor et al., 1987, 1989; Miller et al., 1993b). Mesozoic rocks of the Northwest Cascades system and western mélange belt exposed east of the Straight Creek fault lie, at least in part, structurally above the southern margin of the Cascades core (Miller, 1985; Miller et al., 1993a). The Cascades core is bounded on the northeast by the Ross Lake fault zone, which separates the crystalline rocks from the Jurassic-Cretaceous Methow basin (Fig. 1; Misch, 1966).

The Cascades core is bounded on the south by the middle Cretaceous Windy Pass thrust, in which the Ingalls Ophiolite is

thrust over the Chiwaukum Schist (Miller, 1985). Movement on the Windy Pass thrust overlapped in time with intrusion of the 91–96 Ma Mount Stuart batholith (Miller, 1985; Miller and Paterson, 1992; Paterson et al., 1994, 2002) and was synchronous with metamorphism of the Cascades core. Regional metamorphic grade is amphibolite facies in the northernmost part of the Ingalls Ophiolite, but most of the ophiolite lies outside the contact aureole of the Mount Stuart batholith and experienced only sub-greenschist facies regional metamorphism.

The Ingalls Ophiolite is part of a belt of ca. 160–170 Ma ophiolites in the western United States. This belt includes the Fidalgo Ophiolite of northwest Washington (Brown et al., 1979; Gusey and Brown, 1987) and the Josephine Ophiolite of northwestern California and southwestern Oregon (Harper et al., 1994, 2002).

INGALLS OPHIOLITE COMPLEX

The Ingalls Ophiolite is the largest (450 km^2) and most complete of several Middle to Late Jurassic ophiolites in northwestern Washington, and represents a northward extension of an ophiolite belt that includes the Josephine ophiolite of California and Oregon (e.g., Miller et al., 1993a; Saleeby, 1992; Metzger et al., 2002). Although lithologies in the Ingalls Ophiolite are typical of ophiolites in general, including 161 Ma gabbro-dike-pillow complexes (Esmeralda Peaks unit, Figs. 2, 3, 4; Miller et al., 2003), large areas consist of serpentinite mélange (Fig. 2; Miller, 1985). In addition, much of the southern part of the ophiolite appears to have an early Mesozoic basement (Miller et al., 1993a; Metzger et al., 2002; MacDonald et al., 2002).

A sedimentary sequence is included as part of the ophiolite complex since it is interlayered with Late Jurassic pillow lavas and ophiolitic breccias, although most of it probably postdates these pillow lavas. In general, the sedimentary sequence dips steeply toward the north.

The presence of older basement in the Ingalls Ophiolite Complex suggests it is, at least in its southern part, a polygenetic ophiolite formed over an extended period of time. Most likely it represents a rift-edge facies similar to the Preston Peak ophiolite and Devils Elbow remnant of the Josephine ophiolite in the Klamath Mountains (Saleeby et al., 1982; Wyld and Wright, 1988). Geochemistry of the Late Jurassic mafic rocks of the ophiolite and the presence of a Late Jurassic arc complex exposed in outliers southwest of the Ingalls Ophiolite Complex (Fig. 1) suggest it formed in a backarc or forearc basin (Miller et al., 1993a; Metzger et al., 2002). The presence of a high-T mantle shear zone separating harzburgite tectonite from lherzolite tectonite (Fig. 2; Miller and Mogk, 1987) is compatible with formation of the ophiolite in a fracture zone. In as much as many fracture zones are characterized by basement uplifts and abundant talus, the presence of abundant ophiolite breccias (discussed below) within the Ingalls Ophiolite Complex is consistent with a fracture zone origin.

Below we briefly describe the units of the complex from oldest to youngest.

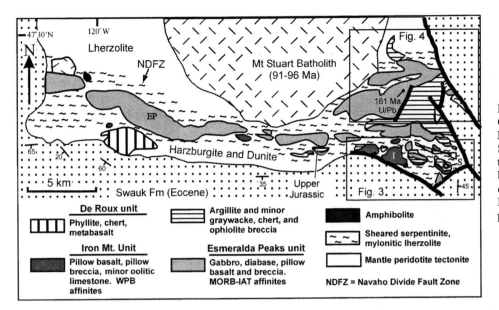

Figure 2. Map of the Ingalls Ophiolite Complex and surrounding units. EP—Esmeralda Peaks; I—Iron Mountain; S—Sheep Mountain. IAT—island-arc tholeiite; MORB—mid-ocean ridge basalt; WPB—within-plate basalt. Indicated Late Jurassic age is based on Radiolaria in chert (E. Pessagno, 2002, personal commun.).

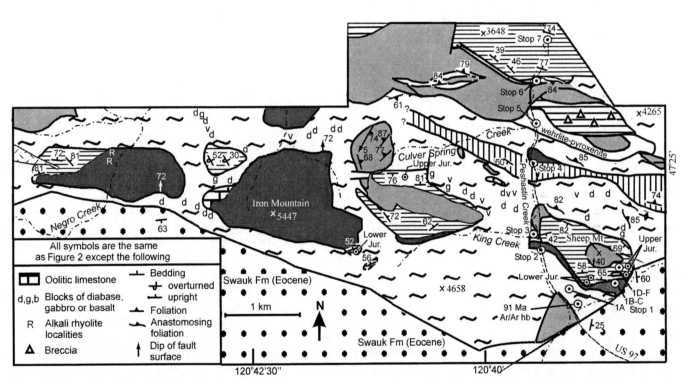

Figure 3. Geologic map of the southeastern part of the Ingalls Ophiolite Complex. Jurassic ages are based on radiolaria (Miller et al., 1993a; E. Pessagno, 1999, personal commun.; C. Blome, 1992, personal commun.). See key and location of map in Figure 2.

De Roux Unit

The De Roux unit is an undated assemblage of mostly pillow basalt, chert, and phyllite intercalated with reworked water-lain airfall tuffs. In addition, pods of limestone, serpentinite, and local amphibolite, and gabbro and graywacke are scattered throughout the unit (Miller, 1985). The De Roux unit

shows many lithologic similarities with the Ingalls Ophiolite Complex, but deformation under greenschist-facies conditions is more pervasive and ductile than in the ophiolite and numerous limestone pods and tuffs are atypical of the ophiolite. Geochemistry of a limited number of metabasalts from the De Roux unit are all normal mid-ocean ridge basalts (N-MORB; MacDonald et al., 2003), distinct from the dominantly transitional

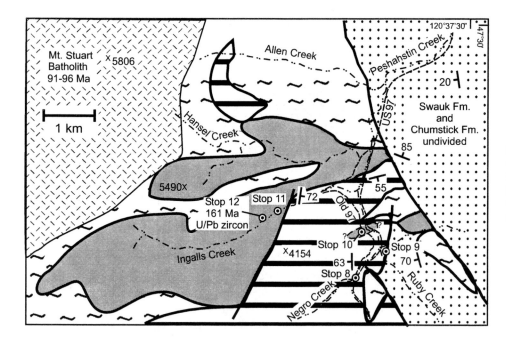

Figure 4. Geologic map of the northeastern part of the Ingalls Ophiolite Complex modified from Tabor et al. (1982). U/Pb zircon age is from Miller et al. (2003). See key and location of map in Figure 2.

MORB-IAT (island-arc tholeiite) affinities of the Late Jurassic Esmeralda Peaks mafic unit and the within-plate basalt (WPB) affinities of the Early Jurassic Iron Mountain unit.

The type area of the De Roux unit is in the southwestern part of the map area of Figure 2, where it is in fault contact with ultramafic rocks of the ophiolite complex (Fig. 2). Recent mapping has revealed the presence of De Roux–like rocks in the eastern part of the Ingalls Ophiolite Complex directly underlying the Iron Mountain unit (Stop 4; Fig. 5B) and extending for at least 6 km along strike (Fig. 3). We infer that the De Roux unit along with the Lower Jurassic Iron Mountain unit represent older basement that underlies at least the southern part of the Ingalls Ophiolite Complex.

Iron Mountain Unit

The Iron Mountain unit consists of a coherent sequence of basalt and rare rhyolite along with minor sedimentary rocks that generally dip steeply north and extend for ~24 km along the southern margin of the Ingalls Ophiolite Complex (Figs. 2, 3). Its thickness ranges from ~100 to >500 m. It structurally overlies serpentinized peridotite (Stops 1, 2; Figs. 5A, 6) or rocks correlated with the De Roux unit (Stop 4; Fig. 5B). The Iron Mountain unit is assigned an Early Jurassic age based on radiolaria from chert interbedded with pillow basalt near the base of the unit on the eastern flank of Iron Mountain (Fig. 3; Miller et al., 1993a). Recently, a 192.1 ± 0.3 Ma Pb/U zircon age was obtained for a rhyolite (J.S. Miller, 2003, personal commun.), consistent with the Early Jurassic age based on radiolaria. Lower Jurassic chert also occurs interbedded with mudstone that overlies the Iron Mountain unit on Sheep Mountain (Fig. 6). The Iron Mountain unit is distin-

guished from the De Roux unit by its lack of ductile deformation, lack of felsic tuffaceous rocks and amphibolite, and distinctive within-plate affinities (Fig. 7). It is distinguished from the 161 Ma Esmeralda Peaks unit by vesicularity of basalts, very common interpillow limestone, older age, and within-plate affinities.

The volcanic rocks of the Iron Mountain unit are mostly vesicular basalts, including both pillow lavas and pillow breccias. Also present are massive basalt and hyaloclastite (some bedded; Stop 2B). Rhyolite is exposed west of Iron Mountain (Fig. 3), along a road adjacent to Negro Creek. Plagioclase microphenocrysts are common in both the basalt and rhyolite. Relict textures in basalts are well preserved, but the rocks are altered to albite, chlorite, epidote, pumpellyite, and sphene. Vesicles are typically filled with calcite, commonly intergrown with epidote and pumpellyite.

All Iron Mountain basalts have WPB magmatic affinities on the Ti-Y-Zr diagram of Pearce and Cann (1973), which is the most diagnostic diagram for recognizing WPB (Pearce, 1996). This is consistent with where samples plot on both the Th-Hf-Ta and Ti/V diagrams (Fig. 7; Metzger et al., 2002; MacDonald et al., 2002). Ta/Yb ratios indicate they are transitional between tholeiitic and alkali basalts. Rhyolite from the Iron Mountain unit plots in the same field on the Ta-Hf-Th diagram as that defined by the basalts (Fig. 7), consistent with its formation by fractionation of the within-plate mafic magma (fractionation does not significantly change where a magma plots on the Ta-Hf-Th diagram).

Limestone and less common chert occur as interpillow sedimentary rocks. The former are present in most outcrops of pillow basalts, occurring as pods up to 1 m in diameter. On Iron Mountain, well-bedded Early Jurassic radiolarian chert is interlayered with pillow basalt, and massive oolitic limestone (Fig. 8) up to 70 m

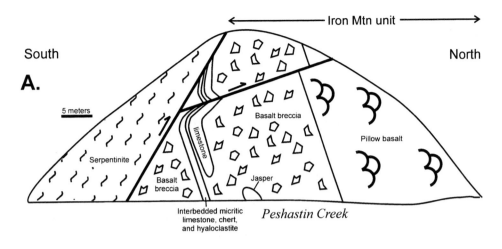

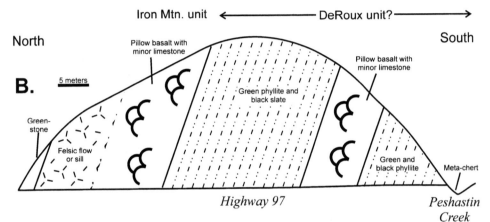

Figure 5. A: Sketch of section exposed at Stop 2B showing Iron Mountain unit overlying serpentinite (faulted contact). B: Sketch of section exposed at Stop 4 showing Iron Mountain unit overlying rocks correlated with the undated De Roux unit. C: Sketch of section at Stop 6 showing ophiolitic breccias, including detrital serpentinite.

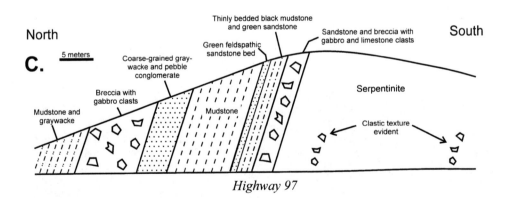

thick occurs at a stratigraphically higher level. Within a basalt breccia unit at Stop 2B, thinly bedded pink micritic limestone and red radiolarian chert interlayered with calcite-cemented hyaloclastite is overlain by breccia containing decimeter-scale oolitic limestone clasts, which is in turn overlain by a large, massive, discontinuous limestone (breccia clast?; Fig. 5A). The association of WPB and oolitic (shallow water) limestone suggests the Iron Mountain unit formed as a seamount (MacDonald et al., 2002).

Esmeralda Peaks Unit

The Esmeralda Peaks unit consists of gabbro, diabase dikes (some sheeted), pillowed and massive basalt, and minor chert and jasper. In addition, sedimentary breccias having abundant clasts of basalt with minor diabase and plagiogranite, or of diabase and less common gabbro, are present and probably represent talus from submarine fault scarps (Miller, 1985).

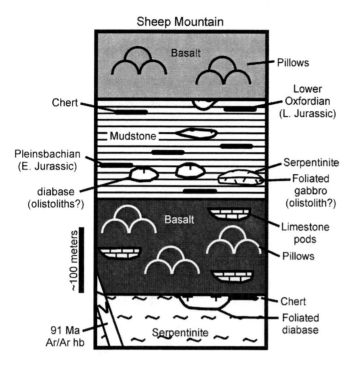

Figure 6. Schematic section of Sheep Mountain (Stop 1). Patterns are the same as those in Figure 2. Radiolarian ages are from Miller et al. (1993a), E. Pessagno (1999, personal commun.), and C. Blome (1992, personal commun.). 91 Ma hornblende Ar/Ar age is from M. Heizler (2003, personal commun.).

The Esmeralda Peaks unit occurs essentially as large blocks within the serpentinite mélange (Fig. 2). The age of a gabbro from this unit (Stop 12) was recently dated by Pb/U zircon at 161 ± 1 Ma (Miller et al., 2003), significantly older than a previous, apparently discordant 155 ± 2 Ma zircon age from a nearby gabbro (Miller et al., 1993a). Upper Jurassic chert, consistent with this age, occurs immediately above pillow basalt at Navaho Divide (Fig. 2, south central part of area; E. Pessagno, 2002, personal commun.).

Gabbros commonly have a magmatic foliation, but only rarely display cumulus layering. Diabase dikes are common in the gabbro, and a transition from gabbro to sheeted dikes, as well as from sheeted dikes to pillow lavas, is exposed in the Esmeralda Peaks area and the Three Brothers area. Ultramafic cumulates (wehrlite and clinopyroxenite) form a >3-km-long belt in the eastern part of the Ingalls Ophiolite Complex (Stop 5). Some of these cumulates have undergone high-T subsolidus crystallization, but others are nearly undeformed.

Geochemically, dikes and basalts from the Esmeralda Peaks unit are mostly transitional between MORB-IAT. Nearly all samples plot in the MORB field on a Ti/V diagram (Fig. 7A), but they overlap the field between N-MORB and island-arc tholeiite on a Ta-Hf-Th diagram (Fig. 7B). These characteristics indicate dominantly transitional MORB-IAT magmatic affinities, similar to that observed in several backarc basins (e.g., Pearce et al., 1994; Fretzdorff et al., 2002). A few basalt and diabase samples have a much greater arc component, ranging from transitional IAT-CAB (calc-alkaline basalt) to boninitic affinities (Fig. 7B). Many gabbros and three plagiogranites also have transitional IAT-CAB affinities (Fig. 7B). The crystallization order of

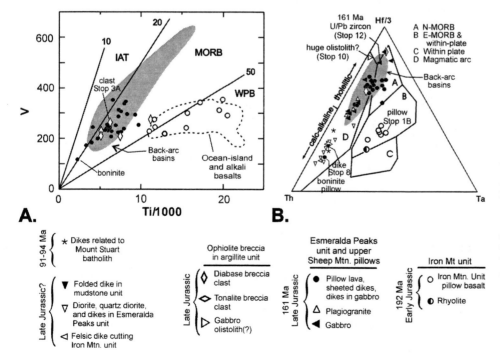

Figure 7. A: Ti/V discriminant diagram (Shervais, 1982) showing Esmeralda Peaks and Iron Mountain dikes and lavas. Only mafic rocks are shown, and gabbros are excluded, as many of them appear to be cumulates. Calc-alkaline basalts from are not plotted because this diagram is not useful for distinguishing them (Ti/V changes with fractionation). B: Ta-Hf-Th discriminant diagram of Wood (1980). All rock compositions are plotted on this diagram because fractionation and crystal accumulation does not significantly change ratios of these elements.

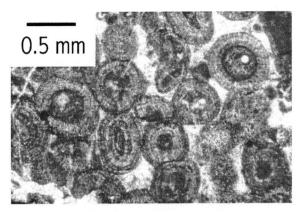

Figure 8. Photomicrograph of oolites in limestone within the Iron Mountain unit.

olivine → clinopyroxene → plagioclase in ultramafic cumulates also indicates arc magmatic affinities (e.g., Pearce et al., 1984). The presence of a boninite pillow lava is important because boninites only occur in suprasubduction zone settings, mostly in forearcs, and indicate unusually hot melting at shallow levels in the mantle (e.g., Cameron, 1989; Murton et al., 1992).

Pillow basalts at the top of the sequence at Sheep Mountain (Fig. 6), and related rocks in the Kings Creek drainage to the west, are apparently related to the Esmeralda Peak unit based on similar MORB-IAT geochemistry and age (Metzger et al., 2002).

Mantle Tectonite Unit

Ultramafic tectonites, representing residual peridotite left after extraction of a mafic melt, dominate the western two-thirds of the ophiolite. Lherzolite (orthopyroxene + clinopyroxene peridotite) and cpx-bearing harzburgite (orthopyroxene peridotite), with minor plagioclase peridotite and dunite, occur north of the Navaho Divide shear zone (Fig. 2; Miller, 1985; Miller and Mogk, 1987). In contrast, harzburgite (~80%) cut by numerous bodies of dunite (~20%), some containing podiform chromitites, occur south of the shear zone (Miller and Mogk, 1987). The Navaho Divide shear zone is up to 2.3 km wide and is comprised of well foliated to mylonitic lherzolite and hornblende peridotite. The high temperatures (700 to ≥900 °C) recorded by assemblages in this shear zone suggest that it formed in the mantle (Miller and Mogk, 1987).

Mineralogy and Cr-spinel compositions indicate that the harzburgite is more depleted than the lherzolite, representing residue of transitional MORB-IAT magma, whereas the lherzolite is a residue of MORB magma (Miller and Mogk, 1987; Metzger et al., 2002). Cr-spinels in dunites have much higher Cr/(Cr + Al), suggesting the dunites formed when depleted IAT or boninitic magmas passed through the harzburgite (Metzger et al., 2002). Thus, the harzburgite and dunite both indicate a suprasubduction zone origin, consistent with the magmatic affinities of the Esmeralda Peaks unit. In contrast, the lherzolite is a residue of MORB

magmas and thus could have formed either beneath a backarc spreading center or mid-ocean ridge (if older than Esmeralda Peaks magmatism).

The mantle peridotite and dunite vary from fresh to completely serpentinized. Gabbro and diabase in serpentinized peridotite are typically altered to rodingite—a white rock comprised largely of hydrogrossular and formed by metasomatism associated with serpentinization (e.g., Coleman, 1967). Highly serpentinized peridotites are typically sheared. The Navaho Divide shear zone (Fig. 2) has been strongly affected by serpentinization and shearing, resulting in mélange. The age of mélange formation is constrained to predate the middle Cretaceous Mount Stuart batholith (Miller, 1985) and probably predates, or was synchronous with, deposition of the Late Jurassic mudstone unit based on the presence of intercalated ophiolitic sedimentary breccias (including serpentinite) and the coincidence of the mélange with the zone of mantle mylonites.

Sedimentary Rocks

Mudstone-rich sedimentary rocks overlie the Esmeralda Peaks unit (including the upper pillow basalt unit of Sheep Mountain), separate the Lower Jurassic Iron Mountain basalts from the Upper Jurassic Sheep Mountain basalts (Fig. 6), and locally occur within the Iron Mountain unit. Most of the mudstone is massive. Minor graywacke, pebble conglomerate, pebbly mudstone, and chert are interbedded with the mudstone (Miller, 1985). The graywacke varies from thin beds showing partial Bouma sequences interbedded with mudstone, indicative of deposition by turbidity currents, to very thick massive beds. Most pebble conglomerate occurs as thick or very thick massive beds (Fig. 5C), in some places overlain by thinly bedded turbidites (Stop 7). The thick massive sandstones and pebble conglomerates likely represent submarine fan channel deposits, and the thinly bedded sandstone and mudstone overbank deposits.

Radiolarians from chert in the Sheep Mountain sequence indicate that the lower part of the mudstone unit is Pliensbachian (Lower Jurassic), whereas the upper part is Lower Oxfordian (Upper Jurassic; Metzger et al., 2002; E. Pessagno, 1999, personal commun.). E. Pessagno found that radiolaria from the Late Jurassic chert are identical in age and inferred paleolatitude as those directly overlying the Josephine Ophiolite (e.g., Pessagno et al., 2000).

Detrital zircon from a graywacke (Stop 7) yielded two distinct age peaks ca. 153 and 223 Ma (Fig. 9; Miller et al., 2003). Because the graywacke is rich in volcanic rock fragments and compositionally immature, it is likely that the 153 Ma peak approximates the age of deposition. A well-dated Late Jurassic graywacke from above the ca. 162 Ma Josephine Ophiolite gave two identical peaks in detrital zircon ages (Miller et al., 2003). Graywacke and pebble conglomerate clasts consist mostly of chert, mafic to felsic volcanic rock fragments, quartz, and plagioclase, in addition to minor siltstone, diabase, gabbro, felsic plutonic rock, and chromite (Southwick, 1974; Miller, 1985).

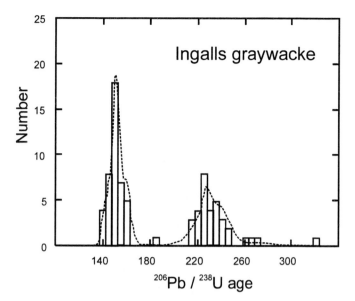

Figure 9. Detrital zircon age histogram for a graywacke overlying Ingalls ophiolite (Stop 7) with superimposed relative-probability curve (Miller et al., 2003). Note age peaks at ca. 153 and 227 Ma. The younger peak probably approximates the age of deposition.

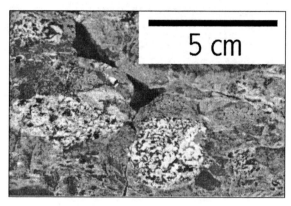

Figure 10. Ophiolitic breccia consisting of gabbro, diabase, and basalt clasts. Sample from talus along old Highway 97 east of bridge over Ingalls Creek.

These clast types are most compatible with derivation from a partially unroofed magmatic arc and a chert-rich ophiolitic terrane (arc basement?).

Miller (1985) noted the presence of minor ophiolitic breccias in the Esmeralda Peaks unit in the central part of the Ingalls Ophiolite Complex. Our recent mapping in the eastern part of the Ingalls Ophiolite Complex has revealed the presence of abundant ophiolitic breccias intercalated with the mudstone unit (Fig. 10; Mlinarevic et al., 2003). These occur as lenses <1 km in length and tens of meters in thickness. Isolated outcrops of diabase or gabbro observed within the mudstone unit are probably olistoliths (blocks in submarine landslide). One gabbro block is >300 m in length (Fig. 11; Stop 10). The ophiolite breccias consist mainly of either mafic igneous or serpentinite clasts, although some consist entirely of basalt (Stop 3A). Mafic breccias are generally poorly sorted and nonstratified, with angular to rounded clasts of diabase, gabbro, basalt, metagabbro (amphibolite), and less common quartz diorite and felsic volcanic rocks. The mafic clasts have magmatic affinities that are, for the most part, similar to the Esmeralda Peaks unit (Figs. 7A, B; Mlinarevic et al., 2003). One breccia consists of tonalite clasts that plot in the calc-alkaline field (Fig. 7B).

Detrital serpentinites are mostly unsorted breccias with angular to subangular clasts (Fig. 12A; Stop 6), but well-sorted and graded serpentinite sandstone and pebble conglomerate beds were found near Navaho Divide (Fig. 12B). Some serpentinite breccias have clasts of metagabbro tectonite (amphibolite) or diabase.

We interpret the association of ophiolitic sedimentary breccias with mudstones locally interbedded with turbidites to reflect a mixture of locally derived uplifted oceanic crust, probably

exposed on fault scarps, with distal deep-marine terrigenous sediments (Mlinarevic et al., 2003). Topographic relief caused by fracture zone-ridge interaction provides a unique environment in which rock from fault scarps are locally shed into the basin. In the Ingalls Ophiolite Complex, locally derived sediment from such fault scarps apparently periodically interrupted turbiditic and pelagic sedimentation.

Calc-alkaline Intrusives

A few plutonic rocks and dikes in the gabbro part of the Esmeralda Peaks unit (e.g., Stop 11), a dike cutting the mudstone unit (Fig. 13; Stop 8), and a dike cutting the Iron Mountain unit have calc-alkaline affinities on the Ta-Hf-Th plot (Fig. 7B). These intrusive rocks are mostly basaltic andesite to andesite in composition, and most have relict igneous brown hornblende.

Two hornblende-plagioclase andesite porphyries, one from near Stop 1 and a second from the Three Brothers area (central part of area in Fig. 2; Miller, 1985), also plot in the calc-alkaline field (Fig. 7B). Ar/Ar hornblende ages on these dikes indicate they are 91 and 94 Ma, respectively, and thus related to the Mount Stuart batholith (M. Heizler, 2003, personal commun.). These dikes are common on Sheep Mountain (Stop 1) and abundant to the west of Sheep Mountain (west of Highway 97; Southwick, 1974).

It is uncertain how many, if any, of the other calc-alkaline intrusions might also be middle Cretaceous in age. Most are different from the two dated samples in that they are much more altered and tend to be more depleted in heavy rare earth elements (REEs) at a given Cr level (fractionation index), sufficiently so that many, if not most, are boninitic. Most calc-alkaline intrusions are probably older, perhaps Late Jurassic age. Folding of a dike in the mudstone unit (Fig. 13) also suggests a relatively old age. Calc-alkaline intrusive rocks were certainly present during the Late Jurassic based on calc-alkaline tonalite clasts in a breccia within the mudstone unit (Fig. 7B); the clast sampled is very depleted in heavy REE and may thus be a boninite fractionate.

Figure 11. Gabbro block within mudstone unit, interpreted as a huge olistolith (block in submarine landslide). This is the same gabbro as that at Stop 10, from where this photo was taken. The mudstone unit in this area is strongly deformed. The clasts are highly stretched, but this surface is oriented normal to the stretching lineation.

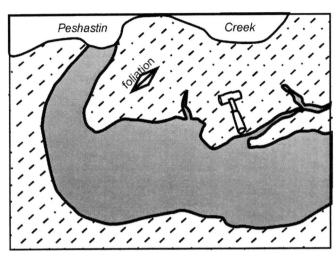

Figure 13. Folded dike in mudstone unit (Stop 8). The mudstone unit is metamorphosed and deformed in this area and has a well-defined cleavage.

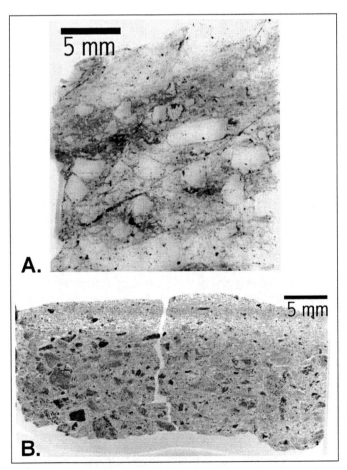

Figure 12. A: Photomicrograph of detrital serpentinite at Stop 6. B: Photomicrograph of a graded bed of detrital serpentinite.

Teanaway Dikes

Eocene (Teanaway) dikes are very common throughout the Ingalls Ophiolite Complex and the overlying Eocene Swauk Formation (e.g., Tabor et al., 1982). They are easily recognized in the field by their distinctive reddish-brown weathering color, gray fresh color, abundant fresh cleavage faces in hand sample, dike-normal jointing, and thickness (generally >5 m).

CONCLUSIONS

The Ingalls Ophiolite Complex is polygenetic, at least in its southern part, as indicated by the presence of the early Mesozoic Iron Mountain unit (fossil seamount) and underlying De Roux unit of unknown age. Gabbro-dike-pillow complexes of the Esmeralda Peaks unit and correlative pillow lavas overlying the Iron Mountain unit are early Late Jurassic and formed either during backarc or forearc rifting. The mudstone unit is mostly Late Jurassic, and detrital zircon from a graywacke indicates source areas of ca. 153 and 227 Ma. The age of the Esmeralda Peaks unit, radiolarian chert, and detrital zircons indicates these units are correlative with the Josephine Ophiolite and overlying Galice Formation in the Klamath Mountains (Miller et al., 2003). Ophiolitic breccias, including detrital serpentinite, are abundant in the mudstone unit and indicate uplift and exposure of ocean crust during deposition. This and the presence of a high-T mylonitic shear zone in the mantle tectonite unit suggest the Ingalls Ophiolite Complex formed in a fracture zone setting. The MORB-IAT magmatic affinities of the Esmeralda Peaks unit, the presence of Late Jurassic arc rocks in inliers to the southwest (Fig. 1), and the presence of arc detritus indicate the Ingalls Ophiolite Complex was in a suprasubduction zone setting during the Late Jurassic.

ROAD LOG

The road log is in miles measured northward from Blewett Pass. Numbers in parentheses that start with "10 T" are UTM (Universal Transverse Mercator) coordinates.

Miles	Description
0.0	Blewett Pass, U.S. Highway 97.
0.9	View of ridge of lacustrine facies of Eocene Swauk Formation in distance (see field guide by Evans (1994) for details.
4.0	View of 91–96 Ma Mount Stuart batholith on skyline.
6.4	Eocene Teanaway dike cutting Swauk Formation in roadcut.
7.2	Eocene Teanaway dike cutting Swauk Formation in roadcut.
7.5	Multiple Teanaway dikes cutting Swauk Formation in roadcut.
7.6	First view of Ingalls Ophiolite Complex to left in distance. Pillow lavas farthest left; serpentinite underlies grassy areas.
8.0	Outcrop on right (east) side of road shows many Teanaway dikes cutting basal Swauk Formation conglomerate. Serpentinite of the Ingalls Ophiolite Complex is exposed on the north side of outcrop.
8.1	Turn right off Highway 97 at the end of above roadcut onto gravel road up Magnet Creek. Beginning at crossing of road over small drainage, Eocene Teanaway dikes are exposed (jointed and brown weathering). Also exposed is a whitish weathering andesite dike (10 T 677329, 5252039) having phenocrysts of hornblende and plagioclase. Hornblende from this dike gave an Ar/Ar age of 91 Ma (M. Heizler, 2003, personal commun.), indicating it is related to the Mount Stuart batholith and much younger than the ophiolite. We will not stop here, as many similar dikes are exposed farther up the road where we will be walking.
8.5	Park at first switchback and walk 215 m up the road (east) to small outcrop of serpentinite in roadcut. Continue up road ~310 m onto the next, better outcrop of serpentinite. Some float of orange weathering silica-carbonate rock (magnesite-quartz) is also present and represents metasomatized serpentinite.

Stop 1A (10 T 677446 5252187)

Outcrop of massive and sheared serpentinized peridotite. This is the lowest unit exposed in the Sheep Mountain section (Fig. 6). Elsewhere it can be seen to consist of partially serpentinized peridotite separated by zones of sheared serpentinite.

Approximately 170 m farther up the road from Stop 1A is a roadcut of diabase displaying a magmatic (?) foliation dipping ~30–60° east, and a steeply plunging lineation. Approximately 90 m farther up the road are NE-dipping, 5–65-cm-thick beds of black chert interbedded with mudstone. Just past this outcrop, the road curves left (NE) and widens into a large flat area. Cliffs

at top of Sheep Mountain are visible from here—these are Late Jurassic pillow lavas having MORB-IAT affinities. The slope below the cliffs is a mudstone-rich unit containing minor chert, graywacke, blocks of diabase and gabbro, and lenses of serpentinite (Fig. 6). These rocks overlie pillow basalts of the Iron Mountain unit, which occur at the level we are standing.

Stop 1B (10 T 677664 5252302)

Just off the corner of the road curve is a natural outcrop of fractured vesicular greenstone belonging to the Iron Mountain unit (Fig. 6). With careful observation, pillow structures up to 1 m in diameter can be seen on the near-vertical outcrop surface. At the top of the outcrop is an ~1 x 0.5 m limestone pod within the greenstone. A sample from this outcrop (BL-14-1) has WPB affinities (Metzger et al., 2002) characteristic of the Iron Mountain unit (Fig. 7). Vesicular WPB pillow lavas and limestone are characteristics of the Iron Mountain unit that distinguish it from the Late Jurassic pillow lavas.

Continue up the road (NE) for another ~200 m to a serpentinite outcrop. In this interval are roadcuts of Iron Mountain greenstone and mudstone.

Stop 1C (10 T 677824 525234)

A steeply NE-dipping sheared serpentinite is exposed in a roadcut. The serpentinite has a NW-striking fabric with composite foliations, one of which is penetrative (~82° dip) and the other comprising discrete surfaces (~58° dip) with centimeter-scale spacing. These appear to define S surfaces (foliation) and shear bands, respectively, and their orientations imply reverse faulting (NE side up). The serpentinite is in contact on the northeast with ~1 m of mudstone-matrix breccia overlain by greenstone that may be pillowed. The breccia is ophiolitic, consisting of clasts of mafic volcanic rock (some variolitic), diabase with variably colored amphibole (hydrothermal), and plagioclase-phyric leucocratic volcanic rock (looks like pale green chert in hand sample). The greenstone adjacent to the breccia, as well as the gabbro above (next stop) may be large clasts within this breccia, and thus the breccia may be an olistostrome (submarine landslide deposit).

From the east end of the greenstone outcrop on the road, walk up hill ~100 m, bearing left (west). Argillite float is evident in the soil.

Stop 1D (10 T 677781 5252436)

An ~9-m-thick serpentinite pod occurs within mudstone in a grassy area at ~3200 ft elevation. An outcrop along the lower margin of the serpentinite is gabbro displaying a magmatic foliation that dips steeply north. Blocks of serpentinite and diabase are found locally elsewhere on Sheep Mountain within the mudstone. We suspect that these are olistoliths, but we cannot rule out a tectonic origin.

Walk uphill along bearing 023°, ~120 m to the next stop. The base of the upper, Late Jurassic pillow lavas begins at

~3290–3440 ft elevation and is evident from prominent, but low, outcrops. Continue to an ~3-m-high outcrop in brush at ~3440 ft elevation.

Stop 1E (10 T 677822 5252549)

Pillow basalt of upper Late Jurassic basalt that caps Sheep Mountain. Samples from outcrops at higher elevation have transitional MORB-IAT affinities (BL-69 in Metzger et al., 2002). Maximum diameter of pillows here is ~1 m.

Walk 185 m ENE to prominent outcrop of chert at ~3400 ft elevation. Note mudstone float.

Stop 1F (10 T 677960 5252648)

Chert outcrops trending downhill at ~136°. The chert is pale greenish gray, and Radiolaria (dark spheres) are visible with a hand lens. The chert is pure with very few shale interbeds even though mudstone float is evident on either side of the chert. Bedding is difficult to see, but dips steeply to the NE. Emile Pessagno (1999, personal commun.) has extracted Radiolaria that indicate an Early Oxfordian (early Late Jurassic) age.

Hike downhill to road. Follow road back (SW) to parked vehicles. Drive back to Highway 97, turn right (NW), and drive 0.6 mi. Park in pullout on right (NE) side of road. Walk ~300 m back down road (SE) from pullout to Stop 2A.

Miles	Description
8.9	Stop 2A.

Stop 2A (10 T 676944 5252425)

Partially serpentinized mantle peridotite separated by sheared serpentinite (poorly exposed in gullies). A foliation defined by aligned pyroxene is evident on some brownish weathered surfaces. Weakly serpentinized tectonite peridotite transitional from harzburgite to lherzolite lies a few kilometers on strike to the west, south of the Iron Mountain unit.

Return to pullout and cross highway to look at exposure on western side of Peshastin Creek (Fig. 5A).

Stop 2B (10 T 676780 5252714)

Sheared serpentinite on the south end of outcrop is faulted against Iron Mountain unit (Fig. 5A). The serpentinite has a shear foliation oriented parallel to the E-W-striking, south-dipping fault; steeply plunging slickenside striae are evident on some of the shear surfaces. The slickenside orientation, drag of bedding (Fig. 5A), and small associated fault (Fig. 5A) indicate reverse displacement (south side up).

Cross Peshastin Creek to view ~1.5-m-thick massive gray limestone lens underlain by NNE-dipping, red radiolarian chert, thin-bedded red micritic limestone, and calcite-cemented hyaloclastite (Fig. 5A). An ~10-cm-diameter oolitic limestone clast

occurs in breccia immediately below the massive limestone, suggesting the massive limestone might be a breccia clast (slide block) rather than a bed. Clast size in the basalt breccias here and upstream varies from sand size up to ~30 cm. Centimeter-size clasts are most common in the lower part of the breccia sequence, whereas decimeter-size clasts are most common in the upper part of the sequence. A pod of jasper, probably hydrothermal in origin, is exposed at creek level (Fig. 5A).

Return to highway side of creek and proceed ~26 m downstream (north). If possible, cross the creek on the north side of the outcrop to view pillow basalts (shapes visible from highway). Pillows range from ~30 cm to 1.2 m in diameter.

Return to vehicles and continue driving north on Highway 97.

Miles	Description
9.8	Turn left onto Kings Creek Road and park in front of gate. Walk to the intersection of Kings Creek Road with Highway 97, and then descend to the creek and cross it.

Stop 3A (10 T 676733 5252952)

Mafic breccia consisting of aphyric basalt is well exposed along the creek. Clasts range up to 13 cm in diameter. They are mostly angular, but a few are rounded. At the north end of the outcrop near water level are two steeply NE-dipping bedding (?) surfaces defined by lighter-green angular centimeter-scale clasts. Analysis of one of the larger clasts from this outcrop indicates magmatic affinities similar to the Late Jurassic pillow lavas in the upper part of the Sheep Mountain section and the Esmeralda Peaks unit (Fig. 7A). This breccia appears to be in a similar stratigraphic position as gabbro and diabase olistoliths (?) on the south side of Sheep Mountain (i.e., above Iron Mountain unit but below upper pillow basalt unit).

Stop 3B (Optional. 10 T 676803 5252902)

Walk south of Kings Creek Road to outcrops on west side of Highway 97 where SE-dipping graywacke is interbedded with mudstone. On the east side is an outcrop of sheared mudstone containing limestone pods up to 50 cm in diameter. We suspect these pods are olistoliths. The higher part of the outcrop is massive graywacke that extends to road level at the south end of the outcrop. At the north end of this outcrop is a 7-m-thick breccia interbedded with the mudstone that consists of unsorted mafic volcanic clasts up to 12 cm in diameter. This breccia is similar to that just observed on Peshastin Creek (Stop 3A).

Stop 3C

Walk back north to the King Creek Road junction and cross main highway to the large outcrop on the east side, which is probably the continuation of the breccias in the creek (Stop 3A). A thin section from this outcrop displays fresh clinopyroxene phenocrysts, smaller plagioclase phenocrysts, and numerous prehnite veins.

From the highway, climb up the slope on the south side of the outcrop (use animal trails through brush) for ~100 m to top of outcrop where an obvious NW-striking, subvertical fault contact is exposed between the basalt breccia and serpentinite (10 T 0676853, 5253292). The breccia is extensively veined, and the serpentinite encloses rodingitized mafic pods. Walk up the ridge to the east to view WNW-ESE striking, subvertical foliation and local pyroxene-rich bands in mantle peridotite at ~2680 ft elevation (10 T 0676923, 5253321). This ultramafite lies within a belt of high-T, mylonitic plagioclase peridotite and lherzolite that is characteristic of a mantle shear zone, best documented farther west (Miller, 1985; Miller and Mogk, 1987). This shear zone is interpreted as a fossil transform (Miller and Mogk, 1987). Less serpentinized samples farther east and up the ridge display enstatite and less common olivine and diopside porphyroclasts set in a fine-grained mosaic of olivine, diopside, plagioclase, and enstatite. The plagioclase forms elongate isolated grains and highly elongate aggregates in the foliation. Most of the scattered rodingites on the walk up through the ultramafites are probably altered diabase dikes. Dikes analyzed to the west that cut the mylonite have transitional MORB-IAT chemistry typical of the Esmeralda Peaks unit.

Retrace route back down ridge to the vehicles and continue driving north on Highway 97.

Miles	Description
10.0	Outcrop of rodingite (white) in serpentinite on east side of Highway 97.
10.2	Highway crosses Peshastin Creek. Park in pullout on right (east) side of highway. Walk down to creek.

Stop 4A (10 T 676742, 5253787)

Ribbon chert is exposed on the creek. This chert is distinct from that within the Iron Mountain unit and overlying mudstone sequence in that it is more deformed and recrystallized (note sugary texture with hand lens). These rocks are very similar to the De Roux unit farther west in the Ingalls Ophiolite Complex (Fig. 2).

Return to the highway and walk north ~185 m to view N-dipping section exposed on a roadcut on the east side of the highway (Fig. 5B), just after the highway crosses Peshastin Creek.

Stop 4B (10 T 676659 5253948)

This N-dipping section, with the deformed chert at the base, appears to represent a repetition of the Iron Mountain unit and its underlying basement. Instead of serpentinite like we saw at Sheep Mountain, the basement here is greenschist-facies metachert, phyllite, metatuff, and pillow basalt (Fig. 5C). These rocks are more deformed and metamorphosed than other rocks in this area, but are very similar to those of the type area of the De Roux unit exposed farther west (Fig. 2; Miller, 1985).

Drop down to the chert outcrop on the northeast side of the creek, ~30 m upstream of the highway. This chert is very similar to that observed at the last stop. Proceed up the stream bank to observe strongly foliated gray phyllite and greenschist (metatuff). Some of the latter is wispy on a centimeter scale, similar to metatuffs of the De Roux unit. The metatuffs display an intense foliation, which is axial-planar to local rootless folds on the thin section scale. Proceed to roadcut where the greenschist and phyllite are overlain by greenstone that contains limestone and appears to be pillowed. A sample from this greenstone (BL-179-1) has N-MORB affinities, which is the same as that for analyzed metabasalts from the De Roux unit (MacDonald et al., 2003). The pillow lavas are overlain by green metatuff and black slate (Fig. 5B). These De Roux (?) rocks are overlain by undeformed variolitic pillow lava containing a few limestone pods up to 10 cm in diameter. The limestone has spherical grains that may be oolites. These pillow basalts are considered to belong to the Iron Mountain unit based on the WPB affinity of a sample collected from here (BL-179-5) and the presence of oolitic (?) limestone. This unit is overlain by a felsic flow or sill, possibly correlative with rhyolite locally found in the Iron Mountain unit NE of Iron Mountain (Fig. 3). An outcrop of greenstone occurs above the rhyolite at the north end of the roadcut.

Return to vehicles. Continue driving north on Highway 97 to Stop 5.

Miles	Description
10.6	Stop 5.

Stop 5

Turn right onto gravel road and park ~25 m from main highway where this road crosses old Highway 97. Walk up logging road to east, which shortly switches back toward the north parallel to Highway 97. At ~0.1 mi along this road (10 T 0676765 5254372) another spur takes off to the right, back toward an E-W-oriented drainage. At the road junction is exposed sheared serpentinite and pyroxenite. Turn right onto the old logging road and walk a short distance to where there is an open slope on the left of mostly serpentinite. Most of the blocks are gabbro, but a weakly layered wehrlite is present along the road (10 T 0676843 5254372; elevation ~2445 ft). The ultramafic cumulates also occur in a roadcut on Highway 97 immediately north of the turnoff to this stop. The wehrlite and pyroxenite are part of a >3-km-long belt of cumulate ultramafic rocks. Some of the wehrlites are almost undeformed, whereas others show evidence of high-T subsolidus recrystallization.

Retrace route back to the vehicles and look for a large talus block of ophiolitic sedimentary breccia, derived from the ridge to the NE. Turn right back onto Highway 97 and continue driving north.

Miles	Description
10.8	Historic site where mining town of Blewett was located.
11.1	Park on right side of highway (10 T 676812 5255170). Walk back down road (south) to the farthest outcrop, ~50 m past the section shown in Figure 5C.

Stop 6 (10 T 676720 5254990)

Outcrop of serpentinite. Close observation shows this rock has a clastic texture with sand- to pebble-size angular grains. Sorting is poor. The clastic texture is also evident in thin section (Fig. 12A). Walk back (north) to the section shown in Figure 5C (note float of gabbro-clast breccia) to next exposure of serpentinite having similar clastic texture. Some shearing in the serpentinite is also evident.

At the base of the section is 5 m of very coarse-grained sandstone and breccia containing clasts that include gabbro, chert, and limestone. This is in turn overlain by ~1 m of interbedded black mudstone, pale-green siltstone, and green feldspathic silicic sandstone (or tuff) having graded bedding, followed by ~10 m of mudstone, and 5 m of dark gray coarse graywacke and pebble conglomerate. Overlying this is ~1.5 m of breccia containing clasts of gabbro and quartz diorite; some of the clasts are tectonites as evident in hand sample and thin section.

Return to vehicles and continue driving north on Highway 97.

Miles	Description
11.4	Park on right side of highway.

Stop 7 (10 T 676862 5255553)

View outcrops of interbedded mudstone and graywacke. Graywacke beds are 2–20 cm thick and show graded bedding and ripple cross lamination characteristic of turbidites (Bouma C-E). The steep north dip is typical of sedimentary rocks in the Ingalls Ophiolite Complex. Cross the highway to see very coarse-grained graywacke and pebble conglomerate. Detrital zircons separated from a graywacke (BL-138) collected here were dated using the Stanford sensitive high-resolution ion microprobe (SHRIMP). A striking bimodal age distribution is evident with peaks at ca. 153 and 227 Ma (Fig. 9). The pebble conglomerate and massive graywacke may be a submarine channel deposit and the overlying, thinly bedded turbidites representing overbank deposits.

Walk to next roadcut to the south. The northern part of the outcrop consists of massive coarse-grained sandstone underlain by mudstone. The central part of the outcrop is an ~25-m-thick felsic sill or flow, which is sheared in several places. Below this is argillite and a few thin graywacke interbeds; a tight anticline is exposed several meters south of the felsic unit and is best seen from the other side (west) of the highway. A weak, E-W-striking, approximately axial-planar fabric is evident in the argillite.

Return to vehicles and continue driving north on Highway 97 to Stop 7.

Miles	Description
12.3	Stop and park on the left (west) side of the highway, across from the mouth of Negro Creek. Descend from the road and walk ~20 m north along the east bank of Peshastin Creek.

Stop 8 (10 T 676379 5256830)

Note large (transported) boulder of porphyroclastic peridotite mylonite, transported from the high strain zone in the Ingalls peridotite unit (oceanic fracture zone).

Cross Peshastin Creek (if possible; if not, outcrop is visible from east side). A folded dike occurs within foliated argillite (Fig. 13). This dike (IO-24) is mafic, has high Cr and Ni contents indicating it is primitive (unfractionated), and has calc-alkaline affinities. We do not know the age of this dike, but since it is so deformed we assume it is relatively old, perhaps Late Jurassic (i.e., just younger than the argillite).

Return to vehicles and continue driving north on Highway 97.

Miles	Description
12.8	At north end of outcrop is a road sign indicating turn to Ruby Creek Road. Stop 9.

Stop 9. (Optional. 10 T 676777 5257413)

This is an optional stop that we will make if we are unable to cross Peshastin Creek to see the dike at Stop 8. Large outcrop consists of massive, featureless argillite cut by large andesitic dike. The dike is irregular and may be folded.

Return to vehicles and continue driving north on Highway 97.

Miles	Description
13.1	Intersection with Ruby Creek Road.
13.5	Stop and park on right (east) side of highway. Looking southeast, a prominent knob is a huge gabbro block in the argillite unit that we will visit at the next stop. It is likely that this gabbro is a large olistolith.
14.0	End of narrow canyon (Peshastin Canyon).
15.4	Turn left onto Ingalls Creek Road. Drive to end of road and park at gated bridge over Ingalls Creek (10 T 675449 5259039). Note that a U.S. Forest Service (USFS) permit is required to park here (can be obtained in Leavenworth at the USFS station). Cross bridge and hike along old highway 600 m to Stop 10.

The argillite unit along this road is metamorphosed to higher grade and is more deformed than elsewhere. The cause and age of this locally higher-grade metamorphism is uncertain, but it may reflect the increase in metamorphic grade from south to north during middle Cretaceous regional metamorphism of the Cascades core (Paterson et al., 1994). Alternatively, an apophysis of the Mount Stuart batholith underlies this area, resulting in dynamic contact metamorphism.

Talus blocks in the first roadcuts after crossing the bridge are deformed gabbro and diabase. It is evident in some of the blocks that the gabbro and diabase are clasts in ophiolitic breccia (Fig. 10). Clasts are strongly elongated parallel to a mineral lineation. Farther up the road are outcrops of gabbro with steeply dipping intercalations of biotite-rich metagraywacke. These are followed uproad (east) by talus of phyllite, then gabbro, and finally a large outcrop of gabbro (Stop 10). Pelitic and psammitic rocks along the road are well foliated and lineated, and foliation in pelitic rocks wraps porphyroblasts of altered cordierite.

Stop 10 (10 T 676609 5257974)

Large gabbro block in argillite unit that was viewed from the highway. Pelitic rocks are exposed at the upper (east) end of outcrop. The shape of the gabbro block is evident in the large roadcut on Highway 97 where the gabbro block can be seen to be bounded on either side by dark argillite (Fig. 11). The boundaries of the block are steeply dipping, similar to the regional dip of bedding in the argillite unit. A minimum along-strike length of this block is ~300 m from the roadcut to outcrops above where we are standing. We interpret this block to be a very large olistolith in the argillite unit. This interpretation is consistent with the common occurrence of ophiolite breccias containing gabbro clasts in the argillite unit.

The gabbro is dominantly medium grained, but is markedly heterogeneous. It locally has a strong magmatic foliation. The dominant gabbro is cut by dikes and pods of hornblende pegmatite. Steeply dipping serpentinite bodies several meters wide occur within the gabbro outcrops; little texture is evident with a hand lens, but they are probably ultramafic cumulates. The gabbro and gabbro pegmatite are both cut by ductile shear zones up to 30 cm thick. Thin hornblende veins and white veins developed along small faults are probably hydrothermal in origin (ocean-ridge metamorphism). A geochemical analysis of the dominant gabbro indicates arc magmatic affinities (Fig. 7B). High Ni content (328 ppm) suggests olivine accumulation, but a REE pattern does not show a positive Eu anomaly characteristic of plagioclase accumulation. Very low values of incompatible trace elements (e.g., heavy REE, Ti, Y) and a positive Zr anomaly on a MORB-normalized diagram are suggestive of boninitic affinities, although the former might be due to dilution by olivine accumulation.

Walk back down the road to the parking lot where vehicles are parked. Hike west 0.6 mi (0.98 km) along Ingalls Creek trail (on north side of creek), then follow a side trail that leads down to the creek.

Stop 11 (10 T 674633 5258633)

On the side of the outcrop that faces the creek, at the edge of the creek, is a fine-grained hornblende quartz diorite that is cut by an irregular white dike of quartz monzonite, both of which are cut by a greenish-gray mafic dike. The quartz diorite was sampled for U/Pb dating but no zircon was recovered during mineral separation.

This outcrop is just upstream from the (unexposed) contact with the argillite unit and appears to be part of the gabbro belonging to the Esmeralda Peaks unit (Fig. 4). The quartz diorite,

quartz monzonite, and mafic dike all have calk-alkaline affinities, which is distinct from the dominantly IAT-MORB affinity of the sheeted dikes and pillow lavas of the Esmeralda Peaks unit. We infer that they are older than the Mount Stuart batholith because they are strongly recrystallized and are locally cut by ductile shear zones evident in nearby exposures in the creek. They probably postdate gabbro of the Esmeralda Peaks unit as a calc-alkaline dike cuts gabbro on the slope northwest of here. As for the folded calc-alkaline dike cutting the argillite unit (Stop 8), we postulate the rocks at this stop are Late Jurassic in age.

Return to main trail and continue west for 0.4 mi (0.64 km) to where the trail crosses a prominent talus slope.

Stop 12 (10 T 674110 5258258)

Gabbro, pegmatitic hornblende gabbro, and diabase occur as blocks in the talus. Some blocks show chilled dike margins of diabase cutting gabbro. These rocks belong to the gabbro part of the Esmeralda Peaks unit. The talus blocks are derived from prominent cliff exposures ~100 m upslope.

A pegmatitic hornblende gabbro block from here was dated by U/Pb zircon as 161 ± 1 Ma (Miller et al., 2003). This gabbro has transitional IAT-MORB magmatic affinities (Fig. 7B). It has a positive Eu spike on an REE diagram, suggesting plagioclase accumulation. A mafic dike collected upslope from here, and another dike 350 m to the west, have calc-alkaline affinities.

Return to vehicles and drive back to Seattle.

ACKNOWLEDGMENTS

This research was supported by National Science Foundation (NSF) grant EAR-000344 to G.D. Harper, NSF grant EAR-0087829 to R.B. Miller and J.S. Miller, and U.S. Geological Survey EDMAP award 03HQAG0066 to R.B. Miller. J.H. MacDonald was funded in part by GSA Research Grant 6951-01, a Sigma Xi Grant-in-Aid of Research, and the Department of Earth and Atmospheric Sciences, State University of New York, Albany. We thank Ron Karpowicz, Scott McPeek, Cindy Schultz, and Mike Siudy for assistance during fieldwork. We thank Terry Swanson for his review and for editing the field trip volume.

REFERENCES CITED

Brandon, M.T., Cowan, D.S., and Vance, J.A., 1988, The Late Cretaceous San Juan thrust system, San Juan Islands, Washington: Boulder, Colorado, Geological Society of America Special Paper 221, 81 p.

Brown, E.H., 1987, Structural geology and accretionary history of the Northwest Cascades system, Washington and British Columbia: Geological Society of America Bulletin, v. 99, p. 201–214.

Brown, E.H., Bradshaw, J.Y., and Mustoe, G.E., 1979, Plagiogranite and keratophyre in ophiolite on Fidalgo Island, Washington: Geological Society of America Bulletin, v. 90, p. 493–507.

Cameron, W.E., 1989, Contrasting boninite-tholeiite associations from New Caledonia, *in* Crawford, A.J., ed., Boninites and Related Rocks: Unwin Hyman, London, p. 314–338.

Coleman, R.G., 1967, Low-temperature reaction zones and alpine ultramafic rocks of California, Oregon, and Washington: U.S. Geological Survey Bulletin, B 1247, 49 p.

Davis, G.A., Monger, J.W.H., and Burchfiel, B.C., 1978, Mesozoic construction of the Cordilleran "collage," central British Columbia to central California, *in* Howell, D.G., and McDougall, K., eds., Mesozoic Paleogeography of the Western United States, Pacific Coast Paleogeography Symposium, v. 2: Society of Economic Paleontologists and Mineralogists, p. 1–32.

Evans, J.E., 1994, Tectonics and sedimentation of the Eocene Chumstick Formation, central Washington State: Society of Economic Geologists Guidebook Series, v. 20, p. 18–30.

Fretzdorff, S., Livermore, R.A., Devey, C.W., Leat, P.T., and Stoffers, P., 2002, Petrogenesis of the backarc East Scotia Ridge, South Atlantic Ocean: Journal of Petrology, v. 43, p. 1435–1467.

Gusey, D., and Brown, E.H., 1987, The Fidalgo Ophiolite, Washington, *in* Hill, M.L., ed., Cordilleran Section of the Geological Society of America: Boulder, Colorado, Geological Society of America, Centennial Field Guide, v. 1, p. 389–392.

Harper, G.D., Saleeby, J.B., and Heizler, M., 1994, Formation and emplacement of the Josephine ophiolite and the age of the Nevadan Orogeny in the Klamath Mountains, California-Oregon: U/Pb zircon and $^{40}Ar/^{39}Ar$ geochronology: Journal of Geophysical Research, v. 99, p. 4293–4321.

Harper, G., Giaramita, M., Kosanke, S., 2002, Field guide to the Josephine and Coast Range Ophiolites, Oregon and California: Oregon Department of Geology and Mineral Industries Special Paper 36, p. 1–22.

MacDonald, J.H., Jr., Harper, G.D., Miller, R.B., and Miller, J.S., 2002, Within-plate magmatic affinities of a lower pillow unit in the Ingalls Ophiolite Complex, Northwest Cascades, Washington: Geological Society of America Abstracts with Programs, v. 34, no. 5, p. 22.

MacDonald, J.H., Jr., Harper, G.D., and Miller, R.B., 2003, The De Roux unit of the central Cascades, Washington: Geochemistry, tectonic setting, and possible correlations: Geological Society of America Abstracts with Programs, v. 35, no. 6, p. 513.

Metzger, E.P., Miller, R.B., Harper, and G.D., 2002, Geochemistry and tectonic setting of the ophiolitic Ingalls Complex, North Cascades, Washington; implications for correlations of Jurassic Cordilleran ophiolites: Journal of Geology, v. 110, p. 543–560.

Miller, J.S., Miller, R.B., Wooden, J.L., and Harper, G.D., 2003, Geochronologic links between the Ingalls Ophiolite, North Cascades, Washington and the Josephine Ophiolite, Klamath Mts., Oregon and California: Geological Society of America Abstracts with Programs, v. 35, no. 6, p. 113.

Miller, R.B., 1985, The ophiolitic Ingalls Complex, north-central Cascade Mountains, Washington: Geological Society of America Bulletin, v. 96, p. 27–42.

Miller, R.B., Mattinson, J.M., Funk, S.G., Hopson, C.A., Treat, C.L., 1993a, Tectonic evolution of Mesozoic rocks in the southern and central Washington Cascades, *in* Dunn, G., and McDougall, K., eds., Mesozoic Paleogeography of the Western United States II, Book 71: Pacific Section Society of Economic Paleontologists and Mineralogists, p. 81–98.

Miller, R.B., and Mogk, D.W., 1987, Ultramafic rocks of a fracture-zone ophiolite, North Cascades, Washington: Tectonophysics, v. 142, p. 201–289.

Miller, R.B., and Paterson, S.R., 1992, Tectonic implications of syn- and post-emplacement deformation of the Mount Stuart Batholith for Mid-Cretaceous orogenesis in the North Cascades: Canadian Journal of Earth Sciences, v. 29, p. 479–485.

Miller, R.B., Whitney, D.L., and Geary, E.E., 1993b, Tectonostratigraphic terranes and the metamorphic history of the northeast part of the Cascades crystalline core—Evidence from the Twisp Valley Schist: Canadian Journal of Earth Science, v. 30, p. 1306–1323.

Misch, P., 1966, Tectonic evolution of the northern Cascades of Washington State—a west-Cordilleran case history: Canadian Institute of Mining and Metallurgy, Special Volume 8, p. 101–148.

Misch, P., 1977, Dextral displacements at some major strike faults in the North Cascades: Geological Association of Canada Program with Abstracts, v. 2, p. 37.

Mlinarevic, A.N., Miller, R.B., Harper, G.D., MacDonald, J.H., Jr., and Miller, J.S., 2003, Nodal basin(?) sedimentation in an ancient oceanic fracture zone, Ingalls Ophiolite Complex, Washington: Geological Society of America Abstracts with Programs, v. 35, no. 6, p. 513.

Murton, B.J., Peate, D.W., Arculus, R.J., Pearce, J.A., and van der Laan, S.R., 1992, Trace element geochemistry of volcanic rocks from Site 786; the Izu-Bonin forearc, *in* Dearmont, L.H., Mazullo, E.K., Stewart, N.J., and Winkler, W.R., eds., Proceedings of the Ocean Drilling Program, Scientific Results 125: College Station, Texas, Ocean Drilling Program, p. 211–235.

Paterson, S.R., Miller, R.B., Anderson, J.L., Lund, S., Bendixen, J., Taylor, N., and Fink, T., 1994, Emplacement and evolution of the Mt. Stuart Batholith, *in* Swanson, D.A., and Haugerud, R.A., eds., Guides to Field Trips, Chapter 2F: Geological Society of America Annual Meeting, Seattle, p. 2F1–2F48.

Paterson, S.R., Matzel, J.P., and Miller, R.B., 2002, Spatial and temporal evolution of magmatic systems in a continental margin arc; the Cascades Core, Washington: Geological Society of America Abstracts with Programs, v. 34, no. 5, p. 96.

Pearce, J.A., 1996, A user's guide to basalt discrimination diagrams, *in* Wyman, D.A., ed., Trace Element Geochemistry of Volcanic Rocks: Applications for Massive Sulphide Exploration, Geological Association of Canada Short Course Notes, v. 12: Memorial University, St. John's, Newfoundland, p. 79–114.

Pearce, J.A., and Cann, J.R., 1973, Tectonic setting of basic volcanic rocks determined using trace element analyses: Earth and Planetary Science Letters, v. 19, p. 290–300.

Pearce, J.A., Ernewein, M., Bloomer, S.H., Parson, L.M., Murton, B.J., and Johnson, L.E., 1994, Geochemistry of Lau Basin volcanic rocks: influence of ridge segmentation and arc proximity, *in* Mellie, J.L., ed., Volcanism Associated with Extension at Consuming Plate Margins: Geological Society [London] Special Publication 81, p. 53–75.

Pearce, J.A., Lippard, S.J., and Roberts, S., 1984, Characteristics and tectonic significance of suprasubduction zone (ssz) ophiolites, *in* Kokellar, B.P., and Howell, M.F., eds., Marginal Basin Geology: Geological Society [London] Special Publication 16, p. 77–94.

Pessagno, E.A., Jr., Hull, D.M., and Hopson, C.A., 2000, Tectonostratigraphic significance of sedimentary strata occurring within and above the Coast Range ophiolite (California Coast Ranges) and the Josephine ophiolite (Klamath Mountains), northeastern California: Boulder, Colorado, Geological Society of America Special Paper 349, p. 383–394.

Saleeby, J.B., 1992, Petrotectonic and paleogeographic settings of U.S., Cordilleran ophiolites, *in* Burchfiel, B.C, Lipman, P.W., and Zoback, M.L., eds., The Cordilleran Orogen; Conterminous U.S.: Boulder, Colorado, Geological Society of America, Geology of North America, G-3, p. 653–682.

Saleeby, J.B., Harper, G.D., Snoke, A.W., and Sharp, W., 1982, Time relations and structural-stratigraphic patterns in ophiolite accretion, west-central Klamath Mountains, California: Journal of Geophysical Research, v. 87, p. 3831–3848.

Shervais, J.W., 1982, Ti-V plots and the petrogenesis of modern and ophiolitic lavas: Earth and Planetary Science Letters, v. 59, p. 101–118.

Southwick, D.L., 1974, Geology of the alpine-type ultramafic complex near Mount Stuart, Washington: Geological Society of America Bulletin, v. 85, p. 391–402.

Tabor, R.W., 1994, Late Mesozoic and possible early Tertiary accretion in western Washington State; the Helena-Haystack mélange and the Darrington–Devils Mountain fault zone: Geological Society of America Bulletin, v. 106, p. 217–232.

Tabor, R.W., Frizzell, V.A., Jr., Vance, J.A., and Naeser, C.W., 1984, Ages and stratigraphy of lower and middle Tertiary sedimentary and volcanic rocks of the central Cascades, Washington; application to the tectonic history of the Straight Creek Fault: Geological Society of America Bulletin, v. 95, p. 26–44.

Tabor, R.W., Frizzell, V.A., Jr., Whetten, J.T., Waitt, R.B., Jr., Swanson, D.A., Byerly, G.R., Booth, D.B., Hetherington, M.J., and Zartman, R.E., 1987, Geologic map of the Chelan 30' by 60' Quadrangle, Washington: U.S., Geological Survey Miscellaneous Investigations Series, I-1661, 33 p.

Tabor, R.W., Haugerud, R.H., Brown, E.H., Babcock, R.S., and Miller, R.B., 1989, Sedimentation and tectonics of western North America, *in* Hanshaw, P.M., ed., Accreted Terranes of the North Cascades Range, Washington, Volume 2: Washington, D.C., American Geophysical Union, Field Trips for the 28th International Geological Congress, Field Trip Guidebook T307, 62 p.

Tabor, R.W., Waitt, R.B., Jr., Frizzell, V.A., Jr., Swanson, D.A., Byerly, G.R., Bentley, R.D., 1982, Geologic map of the Wenatchee 1:100,000 Quadrangle, central Washington: U.S. Geological Survey Miscellaneous Investigations Series, I-1311, 26 p.

Wood, D.A., 1980, The application of a Th-Hf-Ta diagram to problems of tectonomagmatic classification and to establishing the nature of crustal contamination of basaltic lavas of the British Tertiary Volcanic Province: Earth Planet Science Letters, v. 50, p. 11–30.

Wyld, S.J., and Wright, J.E., 1988, The Devils Elbow ophiolite remnant and overlying Galice Formation: new constraints on the Middle to Late Jurassic evolution of the Klamath Mountains, California: Geological Society of America Bulletin, v. 100, p. 29–44.

Geological Society of America
Field Guide 4
2003

Quaternary geology of Seattle

Kathy Goetz Troost*
Derek B. Booth*
University of Washington, Department of Earth and Space Sciences, Box 351310, Seattle, Washington 98195-1310, USA

William T. Laprade*
Shannon & Wilson, Inc., Box 300303, Seattle, Washington 98103, USA

ABSTRACT

Seattle lies within the Puget Sound Lowland, an elongate structural and topographic basin bordered by the Cascade and Olympic Mountains. The geology of the Seattle area is dominated by a complex, alternating, and incomplete sequence of glacial and interglacial deposits that rest upon an irregular bedrock surface. The depth to bedrock varies from zero to several kilometers below the ground surface. Bedrock outcrops in an east-west band across the lowland at the latitude of south Seattle and also around the perimeter of the lowland. Numerous faults and folds have deformed both the bedrock and overlying Quaternary sediments across the lowland, most notably the Seattle fault. During an earthquake on the Seattle fault ca. 980 A.D., 8 m of vertical offset occurred.

The Seattle area has been glaciated at least seven times during the Quaternary Period by glaciers coalescing from British Columbia. In an area where each glacial and interglacial depositional sequence looks like its predecessor, accurate stratigraphic identification requires laboratory analyses and age determinations. The modern landscape is largely a result of repeated cycles of glacial scouring and deposition, and recent processes such as landsliding and river action. The north-south ridges of the lowland are the result of glacial scouring and subglacial stream erosion. The last glacier reached the central Puget Sound region ca. 15,000 years ago and retreated past this area by 13,650 ^{14}C yr B.P. Post-glacial sediments are poorly consolidated, as much as 300 m thick in deep alluvial valleys, and susceptible to ground failure during earthquakes.

Keywords: Quaternary, stratigraphy, Puget Lowland, Seattle, tectonics.

INTRODUCTION

This field trip includes seven stops to explore the Quaternary geology of Seattle, as summarized in Table 1. Seattle lies within a glaciated basin exhibiting a wide range of subglacial erosional features (Fig. 1A). Although the stops are centered around Seattle (Fig. 1B), they provide good proxies for the Quaternary record throughout the Puget Lowland. Figure 2 provides a conceptual view of the relationships between the field trip stops and the range of geologic contacts. The remainder of this section provides some background for the field stops and their deposits.

Tectonic Setting and Bedrock Framework

The rocks and unconsolidated deposits of Washington record more than 100 million years of earth history. Knowledge of this history has been gained from over a century of careful study, and the story is still unfolding as a result of continued geologic research. The foundation of this landscape is incompletely displayed by rocks now exposed in the North Cascades along the eastern boundary of the Puget Lowland. They record a history of oceans, volcanic island arcs, and subduction zones, mostly of Mesozoic age (the geologic period 220–65 Ma), but including some late Paleozoic components (in western Washington, as old as 275 Ma) (Frizzell et al. 1987; Tabor and Haugerud 1999; Tabor et al. 2001).

*E-mail: Troost—ktroost@u.washington.edu; Booth—dbooth@u.washingon.edu; Laprade—wtl@shanwil.com.

Troost, K.G., Booth, D.B., and Laprade, W.T., 2003, Quaternary geology of Seattle, *in* Swanson, T.W., ed., Western Cordillera and adjacent areas: Boulder, Colorado, Geological Society of America Field Guide 4, p. 267–284. For permission to copy, contact editing@geosociety.org. © 2003 Geological Society of America.

TABLE 1. FIELD TRIP ITINERARY

Stop	Location	Time	Activity
	WSCTC	7:45 am	Pick up guidebooks and name tags, load bus.
		8:00–8:30	Drive to Stop 1.
1	Mee Kwa Mooks Park	8:30–9:00	Walk out onto beach from lower walkway; view of folded Olympia peat beds, MIS 3 deposits.
		9:00–9:15	Drive to Stop 2.
2	Alki Point	9:15–9:45	View dipping Tertiary beds; first discussion of the Seattle fault zone and structural setting.
		9:45–10:20	Drive to Stop 3, bathroom stop, pick up permit.
3	South Beach, Discovery Park	10:20–12:00	Walk south onto beach to view the Olympia and Vashon type sections and landslide impacts.
	Near parking lot	12:00–12:30	Lunch on the beach.
		12:30–12:45	Drive to Stop 4.
4	North slope, Discovery Park	12:45–1:20	Short walk up the road to a landslide scarp and outcrop of Vashon till.
		1:20–1:45	Bathroom stop; drive to Stop 5.
5	Perkins Lane, Magnolia	1:45–2:15	Gather at south end of fence for overview of large landslide at till/lacustrine clay contact.
		2:15–2:30	Drive to Stop 6.
6	West side, Interbay	2:30–3:00	Walk up to till/outwash contact, visible because of a shallow slide. Watch for glass.
		3:00–3:45	Drive to Stop 7.
7	Jose Rizal overlook	3:45–4:30	View Duwamish fill, Vashon advance outwash plain, and the Seattle fault zone. Bathrooms.
End	WSCTC	4:30–5:00	Return to WSCTC.

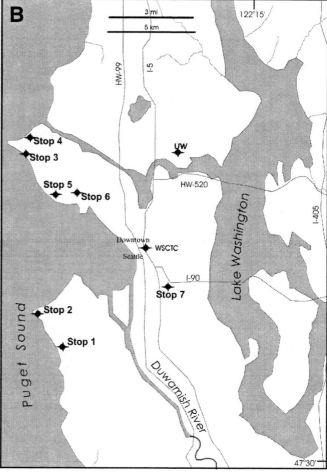

Figure 1. Field trip index maps. A: Physiographic map of the Puget Lowland with the Vashon-age ice limit, B: Location map of the field trip sites in the city of Seattle.

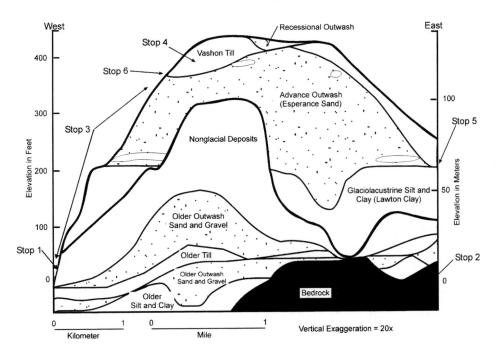

Figure 2. Diagrammatic cross section through a hill in the Puget Lowland, emphasizing the variety and irregularity of deposits, and the abundance of erosional unconformities. The stratigraphic positions of field trip Stops 1–6 are shown.

Middle and late Eocene (ca. 50–42 Ma) sandstone and volcanic rocks overlie these older rocks. During this time, large rivers flowed across an extensive (and subsiding) coastal plain that lay west of the modern Cascade Range and east of the modern Olympic Mountains (and probably east of Puget Sound as well). This ancient river system produced the rocks of the Puget Group, whose relatively good resistance to erosion is responsible for the prominence of the Newcastle Hills in the central Puget Lowland, and whose abundant plant debris resulted in coal deposits that helped shape the nineteenth century economy and history of the region.

Subsequent reorganization of tectonic plates in the northeastern Pacific Ocean resulted in renewed plate convergence, subduction, and volcanism along the Cascade Arc in earliest Oligocene time (ca. 35 Ma), followed by sedimentary deposition of the marine Blakeley Formation across the central Puget Lowland on what is now Seattle and Bainbridge Island. Marine deposition was followed by terrestrial deposition of the Blakely Harbor Formation, ca. 10 Ma, from erosion of the proto–Cascade Range. The modern-day form of the Cascade Range is not a direct descendant of this interval of tectonic and volcanic activity—the modern mountain range was uplifted less than 4 Ma (Cheney, 1997), but it expresses a similar style of tectonic activity that has been episodically active over the past several tens of millions of years.

Subduction continues today, along the Juan de Fuca plate (Fig. 3), and establishes a compressional regime in the Puget Lowland (Fig. 4), resulting in a series of basins and uplifts bounded by

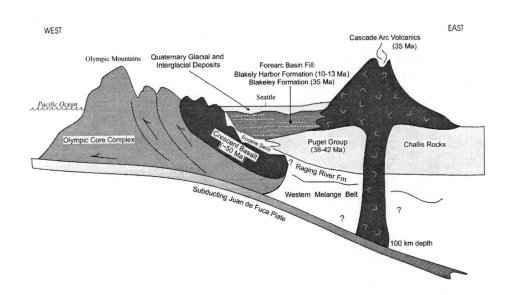

Figure 3. Diagrammatic W-E cross section across western Washington, emphasizing the major Tertiary rock units and their associated depositional ages. The Puget Lowland basin is filled with as much as 9 km of young volcaniclastic and sedimentary rocks and Quaternary deposits.

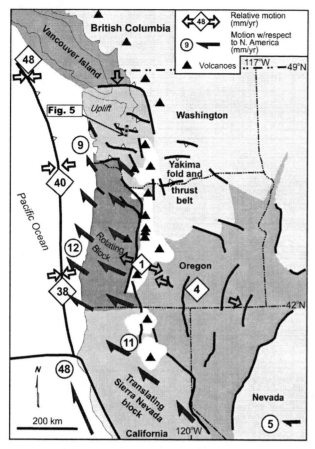

Figure 4. Relative motion of the western United States as transferred to western Washington (modified from Wells et al., 1998). The location of the cross section in Figure 5 is shown.

reverse faults. The Seattle fault zone is the border between the Seattle uplift to the south, and the Seattle basin to the north (Fig. 5).

Glaciations

Although the evolution of the tectonic and bedrock framework of the Puget Lowland continues to the present day, most

of these influences had been established by the beginning of the Quaternary Period, during which glacial ice has advanced into the Puget Lowland at least seven times, leaving a discontinuous record of Pleistocene glacial and interglacial intervals (Blunt et al., 1987; Troost et al., 2004). Originating in the mountains of British Columbia, this ice was part of the Cordilleran Ice Sheet of northwestern North America (Fig. 6). During each successive glaciation, ice advanced into the lowland as a broad tongue first called the "Puget lobe" by Bretz (1913). All but the most recent of these advances are older than can be dated with the ^{14}C technique, and so their detailed history is somewhat indeterminate (Fig. 7).

This most recent ice-sheet advance into western Washington was named the Vashon Stade of the Fraser Glaciation by Armstrong et al. (1965). Ice occupied the Puget Lowland ca. 18,000–15,000 cal. yr B.P.; at its maximum, Seattle was buried by at least 900 m (3000 ft) of ice (Booth, 1987; Porter and Swanson, 1998). Most Puget Lowland topography is a direct product of, or at least shows a strong imprint from, this period of ice-sheet occupation.

Most, although not all, of the glacial deposits exposed at the ground surface in the Puget Lowland are products of the advance and retreat of the Puget lobe during the Vashon Stade. These sediments, collectively named Vashon Drift, are divided into several units: advance deposits—lacustrine silt and clay deposited into proglacial lakes, followed by well-sorted sand and gravel carried by streams flowing from the ice sheet as it spread south; till—unsorted sand, gravel, and silt deposited beneath the ice sheet; ice-contact and ice-marginal deposits; and recessional deposits—well-sorted sand and gravel deposited by streams draining from the ice as the ice-front receded, as well as silt and clay deposited in lakes dammed by the receding ice. In the central and northern Puget Sound, extensive deposits of glaciomarine drift—nonsorted sediment deposited off the front or the base of the ice sheet into marine waters—are also found.

Distribution of Deposits in the Puget Lowland

Although a great variety of Quaternary deposits are found throughout the Puget Lowland, the characteristics of glacial and

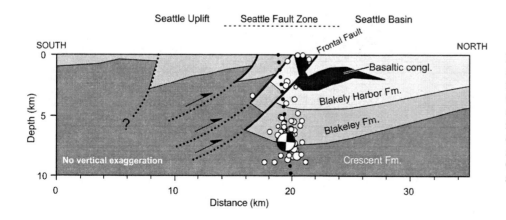

Figure 5. The Seattle fault zone shown as a series of south-dipping reverse faults (modified by B. Sherrod [U.S. Geological Survey] from Johnson et al., 1994) and showing the north-verging compressional motion and resultant displacement across the Seattle fault zone. Circles indicate earthquake foci. Quaternary cover not shown but locally as thick as 1 km over the Seattle basin.

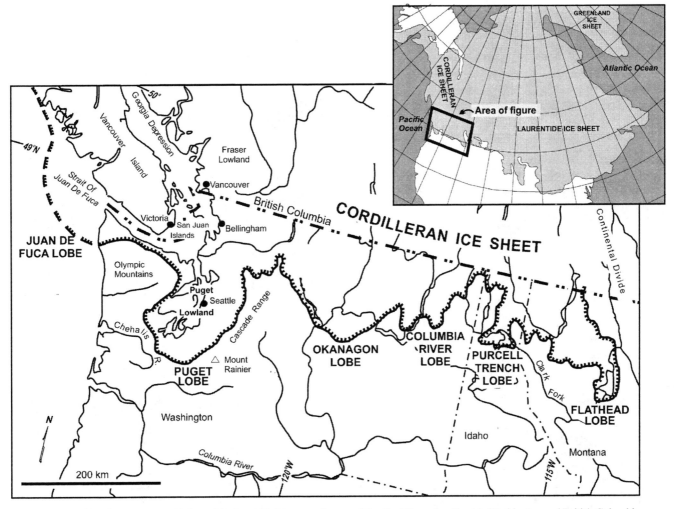

Figure 6. Map of southern extent and lobes of the latest Pleistocene advance of the Cordilleran Ice Sheet in Washington and British Columbia.

postglacial erosion and deposition have given rise to broadly predictable patterns of geologic materials across the modern landscape. In plan view, the landscape is mostly mantled by Vashon till. Locally the till is thin or absent, revealing the underlying deposits, most commonly the sandy Vashon advance outwash (but locally some older and generally less permeable sediments as well). The Vashon advance outwash formed an apron of outwash that filled the lowland (Fig. 8). Extensive advance outwash deposits are now exposed in channels that were cut through the till, most commonly during ice-sheet retreat, when voluminous meltwater discharges off of the ice sheet combined with alpine rivers draining the Cascade Range and Olympic Mountains.

Glaciomarine drift is also broadly exposed, but its geographic extent is more limited, a consequence of the late-glacial history of the lowland. This deposit, a product of subaqueous melting of the terminus of an ice sheet or the bottom of a floating ice shelf, is common, but only in areas that were below sea level during ice retreat. Although global sea level continued to rise following deglaciation of the lowland, Earth's crust in the Puget Lowland

also rebounded after deglaciation (Thorson, 1989) in response to the weight lost by melting hundreds to thousands of meters of ice, just as it had depressed several thousand years earlier, when the ice first advanced. The initial depression (and so also the subsequent uplift) was greater to the north than to the south because the ice was thicker to the north. At about the latitude of Seattle, isostatic rebound equaled eustatic sea-level rise. Thus, marine deposits dating from the first opening of Puget Sound to marine waters after deglaciation are common only in the northern Puget Lowland and are not recognized south of Seattle (Pessl et al., 1989; Yount et al., 1993).

As the ice sheet retreated from its maximum position, meltwater drained into the axis of the lowland but could not follow what would become its modern drainage path north and west out the Strait of Juan de Fuca, because the strait was still filled with many hundreds or even thousands of meters of ice. Instead, meltwater was diverted south along the margins of the retreating ice sheet, coalescing into ever-broader rivers. Channels and locally broad plains of Vashon recessional outwash now form much of

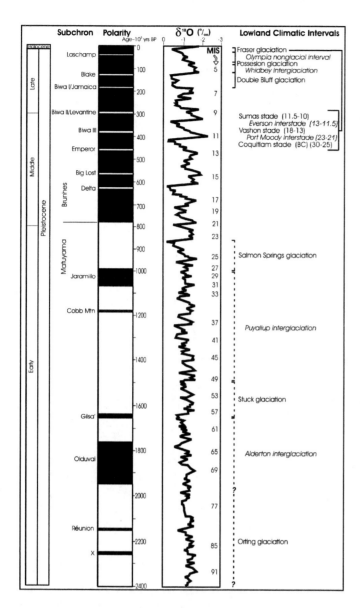

Figure 7. Comparison of the marine oxygen-isotope curve stages (MIS) using the deep-sea oxygen-isotope data for ODP677 from Shackelton et al. (1990), global magnetic polarity curve (Mankinnen and Dalrymple, 1979; Barendregt, 1995; Cande and Kent, 1995), and ages of climatic intervals in the Puget and Fraser Lowlands. Ages for deposits of the Possession Glaciation through Orting Glaciation from Easterbrook et al. (1981), Easterbrook (1986), Blunt et al. (1987), and Easterbrook (1994). Additional ages from deposits of the Puyallup Interglaciation from R.J. Stewart (1999, personal commun.). Ages for the Olympia nonglacial interval from Armstrong et al. (1965), Mullineaux et al. (1965), Pessl et al. (1989), and Troost (1999). Ages for the Coquitlam Stade from Hicock and Armstrong (1985); ages for the Port Moody interstade from Hicock and Armstrong (1981). Ages for the Vashon Stade from Armstrong et al. (1965) and Porter and Swanson (1998). Ages for the Everson interstade from Dethier et al. (1995) and Kovanen and Easterbrook (2001). Ages for the Sumas Stade from Clague et al. (1997), Kovanen and Easterbrook (2001), and Kovanen (2002). Many other glacial and interglacial depositional periods are likely represented but not yet identified in the Puget Lowland.

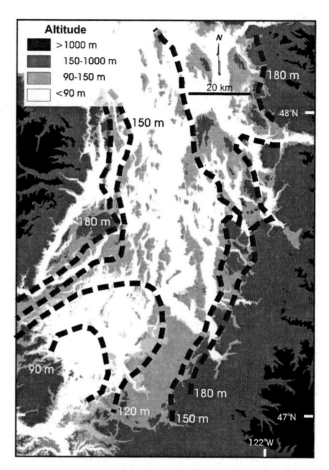

Figure 8. Topographic shading of the Puget Lowland, from U.S. Geological Survey 10 m digital elevation model. Contours show generalized topography of the Great Lowland Fill (modified from Booth, 1994), with a modern altitude between 120 and 150 m across most of the lowland as reconstructed from the altitude of modern drumlin tops and subsequently incised by both subglacial channels and modern river valleys.

the landscape in these ice-marginal locales and recessional river valleys (Fig. 9). These landforms can be traced downstream to their glacial-age spillway out of the Puget Lowland, south through the valley of Black Lake near Olympia, and then along the valley of the modern Chehalis River west to the Pacific Ocean.

The vertical distribution of sediments is also broadly predictable. Vashon till is laterally extensive but commonly just 1 or 2 m thick. Below it, Vashon advance outwash provides the bulk of the modern landscape lying above sea level, commonly with thickness of many tens of meters. The top of the advance outwash is relatively flat, notwithstanding abundant superimposed channels and elongated hills; it records the level of the vast braided-river outwash plain formed during the advance of the Cordilleran Ice Sheet into the Puget Lowland (Booth, 1994). Beneath the advance outwash, deposits are less commonly exposed and more variable in origin and in character. In the Seattle area, early Vashon-age lake deposits (the Lawton Clay) are common and can be found up to elevations

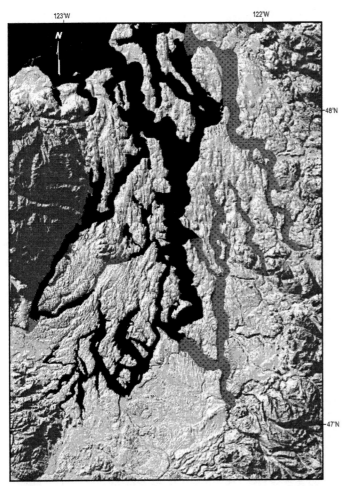

Figure 9. Shaded topography of the Puget Lowland from U.S. Geological Survey 10 m digital elevation model, displaying the major subglacial drainage channels of the Puget lobe. Most are now filled by marine waters (black), and others are filled by late glacial and Holocene alluvium and mudflows (dark gray stipple).

of ~60 m; farther south, these sediments are almost entirely absent. These Vashon sediments overlie pre-Vashon deposits, in places conformably, and in other places unconformably.

Postglacial Processes and Deposits

Rivers, landslides, waves, and volcanic mudflows have continued to modify the Puget Lowland landscape and to create deposits younger than those of the last glacial advance. Although not influential at the same scale as continental glaciations, these processes and their resulting products are locally significant. The sediments are uniformly unconsolidated but permeabilities vary widely, reflecting the variety of depositional process and parent materials.

The most dramatic of these processes has been the Osceola mudflow near the town of Enumclaw, deposited after the catastrophic collapse of the northeast side of Mount Rainier ca. 5700 yr ago. Subsequent erosion and redeposition of the abundant post-Osceola supply of volcanic debris eventually filled the previously marine Green/Duwamish River valley (Dragovich et al., 1994) from about the present city of Auburn north to Seattle's Elliott Bay.

ROAD LOG AND STOP DESCRIPTIONS

Miles	Description
0.0	Mileage begins at Interstate 5 and Spring Street.
2.3	Drive south on I-5; exit at West Seattle Bridge.
4.7	Follow West Seattle Bridge; take the Admiral Way exit.
5.7	Head north on Admiral Way; note view of downtown Seattle.
7.35	Follow Admiral Way; drop onto uplifted beach terrace at 60th Ave. SW.
7.6	Turn left onto 63rd Ave. SW.
8.6	Follow 63rd Ave., bear left onto Beach Drive SW, head south, and park near SW Oregon St.

Stop 1. Mee Kwa Mooks Park

The field trip begins in west Seattle with two of the oldest exposed deposits in Seattle (Fig. 10), a logistical choice determined by the day's morning low tide. At this stop, peat layered with silt is well exposed on the low-tide beach (Fig. 11). The material has been radiocarbon dated at 27,810 ± 130 yr B.P. (B-136967), and so it is part of the Olympia nonglacial interval. Deposits of this interval accumulated during the interglacial time preceding the Vashon Glaciation, 15,000–60,000 yr B.P. (MIS 3). The climate was cooler than today, and the sea level was at least 50 m (150 ft) lower than today.

Folds are evident in the deposits here across the several hundred yards of exposure (Fig. 11A). Figure 11B depicts one of the doubly plunging synclines visible on the beach at or below –2′ tide (and so will not be visible today). At this site we are standing within the Seattle fault zone; the folds here are aligned with the trend of deformed beds observed in offshore seismic data along one of the southern strands of the Seattle fault (Johnson et al., 1999). The relationship of this folding to fault movement is unknown. Note the lack of other obvious surface expression of the fault zone (e.g., prominent scarps). More subtle evidence, however, is apparent throughout this zone, notably stratigraphic offsets of Vashon age (i.e., <15,000 yr) and a small scarp that crosscuts Beach Drive SW just south of SW Jacobsen Street.

Miles	Description
8.8	Drive south across the south end of the uplifted beach terrace at SW Jacobsen St.
8.85	Turn around at SW Angeline St. and head north on Beach Drive SW.
9.8	Bear left at 63rd Ave. SW, staying on Beach Drive SW.
10.1	Park near treatment plant on Beach Drive SW; walk down to the beach.

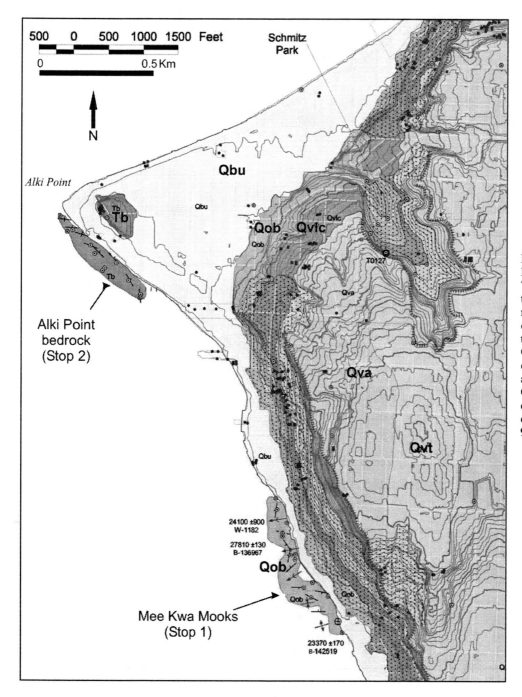

Figure 10. Geologic map of Alki Point and environs, west Seattle (from Troost et al., 2004) showing the locations of Stops 1 and 2. Tb—Tertiary marine volcaniclastic rock, may be correlative with the Blakeley Formation. Qob—Olympia beds of MIS 3. Qvlc—Vashon Lawton Clay, proglacial silts and clays. Qva—Vashon advance outwash, Esperance Sand. Qvt—Vashon till, mostly subglacial in origin. Qbu—uplifted beach deposits; uplifted 7 m during the 980 A.D. Seattle fault earthquake.

Stop 2. Alki Point

Bedrock, persumably of the Blakeley Formation, is well exposed in the beach at this stop (Fig. 10). The rock is ca. 40 m.y. old and now dips primarily to the east, although a broad NE-plunging anticline is exposed toward the point. The rock here consists of interbedded, bioturbated, very fine-grained sandstone and siltstone, and graded sandstone, representative of a nearshore marine sandy shelf environment. Clast composition is chiefly tuff, mudstone, and sandstone. Some beds contain shell fossils and some contain wood debris. East of Seattle, the presumed Blakeley Formation contains abundant volcanic material and becomes less competent than the rock seen here at Alki Point. The bedrock at the south end of the exposure is sheared, suggestive of fault movement some time in the past. Just north of this exposure is one of the traces of the Seattle fault, which has uplifted this block of Earth's crust several kilometers relative to the block just north of Alki Point over the past several million years (and by an additional 7 m in the year 980 A.D.; Fig. 12).

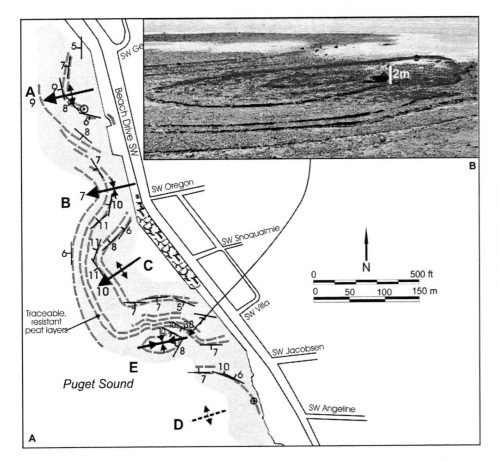

Figure 11. Structure of folded Olympia beds, across from Mee Kwa Mooks Park, west Seattle. Inset photograph taken at –2.5′ tide. The resistant beds are the highly compressed peat layers that are interbedded with organic silt layers. Rooted trees and abundant branches are dispersed throughout the floodplain deposit.

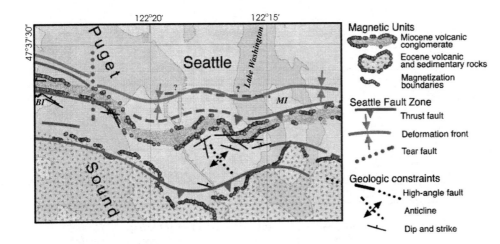

Figure 12. Geologic interpretation of aeromagnetic data across the Seattle fault zone, east Bainbridge Island (BI) to Mercer Island (MI) (from Blakely et al., 2002).

Miles	Description

10.2 Continue on Beach Drive SW; it turns into Alki Ave. SW; continue northeast.

10.7 Restrooms opposite 63rd Ave. SW; note Statue of Liberty opposite 61st Ave. SW.

11.2 Follow Alki Ave. SW and cross the north end of the uplifted beach terrace.

11.7 Note steep landslide scarps behind condominiums (Fig. 13).

12.3 Drive around Duwamish Head, great views of downtown Seattle.

12.6 Restrooms at boat launch, east side of Alki Ave. SW.

12.7 Note large landslide on California Way SW, west side of road (Fig. 14).

14.2 Turn left at stoplight under the overpass. Move into right lane to access the West Seattle Bridge, heading east.

15.4 Take the Highway 99 north exit.

17.4 Landmarks include the Mariners and Seahawks stadiums, ferry terminal.

18.2 Drive north on HW 99 on "Alaskan Way Viaduct."

18.9 Take the Western Ave. exit.

19.8 Western Ave. merges into Elliott Ave. W; drive on fill on former tidelands.

21.0 Elliott Ave. turns into 15th Ave. W; drive past the Interbay landfill.

21.9 Take the Dravus St. exit and go left on Dravus St., back across 15th Ave. W.

22.3 Turn right onto 20th Ave. W; it then merges into Gilman Ave. W.

23.5 Gilman Ave. W turns into W Government Way.

24.0 Follow W Government Way to the entrance to Discovery Park.

24.1 Continue into park; stop at Visitor's Center, bathrooms, parking permits. Drive to drop-off point at parking lot near lighthouse at the beach (mileage temporarily stops).

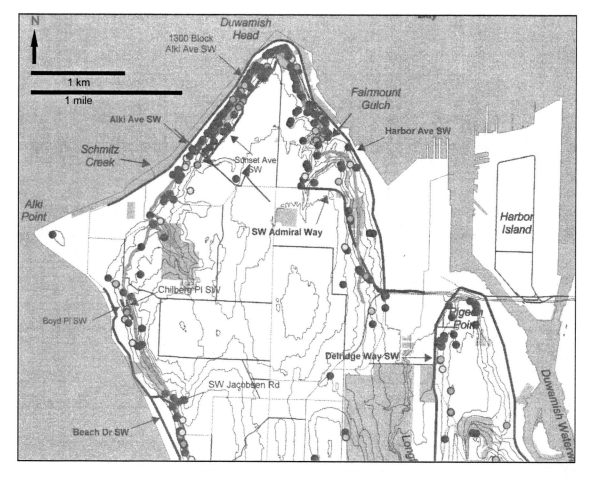

Figure 13. Distribution of historical landslides in west Seattle, compiled by the city of Seattle and Shannon & Wilson, Inc. 1-mm-wide medium gray (red in online color version) line shows approximate location of Esperance Sand-Lawton Clay contact as mapped by Waldron et al. (1962). The color version shows landslides by decade: white—1890–1900; dark green—1901–1920; blue-green—1921–1930; dark blue—1931–1950; light blue—1951–1960; pink—1961–1970; orange—1971–1980; red—1981–2000.

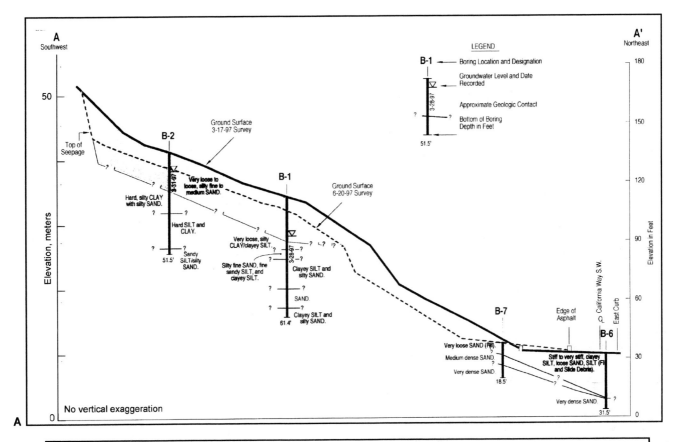

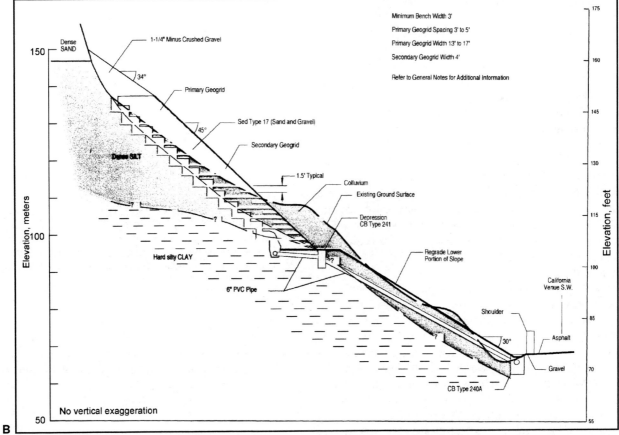

Figure 14. A: Profile A–A' through the middle of the California/Ferry landslide. The upper layer of colluvium was removed above elevation ~90 ft and recompacted in place after interceptor drains were installed at the colluvium/sand contact. The lower slope was mixed with soil cement to allow it to maintain the angle at which it now stands. B: Geogrid section for the California Way landslide repair, which consisted of interceptor drains at the sand/clay contact and then geogrid-reinforced slope to allow it to be built at a 45° angle and to stay within park property lines.

Stop 3. South Beach, Discovery Park

This stop, as shown on Figure 15, displays the type sections of the Vashon advance outwash (the Esperance Sand Member of the Vashon Drift), the Lawton Clay Member of the Vashon Drift (both from Mullineaux et al., 1965), and the Olympia beds (Fig. 16), first seen today at Mee Kwa Mooks Park. Here, Olympia beds (oxidized sand and silt beds) are exposed at beach level at a bluff point. Freshwater mollusk fossils and sedimentary structures indicate a floodplain environment (Fig. 16B). Radiocarbon dating of wood has yielded ages of 18.1–22.4 ka.

The climate cooled, closing the Olympia interval, and the Vashon-age Puget lobe (Bretz, 1913), the southwestern-most extension of the Cordilleran Ice Sheet, approached from the north after crossing the Canadian border ca. 17,000 yr B.P. The ice dammed the Strait of Juan de Fuca, forming a lake in the Puget Lowland. Deposition in that lake formed the Lawton Clay. The contact between the Lawton Clay and Olympia beds is therefore gradational. By definition at this locality, the Lawton Clay includes only the silt and clay deposits of the proglacial lake; gradational zones are included in the underlying unit.

The advance outwash (Esperance Sand) is the subsequent deposit of sand and gravel carried by rivers emerging from the snout of the approaching ice sheet. It can be as much as 150 m thick, and in places it serves as a major sand and gravel resource. It was subsequently overridden by the glacier and so is dense to very dense. At its base, the advance outwash commonly interfingers with silt of the Lawton Clay; this transition zone is defined as the basal deposits of the advance outwash (and not part of the Lawton Clay). Locally, the advance coarsens upward to gravel and includes gravelly interbeds.

Landslides are common in the upper part of this exposure, because groundwater readily penetrates through the permeable advance outwash and is blocked by the much less permeable Lawton Clay. The buildup of groundwater increases pore-water pressures, which in turn buoys up the sand grains and allows them to slide over the silt-clay toward the open bluff face as (disintegrating) block glides. Even larger rotational failures that cut through both the advance outwash and Lawton Clay, along with much smaller debris flows and slurries, are also visible here and along many of the bluff faces of the major river valleys and Puget Sound.

Although our schedule precludes much discussion of the Holocene history of West Point, the treatment plant sits on an archaeological site showing evidence of a tsunami and fault subsidence. The West Point spit accreted over the past 5000 yr, building outward with longshore drift and upward with sea-level rise. Prior to 1000 yr ago, perimeter beach berms protected a back beach tidal marsh. Native Americans occupied dry ground around the marsh and used the land as a seasonal shellfish processing area. The major earthquake on the Seattle fault, 1000 yr ago, caused 1 m of subsidence and deposited a tsunami sand layer. Rapid sedimentation buried the tsunami sand until excavation for the treatment plant in the early 1990s exposed the layer.

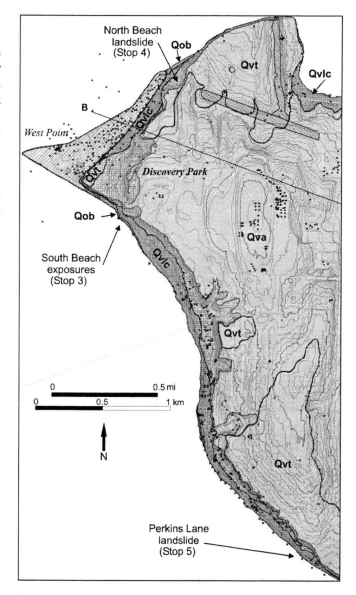

Figure 15. Geologic map of the Magnolia and Discovery Park area, northwest Seattle (from Booth et al., 2004) showing locations of Stops 3, 4, and 5. Qob—Olympia beds of MIS 3. Qvlc—Vashon Lawton Clay, proglacial silts and clays. Qva—Vashon advance outwash, Esperance Sand. Qvt—Vashon till, mostly subglacial in origin. The beach at West Point subsided 1 m during the 980 A.D. Seattle fault earthquake.

Miles	Description
24.1	Return to Visitor's Center; return parking permit (resume mileage). Go left out of the parking lot and follow signs to Daybreak Star Cultural Center.
24.4	Go past the turnoff to South Beach and its restricted parking.
24.7	Bear right at parking lot.
25.3	Take an immediate left and follow road to Daybreak Star; park on right.

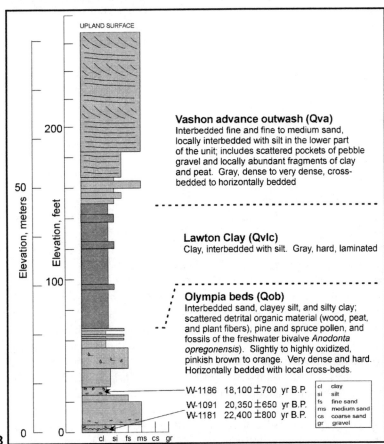

Figure 16. South Beach bluff exposure at West Point, northwest Seattle (Stop 3). Photograph (A) displays the type locality of the Esperance Sand (Vashon advance outwash), Lawton Clay, and deposits of the Olympia non-glacial interval; measured section (B) is from Booth et al. (2004), adapted from Mullineaux et al. (1965).

Stop 4. North Slope, Discovery Park

Vashon till mantles most of the upland surfaces within the Puget Lowland. Despite the great extent of mapped Vashon till, good outcrops are few and far between. The till is exposed here because of landsliding (Fig. 15). Although Vashon till is typically a very dense, matrix-supported gravelly, silty sand or sandy silt, frequent discontinuities within the till increase its permeability by several orders of magnitude. Depending on subglacial conditions, the discontinuities may consist of intercalated sand and silt layers, joints, and bedding.

The prominent landslide here involves the Vashon till. This condition is unusual but not unprecedented in the region; normally, we expect the till to form a coherent and generally quite stable slope or bluff. At this site, however, two conditions have reduced the strength of this material. First, the normally competent till has been weakened here by numerous lenses of sand and joints, which locally transmit groundwater quite effectively. And secondly, the till directly overlies the Lawton Clay; the Esperance Sand is entirely absent.

Known as "hardpan" by the construction trade, till is one of the most stable substrates in the region. It is rarely implicated in landsliding (although this stop and the next are exceptions). Rapid surface-water runoff and surface erosion, however, is common over the till where overlying soil has been stripped or compacted.

Miles	Description
26.4	Retrace steps to Visitor's Center.
26.5	Leave park; turn right onto 36th Ave. W.
26.8	Turn right onto W Emerson St.
27.0	Turn left onto Magnolia Blvd. W.
28.8	Park near fence at top of scarp.

Stop 5. Perkins Lane, Magnolia

This site displays an impressive variety of landslide types, with a variety of attendant hazards and property damage. The most prominent landsliding here has involved the uppermost unit, Vashon till, that has detached in large blocks from the upper scarp (Fig. 17A). At this site, the till has numerous lenses of sand and joints that locally transmit groundwater quite effectively, and that have created zones of weakness through the otherwise strong material (Fig. 17B). This site also lies at the edge of an older, larger rotational landslide that has involved not only the failed material at the foot of the slope but also several lots to the northwest.

Despite the magnitude of landslide motion, this locality does *not* have the "classic" Seattle landslide stratigraphy of Esperance Sand over Lawton Clay. Instead, the till directly overlies the clay, and the Esperance Sand is entirely absent. Thus, a landslide-hazard map predicated exclusively on the location of the sand-clay contact would not have identified this site as a zone of elevated hazard.

Miles	Description
29.2	Continue south on Magnolia Blvd. W, right onto W Howe St.; cross bridge.
29.25	Turn right onto Magnolia Blvd.
29.6	Merge with W Galer St.
29.7	Turn left onto Thorndyke Ave. W.
30.6	Turn right onto 21st Ave. W, heading south.
30.7	Park on side of road.

Stop 6. West Side, Interbay

This site displays the two most common surficial geologic materials in the central Puget Lowland: Vashon till, and its associated advance outwash. A sharp contact relationship between them is nicely displayed in a face of a shallow colluvial slide. The outcrop as a whole appears to have been slumped from a higher elevation on the east side of Magnolia Hill, given its anomalous position relative to older sediments to the south.

Miles	Description
30.9	Make a U-turn and head north on 21st Ave. W; turn right onto Thorndyke W.
21.15	Turn right onto W Dravus St.
31.4	Turn right onto 15th Ave. W (poorly marked), heading south.
32.5	Bear left onto Elliott Ave. W at W Galer St.
33.6	Bear right from Elliott onto Western Ave.
34.5	Turn right onto southbound HW 99.
35.6	Take a left-hand exit marked "to ferries."
36.5	Continue south on 1st Ave. S, in the SODO district; turn left onto S Holgate St.
37.0	Follow Holgate in left lane to go over I-5 and up a hill.
37.55	Turn left onto 14th Ave. S.
38.2	Turn left on S Judkins St.
38.3	Turn right on 12th Ave. S.
38.4	Turn left into parking lot at the overlook.

Stop 7. Jose Rizal Overlook

This stop offers a view across four landscape elements of the Puget Lowland:

1. Duwamish Fill

From this site we can view the extensive fill deposits around the mouth of the Duwamish River. The flat, industrial area surrounding Safeco Field is an extensive fill deposit that rests on soft tideflat deposits. Much of the fill was derived from regrading projects in the early 1900s and dredging of Elliott Bay. In places the fill is as much as 15 m thick, resting on 15 m of soft and loose native deposits. Upvalley, the Duwamish River flows over more than 100 m of loose alluvial deposits. With the shallow water table and sandy soils, this area is highly susceptible to liquefaction. Where fill overlies tideflats, extensive liquefaction occurred from the Nisqually earthquake on February 28, 2001. Filling efforts were extensive in the early 1900s (Galster and Laprade, 1991).

Figure 17. Slide scarp above the south end of Perkins Lane (A) (photo taken in 1998). Stop 5 overlook is above and to the right of the main scarp. Vashon till (B) well exposed in boxed area of photo; dark areas in close-up view are voids caused by groundwater piping through subhorizontal sand lenses.

2. The Great Lowland Fill

This viewpoint displays some of the variety of the city's topography, which in turn reflects important elements of the underlying geology. Particularly prominent is the surface of the Great Lowland Fill, a planar surface that forms the upland plateaus and promontories of the Puget Lowland (Booth, 1994). It is a surface defined by the top of the Vashon advance outwash, locally mantled but barely increased by at most a few meters of overlying till.

Where the land surface of the lowland rises *above* this level, it is a result of local ice-recessional and ice-contact deposits. Where the land surface drops *below* this level, subglacial erosion is the most common cause (such as the channels of Puget Sound and Lake Washington) (Booth and Hallet, 1993). The borders of the lowland can be defined by the emergence of bedrock from beneath the blanket of glacial deposits. The rock forms the higher surfaces of the Cascade and Olympic foothills, and the more localized promontories south of the Seattle fault (the Newcastle Hills to the east, Green and Gold Mountains to the west).

3. Seattle Fault Zone

We are standing within the Seattle fault zone. Note the lack of lowland surface expression of the fault zone. Although the Seattle fault has little surface expression and no historic activity, the region as a whole has experienced many historic earthquakes; most recently, the magnitude 6.8 Nisqually earthquake occurred on February 28, 2001. The shaking from the earthquake caused ground failures throughout the Puget Lowland, particularly in Olympia and Seattle. Intensive field observations were made over a two-week period after the earthquake. Scientists from many state and federal agencies covered the areas most likely to fail (shallow water table with loose, sandy soil). Extensive access was obtained to private property. Ground failures included settlements, ground cracks, landslides, and liquefaction-induced lateral spreading and loss of bearing strength.

The geologic environment strongly controlled ground failures. For example:

- Ground failures were more prevalent over reclaimed tidelands.
- The Duwamish Valley alluvial fill amplified shaking.
- Methods of fill placement strongly controlled ground failures, and thus historical filling records may better delineate high-risk areas:
  - Old, poorly compacted fill failed more commonly.
  - Recent surface compaction minimized occurrences.
  - Inconsistencies in fill placement are evident (e.g., Boeing Field).

4. Hills of Seattle

This is a good location to note the differences in the cores of the various hills of Seattle, as depicted in Figure 18. Extensive drilling programs for tunnels have provided opportunities to compare the cores of many hills. Some drumlins and hills are cored with pre-Vashon deposits, pre-existing topographic highs. Some drumlins and hills are topped by thick, relatively tabular beds of Vashon drift that were subsequently scoured into drumlin shapes. The subglacial channels carved by the Vashon Glaciation do not coincide completely with pre-existing channels. This pattern of buried channels and channel stacking (Fig. 18) has been repeated many times during the Quaternary, giving rise to a series of unconformities and buried topographies.

Miles	Description
38.5	Turn left out of parking lot; go left onto Golf Dr. S.
39.0	Cross the Jose Rizal Bridge, over I-90; then turn left onto Boren Ave.
39.7	Turn left onto Madison St.
40.0	Turn right onto 6th Ave.
40.1	I-5 southbound ramp is at Spring St.
40.2	I-5 northbound ramp is at University St.

REFERENCES CITED

Armstrong, J.E., Crandell, D.R., Easterbrook, D.J., and Noble, J.B., 1965, Late Pleistocene stratigraphy and chronology in southwestern British Columbia and northwestern Washington: Geological Society of America Bulletin, v. 76, p. 321–330.

Barendregt, R.W., 1995, Paleomagnetic dating methods, in Rutter, N.W., and Catto, N.R., eds. Dating methods for Quaternary deposits: Geological Association of Canada, GEOtext2, 308 p.

Blakely, R.J., Wells, R.E., Weaver, C.S., and Johnson, S.Y., 2002, Location, structure, and seismicity of the Seattle fault zone, Washington: Evidence from aeromagnetic anomalies, geologic mapping, and seismic-reflection data: Geological Society of America Bulletin, v. 114, p. 169–177.

Blunt, D.J, Easterbrook, D.J., and Rutter, N.W., 1987, Chronology of Pleistocene sediments in the Puget Lowland, Washington: Washington Division of Geology and Earth Resources Bulletin, v., 77, p. 321–353.

Booth, D.B., Troost, K.G., and Shimel, S.A., 2004, Geologic map of the Seattle NW quadrangle: U.S. Geological Survey Miscellaneous Field Investigations Map, scale 1:12,000 (in press).

Booth, D.B., 1987, Timing and processes of deglaciation along the southern margin of the Cordilleran ice sheet, in Ruddiman, W.F., and Wright, H.E., Jr., eds., North America and adjacent oceans during the last deglaciation: Boulder, Colorado, Geological Society of America, The Geology of North America, v. K-3, p. 71–90.

Booth, D.B., 1994, Glaciofluvial infilling and scour of the Puget Lowland, Washington, during ice-sheet glaciation: Geology, v. 22, p. 695–698.

Booth, D.B., and Hallet, B., 1993, Channel networks carved by subglacial water: Observations and reconstruction in the eastern Puget Lowland of Washington: Geological Society of America Bulletin, v. 105, p. 671–683.

Bretz, JH., 1913, Glaciation of the Puget Sound region: Washington Geological Survey Bulletin No. 8, 244 p.

Cande, S.C., and Kent, D.V., 1995, Revised calibration of the geomagnetic polarity time scale for the late Cretaceous and Cenozoic: Journal of Geophysical Research, v. 100, p. 6093–6095.

Cheney, E.S., 1997, What is the age and extent of the Cascade magmatic arc?: Washington Geology, v. 25, p. 28–32.

Clague, J.J., Mathewes, R.W., Guilbault, J.P., Hutchinson, I., and Ricketts, B.D., 1997, Pre-Younger Dryas resurgence of the southwestern margin of the Cordilleran ice sheet, British Columbia, Canada: Boreas, v. 26, p. 261–278.

Dethier, D.P., Pessl, F., Jr., Keuler, R.F., Balzarini, M.A., and Pevear, D.R., 1995, Late Wisconsinan glaciomarine deposition and isostatic rebound, northern Puget Lowland, Washington: Geological Society of America Bulletin, v. 107, p. 1288–1303.

Dragovich, J.D., Pringle, P.T., and Walsh, T.J., 1994, Extent and geometry of the mid-Holocene Osceola Mudflow in the Puget Lowland: Implications for Holocene sedimentation and paleogeography: Washington Geology, v. 22, p. 3–26.

Easterbrook, D.J., 1986, Stratigraphy and chronology of Quaternary deposits of the Puget Lowland and Olympic Mountains of Washington and the Cascade Mountains of Washington and Oregon: Quaternary Science Reviews, v. 5, p. 145–159.

Easterbrook, D.J., 1994, Chronology of pre-late Wisconsin Pleistocene sediments in the Puget Lowland, Washington, in Lasmanis, R., and Cheney, E.S., conveners, Regional geology of Washington State: Washington Division of Geology and Earth Resources Bulletin, v. 80, p. 191–206.

Easterbrook, D.J., Briggs, N.D., Westgate, J.A., and Gorton, M., 1981, Age of the Salmon Springs Glaciation in Washington: Geology, v. 9, p. 87–93.

Frizzell, V.A., Jr., Tabor, R.W., Zartman, R.E., and Blome, C.D., 1987, Late Mesozoic or early Tertiary mélanges in the western Cascades of Washington, in Schuster, J.E., ed., Selected papers on the geology of Washington: Washington Division of Geology and Earth Resources Bulletin, v. 77, p. 129–148.

Galster, R.W., and Laprade, W.T., 1991, Geology of Seattle, Washington, United States of America: Bulletin of the Association of Engineering Geologists, v. 28, no. 3 (supplement), p. 239–302.

Hicock, S.R., and Armstrong, J.E., 1981, Coquitlam Drift—a pre-Vashon Fraser glacial formation in the Fraser Lowland, British Columbia: Canadian Journal of Earth Sciences, v. 18, p. 1443–1451.

Hicock, S.R., and Armstrong, J.E., 1985, Vashon drift—definition of the formation in the Georgia Depression, southwest British Columbia: Canadian Journal of Earth Sciences, v. 22, p. 748–757.

North Capitol Hill Area

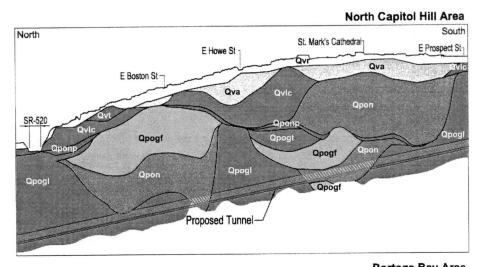

Portage Bay Area

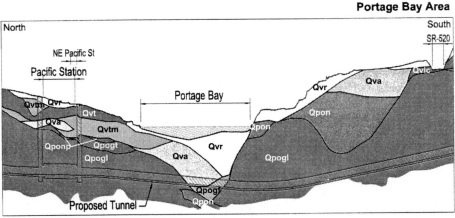

Figure 18. N-S cross sections through the North Capitol Hill and Portage Bay areas depicting the geology along the proposed Sound Transit tunnel alignment. Note multiple unconformities and channel fills in the subsurface.

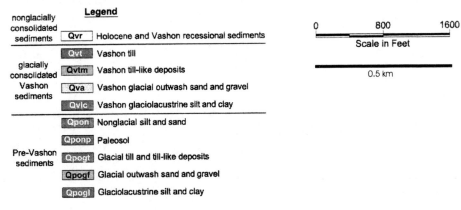

Legend

nonglacially consolidated sediments	Qvr	Holocene and Vashon recessional sediments
glacially consolidated Vashon sediments	Qvt	Vashon till
	Qvtm	Vashon till-like deposits
	Qva	Vashon glacial outwash sand and gravel
	Qvlc	Vashon glaciolacustrine silt and clay
Pre-Vashon sediments	Qpon	Nonglacial silt and sand
	Qponp	Paleosol
	Qpogt	Glacial till and till-like deposits
	Qpogf	Glacial outwash sand and gravel
	Qpogl	Glaciolacustrine silt and clay

Scale in Feet
0   800   1600

0.5 km

Johnson, S.Y., Dadisman, C.V., Childs, J.R., and Stanley, W.D., 1999, Active tectonics of the Seattle fault and central Puget Sound, Washington—Implications for earthquake hazards: Geological Society of America Bulletin, v. 111, p. 1042–1053.

Johnson, S.Y., Potter, C.J., and Armentrout, J.M., 1994, Origin and evolution of the Seattle fault and Seattle basin, Washington: Geology, v. 22, p. 71–74.

Kovanen, D.J., 2002, Morphologic and stratigraphic evidence for Allerod and Younger Dryas age glacier fluctuations of the Cordilleran Ice Sheet, British Columbia, Canada, and northwest Washington, U.S.A.: Boreas, v. 31, p. 163–184.

Kovanen, D.J., and Easterbrook, D.J., 2001, Late Pleistocene, post-Vashon, alpine glaciation of the Nooksack drainage, North Cascades, Washington: Geological Society of America Bulletin, v. 113, p. 274–288.

Mankinnen, E.A., and Dalrymple, G.B., 1979, Revised geomagnetic polarity time scale for the interval 0–5 m.y. b.p.: Journal of Geophysical Research, v. 84, p. 615–626.

Mullineaux, D.R., Waldron, H.H., and Rubin, M., 1965, Stratigraphy and chronology of late interglacial and early Vashon time in the Seattle area, Washington: U.S. Geological Survey Bulletin, p. 1–10.

Pessl, F., Jr., Dethier, D.P., Booth, D.B., and Minard, J.P., 1989, Surficial geology of the Port Townsend 1:100,000 quadrangle, Washington: U.S. Geological Survey Miscellaneous Investigations Map I-1198F.

Porter, S.C., and Swanson, T.W., 1998, Radiocarbon age constraints on rates of advance and retreat of the Puget lobe of the Cordilleran ice sheet during the last glaciation: Quaternary Research, v. 50, p. 205–213.

Shackelton, N.J., Berger, A., and Peltier, W.R., 1990, An alternative astronomical calibration of the lower Pleistocene time-scale based on ODP site 677: Transactions of the Royal Society of Edinburgh, Earth Sciences, v. 81, p. 251–261.

Tabor, R.W., and Haugerud, R.A., 1999, Geology of the North Cascades—A mountain mosaic: Seattle, WA, The Mountaineers, 143 p.

Tabor, R.W., Booth, D.B., Vance, J.A., and Ford, A.B., 2001, Geologic map of the Sauk River 1:100,000 quadrangle, Washington: U.S. Geological Survey Miscellaneous Investigations Map I-2592.

Thorson, R.M., 1989, Glacio-isostatic response of the Puget Sound area, Washington: Geological Society of America Bulletin, v. 101, p. 1163–1174.

Troost, K.G., Booth, D.B., and Shimel, S.A., 2004, Geologic map of the Seattle SW quadrangle: U.S. Geological Survey Miscellaneous Field Investigations Map, scale 1:24,000 (in press).

Troost, K.G., Booth, D.B., and Wells, R.E., 2004, Geologic map of the Gig Harbor 7.5-minute quadrangle: U.S. Geological Survey Miscellaneous Field Investigations Map, scale 1:24,000 (in press).

Troost, K.G., 1999, The Olympia nonglacial interval in the southcentral Puget Lowland, Washington [M.S. thesis]: Seattle, University of Washington, 123 p.

Waldron, H.H., Liesch, B.A., Mullineaux, D.R., and Crandell, D.R., 1962, Preliminary geologic map of Seattle and vicinity, Washington: U.S. Geological Survey Miscellaneous Investigations Map I-354.

Wells, R.E., Weaver, C.S., and Blakely, R.J., 1998, Fore-arc migration in Cascadia and its neotectonic significance: Geology, v. 26, p. 759–762.

Yount, J.C., Minard, J.P., and Dembroff, G.R., 1993, Geologic map of surficial deposits in the Seattle 30' × 60' quadrangle, Washington: U.S. Geological Survey Open-File Report 93-233, 2 sheets, scale 1:100,000.